Handbook of Environmental Engineering

Handbook of Environmental Engineering

Edited by Chloe Miller

MURPHY & MOORE
www.murphy-moorepublishing.com

Published by Murphy & Moore Publishing,
1 Rockefeller Plaza,
New York City, NY 10020, USA
www.murphy-moorepublishing.com

Handbook of Environmental Engineering
Edited by Chloe Miller

International Standard Book Number: 978-1-63987-278-7 (Hardback)

Cataloging-in-Publication Data

Handbook of environmental engineering / edited by Chloe Miller.
p. cm.
Includes bibliographical references and index.
ISBN 978-1-63987-278-7
1. Environmental engineering. 2. Sustainable engineering. 3. Environmental protection. I. Miller, Chloe.
TA170 .H36 2022
628--dc23

TABLE OF CONTENTS

PREFACE

Every book is a source of knowledge and this one is no exception. The idea that led to the conceptualization of this book was the fact that the world is advancing rapidly; which makes it crucial to document the progress in every field. I am aware that a lot of data is already available, yet, there is a lot more to learn. Hence, I accepted the responsibility of editing this book and contributing my knowledge to the community.

Environmental engineering is a professional engineering discipline that focuses on improving the quality of the environment and protecting the health of living organisms through the application of scientific and engineering principles. It falls under the domain of civil engineering and chemical engineering. An understanding of how technological advances affect the environment, and the local and global environmental effects of global warming, water and air pollution, and acid rain is vital to environmental engineering. This field enables solutions for water and air pollution control, waste disposal, wastewater management, recycling and public health. It also focuses on the evaluation of hazardous waste management systems, design of industrial wastewater treatment plants and municipal water supply systems, as well as the improvement of sanitation in rural, urban and recreational areas. This book brings forth some of the most innovative concepts and elucidates the unexplored aspects of environmental engineering. While understanding the long-term perspectives of the topics, the book makes an effort in highlighting their impacts as a modern tool for the growth of the discipline. The readers would gain knowledge that would broaden their perspective about environmental engineering.

While editing this book, I had multiple visions for it. Then I finally narrowed down to make every chapter a sole standing text explaining a particular topic, so that they can be used independently. However, the umbrella subject sinews them into a common theme. This makes the book a unique platform of knowledge.

I would like to give the major credit of this book to the experts from every corner of the world, who took the time to share their expertise with us. Also, I owe the completion of this book to the never-ending support of my family, who supported me throughout the project.

Editor

1

PROTECTING BIODIVERSITY OUTSIDE PROTECTED AREAS: CAN AGRICULTURAL LANDSCAPES CONTRIBUTE TO BIRD CONSERVATION ON NATURA 2000 IN ITALY?

Mattia Cai[1], Davide Pettenella[2]

Department of Land, Environment, Agriculture and Forestry, University of Padova, Agripolis, V.le dell'Università, 16, 35020 Legnaro (PD), Italy

E-mail: [1]mattia.cai@unipd.it (corresponding author); [2]davide.pettenella@unipd.it

Abstract. Even when a Protected Area (PA) is adequately managed within its boundaries, its ecological functioning may be influenced by human activities in the surrounding landscape through a number of mechanisms. This paper investigates how conservation outcomes are affected by land use change in neighboring areas on a cross-section of protected sites, the Italian portion of the EU's Natura 2000 network. We exploit variation in space to relate indicators of success of conservation efforts to measures of landscape structure. We focus on features of the landscape that pertain to agriculture because this is the dominant form of land use in many areas of the country, and the major ongoing trends in land use dynamics involve farmland. Our results suggest that the patterns and rates of land use change currently observed in Italy have the potential to result in non-negligible adverse consequences for wildlife.

Keywords: landscape management, protected areas, agricultural landscapes, land use change, Natura 2000.

Introduction

Ever since concerns began to emerge over the loss of biodiversity, protected areas (PAs) have represented the cornerstone of conservation strategies throughout the world. Over the past three decades, about 26,000 PAs – which are now collectively known as the Natura 2000 network – have come into existence in the European Union, covering about 20% of its territory (European Commission 2008).

The designation of a site as a PA typically implies restricting the acceptable types of land use in order to preserve species or habitats that are deemed of particular interest. Especially in regions with a long history of habitat conversion, a significant proportion of the species of conservation concern are dependent on PAs for their survival. In Britain, for example, it is estimated that more than a half of such species occur largely or entirely within PAs (Jackson, Gaston 2008).

Even adequate management of a PA within its borders may not be enough to ensure that valuable species and habitats will continue to exist. In general, the outcome of conservation efforts within PAs will depend also on influences coming from the surrounding landscapes, where human activities are not subjected to special regimes of protection. A number of ecological mechanisms have been proposed through which biodiversity within a PA may be affected by land use change in the neighboring areas (Dunning *et al.* 1992; Hansen, DeFries 2007). For instance, if the effective size of the ecosystem is larger than the PA, loss of habitats in its unprotected portions will generally result in smaller populations and higher risk of extinction. It is likely that, as the effective size of the ecosystem shrinks, it is large predators that disappear first, which in turns impacts the trophic structure of the ecosystem. Also, habitats that are seasonally important may be located outside the PA's boundaries. Furthermore, different uses of the lands surrounding a PA entail different levels of water and air pollution, different regimes of fire and flood disturbance, and different levels of exposure to humans.

The effects of processes occurring off-site can be unexpectedly strong on a PA's conservation status. For example, the extinction rates of large mammals in national parks of the western United States correlate more strongly with the density of human population in the surrounding area than they do with park size (Park, Harcourt 2002).

Indeed, conservationists have long recognized the importance of managing not only PAs but also the neighboring lands (Schonewald-Cox *et al.* 1992; Shafer 1999). To a significant extent, the literature on ecological networks and habitat fragmentation is inspired by similar concerns (Harrison, Bruna 1999; Jongman *et al.* 2004; Fischer, Lindenmayer 2007).

Even so, there has been relatively little empirical research on how conservation efforts within PAs are affected by land use change in the surroundings, especially in Europe. Most existing research focuses on individual large reserves located in North America (Hansen, Rotella 2002; Gude *et al.* 2007) or in developing countries (Vester *et al.* 2007; Viña *et al.* 2007). At a much coarser scale, some recent studies have called attention to the threats posed to conservation by the current trends in residential development around PAs in the United States (Wade,

Theobald 2010), Europe (Di Giulio *et al.* 2009) and other regions of the world (Macdonald *et al.* 2008).

In this paper we use a spatially explicit dataset assembled from a variety of sources to investigate the relationship between the conservation status of Italy's Natura 2000 sites and the structure of the landscapes in which they are situated. Because they typically have small surface areas (the median in our sample is 11 km^2), these PAs are potentially highly vulnerable to pressures arising in their surroundings. Indeed, it has been argued that Italy's Natura 2000 network is not sufficient to preserve many of the species for which it was designed, and that attaining the conservation goals will not be possible unless the surrounding matrix is managed as a functional part of the system (Maiorano *et al.* 2007). Furthermore, even though PAs in Italy have generally been effective at protecting the ecosystems within their borders from land use change, in many cases there is significant and widespread land use pressure in the neighboring areas (Maiorano *et al.* 2008).

We pay particular attention to landscape characteristics related to agriculture, as in most cases farming is the dominant form of land use around PAs. On average, about half the land within 20 km of a site in our sample is used for agriculture. In 42% of sample PAs, indeed, managers report concerns that conservation may be negatively affected by changes in agriculture occurring off-site. Furthermore, agriculture is involved in two important long-term trends in land use dynamics. Firstly, as the profitability of farming deteriorated, woodlands have gradually been replacing pastures and extensively cultivated lands in many marginal areas. In several mountainous or hilly parts of the country, this phenomenon has continued for decades (Falcucci *et al.* 2007). Secondly, on the plains and along the coasts significant portions of land have been switching from farming to urban development, as common throughout Europe (e.g. Bauža 2007). For example, it has been calculated that, between 1999 and 2005, 2.3% of Lombardy's agricultural land was converted to artificial uses (Anon 2009). As is clear from Fig. 1, there is a widespread tendency for the proportion of Italian landscapes used for agriculture to shrink, albeit at differing rates. Based on Corine Land Cover data, we compute that between 1990 and 2000, agricultural surface area declined in about 39% of Italy's 8,101 municipalities (the smallest administrative units). Increases only appear to have taken place in less than 4% of municipalities. This is a cause of concern because changes in farming systems have been shown to produce potentially dramatic effects on biodiversity. On the one hand, the widespread declines in wildlife observed in Europe over the past few decades have been widely blamed on the intensification of farming (Robinson, Sutherland 2002; Wickramasinghe *et al.* 2003; Báldi, Faragó 2007; Hanley *et al.* 2009), and a number of common farming practices have been linked with adverse effects on various components of biodiversity (see, for example, Tilman 1984; Freemark, Boutin 1995; McLaughlin, Mineau 1995; De Snoo 1999; Kladivko 2001). Indeed, intensively farmed land temporarily set-aside from production – as mandated by the European Commission schemes – has been unequivocally shown to enhance biodiversity (Van Buskirk, Willi 2004). The negative implications of intensifying farming are particularly well documented in the case of birds (Chamberlain *et al.* 2000; Donald *et al.* 2001; Benton *et al.* 2002). Yet, reductions in agricultural area should not be expected to unambiguously benefit wildlife. For a start, in several parts of Italy there has been a conversion of farmed land into uses, such as urban development, that are at best no more beneficial to wildlife than agriculture. In addition, low intensity agriculture and traditional farming practices often support high levels of bird (Farina 1997; Fox 2004; Laiolo *et al.* 2004; Verhulst *et al.* 2004), plant (Fischer, Wipf 2002; Luoto *et al.* 2003; Sendžikaite, Pakalnis 2006) and other (e.g. Zervas 1998; Giupponi *et al.* 2006) biodiversity. European landscapes have been shaped by agriculture over centuries and it has been speculated that at least 50% of Europe's most valuable biotopes are found on farmland that is managed extensively (Bignal, McCracken 1996). In the Mediterranean basin, where only 4.7% of the primary vegetation remains, high environmental diversity is the result of anthropogenic disturbances over millennia (Falcucci *et al.* 2007). Abandoning farming could therefore place the local survival of many species at risk.

In fact, not only is land abandonment widespread in Europe and typically associated with negative consequences for the environment (MacDonald *et al.* 2000), but there also seems to be more of it ahead. Deteriorating farm profitability as a result of Common Agricultural Policy reform is likely to result in more land being abandoned in areas of high significance for wildlife but low agricultural productivity (Oñate *et al.* 2007). Demography, international trade and technological change are also expected to contribute to this trend (Verburg *et al.* 2006).

The conservation implications of land use trends in Italy's landscapes feature as an important issue in the ongoing debate on the definition of a national biodiversity strategy for the next decade (Ministero dell'Ambiente 2010). While a substantial proportion of the residual area of valuable habitats in the country is probably already under protection, this does not make it immune from the influence of land use change in the vicinity. This paper attempts to assess to what extent the existence of a buffer zone around valuable natural sites contributes to maintaining that nature in good conservation status and how important agricultural forms of land use around Natura 2000 PAs are in providing support for wildlife. Our analysis exploits spatial variation across PAs to relate indicators of success of conservation efforts to site characteristics and measures of the structure of the surrounding landscape. We show that valuable natural habitats inside a PA generally have a better conservation status when the PA includes some kind of internal buffer area within its borders. Focusing on birds, we also find that – controlling for a variety of confounding factors – wildlife is richer on PAs situated in landscapes that are largely used for low-intensity agriculture.

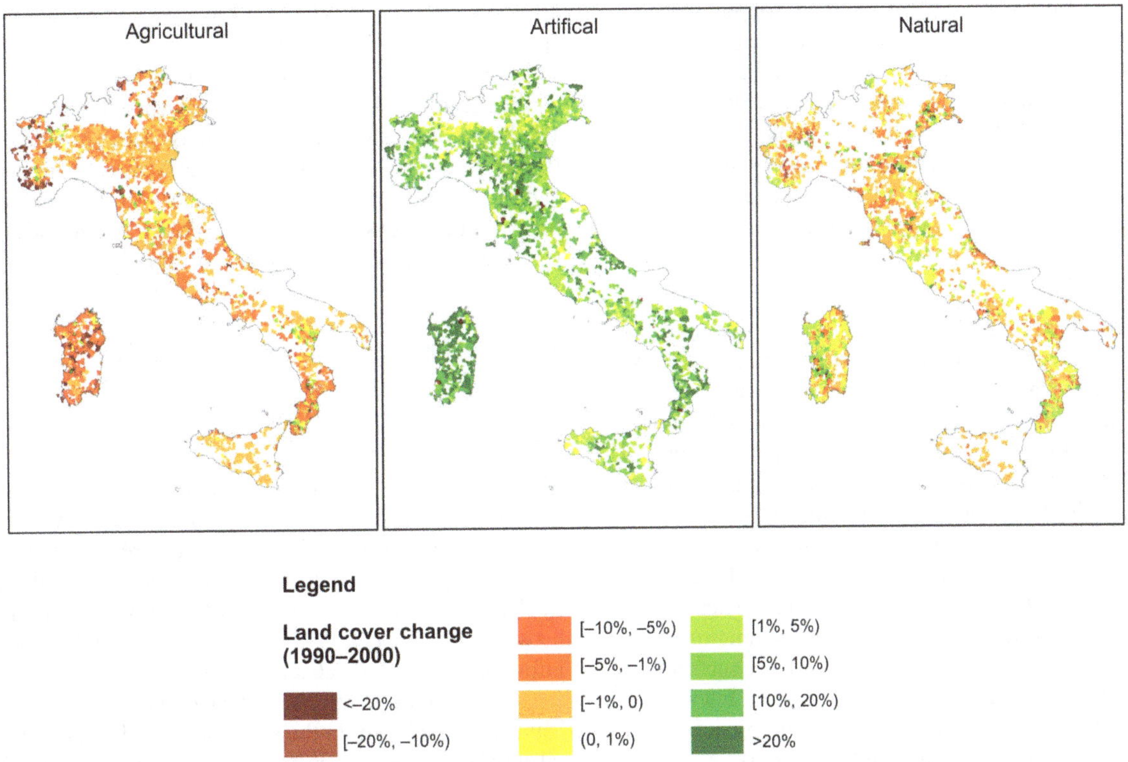

Fig. 1. Percent land cover change by municipality (1990–2000); smoothed rates using locally weighted estimator with 5-km radius; areas represented in white experienced no change

1. Methods

1.1. Methodological framework

The studied PAs are located throughout Italy and belong to the Natura 2000 network established in compliance with EU Directives 79/409/EEC ('Bird Directive') and 92/43/EEC ('Habitat Directive'). Using the cross-sectional dataset described in section 2.2, we exploit variation across sites to perform two types of statistical analysis.

1.1.1. Buffer zones and habitat conservation status

As a first step, we examine the relationship between the conservation status of the habitats a PA is designed to protect and the amount of additional 'buffer' zones that are included within the PA's boundaries. The conservation indicator that we use as the dependent variable is the share of valuable habitat surface whose conservation status is reported by PA managers as not compromised. Using ordinary least squares (OLS) regression, we relate that indicator to a number of bio-geographic site characteristics and a variable measuring the share of PA surface that is not occupied by habitats of conservation concern (MARGIN). We take this variable to indicate the extent to which a PA may include within its boundaries some kind of buffer zone around its most valuable areas, and attempt to evaluate its statistical and practical significance for PA conservation.

1.1.2. Bird species richness and landscape structure

The conservation status of a PA, we postulate, depends on both its local features and the structure of the surrounding landscape matrix. In the second step of the analysis, a linear regression model is estimated in which (the natural logarithm of) bird species richness is explained as a function of characteristics of both the PA and the landscape beyond its boundaries. Our aim is to assess the consequences for biodiversity of different types of land use and different levels of human disturbance in the surroundings of a PA, while controlling for the presence of potentially confounding factors.

1.1.3. Site selection

All the analyses focus on birds and on those Natura 2000 areas – known as Special Protection Areas (SPAs) – that were designed with an explicit emphasis on their conservation. Because birds perceive the landscape at a broader spatial scale than less vagile taxa, the issue of offsite influences on conservation outcomes within PAs appears both particularly relevant and easier to assess at the coarse spatial scale that we use. Even though in general it has proved difficult to identify taxa that can be used as predictors of general biodiversity patterns (Billeter *et al.* 2008), there is some evidence that bird indicators can provide a useful surrogate for information on other elements of biodiversity in agricultural landscapes (Gregory *et al.* 2005). In some countries, indeed, the government routinely uses indicators based on bird populations to monitor environmental quality in the countryside (e.g. Anon 2007). Irrespective of their relevance as a proxy for broader biodiversity, from an economic standpoint birds are associated with non-negligible cultural and recreational benefits, a classic example being bird-watching (e.g. Woodward, Wui 2001).

In addition, most SPAs were created earlier than other sites in the Natura 2000 network and more reliable data are typically available for them. As opposed to PAs of more recent creation, SPAs have a longer history of being managed to preserve or restore valuable habitats, and the effects should be visible. Thus the risk of mistaking the effect of deteriorated local habitat for negative influences arising in the surrounding landscape is small. In fact, there is a substantial amount of overlap between SPAs and the remaining Natura 2000 sites.

1.1.4. Covariates

For convenience, the explanatory variables that appear in the various stages of the analysis are divided into two groups: 'site characteristics' that describe the local features of the PA (either pertaining to its natural environment or to the way it is managed); and 'landscape characteristics' that describe aspects of the landscape beyond the PA boundaries (its structure and the intensity of land use) that represent the main object of our interest.

While its actual content varies among the different analyses, the vector of general site characteristics is composed of the following variables: latitude, longitude, elevation above sea level, site area, share of PA surface used for agriculture, and a set of binary indicators for wetland, coastal area, location in the Po Valley, and biogeographic region (Alpine, continental, Mediterranean). A dummy variable for national park or reserve is also included in order to control for the higher level of protection granted to some sites and their longer history.

Potentially, biodiversity and conservation outcomes at a given location can be affected by a number of other factors that cannot be accounted for here (temperatures, rainfall, moisture, and so forth; see, for an example, Okzan *et al.* 2009). In practice, however, many of the environmental factors that influence biodiversity are often highly collinear (Gaston, Williams 1996). At a sufficiently large scale, a substantial proportion of spatial variation in species richness can be explained statistically in terms of a few environmental variables (Gaston 2000).

Landscape characteristic variables are meant to account for the composition of the landscape in a buffer area around the PA and the intensity of land use or other disturbance by humans. For the purpose of this analysis, we distinguish three classes of land use/land cover: agricultural, artificial, and a residual class that includes woodlands and other natural areas, which is labeled 'natural' for convenience.

In addition to the proportion of buffer area used for agriculture (FARM) and artificial purposes (URBAN), our vector of landscape variables includes the natural logarithm of the distance to the nearest highway (HIGHWAY). Inclusion of this explanatory variable is meant both to account for the potential adverse effects of closeness to road networks (Csereklye 2010) and to serve as a proxy for the intensity of disturbance from human presence. Also, in order to allow the effect of agriculture to depend on the intensiveness of farming, a term is also included that is the interaction of FARM with an indicator of pesticide use (FARM×PEST).

An important issue in our analysis is how far from a PA's boundaries can land use change affect conservation. This question appears to have no general answer, as the appropriate buffering distance seems to vary substantially across locations and taxonomic groups. For instance, Park and Harcourt (2002) use zones of influence that are 50- and 100-km wide to study large mammals. An investigation by Bolger *et al.* (1997) of the relative influence on bird abundance of local versus landscape factors works with buffering distances between 250 m and 3 km, but sensitivity to landscape characteristics appears to vary significantly even among bird species (Flather, Sauer 1996). We consider a 5-km buffer around each PA, but all the analyses were also performed for longer distances (10, 15, and 20 km) to verify the sensitivity of the results to this assumption.

1.2. Data

The data used in the analysis were collected from a variety of administrative sources. Information about Italy's Natura 2000 areas was obtained from a database maintained by the Ministry of the Environment. The data were submitted by PA managers over the course of some years, but mostly between 2004 and 2007. They include digital maps and information on general site characteristics, the presence and conservation status of protected species and types of habitat, land use within the site, human activities and natural processes that may influence the site's ecological functioning. More detailed information about the data collection process for Natura 2000 – including the questionnaire – is available from the European Commission (Commission Decision No. 97/266/EEC, O.J. L 107/1).

PA managers were required to report what types of habitats protected under EU legislation were present on the site and what surface area each of them occupied, and to rate the conservation status of each on a three-point scale ('excellent', 'good', 'average or reduced'). We used that information to compute for each PA the proportion of valuable habitat surface whose conservation status is not compromised (i.e. that did not receive the lowest grade), which represents the dependent variable in our first analysis. The natural logarithm of bird species richness – the dependent variable in our second analysis – was calculated as the total count of protected bird species that are found on the PA. Most information about site characteristics (geographical coordinates, elevation a.s.l., and so forth) was also obtained from this dataset.

Data about the structure of the landscapes surrounding the PAs were extracted using GIS software (ESRI ArcInfo) from the land cover maps produced by the Corine Land Cover (CLC) project for the year 2000. Our definitions of the land-use types 'artificial' and 'agricultural' match CLC classes 1 and 2, respectively. The residual type 'natural' includes all remaining CLC classes. Additional information was obtained from Italy's bureau of statistics (ISTAT), institute of agricultural economics (INEA), and environmental protection agency (APAT). In particular, cereal yields (YIELD) and land prices (LLANDVAL) in the surroundings of the PAs were estimated from official statistics aggregated at the province level. Table 1 reports a brief description of all the variables that are used at some point in the empirical analysis, along with summary statistics and the source of the data.

Table 1. Descriptive statistics (N. Obs = 300; only 289 obs. on HABCONS)

Variable	Description	Mean	Std. Dev.	Source
Conservation indicators				
LBIRD	Log bird species richness	3.40	1.03	N2000
HABCONS	Proportion of protected habitat surface in good conservation status	0.93	0.13	N2000
Site characteristics				
LAREA	Log PA surface (ha)	6.84	1.59	N2000
ELEV	Mean site elevation (100 m a.s.l.)	2.36	2.78	N2000
LAT	Latitude (degr. north)	43.5	2.30	N2000
LON	Longitude (degr. east)	11.6	2.20	N2000
MARGIN	Proportion of site area not occupied by valuable habitats	0.55	0.31	N2000
MEDIT	=1 if Mediterranean biog. region	0.42	0.49	N2000
ALPINE	=1 if Alpine biog. region	0.07	0.25	N2000
PO	=1 if in Po watershed	0.33	0.47	APAT
WET	=1 if wetland	0.30	0.46	N2000
PARK	=1 if national park or reserve	0.45	0.50	N2000
Landscape characteristics				
FARM	Share of 5-km buffer surface occupied by agricultural areas	0.50	0.29	CLC2000
URBAN	Share of 5-km buffer surface occupied by artificial surfaces	0.08	0.09	CLC2000
HIGHWAY	Log distance to nearest highway (km)	2.59	1.23	APAT
LPEST	Log pesticide use (kg a.i./ha ag. land)	1.62	0.82	ISTAT
Other variables				
YIELD	Average cereal yield (100 kg/ ha, avg. 2000–2004)	55.6	25.0	ISTAT
LLANDVAL	Log land price (1,000/ha, avg. 1992–2000)	2.61	0.61	INEA

As mentioned in previous sections, we focus on SPAs, the PAs in Natura 2000 designated for bird conservation. PAs that are located at very high elevation (>1000 m a.s.l.), purely marine areas, and sites with extremely large or extremely small surfaces were excluded from the analysis. The resulting sample is comprised of 300 observations out of 588 SPAs.

All statistical computations were performed in Stata (StataCorp 2007), except for the spatial analyses that were conducted using R (R Development Core Team 2009) and the routines described in Bivand *et al.* (2008).

2. Results

2.1. Buffer zones and habitat conservation status

The first step of the analysis is concerned with the health status within PAs of the habitats of conservation interest. Using ordinary least squares (OLS), we studied its relationship with a measure of the extent to which PA design allows for a buffer area around those habitats (MARGIN). Table 2 reports two sets of results which only differ by whether or not MARGIN is included in the model. While it appears that there is a substantial amount of variation in conservation outcomes which our model is not able to able to account for, the covariates we use are jointly highly significant in either specification ($F_{6.282}$ = 4.18 and $F_{7.281}$ = 4.38, respectively). When MARGIN is added to the model (column 2), its coefficient is positive and highly significant, and the overall fit of the model improves remarkably. Including the landscape characteristics FARM, URBAN and HIGHWAY (column 3) further improves fit, but these variables – although signed consistently with the results of subsequent analyses – are not significant either individually or jointly ($F_{3.278}$ = 2.04).

From a methodological standpoint, a potential issue with the estimates in Table 2 is the presence of spatial correlation. To the extent that nearby PAs resemble one another more than PAs located farther away with respect to unobserved factors, OLS estimation may not be the most satisfying option. We tested for the presence of spatial autocorrelation using Moran's I statistic for regression residuals and found little evidence of it (p = .212).

As expected, we observed that the share of PA valuable habitat area that is in satisfactory conditions of conservation tends to be larger on PAs that also include additional 'buffer' zones within their boundaries. This observation is consistent with the notion that the areas surrounding valuable natural sites can have an important influence on their conservation, and that allocating those areas to more nature-friendly types of land use can contribute to the viability of a PA. In fact, the effect of MARGIN seems modest: the results in column 2 imply that, at the median of the data, adding 100 hectares of buffer zone to a PA – which represents 9% and 15% increases in total area and in buffer zone, respectively – is associated with an average increase of about .4 percentage points in HABCONS.

Table 2. Share of habitat area with conservation status not compromised: OLS estimates (robust standard errors in brackets). Legend: * $p < .1$; ** $p < .05$; *** $p < .01$

	(1)	(2)	(3)
LAREA	0.009*	0.010**	0.009*
	(0.005)	(0.005)	(0.005)
ELEV	0.008**	0.009**	0.010**
	(0.004)	(0.004)	(0.004)
MEDIT	–0.031	–0.016	–0.020
	(0.022)	(0.022)	(0.023)
ALPINE	–0.092**	–0.081**	–0.071*
	(0.042)	(0.039)	(0.040)
WET	0.034*	0.037*	0.038**
	(0.019)	(0.019)	(0.019)
PO	0.063***	0.045**	0.044**
	(0.022)	(0.020)	(0.019)
MARGIN		0.101***	0.105***
		(0.029)	(0.032)
FARM			0.039
			(0.034)
URBAN			–0.036
			(0.126)
HIGHWAY			0.012
			(0.007)
Intercept	0.841***	0.770***	0.728***
	(0.044)	(0.053)	(0.070)
R^2	0.096	0.136	0.153
N	289	289	289

2.2. Bird species richness and landscape structure

As a second step, we examined how a site's bird species richness relates to the features of the site itself and the landscape in which it is situated. The estimates from OLS regression of the natural logarithm of bird species richness on site and landscape characteristics are displayed in the first two columns of Table 3.

In the first specification (column 1), landscape characteristics are excluded from the model. All coefficients have plausible signs and the overall fit is acceptable. When included in the model (column 2), FARM, FARM×PEST, URBAN and HIGHWAY turn out significant both jointly ($F_{4,287}$ = 7.15) and individually, and improve the model fit considerably. Species richness on PAs appears to be positively associated with the proportion of the surrounding landscape that is used for agriculture. As expected, landscapes where farming is more intensive appear less welcoming to wildlife than those farmed at lower intensities. Also, species richness tends to be greater away from highways. On the other hand, the sign on URBAN is disturbingly positive and significant.

As in the previous section, we tested for the presence of spatial autocorrelation among PAs using Moran's I. Because in this case independence of the observations is rejected (p < .001), an alternative specification of the model is estimated which allows for spatial correlation. Column 3 in Table 3 reports the estimates from a spatial autoregressive error model (SAR) that uses an inverse-distance weight matrix (Anselin 1988). While the results remain qualitatively unchanged, the newly estimated coefficients on landscape characteristics have generally smaller magnitudes. This is especially true of URBAN, which indeed is no longer significant. Statistical significance is also lost for FARM×PEST, which however retains the appropriate sign.

Table 3. Log bird species richness: estimates from alternative specifications (Legend: * $p < .1$; ** $p < .05$; *** $p < .01$; standard errors in brackets)

	OLS (1)	OLS (2)	SAR (3)	2SLS (4)
LAREA	0.225***	0.224***	0.233***	0.227***
	(0.031)	(0.030)	(0.026)	(0.030)
ELEV	–0.093***	–0.053**	–0.074***	–0.050**
	(0.020)	(0.023)	(0.022)	(0.023)
LAT	–0.077**	–0.083**	–0.010	–0.141**
	(0.035)	(0.035)	(0.053)	(0.057)
PARK	0.211**	0.263***	0.406***	0.270***
	(0.096)	(0.095)	(0.083)	(0.094)
MEDIT	–0.636***	0.783***	–0.460*	0.691***
	(0.163)	(0.164)	(0.239)	(0.178)
ALPINE	0.723***	0.814***	0.736***	0.886***
	(0.205)	(0.213)	(0.227)	(0.218)
WET	0.612***	0.608***	0.570***	0.589***
	(0.113)	(0.109)	(0.098)	(0.109)
PO	0.814***	0.733***	0.686***	0.672***
	(0.127)	(0.124)	(0.165)	(0.131)
FARM		1.254***	0.719**	1.168***
		(0.307)	(0.331)	(0.311)
URBAN		1.485**	0.912	1.014
		(0.593)	(0.593)	(0.695)
HIGHWAY		0.132***	0.121**	0.138***
		(0.043)	(0.050)	(0.043)
FARM×LPEST		–0.245**	–0.120	–0.265**
		(0.120)	(0.134)	(0.119)
YIELD (indicator)				0.010
				(0.008)
Intercept	5.113***	4.474***	1.366	6.443***
	(1.582)	(1.567)	(2.388)	(2.202)
lambda			0.497***	
log-likelihood			–305.4	
R^2	0.433	0.484		0.475
N	300	300	300	300

At the median of the data, the estimates in Table 3 suggest that, when 1% of the landscape area surrounding a PA (roughly 160 ha) is converted away from agriculture, bird species richness declines on average by a non-negligible .5% to .9%, (depending on which set of estimates is used). In other words, it would be sufficient that 4% to 6.5% of the buffer zone is withdrawn from agriculture for the change to result in the local disappearance of 1 species. It might be tempting to compare the estimated effects of changes in the landscape matrix to the coefficient on PARK, a binary indicator of the site's status as national park or reserve, which is typically associated with both a longer history of habitat protection and stricter management rules. On average, sites granted protection at the national level appear to enjoy approximately 50% more bird species. As this result is due to better local habitat conditions resulting from longer and more effective conservation efforts, it suggests that, although appropriate interventions in the landscape matrix may contribute to conservation, careful management of the lands within PA boundaries is likely to produce more conspicuous effects. In fact, the large coefficient on PARK is likely to reflect not only the biodiversity consequences of better local habitats, but also the principles that informed the selection of national parks: higher protection was presumably given to richer sites in the first place (as in Paiders 2008).

An analysis very similar to the one presented in this section was also performed on a measure of bird conservation status analogous to the indicator used for habitats in section 2.1. In this case, the dependent variable of the regression model is the proportion of protected bird species whose conservation status is described as not compromised. Because the data on bird conservation status from which our indicator is computed were often incomplete, estimates from the analysis are not presented here. Yet, the results are consistent with those in Table 3.

2.3. Estimation issues

A potentially serious issue with the results of the previous section is the possibility that some of the site and landscape characteristics that we examined are correlated with unobserved determinants of biodiversity. Suppose, for example, that whatever makes an area welcoming to biodiversity also causes it to be attractive to humans. On the one hand, an area that is comparatively more attractive in this sense would support relatively high levels of wildlife, and would perhaps be more resistant or resilient in the face of human disturbance. On the other hand, such an area would tend to have a larger anthropic component: because land is more valuable, smaller portions of the landscape would be set aside for conservation, and larger portions would be farmed or developed. If there is some factor – which, for concreteness, we call 'resources' – that influences both our biodiversity and land use indicators, it is important that our analyses control for it, lest the associations we observe be spurious.

There are reasons to be concerned that a mechanism of this type might be at work. A significant body of literature has documented positive correlations between human population density and species richness in several taxa (McKinney 2002; Araújo 2003; Chown *et al.* 2003; Evans *et al.* 2006) including birds (Gaston, Evans 2004), with the strength of those associations increasing as the spatial scale of the analysis becomes coarser (Pautasso 2007). There is indeed agreement that the main reason why such positive species-human relationships arise is that both species richness and human population respond similarly to environmental energy availability. Also, the data analysis of the previous sections provides a warning sign that bias from ignoring 'resources' may be an issue: the counterintuitive sign on URBAN in the bird species richness equation.

With regard to the analysis of section 2.1, it should be noted that neglecting to control for 'resources' should not invalidate the main conclusion, namely that the conservation status of valuable habitats can be improved through appropriate management of the surrounding lands. As long as long-term planning choices at the landscape level to some extent reflect the relative costs and benefits of different types of land use, LAREA and MARGIN will tend to be smaller in areas with greater availability of resources. Reasonably, such negative correlations would result in the coefficients on LAREA and MARGIN being biased downwards rather than upwards.

On the other hand, the issue looks more serious for the bird species richness equation of section 2.2. In theory, indeed, it is possible that the observed positive association of biodiversity with human-related types of land use (e.g. with FARM) are entirely due to the latter being positively correlated with lurking 'resources'.

Plausibly, including a variety of control variables in the model should be enough, if not to address, at least to mitigate the problem. As we argued above, it is likely that the effect of many unobserved factors influencing biodiversity can be captured by relatively few environmental macro-descriptors.

In principle it is possible, using instrumental variable (IV) procedures, to test – and correct, if need be – for correlation between landscape characteristics and the unobservables that are dumped in the error term of the equation. Doing so would require identifying the (potentially) endogenous regressors that are correlated with 'resources' and coming up with at least as many instrumental variables that have explanatory power for the endogenous regressors, while being uncorrelated with 'resources'. In our application, however, IV estimation looks impractical, because the list of potentially endogenous regressors is long and data are in short supply on variables that would make for defensible instruments. Therefore, we used a variation of this approach that is sometimes referred to as the multiple indicator solution. The procedure, which requires no assumptions as to which regressors are potentially endogenous, makes use of two indicators of 'resources', one to be included in the equation of interest, the other to be used as an instrument for the first in order to carry out two-stage least squares (2SLS) estimation. Readers are referred to Wooldridge (2002) for a detailed description of the procedure.

The two indicators that we used are average cereal yields (YIELD) and the log of average land prices (LLANDVAL) in the surroundings of PAs. The estimates from multiple indicator estimation are reported in the fourth column of Table 3. Remarkably, all coefficients are left substantially unaffected, with the notable exception of that on URBAN which drops remarkably in magnitude and loses statistical significance. The coefficient on LPOP, the resource indicator, while positive, also falls short of statistical significance, but that should not be too surprising given that two-stage estimation typically produces large standard errors. As appropriate, the key results remain essentially unaffected when the roles of LLANDVAL and YIELD are reversed in the estimation.

Taken together, these results do provide some evidence of a confounding effect of 'resources'. When no attempt is made to account for the effect of resources, that effect appears to be largely picked up by the coefficient on URBAN, which consequently turns out positive and large; once 'resources' are controlled for, the coefficient on URBAN drops in magnitude dramatically. By contrast, our conclusions on the importance of the surrounding landscape matrix – and more specifically of its agricultural component – for PA conservation hold up across model specifications, and do not appear to merely result from neglecting to control for differences in resource availability.

3. Discussion

The network of PAs known as Natura 2000 represents the cornerstone of Europe's biodiversity conservation strategy. In Italy, N2000 sites are often small and situated in landscapes dominated by human presence. In many cases, especially along the coasts and on the plains, the areas around the PAs are undergoing substantial land use change (Maiorano *et al.* 2008). Against this backdrop, it has been proposed that, because small PAs are heavily influenced by what goes on outside their boundaries, for those PAs to be viable in the long run, it is necessary to manage the surrounding landscape matrix as a part of the system (Maiorano *et al.* 2007).

In this paper, we have observed that, when a PA includes within its boundaries some buffer zones in addition to the habitats of conservation interest that motivated its creation, those valuable habitats are generally in a better status of conservation. Also, we have found that, controlling for a variety of potentially confounding factors, the richness of protected bird species is significantly associated with the characteristics of the surrounding landscape. In particular, our measure of bird species richness is positively related to the share of the landscape matrix that is used as agricultural land, the more so the less intensive farming appears to be.

As there is a sizable literature linking expansions of agricultural land use and biodiversity loss, the observed positive associations between birds and farmland require an explanation. One possible interpretation of our results is that, because agriculture has dominated Italy's landscapes for centuries – so that species either adapted or disappeared – many of the remaining species require at least some level of farming activity for their survival (Kleijn, Baldi 2005). On the other hand, it is possible that the observed positive association between birds and agriculture is merely the result of neglecting to control for some factor – such as 'resources' – that simultaneously influences both the level of biodiversity and extent of farmland in a landscape. While it is difficult, given the limited availability of data, to devise an unquestionable statistical setup to control for this possibility, our analysis has attempted to do so in a 2SLS framework.

Assessing at what distance land use change around a PA ceases to have an influence on conservation efforts is difficult. Based on the findings of previous research, a 5-km buffering distance was chosen for the analysis reported in this paper. At this distance, bird species richness is associated with landscape characteristics, whereas the conservation status of protected habitats is not (after allowing for the presence within PA boundaries of an internal buffer zone). Repeating the analyses using longer distances (10, 15 or 20 km) leaves these results qualitatively unaffected. In fact, given the resolution of the data, landscape composition remains relatively stable as the buffering distance increases. Ultimately, how far offsite influences will affect conservation outcomes will vary across locations and taxa.

By and large, this paper has been concerned with birds, vagile organisms for which the issue of changing land use around protected areas seems especially relevant. Focusing on PAs that were designed primarily for bird protection (SPAs) ensured better data availability, and it is a natural choice to use bird indicators to evaluate the conservation outcomes of those PAs. In fact, although it has been proposed that bird indicators may also provide useful information about other components of biodiversity, evidence that those components respond to the landscape around the PAs in similar ways is relatively thin.

With all these caveats, however, the results presented in this paper show that the relationship between N2000 PAs and their surroundings deserves some consideration on behalf of landscape managers and researchers. Our estimates indicate that the effects of changes in agricultural landscapes on bird species richness are non-negligible – especially at very low intensity of farming – and suggest that the patterns and rates of land use change observed in the recent past in many parts of the country have the potential to result in significant adverse consequences for wildlife.

Conclusions

1. Changes in land use in the surroundings of a protected area have an influence on the protected area's conservation status. Allowing for a 'buffer area' around the habitats of conservation concern is found to increase the chances that those habitats are preserved in good conditions.

2. In particular, the presence of agricultural landscapes around a protected area appears to affect positively some components of biodiversity, namely the richness of birds. This effect tapers off as farming becomes more intensive.

3. The magnitude of the landscape effects that we observed for birds suggests that the changes in land use that are taking place in many parts of Italy represent a non-negligible threat to the long-term viability of many small protected areas of the Natura 2000 network. In this case, adoption of specific zoning rules for the lands neighboring with protected areas may be justified.

4. To the extent that other taxonomic groups are found to be affected in a similar way as birds, preserving agricultural landscapes farmed at low intensity may even, under some circumstances, contribute to the preservation of wildlife in a more cost effective way than expanding protected areas.

Acknowledgements

This paper is based on research conducted within the EU-funded EXIOPOL project 'A new environmental accounting framework using externality data and input-output tools for policy analysis'. The Italian Ministry of the Environment - Directorate for Nature Conservation provided the Natura 2000 dataset used in the analysis. Earlier versions of this paper was presented at the 11th Bioecon conference in Venice, September 21–22, 2009, and at the 47th SIDEA annual meeting in Campobasso, September 22–25, 2010. The authors are indebted to two anonymous referees for valuable comments. All errors and omissions are solely ours.

References

Anon 2007. *Sustainable development indicators in your pocket 2007*. London, UK: Department for the Environment, Food and Rural Affairs. 132 p.

Anon 2009. *Osservatorio nazionale sui consumi di suolo*. Primo rapporto 2009. Sant'Arcangelo di Romagna: Maggioli Editore. 128 p. (in Italian)

Anselin, L. 1988. *Spatial econometrics: methods and models*. Dordrecht: Kluwer Academic Publishers. 284 p.

Araújo, M. B. 2003. The coincidence of people and biodiversity in Europe, *Global Ecology and Biogeography* 12: 5–12. http://dx.doi.org/10.1046/j.1466-822X.2003.00314.x

Báldi, A.; Faragó, S. 2007. Long-term changes of farmland game populations in a post-socialist country (Hungary), *Agriculture, Ecosystems and Environment* 118: 307–311. http://dx.doi.org/10.1016/j.agee.2006.05.021

Bauža, D. 2007. Estimation and trends of landscape transformation in the second half of the 20th century, *Journal of Environmental Engineering and Landscape Management* 15(2): 119–124.

Benton, T. G.; Bryant, D. M.; Cole, L.; Crick, H. Q. P. 2002. Linking agricultural practice to insect and bird populations: a historical study over three decades, *Journal of Applied Ecology* 39: 673–687. http://dx.doi.org/10.1046/j.1365-2664.2002.00745.x

Bignal, E. M.; McCracken, D. I. 1996. Low-intensity farming systems in the conservation of the countryside, *Journal of Applied Ecology* 33: 413–424. http://dx.doi.org/10.2307/2404973

Billeter, R.; Liir, J.; Bailey, D.; Bugter, R.; Arens, P.; Augenstein, I.; Aviron, S.; Baudry, J.; Bukacek, R.; Burel, F.; Cerny, M.; De Blust, G.; De Cock, R.; Diekötter, T.; Dietz, H.; Dirksen, J.; Dormann, C.; Durka, W.; Frenzel, M.; Hamersky, R.; Hendrickx, F.; Herzog, F.; Klotz, S.; Koolstra, B.; Lausch, A.; Le Coeur, D.; Maelfait, J. P.; Opdam, P.; Roubalova, M.; Schermann, A.; Schermann, N.; Schimdt, T.; Schweiger, O.; Smulders, M. J. M.; Speelmans, M.; Simova, P.; Verboom, J.; van Wingerden, W. K. R. E.; Zobel, M.; Edwards, P. J. 2008. Indicators for biodiversity in agricultural landscapes: a pan-European study, *Journal of Applied Ecology* 45: 141–150. http://dx.doi.org/10.1111/j.1365-2664.2007.01393.x

Bivand, R. S.; Pebesma, E. J.; Gómez-Rubio, V. 2008. *Applied spatial data analysis with R*. Springer. 378 p.

Bolger, D. T.; Scott, T. A.; Rotenberry, J. T. 1997. Breeding bird abundance in an urbanizing landscape in coastal Southern California, *Conservation Biology* 11(2): 406–421. http://dx.doi.org/10.1046/j.1523-1739.1997.96307.x

Chamberlain, D.; Fuller, R.; Bunce, R.; Duckworth, J.; Shrubb, M. 2000. Changes in the abundance of farmland birds in relation to the timing of agricultural intensification in England and Wales, *Journal of Applied Ecology* 37: 771–788. http://dx.doi.org/10.1046/j.1365-2664.2000.00548.x

Chown, S. L.; van Rensburg, B.; Gaston, K. J.; Rodrigues, A. L.; van Jaarsveld, A. S. 2003. Energy, species richness, and human population size: conservation implications at the national scale, *Ecological Applications* 13(5): 1233–1241. http://dx.doi.org/10.1890/02-5105

Csereklye, E. K. 2010. Monitoring of landscape combinations and concourses in the Hungarian Danube-bend, *Journal of Environmental Engineering and Landscape Management* 18(1): 5–12. http://dx.doi.org/10.3846/jeelm.2010.01

de Snoo, G. R. 1999. Unsprayed field margins: effects on environment, biodiversity and agricultural practice, *Landscape and Urban Planning* 46: 151–160. http://dx.doi.org/10.1016/S0169-2046(99)00039-0

Di Giulio, M.; Holderegger, R.; Tobias, S. 2009. Effects of habitat and landscape fragmentation on humans and biodiversity in densely populated landscapes, *Journal of Environmental Management* 90: 2959–2968. http://dx.doi.org/10.1016/j.jenvman.2009.05.002

Donald, P. F.; Green, R. E.; Heath, M. F. 2001. Agricultural intensification and the collapse of Europe's farmland bird populations, *Proceedings of the Royal Society B* 268: 25–29. http://dx.doi.org/10.1098/rspb.2000.1325

Dunning, J. B.; Danielson, B. J.; Pulliam, H. R. 1992. Ecological processes that affect populations in complex landscapes, *Oikos* 65(1): 169–175. http://dx.doi.org/10.2307/3544901

European Commission 2008. *Management of Natura 2000 sites: best practices* [Online]. [Cited September 2009]. Available from Internet: http://ec.europa.eu/environment/nature/natura2000/

ESRI. ArcInfo® Workstation version 9.2 GIS software.

Evans, K. L.; van Rensburg, B.; Gaston, K. J.; Chown, S. L. 2006. People, species richness and human population growth, *Global Ecology and Biogeography* 15: 625–636. http://dx.doi.org/10.1111/j.1466-8238.2006.00253.x

Falcucci, A.; Maiorano, L.; Boitani, L. 2007. Changes in land-use/land-cover patterns in Italy and their implications for biodiversity conservation, *Landscape Ecology* 22: 617–631. http://dx.doi.org/10.1007/s10980-006-9056-4

Farina, A. 1997. Landscape structure and breeding bird distribution in a sub-Mediterranean agro-ecosystem, *Landscape Ecology* 12: 365–378. http://dx.doi.org/10.1023/A:1007934518160

Fischer, J.; Lindenmayer, D. B. 2007. Landscape modification and habitat fragmentation: a synthesis, *Global Ecology and Biogeography* 16: 265–280. http://dx.doi.org/10.1111/j.1466-8238.2007.00287.x

Fischer, M.; Wipf, S. 2002. Effect of low-intensity grazing on the species-rich vegetation of traditionally mown subalpine meadows, *Biological Conservation* 104: 1–11. http://dx.doi.org/10.1016/S0006-3207(01)00149-5

Flather, C. H.; Sauer, J. R. 1996. Using landscape ecology to test hypotheses about large-scale abundance patterns in migratory birds, *Ecology* 77(1): 28–35. http://dx.doi.org/10.2307/2265651

Fox, A. D. 2004. Has Danish agriculture maintained farmland bird populations?, *Journal of Applied Ecology* 41: 427–439. http://dx.doi.org/10.1111/j.0021-8901.2004.00917.x

Freemark, K.; Boutin, C. 1995. Impacts of agricultural herbicide use on terrestrial wildlife in temperate landscapes: a review with special reference to North America, *Agriculture, Ecosystems and Environment* 52: 67–91. http://dx.doi.org/10.1016/0167-8809(94)00534-L

Gaston, K. J. 2000. Global patterns in biodiversity, *Nature* 405: 220–227. http://dx.doi.org/10.1038/35012228

Gaston, K. J.; Evans, K. L. 2004. Birds and people in Europe, *Proceedings of the Royal Society B* 271: 1649–1655. http://dx.doi.org/10.1098/rspb.2004.2782

Gaston, K. J.; Williams, P. H. 1996. Spatial patterns in taxonomic diversity, Chapter 8 in Gaston, K. J. (Ed.). *Biodiversity: a biology of numbers and difference*. Oxford: Blackwell Science, 202–229 p.

Giupponi, C.; Ramanzin, M.; Sturaro, E.; Fuser, S. 2006. Climate and land use changes, biodiversity and agri-environmental measures in the Belluno province, Italy, *Environmental Science & Policy* 9: 163–173. http://dx.doi.org/10.1016/j.envsci.2005.11.007

Gregory, R. D.; van Strien, A.; Vorisek, P.; Gmelig Meyling, A. W.; Noble, D. G.; Foppen, R. P. B.; Gibbons, D. W. 2005. Developing indicators for European birds, *Philosophical Transaction of the Royal Society B* 360: 269–288. http://dx.doi.org/10.1098/rstb.2004.1602

Gude, P. H.; Hansen, A. J.; Jones, D. A. 2007. Biodiversity consequences of alternative future land use scenarios in Greater Yellowstone, *Ecological Applications* 17(4): 1004–1018. http://dx.doi.org/10.1890/05-1108

Hanley, N.; Tinch, D.; Angelopoulos, K.; Davies, A.; Barbier, E. B.; Watson, F. 2009. What drives long-term biodiversity change? New insights from combining economics, paleoecology, and environmental history, *Journal of Environmental Economics and Management* 57(1): 5–20. http://dx.doi.org/10.1016/j.jeem.2008.03.005

Hansen, A. J.; DeFries, R. 2007. Ecological mechanisms linking protected areas to surrounding lands, *Ecological Applications* 17(4): 974–988. http://dx.doi.org/10.1890/05-1098

Hansen, A. J.; Rotella, J. J. 2002. Biophysical factors, land use, and species viability in and around nature reserves, *Conservation Biology* 16(4): 1112–1122. http://dx.doi.org/10.1046/j.1523-1739.2002.00545.x

Harrison, S.; Bruna, E. 1999. Habitat fragmentation and large-scale conservation: what do we know for sure?, *Ecography* 2(3): 225–232. http://dx.doi.org/10.1111/j.1600-0587.1999.tb00496.x

Jackson, S. F.; Gaston, K. J. 2008. Land use change and the dependence of national priority species on protected areas, *Global Change Biology* 14: 2132–2138. http://dx.doi.org/10.1111/j.1365-2486.2008.01628.x

Jongman, R. H. G.; Külvik, M.; Kristiansen, I. 2004. European ecological networks and greenways, *Landscape and Urban Planning* 68: 305–319. http://dx.doi.org/10.1016/S0169-2046(03)00163-4

Kladivko, E. J. 2001. Tillage systems and soil ecology, *Soil & Tillage Research* 61: 61–76. http://dx.doi.org/10.1016/S0167-1987(01)00179-9

Kleijn, D.; Báldi, A. 2005. Effects of set-aside land on farmland biodiversity: comments on Van Buskirk and Willi, *Conservation Biology* 19(3): 963–966. http://dx.doi.org/10.1111/j.1523-1739.2005.00603.x

Laiolo, P.; Dondero, F.; Cilinto, E.; Rolando, A. 2004. Consequences of pastoral abandonment for the structure and diversity of the alpine avifauna, *Journal of Applied Ecology* 41: 294–304. http://dx.doi.org/10.1111/j.0021-8901.2004.00893.x

Luoto, M.; Rekolainen, S.; Aakkula, J.; Pykälä, J. 2003. Loss of plant species richness and habitat connectivity in grassland associated with agricultural change in Finland, *Ambio* 32(7): 447–452.

MacDonald, D.; Crabtree, J. R.; Weisinger, G.; Dax, T.; Stamou, N.; Fleury, P.; Gutierrez Lapita, J.; Gibon, A. 2000. Agricultural abandonment in mountain areas of Europe: environmental consequences and policy response, *Journal of Environmental Management* 59(1): 47–69. http://dx.doi.org/10.1006/jema.1999.0335

MacDonald, R. I.; Kareiva, P.; Forman, R. T. T. 2008. The implications of current and future urbanization for global protected areas and biodiversity conservation, *Biological Conservation* 141: 1695–1703. http://dx.doi.org/10.1016/j.biocon.2008.04.025

Maiorano, L.; Falcucci, A.; Boitani, L. 2008. Size-dependent resistance of protected areas to land-use change, *Proceedings of the Royal Society* B 275: 1297–1304. http://dx.doi.org/10.1098/rspb.2007.1756

Maiorano, L.; Falcucci, A.; Garton, E. O.; Boitani, L. 2007. Contribution of the Natura 2000 network to biodiversity conservation in Italy, *Conservation Biology* 21(6): 1435–1444. http://dx.doi.org/10.1111/j.1523-1739.2007.00831.x

McKinney, M. L. 2002. Do human activities raise species richness? Contrasting patterns in United States plants and fishes, *Global Ecology and Biogeography* 11: 343–348. http://dx.doi.org/10.1046/j.1466-822X.2002.00293.x

McLaughlin, A.; Mineau, P. 1995. The impact of agricultural practices on biodiversity, *Agriculture, Ecosystems and Environment 55*: 201–212. http://dx.doi.org/10.1016/0167-8809(95)00609-V

Ministero dell'Ambiente 2010. *Italian National biodiversity strategy* [Online]. [Cited September 2011]. Available from Internet: http://www.minambiente.it/export/sites/default/archivio/allegati/biodiversita/estratto_strategia_eng.pdf

Okzan, K.; Senol, H.; Gulsoy, S.; Mert, A.; Suel, H.; Eser, Y. 2009. Vegetation-environment relationships in Mediterranean mountain forests on limeless bedrocks of southern Anatolia, Turkey, *Journal of Environmental Engineering and Landscape Management* 17(3): 154–163. http://dx.doi.org/10.3846/1648-6897.2009.17.154-163

Oñate, J. J.; Atance, I.; Baradají, I.; Llusia, D. 2007. Modelling the effects of alternative CAP policies for the Spanish high-nature value cereal-steppe farming systems, *Agricultural Systems* 94: 247–260. http://dx.doi.org/10.1016/j.agsy.2006.09.003

Paiders, J. 2008. Status of environmental protection as a source of finance for regional economic development: measurement of environmental and regional policy with the *fisher* function, *Journal of Environmental Engineering and Landscape Management* 16(1): 45–55. http://dx.doi.org/10.3846/1648-6897.2008.16.45-55

Park, S. A.; Harcourt, A. H. 2002. Reserve size, local human density, and mammalian extinctions in U.S. protected areas, *Conservation Biology* 16(3): 800–808. http://dx.doi.org/10.1046/j.1523-1739.2002.00288.x

Pautasso, M. 2007. Scale dependence of the correlation between human population presence and vertebrate and plant species richness, *Ecology Letters 10*: 16–24. http://dx.doi.org/10.1111/j.1461-0248.2006.00993.x

R Development Core Team 2009. *R: a language and environment for statistical computing*. Vienna, Austria: R Foundation for Statistical Computing. 1706 p.

Robinson, R. A.; Sutherland, W. J. 2002. Post-war changes in arable farming and biodiversity in Great Britain, *Journal of Applied Ecology* 39: 157–176. http://dx.doi.org/10.1046/j.1365-2664.2002.00695.x

Schonewald-Cox, C.; Buechner, M.; Sauvajot, R.; Wilcox, B. A. 1992. Cross-boundary management between national parks and surrounding lands: a review and discussion, *Environmental Management* 16(2): 273–282. http://dx.doi.org/10.1007/BF02393833

Sendžikaite, J.; Pakalnis, R. 2006. Extensive use of sown meadows – a tool for restoration of botanical diversity, *Journal of Environmental Engineering and Landscape Management* 14(3): 149–158.

Shafer, C. L. 1999. US national park buffer zones: historical, scientific, social, and legal aspects, *Environmental Management* 23(1): 49–73. http://dx.doi.org/10.1007/s002679900167

StataCorp. 2007. *Stata statistical software: release 10.* StataCorp LP, College Station, TX.

Tilman, D. 1984. Plant dominance along an experimental nutrient gradient, *Ecology* 65(5): 1445–1453. http://dx.doi.org/10.2307/1939125

van Buskirk, J.; Willi, Y. 2004. Enhancement of farmland biodiversity within set-aside land, *Conservation Biology* 18(4): 987–994. http://dx.doi.org/10.1111/j.1523-1739.2004.00359.x

Verburg, P. H.; Schulp, C. J. E.; Witte, N.; Veldkamp, A. 2006. Downscaling of land use change scenarios to assess the dynamics of European landscapes, *Agriculture, Ecosystems and Environment* 114: 39–56. http://dx.doi.org/10.1016/j.agee.2005.11.024

Verhulst, J.; Báldi, A.; Kleijn, D. 2004. Relationship between land-use intensity and species richness and abundance of birds in Hungary, *Agriculture, Ecosystems and Environment* 104: 465–473. http://dx.doi.org/10.1016/j.agee.2004.01.043

Vester, H. F. M.; Lawrence, D.; Eastman, J. R.; Turner, B. L.; Calme, S.; Dickson, R.; Pozo, C.; Sangermano, F. 2007. Land change in the Southern Yucatan and Calakmul biosphere reserve: effects on habitat and biodiversity, *Ecological Applications* 17(4): 989–1003. http://dx.doi.org/10.1890/05-1106

Viña, A.; Bearer, S.; Chen, X.; He, G.; Linderman, M.; An, L.; Zhang, H.; Ouyang, Z.; Liu, J. 2007. Temporal changes in giant panda habitat connectivity across boundaries of Wolong nature reserve, China, *Ecological Applications* 17(4): 1019–1030. http://dx.doi.org/10.1890/05-1288

Wade, A. A.; Theobald, D. M. 2010. Residential development encroachment on U.S. protected areas, *Conservation Biology* 24(1): 151–161. http://dx.doi.org/10.1111/j.1523-1739.2009.01296.x

Wickramasinghe, L. P.; Harris, S.; Jones, G.; Vaughan, N. 2003. Bat activity and species richness on organic and conventional farms: impact of agricultural intensification, *Journal of Applied Ecology* 40: 984–993. http://dx.doi.org/10.1111/j.1365-2664.2003.00856.x

Woodward, R. T.; Wui, Y. S. 2001. The economic value of wetland services: a meta-analysis, *Ecological Economics* 37(2): 257–270. http://dx.doi.org/10.1016/S0921-8009(00)00276-7

Wooldridge, J. M. 2002. *Econometric analysis of cross section and panel data.* MIT Press. 752 p.

Zervas, G. 1998. Quantifying and optimizing grazing regimes in Greek mountain systems, *Journal of Applied Ecology* 35: 983–986. http://dx.doi.org/10.1111/j.1365-2664.1998.tb00019.x

Mattia CAI is a post-doc researcher in environmental economics at the University of Padova (Italy), department of Land, Environment, Agriculture and Forestry. His main research interests revolve around the environmental impacts of agriculture, the economic valuation of ecosystem services, and the assessment of climate change impacts on agriculture.

Davide PETTENELLA is associate professor at the University of Padova (Italy). He has published more than 300 papers in the field of environmental economics and nature-based marketing as a result of his research activities and field works carried out within programs financed by the European Commission, FAO, European Forestry Institute, World Bank, and by Italian national and regional institutions.

2

ASSESSMENT OF NITROGEN POLLUTION REDUCTION OPTIONS IN THE RIVER NEMUNAS (LITHUANIA) USING FYRISNP MODEL

Antanas Sigitas Šileika[a], Mats Wallin[b], Kazimieras Gaigalis[c]

[a]*Water Research Institute, Aleksandras Stulginskis University, Parko g. 6, Vilainiai, LT-58102 Kėdainiai, Lithuania*

[b]*Department of Aquatic Science and Assessment, Swedish University of Agricultural Sciences, Box 7050, 750 07 Uppsala, Sweden*

[c]*Alekandras Stulginskis University, Water Research Institute, Mituvos g. 9, LT-50134 Kaunas, Lithuania*

Abstract. The paper quantifies and discusses diffuse and point sources total nitrogen (TN) inputs as well as retention and TN reduction options in the catchment of the main Lithuanian River Nemunas. Modelled average TN export between 2000–2006 from the River Nemunas catchment to the Baltic Sea was 37620 tonnes TN yr^{-1} according to the data oriented FyrisNP model. Loads of TN from diffuse and point sources as well as retention have been estimated for five subcatchments of the River Nemunas including the external load from Belarus. Agriculture contributes 74.6 to 89.5% of the TN load, increasing with the percentage of arable land and load from point sources. The main point source input is poorly treated wastewater at Kaunas city. The contribution from forest land to the TN load increases from 2.2% to 15.8% with an increase in forest land from 28.5 to 56.9% of the total subcatchments area. The highest retention of TN (30.7%) was observed in the Neris river subcatchment with the lowest hydraulic load (5.55 m yr^{-1}). Scenario modelling suggests that the reduction target for Lithuania for nitrogen input to the Baltic Sea by 11700 tonnes can be achieved by installing biological treatment in sewage treatment plants in all district cities and by converting 20% of arable land to pastures or implementation of other equivalent measures in agriculture. Assessment of the FyrisNP model results shows that the model can be successfully applied for river basin management planning in catchments outside the area where the model originally has been developed.

Keywords: water pollution; environmental processes modelling; catchment; nitrogen; concentration; load; retention.

Reference to this paper should be made as follows: Šileika, A. S.; Wallin, M.; Gaigalis, K. 2013. Assessment of nitrogen pollution reduction options in the river Nemunas (Lithuania) using FyrisNP model, *Journal of Environmental Engineering and Landscape Management* 21(2): 141–151. http://dx.doi.org/10.3846/16486897.2012.663088

Introduction

Nitrogen (N) and phosphorus (P) are the two main nutrients responsible for-eutrophication of coastal and marine waters (Fogelberg 2003). Their load to the Baltic Sea has resulted in negative environmental effects including algal blooms and oxygen depletion in deep waters. The Member States of HELCOM adopted an action plan to drastically reduce pollution of the Baltic Sea and restore its good ecological status. To achieve the target for N, Lithuania agreed to a reduction of the yearly N load of 11750 tonnes by 2021 (HELCOM 2007). The EU Water Framework Directive (Anon 2000) prescribes to develop river basin management plans and to achieve good water status by 2015, including coastal waters but not marine waters. Marine waters are, however, covered by the EU Marine Directive with the aim to achieve good environmental status of the EU's marine waters by 2020 (Anon 2008).

Many investigations have been carried out on nutrient concentration trends and loads in rivers in various countries (Keeney, DeLuca 1993; Johnes, Burt 1993; Vuoristo 1998; Stalnacke *et al.* 1999a, b; Vuorenmaa *et al.* 2002; Klavins *et al.* 2001; Hussian *et al.* 2003; Magner *et al.* 2004; Iital *et al.* 2010). The eutrophication of costal lagoons and open seas is analysed by Vitousek *et al.* (2004); Nixon *et al.* (1996); Wulf and Rahm (1988). Despite the actions undertaken by HELCOM countries, the overall eutrophication status of the Baltic Sea is still unacceptable (HELCOM 2008).

Heidi Vuoristo (1998) suggests that water quality in Finnish lakes and rivers subjected predominantly to point source loading has improved, however, in many areas deterioration of water quality is reported to be caused primarily by non-point source pollution, particularly from agriculture. Having investigated water quality in Finish rivers and small agricultural and forest catchments Jussi Vuorenmaa *et al.* (2002) claims that changes in nutrient losses are mainly caused by weather-driven fluctuations in discharge while no or very little impact of changes in agricultural production, structures, or management practices can be observed. After normalization of the Elbe River environmental data, having removed natural fluctuations from the collected data, Mohamed Hussian *et al.* (2003) found N and P loads are strongly influenced by water discharge.

Corresponding author. Antanas Sigitas Šileika
E-mail: sigitas.sileika@asu.lt

Ahti Lepisto *et al.* (2006) claims that in Finland agriculture contributes on average 38% of the TN export, varying between 35–85% in the south-western basins and 0–25% in the northern basins. Forest land contributes on average 9% of the TN export. Of the TN input to Finnish river-systems, 0% to 68% is retained in surface waters and/or peatlands, with a mean retention of 22%. The highest retention of TN (36–68%) has been observed in the basins with the highest percentage of lake.

There is very little data available about the nutrient fate in the Baltic countries where dramatic changes in industry, agricultural production and social life have taken place in the last decade (Jansons *et al.* 2002; Klavins *et al.* 2001; Bagdžiūnaitė-Litvinaitienė 2004; Povilaitis 2008; Iital *et al.* 2010).

Stalnacke *et al.* (1999b) state that the six largest rivers in eastern and southern parts of the Baltic Sea basin are responsible for about half the total riverine export of N and P to the sea. Long-term detailed investigations of nutrient concentration and load in the Nemunas River, the fourth largest river in the Baltic Sea basin, are of the utmost importance for the successful management of Baltic Sea eutrophication. To create cost effective action plans for reducing nutrient loadings to marine areas, it is important to know the contributions from the various sources in the catchment, and the retention in rivers and lakes.

The present study focuses on quantification of total nitrogen (TN) loads from diffuse and point sources as well as retention in the Nemunas River catchment.

The main objective of this study is to determine the measures needed to achieve the 11700 tonnes (30%) nitrogen reduction target in the Baltic Sea Action Plan undertaken by the Lithuanian Government.

1. Study area and data sets

The present study focuses on TN loads to the River Nemunas from the Lithuanian territory at the monitoring site Smalininkai in the River Nemunas and outlets of the main tributaries the Merkys, the Neris, the Nevėžis and the Dubysa (Fig. 1). Monitoring site Smalininkai (110 km from the mouth) was selected because there are no reliable water quality measurements at the river mouth. The modelling includes a calibration period of seven years (2000–2006).

The River Nemunas basin has a total area of 97,864 km^2, of which 46,695 km^2 is in Lithuania, 45, 389 km^2 in Belarus, 3,174 km^2 in Kaliningrad district of Russia, 2,517 km^2 in Poland and 88 km^2 in Latvia (Gailiusis *et al.* 2001). The study area covers 70.4% of the Lithuanian territory (Gailiusis *et al.* 2001).

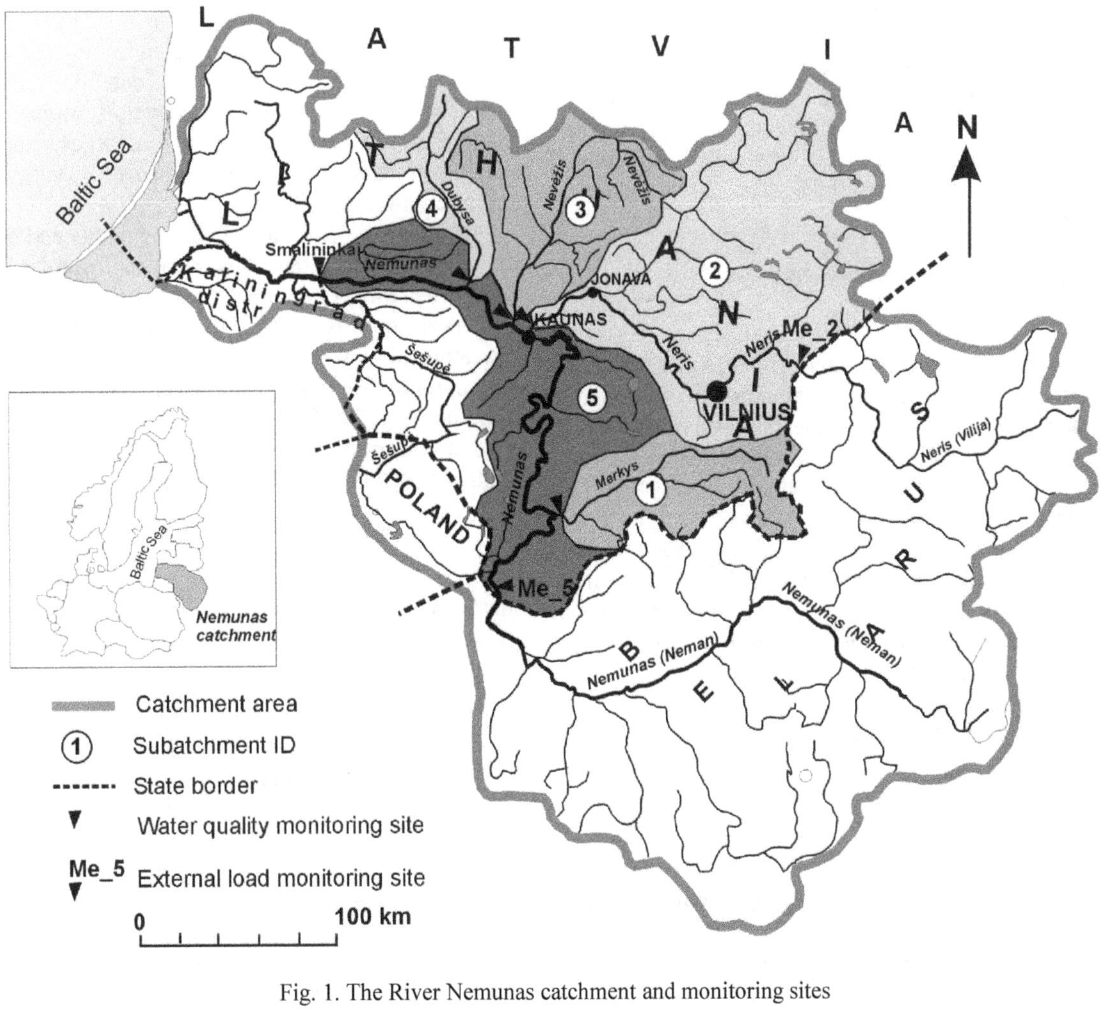

Fig. 1. The River Nemunas catchment and monitoring sites

The River Nemunas basin receives annual precipitation of 520 to 800 mm (Poluckaja 2000). Precipitation is higher at the western hilly areas than in the south-east. Inter year variation in precipitation can be as high as 40%. Snow melt water contributes about 40%, ground water 35% and rain fall 25% of the total River Nemunas runoff. The climate is influenced by the Baltic Sea causing instability of the weather such as frequent thaws in winter and sunny summers being as common as wet ones. Annual average temperature varies between 5.5 and 6.5 °C. The coldest month is January, with temperature –9 to –4.5 °C. Average July temperatures are 17 to 19 °C, with a maximum 34–37 °C (Poluckaja 2000). The long-term mean River Nemunas water discharge to the Baltic Sea is 703 m^3 s^{-1} (Gailiusis *et al.* 2001), it accounts for 19.6% of the total runoff to the Baltic Proper sub basin.

Farming lands cover 56% of the River Nemunas basin, with arable land and grasslands accounting for 73.7 and 26.3% of the agricultural land use. Forest, shrubs and bogs account for the land cover in the remainder of the basin. Sandy loams and sands with a high potential for N leaching account for 63% of the total basin area.

Land use data were obtained from a GIS data base for Lithuania which combines CORINE 2006 land cover for Lithuania (Vaitkuviene *et al.* 2008) and a vector data base LTDBK50000-V (National Land Service 2004). All of the measured water flows and water quality data were obtained from the Lithuanian Hydrometeorological Service and the Lithuanian Environmental Protection Agency (former Lithuanian Joint Research Centre). Leaching coefficients for different land use and precipitation were calculated from monthly measurements in three small agricultural catchments located in western, central and eastern parts of Lithuania. (Sileika *et al.* 2005; Šileika *et al.* 2010).

2. Methods

2.1. Sampling methods and analysis

The measured TN concentration in rivers was obtained from the water quality data base, maintained by Regional Environmental Protection Departments. Twelve samples per year were analysed in all river water quality monitoring sites and point source pollution outlets.

Monthly concentrations of TN (one sample per month) and average monthly values of river water flow were used for calculation of the TN load. Monthly runoff volume was calculated from daily water measurements data. Linear interpolation method (Rekolainen *et al.* 1989) was used for calculation of the nutrient load l_i.

$$l_i = \sum_{m=1}^{n} c(t_m) \cdot q(t_m), \qquad (1)$$

where: l_i – mean load, kg yr^{-1}; $c(t_m)$ – average concentration of the two neighbouring samplings, mg l^{-1}; $q(t_m)$ – runoff volume for the period t_m around the sampling time, m^3; m – sampling number.

Water quality analysis was conducted according to standard methods (Anon 1994) in the chemical analytical laboratory of the Water Research Institute. The laboratory has approbation for the surface water quality analyses form Lithuanian Environmental Protection Agency. The TN was determined by oxidation with peroxodisulphate to nitrate, followed by photometry using phenol disulfoacid. Difference between the determined TN and TN in standard solution should be no more than 10%.

2.2. FyrisNP model and input data

There are a number of tools for nutrient loads calculation and source apportionment in river catchments (Euroharp 2003; Grizzetti *et al.* 2005; SWAT 2005). The mass balance based FyrisNP model is one such tool which calculates the source-apportioned load and transport of TN and TP in rivers (Lindgren *et al.* 2007; Hansson *et al.* 2008b). The main scope of the model is to assess the effects of different nutrient reduction measures on the catchment scale. The time step for the model is one month (or week), and the spatial resolution is on the subcatchment level. Retention, i.e. losses of nutrients in rivers and lakes through sedimentation, up-take by plants and denitrification, is calculated as a function of water temperature, potential TN concentration and lake area, and areas of lakes and streams. The model has two calibrated retention parameters, *kvs* (retention parameter, m yr^{-1}) and c_0 (temperature parameter, dimension less) which are estimated using time series of measured TN concentrations (Hansson *et al.* 2008a).

The data used for calibrating and running the model can be divided into time-dependent data, e.g. time series on observed nutrient (TN) concentration, water temperature, runoff and point source discharges, and time-independent data, e.g. land-use information, lake area and stream length and width (Fig. 2).

Due to sedimentation, uptake by plants and denitrification some TN is retained as it travels from the headwaters downstream. Permanent removal or retention of TN in rivers or lakes is calculated in the model as a function of weather, water temperature, surface water TN concentration, runoff, and lake and river water surface areas (Hansson *et al.* 2008b).

Nitrogen retention is expressed by the coefficient *R* of nitrogen retention in the catchment:

$$R_T = \frac{T_a \cdot kvs}{q_s + kvs}, \qquad (2)$$

where: *kvs* – an empirical coefficient; T_a – a temperature adjustment factor given by:

$$T_a = \begin{cases} 0, & T < 0 \\ c_0 + \frac{T(1-c_0)}{20}, & 0 \le T \le 20, \\ 1, & T > 20 \end{cases} \qquad (3)$$

where: T – the water temperature, °C; c_0 – an empirical calibration parameter.

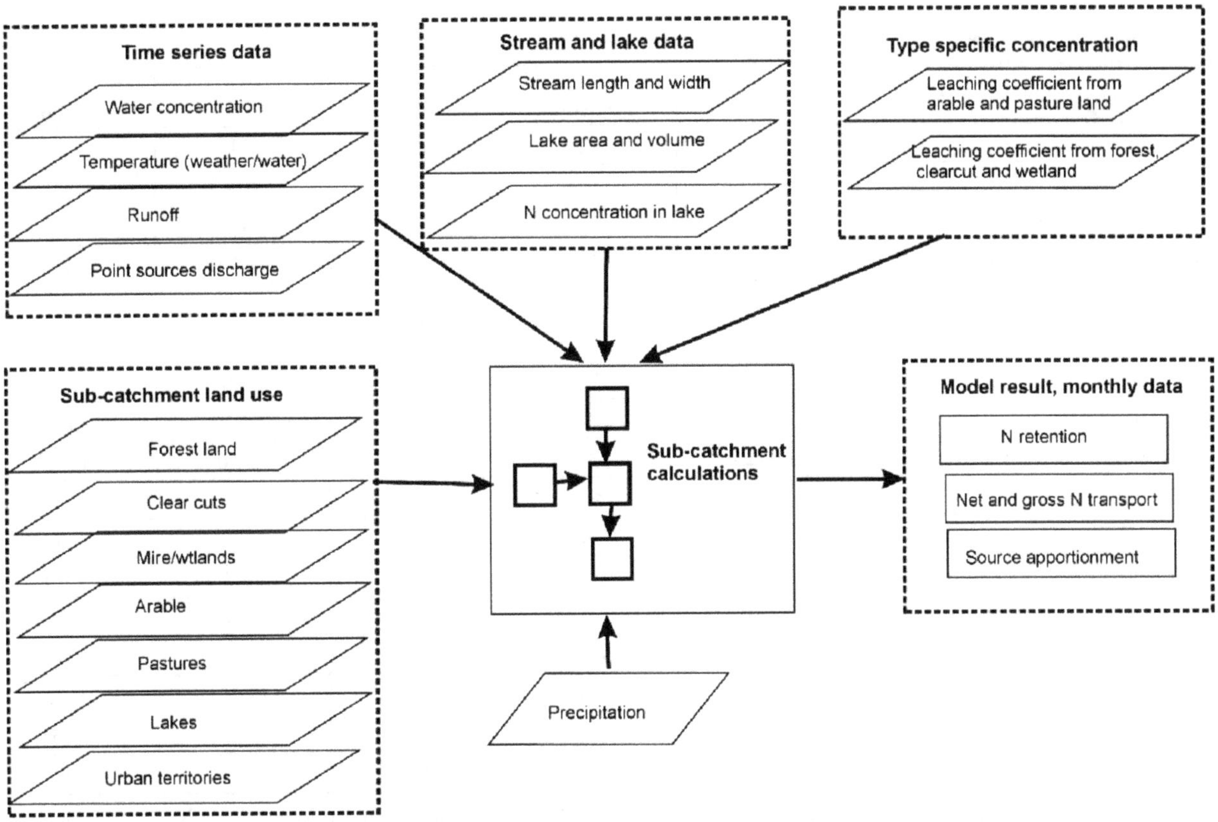

Fig. 2. The structure and characteristics of inputs and outputs of the model

The parameter c_0 determines the level of retention reduction by temperatures between 0 °C and 20 °C. The hydraulic load, q_s (m yr^{-1}), is calculated as follows:

$$q_s = \frac{Q}{A_{lake} - A_{LM} + A_{stream}}, \tag{4}$$

where: Q – water specific runoff from subcatchment, l s^{-1} km^{-2}; A_{lake} – the total surface area of all lakes in the given sub-catchment, m^2; A_{LM} – the area of the lake treated in the separate lake module (if one such exists in the sub-catchment), m^2; A_{stream} – the surface area of all streams in the sub-catchment, m^2.

In order to perform simulations with the FyrisNP model, an Excel-file containing all input data is required (Table 1). The Excel data file contains between eight to ten different worksheets depending on the features used (Hansson *et al.* 2008b). The inclusion of the External load worksheet means that the loading from upstream parts of the river catchment is included in the model without any source apportionment. This worksheet is used to include the TN load to the rivers Nemunas and Neris from Belarus territory.

2.3. Statistical evaluation

In order to evaluate the fit of simulated to observed values, two statistical measures are used in the FyrisNP model: the model efficiency, E and the correlation coefficient, r.

The definition of model efficiency (Nash, Sutcliffe 1970) is:

$$E = 1 - \frac{\sum_{i=1}^{n}(y_i - x_i)^2}{\sum_{i=1}^{n}(y_i - x^{mean})^2}, \tag{5}$$

where: n – the number of observations; x^{mean} – the mean value of all observations, mg l^{-1}; y_i and x_i – the observed and modelled concentrations, respectively, mg l^{-1}.

$E = 1$ implies that the measured and modelled series are identical, and $E = 0$ indicates that the simulation is no better than a straight line representing the average value of the observations (Hansson *et al.* 2008a).

Relative percentage error e was calculated according to Sofia Fogelberg *et al.* (2004):

$$e = \frac{\sqrt{\frac{1}{n}\sum(y_i - x_i)^2}}{\frac{1}{n}\sum y_i} \cdot 100, \tag{6}$$

where: y_i – measured concentration, mg l^{-1}; x_i – calculated concentration, mg l^{-1}; n – number of measurements.

AS well statistical evaluation of the model efficiency model validation was carried out for two independent data series (1997–2001 and 2002–2006) in the river Nevėžis (subcatchment #3). The obtained values of the empirical retention coefficients from calibration were preserved during validation.

Table 1. The River Nemunas catchment (Lithuania) characteristics at the monitoring sites

Subcatchment name	The Merkys	The Neris (Lithuania)	The Nevėžis	The Dubysa	The Nemunas* (Lithuania)
Subcatchment ID	#1	#2	#3	#4	#5
Area, km^2	3792.6	13709.7	6140.5	1965.9	7262.8
Total lakes area, km^2	47.9	455.6	26.2	11.2	265.8
Total streams length, km	1428.0	5123.0	7994.0	1249.0	4386.0
Total stream area, km^2	11.4	22.1	18.3	5.9	21.8
Forest, km^2	2052.0	4916.3	1662.3	545.5	2405.5
Clearcuts, km^2	100.2	259.0	87.5	28.8	121.9
Mire, km^2	45.9	118.4	39.3	31.5	34.6
Arable, km^2	972.1	4835.4	3669.5	991.6	3130.6
Pasture, km^2	510.4	2587.6	447.6	299.3	1005.4
Open, km^2	3.1	23.0	1.5	2.1	6.0
Built, km^2	41.8	367.9	115.3	38.2	201.0
Urban, km^2	7.9	125.3	73.0	11.7	70.2
Altitude, m	80.0	90.0	60.0	60.0	50.0
TN deposition on lakes, kg ha^{-1} yr^{-1})	18.72	18.72	18.72	18.72	18.72
TN leaching from arable land, mg l^{-1}**	5.2	9.4	10.5	5.8	10.5
TN leaching from pasture, mg l^{-1}**	2.2	4.5	4.5	2.8	5.0

* The Nemunas (Lithuania) subcatchment consists the area between the subcatchments of the Merkys, the Neris (Lithuania), the Nevėžis and the Dubysa).

** Annual mean concentration.

3. Results

3.1. Model validation and calibration

Monte Carlo simulations using measured vs modelled TN concentrations was used to find values of parameters c_0 and *kvs*. These parameter values were then used in manual calibration of TN concentration in FyrisNP.

For the model validation the river Nevėžis subcatchment #3 was selected because a rather long and consistent TN data series was available. After Monte Carlo simulations (300 simulations) optimum values of $c_0 = 0.86$ and the coefficient $kvs = 0.78$ best fitted the data series 1997–2001. Model efficiency E = 0.58 and correlation coefficient r = 0.76 were achieved in calibration of TN concentration in FyrisNP. Using the same calibration parameters for the data set 2002–2006 resulted in model efficiency E = 0.43 and correlation coefficient r = 0.68. Model efficiency and correlation coefficient were slightly lower for the validation period but in acceptable agreement with the calibration results. The modelled net load of TN from the River Nemunas catchment includes retention in all upstream subcatchments and in the Nemunas River itself. In the calibrated model set up the temperature adjustment coefficient c_o and calibration parameter *kvs* were the same for all subcatchments and the whole data set 2000–2006. Monte Carlo simulations resulted in optimum retention coefficients $c_0 = 0.65$ and the coefficient $kvs = 0.82$. The model efficiency $E = 0.61$ and the correlation coefficient $r = 0.78$ imply good congruence between observed and modelled data series. E and r are calculated based on all selected pairs of the observed and simulated concentrations lumped before calculation (Hansson *et al.* 2008a). The simulated TN loads plotted against the observed monthly ones during the analysed period (2000–2006) at the monitoring site #5 correspond well (Fig. 3).

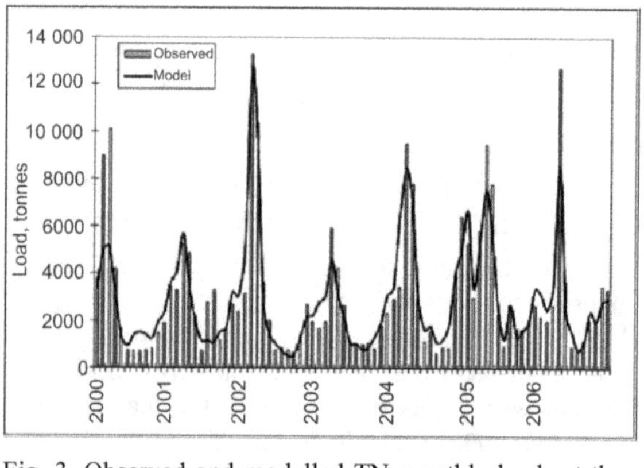

Fig. 3. Observed and modelled TN monthly loads at the outlet of the River Nemunas subcatchment #5

Winter and early spring exhibit peak values where observed values exceeds model results for most of the years. One possible reason for a mismatch between modeled and observed data could be temporal variability in TN concentration. Grab samples for water quality analysis are taken once a month and these values are considered representative for the whole month in FyrisNP. Temporal variability in concentrations causes more uncertainty in estimated loads during periods of peak flow than during low or mean flow periods. In order to achieve more accurate value of mean monthly load, the frequency of water sampling as well as flow proportional sampling must be taken into account. Difference between modelled and measured data decrease when yearly mean loads are used instead of monthly concentrations. The discrepancy between the modelled and observed mean TN load values for seven year is only 0.14% (Fig. 4).

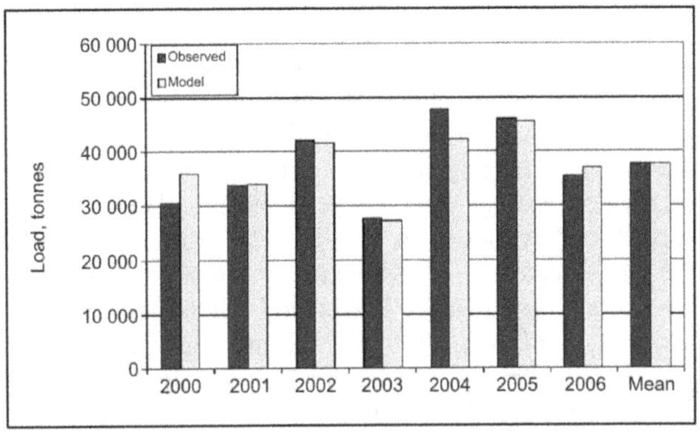

Fig. 4. Observed and modelled TN yearly loads at the outlet of the River Nemunas subcatchment #5

3.2. TN contribution to the watercourses and retention

The internal TN load sources (gross contribution) is calculated for each catchment and summed for the entire simulation period (Fig. 5).

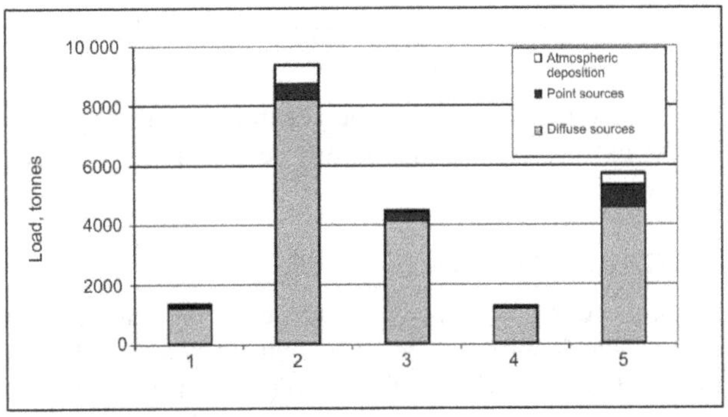

Fig. 5. Internal TN load source apportionment for each subcatchment

The largest total load (93611 tonnes) is to the watercourses of the subcatchment Neris, #2. This subcatchment also has the largest area (24 942 km^2 including 11 100 km^2 in the Belarus territory) and subsequently the largest TN contribution from non-point sources – 81,991 tonnes. The largest point-source contribution (7316 tonnes, 44.6% of the total point-source load) enters to the River Nemunas subcatchment #5. The main source of pollution is Kaunas city with 350.5 thousand inhabitants and a mechanical waste water treatment plant. The second largest point source input of 5360 tonnes (32.7% of the total point-source load) enters to the Neris river, subcatchment #2 and emanates from the Lithuanian capital Vilnius with 554 thousand inhabitants and from the industrial city Jonava (51.7 thousand inhabitants).

The specific TN contribution to the watercourses from diffuse sources is presented in Table 2.

Table 2. The TN gross load to the watercourses from diffuse sources, kg ha^{-1} yr^{-1}

ID	Arable	Pasture	Forest	Mire	Water
#1	12.0	5.1	1.4	2.2	18.7
#2	18.2	8.7	1.2	1.8	18.7
#3	14.8	6.4	0.8	1.2	18.7
#4	12.4	6.0	1.2	1.9	18.7
#5	17.3	8.3	1.0	1.5	18.7
Mean	15.0	6.9	1.1	1.7	18.7

The highest yearly mean TN load to lakes and rivers comes from deposition – 18.7 kg ha^{-1} yr^{-1}. Arable land adds 15.0 kg ha^{-1} yr^{-1}. The least yearly TN load derives from forests – 1.1 kg ha^{-1} yr^{-1}. The average yearly TN loss from pastures is 6.9 kg ha^{-1} yr^{-1} for the simulated period.

The differences between the TN gross contribution to the River Nemunas subcatchments and the net load in the outlets represent the proportion retained in streams and lakes during the investigated period (Fig. 6).

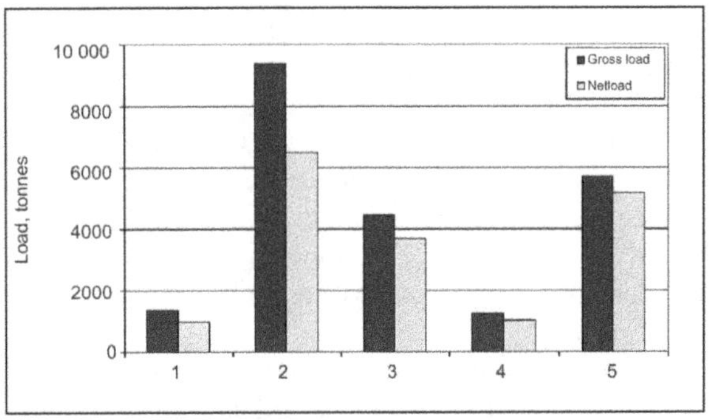

Fig. 6. The TN contribution per subcatchment, before (gross) and after retention (net) within the same subcatchment

The TN retention in the river catchments varies from 9.2 to 30.7% of the total input. The largest proportion, 30.7% of the gross load, is retained in the Neris river, subcatchment #2. The lowest retention 9.2% is in the River Nemunas, subcatchment #5.

3.3. TN export from subcatchments

Using the FyrisNP model, the TN export for 2000–2006 from the River Nemunas catchment to the Baltic Sea was calculated to be 37620 tonnes N yr^{-1}.

Source apportionment for the net load at the outlet of the subcatchments for the entire investigation period is presented in the Table 3.

Table 3. Source apportionment at the outlet of the subcatchments in percent of the total net load

ID	Arable	Pasture	Forest	Mire	Point source	Water
#1	61.0	13.5	15.8	0.5	3.4	5.8
#2	66.0	16.9	4.5	0.2	5.7	6.7
#3	85.1	4.5	2.2	0.1	6.9	1.3
#4	70.4	10.3	15.6	0.3	1.5	1.8
#5	70.3	11.7	4.8	0.1	7.9	5.2

Agricultural load (arable and pasture) varies from 74.5% of the total load in the south-eastern Lithuanian River Merkys (subcatchment #1) outlet to 89.5% in the outlet of the Nevėžis river (subcatchment #3) in the Lithuanian Middle Plain where intensive agriculture dominates. In this basin arable land represents 59.4% of the total catchment area. The TN export from the largest point sources, 7.9% of the total load, is from the River Nemunas (subcatchment #5). The main input enters from Kaunas city with 350.5 thousand inhabitants. The contribution from forest land increases from 2.2% in subcatchment #3 in the Lithuanian Middle Plain, to 15.8% subcatchment #1 in south-eastern part of the country where forest land covers 56.9% of the total catchment area.

3.4. TN load reduction options

Two FyrisNP model scenarios were constructed to evaluate possible measures for reduction of the yearly TN export from Lithuania to the Baltic Sea by 11700 tonnes by 2021, as agreed upon in the new HELCOM Baltic Sea Action Plan (HELCOM 2007). For both scenarios retention coefficients *kvs* and c_0 were set the same as for the calibration period.

The first scenario was performed to evaluate the possible decrease in TN export if major waste water treatments plants (WWTP) were modernised according to HELCOM recommendations (HELCOM 2007) in all district cities. After modernisation the TN concentration in effluents after treatment should not exceed the permitted limit of 10 mg l^{-1}. According to this scenario, TN export for 2015–2021 would decrease by 14.4% or 5436 tonnes.

The second model scenario was the conversion of 20% of arable land to pasture, together with the improvements in WWTP. After this, pastures would make up 41% of total agricultural land. This scenario results in reduction of the yearly TN export of 12000 tonnes (31.8%) by 2021. Modelling results show that reduction of point source pollution only is not enough to reach the target. Conversion of 20% of arable land to pasture may not be realistic if agricultural production should remain unchanged or even increase. Some other measures, thus, have to be applied in agricultural land. These could be e.g. increase of winter crop area, intercropping, aftercropping, fertilisation planning according to crop need, spring manure application, etc.). To be able to run this type of scenarios more detailed information on cropping, fertilization, harvests, soil types, etc. are needed than was available for this study.

4. Discussion

4.1. Evaluation statistics

Besides time series charts for observed and modelled concentration and load, the FyrisNP model produces model efficiency (Nash, Sutclife 1970) and correlation coefficient statistics. Thus, the FyrisNP model offers several possibilities to evaluate correspondence between the observed and modelled results. The time series chart of observed and modelled data for the monitoring site #5 of the River Nemunas catchment shows that many of the observed TN peak loads are higher than the modelled ones (Fig. 3). The same pattern exists for time series from the other four subcatchments. In an-other study it was noted that the HBV-N model (Fogelberg *et al.* 2004), which uses a daily time step, could not manage to capture the TN concentration peaks in the Motala and the Ronea rivers (Sweden).

Arvydas Povilaitis (2008) states that absolute values of deviation between the observed and estimated loads constitutes less than 10% from line of equivalence when testing MESAW model for the Merkys river (subcatchment #1). Sofia Fogelberg *et al.* (2004) showed that the relative percentage error between the calculated and observed loads varies from 13.3 to 47% in the Motala and the Ronea catchments (Sweden) for the period of 1993–1999 when using the HBV-N and MONERIS models. Horst Behrendt *et al.* (2005) reported that the mean deviation between the calculated and observed TN loads in the River Odra is 22% and 12% for the dissolved inorganic nitrogen in the River Danube. The calculated model efficiency, correlation coefficient and relative percentage error for the River Nemunas subcatchment #5 for the study period presented in the Table 4.

Table 4. The FyrisNP model efficiency for 2000–2006

ID	River	Model efficiency, E	Correlation coefficient, r	Relative error, %
#1	Merkys	0.7	0.83	11.7
#2	Neris	0.41	0.64	13.2
#3	Nevėžis	0.36	0.61	15.2
#4	Dubysa	0.5	0.71	13.0
#5	Nemunas	0.6	0.78	7.8

These results confirm the reliability of the FyrisNP model outside the catchments where the model has been developed.

4.2. Gross and net loads

Agriculture is commonly pointed out as the main source of nitrogen input to rivers. Ahti Lepisto *et al.* (2006) estimates that in 30 rivers of Finland (60% of the total land area) the diffuse sources (agriculture, forest land and background) conrtibute 74%, deposition 9% and point sources 17% of the total TN input.

N discharges from diffuse sources, point sources and deposition in the River Odra were 68.9, 25.1 and 3.2%, respectively, in the period of 1998–2002 (Behrendt *et al.* 2005). According to the Swedish report on nutrient loads to HELCOM (Brandt *et al.* 2008) the gross contribution from diffuse sources, point sources and deposition in Swedish part of the Baltic Proper sea-basin catchment was 65.6, 22.9 and 11.5%, respectively, in the period of 1985–2004. The proportion of the net load (after retention) is almost the same as that for the gross load. In the River Nemunas catchment contribution from diffuse sources, point sources and deposition for the study period was: 87.3, 7.4 and 5.3%. Consequently, River Nemunas differ from the examples given above due to the high proportion of arable land which causes significant load from diffuse sources. Arable land in the River Nemunas catchment makes up 41.4% of the total catchment area. In Sweden agriculture makes up 19.8% of the Swedish part of the Baltic Proper sea-basin catchment and in Finland – only 8.2% (Lepisto *et al.* 2006).

Different researchers state very similar TN export coefficient from agriculture. Ahti Lepisto *et al.* (2006) used the the N_EXRET model to apply 15.0 kg ha^{-1} yr^{-1} TN gross area specific load in the South-western Finland. Seppo Rekolainen *et al.* (1995) after analysis of 1986–1990 monitoring data in three small agricultural catchments in Finland suggest that TN gross contribution varied from 10.0–22.0 kg ha^{-1} yr^{-1}.

Having analysed the MESAW model results Povilaitis (2008) gives an estimated TN net export coefficient of 14.4 kg ha^{-1} yr^{-1} for the Merkys river (subcatchment #1 in Nemunas River). Our application of the FyrisNP model give an average gross TN export coefficient of 16.2 kg ha^{-1} yr^{-1} for arable land of the entire River Nemunas catchment and net TN export coefficient 12.0 kg ha^{-1} yr^{-1}. Source apportion calculations for the river Susve (main tributary of the Nevėžis river subcatchment #3) gave an export coefficient from agricultural land after retention of 12.6 kg ha^{-1} yr^{-1} (Kronvang *et al.* 2005). Consequently, independent estimates of TN export coefficients for the Nemunas River indicate results within the same size range.

The share of the total TN load coming from atmospheric deposition varies very widely depending mainly on the lake area in the catchment. The contribution from deposition directly on rivers is generally very small. PLC-5[1] calculations in Sweden report that the N load from deposition in the Swedish part of the Baltic Proper sea basin catchment made up 11.5% of the total load (Brandt *et al.* 2008). Horst Behrendt *et al.* (2005) suggest that nitrogen deposition in the River Odra basin for the period of 1998–2002 made up 3.2% of the total load. The TN load from deposition in Lithuania in the study period made up from 1.3 to 6.7% of the total load depending on lake area in the subcatchments. Despite a quite high inputs from deposition in Lithuania (18.7 kg ha^{-1} yr^{-1}) the contribution to the total load is limited due to the small total lake area in the River Nemunas catchment.

The mean area TN specific gross load from all sources of pollution in the whole River Nemunas catchment is estimated to be 9.6 kg ha^{-1} yr^{-1} for the period of 2000–2006. Beherndt *et al.* (2005a, b), using MONERIS model, reported a TN gross load of 8.9 kg ha^{-1} yr^{-1} for 1998–2002 in the River Odra catchment and 9.4 kg ha^{-1} yr^{-1} for the year 2000 in the River Danube catchment. Lepisto *et al.* (2006) calculated a TN gross load of 5.9 kg ha^{-1} yr^{-1} in 30 Finish catchments. Brandt *et al.* (2008) delivered a TN gross load of 5.1 kg ha^{-1} yr^{-1} to the Swedish part of the Baltic Proper sea basin catchment.

Some investigators present only the value of the area specific export coefficient (net load). Stalnacke *et al.* (1996) reported 6.1 kg ha^{-1} yr^{-1} as an average area-specific export of the TN to the Baltic Proper sea sub basin from the River Vistula in the period of 1980–1993. The area-specific export of TN form the River Elbe was calculated to 7.4 kg ha^{-1} yr^{-1} in 1998–2000 (Hussian *et al.* 2003).

If the TN retention in streams is assumed to be no higher than 20%, the gross load can be estimated to be maximum 7.3 and 8.9 kg ha^{-1} yr^{-1} in the Vistula and the Elbe rivers, respectively. Hence, comparison of the total gross TN loads in the catchments of various European large rivers shows that the area-specific load in the River Nemunas catchment is the highest one.

Previous studies (Sileika *et al.* 2006) have revealed an increase of flow normalised area-specific load of nitrate (NO_3–N) from 1986 to 2000 at almost all the sampling sites in the River Nemunas and its tributaries. These results give evidence that dramatic changes in agricultural production in former Soviet Lithuania have increased NO_3–N loads in the River Nemunas. Changes in land use and extensive ploughing of pastures could have enhanced mineralisation of organic nitrogen and significantly increased the NO_3–N losses to rivers in the catchment. The same trends have been observed in other former Soviet countries (Stalnacke *et al.* 1999b; Klavins *et al.* 2001; Jansons *et al.* 2002; Iital *et al.* 2010). NO_3–N represents a large proportion of TN concentration. Therefore, it can be expected that the TN discharges have also increased.

4.3. Retention

Nutrient gross loads from various pollution sources in the studied catchments almost always exceed the load leaving the catchments. The difference is generally called retention (Harrison *et al.* 2009; Hejzlar *et al.* 2009; Saunders, Kalff 2001). Nitrogen retention is mainly caused by denitrification and sedimentation. Retention is difficult to measure, but it can be estimated using catchment mass balance models (Molot, Dillon 1993; Windolf *et al.* 1996). Various pathways are used for retention calculations in models. The MONERIS model calculates retention on land, in soil, groundwater and in the river systems (Behrendt *et al.* 2005), while the HBV-N model calculates retention in groundwater, streams and lakes (Fogelberg 2003). The MESAW model (Povilaitis 2008) takes into account the total area of the catchment and the N_EXRET model evaluates retention in peat lands and lakes (Lepisto *et al.* 2006). The FyrisNP model simulates retention in rivers and lakes and can incorporate a separate lake module for larger lakes situated close to the subcatchment outlet (Hansson 2008b). Due to different N input, land use and lake areas the estimated retention differs significantly. For example, for the Merkys river subcatchment #1 Povilaitis (2008) reported that 60% of the total TN load is retained before the outlet, while the FyrisNP model simulates 26% retention for the same subcatchment. Having used the EUROHARP-NUTRET tool for retention calculation Kronvang *et al.* (2005) estimated that 13.4% of the TN is retained in the Susve river (main tributary of the Nevėžis river #3), while the FyrisNP model computes 19.3%. Lepisto *et al.* (2006) estimate average total retention (retention in lakes and

[1] PLC-5 HELCOM's fifth Pollution Load Compilation (HELCOM 2011)

peat lands) of 35% for 30 Finish rivers. Having compared retention in two Swedish and two German river catchments Fogelberg (2003) states that there were minimal differences between HBV-N and MONERIS predictions in two of the catchments. However, in other two catchments HBV-N, calculated much lower retention than MONERIS. Retention value depends on the calculation method and N flow pathway assumption, as well as catchment characteristics Kronvang *et al.* (2005). TN retention in FyrisNP depends on many factors but mainly on hydraulic load (Table 5).

Table 5. Hydraulic load and TN retention in the River Nemunas subcatchments

ID	River	Hydraulic load, m yr^{-1}	Retention, % of the total load
#1	Merkys	10.72	26.0
#2	Neris	5.55	30.7
#3	Nevėžis	17.33	17.0
#4	Dubysa	24.84	15.8
#5	Nemunas (excluding tributaries and upper reaches in Belarus)	4.42	9.2

The highest retention (30.7%) was calculated for the Neris River (subcatchment #2) due to low hydraulic load (5.55 m yr^{-1}). Retention in Dubysa river (subcatchment #4) with the highest hydraulic load (24.84 m yr^{-1}) was low – 15.8%. The River Nemunas (subcatchment #5) has the lowest retention (9.2%) of gross TN load despite the lowest hydraulic load (4.42%). It can be because the biggest polluters (Kaunas city and the Nevėžis river outlet) are close to the subcatchment outlet.

Conclusions

1. Agricultural net load varies from 74.6 to 89.5% of the total nitrogen net load depending on the percentage of arable land and the load from point sources in five subcatchments of the River Nemunas. The main input from point sources enters at Kaunas city with poor waste water treatment. Contribution from forest land varies from 2.2% to 15.8% with increasing percentage forest land from 28.5 to 56.9% of the total area of subcatchments. The highest retention of N (30.7%) is observed in the river subcatchment with the lowest hydraulic load (5.55 m yr^{-1}).

2. Scenario modelling suggests that the Lithuanian reduction target of 11,700 tonnes nitrogen by 2021 to the Baltic Sea can be achieved after installation of biological treatment in all district cities and implementation of measures to reduce the TN load in agriculture. By modernisation of waste water treatment plants in all district cities the yearly TN export would decrease by 14.4% or 5436 tonnes. Conversion of 20% of arable land to pasture (or implementation of other equivalent measures in agriculture) together with the improvements in WWTP would give reduction of the TN yearly export of 12,000 tonnes (31.8%).

3. Assessment of the FyrisNP model results shows that the model can be successfully applied for the river basin management planning in catchments outside of the area where the model has been developed.

4. The aggregated target in the national water management plan (reduction of TN load to the Baltic Sea 11 700 tonnes by 2021) should be complemented with a specific reduction target focusing on the main sources of pollution from – agriculture and point sources.

Acknowledgements

This work has been performed within the Harmo Balt Project included in the SIDA/GEF/Worldbank funded Baltic Sea Regional Project. We would like to thank all the persons from the Department of Aquatic Science and Assessment, Swedish University of Agricultural Sciences for introducing us to the FyrisNP model. Special thanks to Klas Hansson and Faruk Djodjic for helping us with the application of the model to the River Nemunas catchment and for valuable comments on the manuscript.

References

Anon 2000. *Directive 2000/60/EC of the European Parliament and of the Council of 23 October 2000 establishing a framework for Community action in the field of water policy (Water Framework Directive)* [online], [cited 23 February 2012]. Available from Internet: http://ec.europa.eu/environment/water/water-framework/index_en.html

Anon 1994. Lietuvos aplinkos ministerija 1994. *Unifikuoti nuotekų ir paviršinių vandenų kokybės tyrimų metodai* [Unified methods for water quality analyses of effluents and surfrace waters]. Vilnius, 9–22 (in Lithuanian).

Anon 2008. Directive 2008/56/EC the European Parliament and of the Council of 17 June 2008 establishing a framework for community action in the field of marine environmental policy (Marine Strategy Framework Directive), *Official Journal of the European Union* 164: 19–40.

Bagdžiūnaitė-Litvinaitienė, L. 2004. Dynamics of nutrient changes in the water of SE Lithuanian Rivers for different periods of water level, *Journal of Environmental Engineering and Landscape Management* 12(4): 146–152.

Behrendt, H.; Opitz, D.; Korol, R.; Stronska, M. 2005. Changes of the nutrient loads of the Odra River during the last century – their causes, in *Proceedings ICID Conference,* 15–19 May 2005, Frankfurt (Oder), (CD).

Brandt, M.; Ejhed, H.; Rapp, L. 2008. *Nutrient loads to the Swedish marine environment in 2006.* Sweden's report for HELCOM's fifth pollution load compilation. Stockholm: Swedish Environmental Protection Agency.

Euroharp 2003. *Review and literature evaluation of quantification tools for the assessment of nutrient losses at catchment scale.* EUROHARP Report 1-2003, Wageningen: ALTERRA.

Fogelberg, S.; Arheimer, B.; Venohr, M.; Behrendt, H. 2004. Catchment modelling of nitrogen flow with two different conceptual models, in *Proceedings of Nordic Hydrologic Conference,* 2004, Tallin, 48: 149–158.

Fogelberg, S. 2003. *Modelling nitrogen retention at the catchment scale.* Report UPTEC W 03 019. Uppsala: Uppsala Universitet, 1–51.

Gailiusis, B.; Jablonskis, J.; Kovalenkoviene, M. 2001. *Lietuvos upės. Hidrografija ir nuotėkis* [Lithuanian rivers. Hydrography and runoff]. Kaunas: Energetikos institutas, 136–139 (in Lithuanian).

Grizzetti, B.; Bouraoui, F.; Marsily, G.; Bidoglio, G. 2005. A statistical method for source apportionment of riverine nitrogen loads, *Journal of Hydrology* 304: 302–315. http://dx.doi.org/10.1016/j.jhydrol.2004.07.036

Hansson, K.; Wallin, M.; Djodjic, F.; Orback, K. 2008a. *The FYRIS model Version 3.1. A tool for catchment-scale modelling of source apportioned gross and net transport of nitrogen and phosphorus in rivers.* A user´s manual, Version 3.1. Uppsala: Swedish University of Agricultural Sciences 18: 6–28.

Hansson, K.; Wallin, M.; Djodjic, F.; Orback, K. 2008b. *The FYRIS model Version 3.1. A tool for catchment-scale modelling of source apportioned gross and net transport of nitrogen and phosphorus in rivers.* Technical description, Version 3.1. Uppsala: Swedish University of Agricultural Sciences 19: 6–17.

Harrison, L. A.; Maranger, R. J.; Alexander, R. B.; Giblin, A. E.; Jacinte, P.-A.; Mayorhga, E.; Seitzinger, S. P.; Sobota, D. J.; Wollheim, W. M. 2009. The regional and global significance of nitrogen removal in lakes and reservoirs, *Biogeochemistry* 93: 143–157. http://dx.doi.org/10.1007/s10533-008-9272-x

Hejzlar, J.; Anthony, S.; Arheimer, B.; Behrendt, H.; Bouraoui, F.; Grizetti, B.; Groenendijk, P.; Jeiken, M. H. J. L.; Johnsson, H.; Lo Porto, A.; Kronvang, B.; Panagopoulos, Y.; Siderius, C.; Silgram, M.; Venohr, M.; Zaloudik, J. 2009. Nitrogen and phosphorus retention in surface waters: an inter-comparison of predictions by catchment models of different complexity, *Journal of Environmental Monitoring* 11: 584–593. http://dx.doi.org/10.1039/b901207a

HELCOM 2011. *The fifth Baltic Sea pollution load compilation (PLC-5)* [online], [cited 23 February 2012]. Available from Internet: www.helcom.fi/stc/files/Publications/ Proceedings/BSEP128.pdf, 2011.

HELCOM 2007. *HELCOM Baltic Sea action plan* [online]. Adopted on 15 November 2007 in Krakow, Poland by the HELCOM Extraordinary Ministerial Meeting [cited 23 February 2012]. Available from Internet: http://www.helcom.fi/stc/files/BSAP/BSAP_Final.pdf

HELCOM 2008. *HELCOM activities 2008 overview*. Helsinki, 19–20.

Hussian, M.; Grimvall, A.; Petersen, W. 2003. Estimation of the human impact on nutrient loads carried by the Elbe River, *Environmental Monitoring and Assessment* 96: 15–33. http://dx.doi.org/10.1023/B:EMAS.0000031722.88972.62

Iital, A.; Pachel, K.; Loigu, E.; Pihlak, M.; Leisk, U. 2010. Recent trends in nutrient concentrations in Estonian rivers as a response to large-scale changes in land-use intensity and life-styles, *Journal of Environmental Monitoring* 12: 178–188. http://dx.doi.org/10.1039/b912923e

Jansons, V.; Vagstad, N.; Sudars, R.; Deelstra, J.; Dzalbe, I.; Kirsteina, D. 2002. Nutrient losses from point and diffuse agricultural sources in Latvia, *Landbauforsch Volkenrode* 1(52): 9–17.

Johnes, P. J.; Burt, T. P. 1993. Nitrate in surface waters, in Burt, T. P., *et al.* (Eds.). *Nitrate: Processes, Patterns and Management*. Chichester, 269–317.

Keeney, D. R.; DeLuca, T. H. 1993. Des Moines River nitrate in relation to watershed agricultural practices: 1945 versus 1980's, *Journal of Environmental Quality* 20: 123–128.

Klavins, M.; Rodionov, V.; Kokorite, I.; Klavina, I.; Apsite, E. 2001. Long-term and seasonal changes in chemical composition of surface waters in Latvia, *Environmental Monitoring and Assessment* 66: 233–251. http://dx.doi.org/10.1023/A:1006352015230

Kronvang, B.; Hezlar, J.; Boers, P.; Jensen, J. P.; Behrendt, H.; Anderson, T.; Arheimer, B.; Venohr, M.; Hoffmann, C. C. 2004. *Nutrient Retention Handbook*, EUROHARP Report 9-2004. Norway: NIVA.

Kronvang, B.; Larsen, S. E.; Jensen, J. P.; Andersen, H. E.; Sileika, A. S. Gaigalis, K. 2005. *Susve, Lithuania. Trend analysis Retention and Source Apportionment*, EUROHARP Report 12-2005. Denmark: NERI.

Lepisto, A.; Granlund, K.; Kortelainen, P.; Raike, A. 2006. Nitrogen in river basins: sources, retention in the surface waters and peatlands, and fluxes to estuaries in Finland, *Science of the Total Environment* 365: 238–259. http://dx.doi.org/10.1016/j.scitotenv.2006.02.053

Lindgren, G. A.; Wrede, S.; Seibert, J.; Wallin, M. 2007. Nitrogen source apportionment modeling and the effect of land-use class related runoff contributions, *Nordic Hydrology* 38: 317–331. http://dx.doi.org/10.2166/nh.2007.015

Magner, J.; Payne, G.; Steffen, L. 2004. Drainage effects on stream nitrate-N and hydrology in south-central Minnesota (USA), *Environmental Monitoring and Assessment* 91: 183–198. http://dx.doi.org/10.1023/B:EMAS.0000009235.50413.42

Molot, L. A.; Dillon, P. 1993. Nitrogen mass balances and denitrification rates in central Ontario Lakes, *Biogeochemistry* 20: 195–212. http://dx.doi.org/10.1007/BF00000787

Nash, J. E.; Sutcliffe, J. V. 1970. River flow forecasting through conceptual models part I – a discussion of principles, *Journal of Hydrology* 10 (3): 282–290. http://dx.doi.org/10.1016/0022-1694(70)90255-6

National Land Service 2004. *Topographic information LTDBK50000-V*. Vilnius (in Lithuanian).

Nixon, S. W.; Ammerman, J. W.; Atkinson, L. P.; Berounsky, V. M.; Billen, G.; Boicourt, W. C.; Boynton, W. R.; Church, T. M.; Ditoro, D. M.; Elmgren, R.; Garber, J. H.; Giblin, A. E.; Jahnke, R. A.; Owens, N. J. P.; Pilson, M. E. Q.; Seitzinger, S. E. 1996. The fate of nitrogen and phosphorus at the land-sea margin of the North Atlantic Ocean, *Biogeochemistry* 35: 141–180. http://dx.doi.org/10.1007/BF02179826

Poluckaja, N. 2000. *Neman* [Nemunas]. Byelorussia: Grodno, 4–5.

Povilaitis, A. 2008. Source apportionment and retention of nutrients and organic matter in the Merkys River basin in southern Lithuania, *Journal of Environmental Engineering and Landscape Management* 16(4): 195–204. http://dx.doi.org/10.3846/1648-6897.2008.16.195-204

Rekolainen, S. 1989. Phosphorus and nitrogen load from forest and agricultural areas in Finland, *Aqua Fennica* 19(2): 95–107.

Rekolainen, S.; Pitkanen, H.; Bleeker, A.; Felix, S. 1995. Nitrogen and Phosphorus fluxes from Finnish agricultural areas to the Baltic Sea, *Nordic Hydrology* 26: 55–72.

Saunders, D. L.; Kalff, J. 2001. Nitrogen retention in wetlands, lakes and rivers, *Hydrobiologia* 443: 205–212. http://dx.doi.org/10.1023/A:1017506914063

Sileika, A. S.; Gaigalis, K.; Kutra, G.; Smitiene, A. 2005. Factors affecting N and P losses from small catchments (Lithuania), *Environmental Monitoring and Assessment* 102(1–3): 359–374. http://dx.doi.org/10.1007/s10661-005-6033-3

Sileika, A. S.; Stalnacke, P.; Kutra, S.; Gaigalis, K.; Berankiene, L. 2006. Temporal and spatial variation of nutrient levels in the Nemunas River (Lithuania and Belarus), *Environmental Monitoring and Assessment* 122: 335–354. http://dx.doi.org/10.1007/s10661-006-9186-9

Šileika, A. S.; Gaigalis, K.; Šmitiene, A.; Baigys, G. 2010. Impact of some environmental and agricultural factors on nitrogen loss in three Lithuanian catchments, *Vandens ūkio inžinerija* [Water Management Engineering] 37(57): 52–63.

Stalnacke, P.; Grimvall, A.; Laznik, M. 1996. Water quality response to the dramatic reduction in the use of fertilizers in Latvia, in *Nutrient Loads to the Baltic Sea:* PhD Thesis. Linkoping: Linkoping University. 1–14.

Stalnacke, P.; Grimvall, A.; Sundblad, K.; Wilander, A. 1999a. Trends in nitrogen transport in Swedish rivers, *Environmental Monitoring and Assessment* 59: 47–72. http://dx.doi.org/10.1023/A:1006007711735

Stalnacke, P.; Grimvall, A.; Sundblad, K.; Tonderski, A. 1999b. Estimation of Riverine Loads of Nitrogen and Phosphorus to the Baltic Sea, 1970–1993, *Environmental Monitoring and Assessment* 58: 173–200. http://dx.doi.org/10.1023/A:1006073015871

SWAT 2005. *Soil and water assessment tool SWAT 2005* [online], [cited 05 March 2005]. Available from internet: http://swatmodel.tamu.edu.

Vaitkuviene, D.; Dagys, M. 2008. CORINE 2006 *land cover, Lithuania.* Implementing CLC2006 project in Lithuania, Report. Vilnius: Institute of Ecology of Vilnius University (in Lithuanian).

Windolf, J.; Jeppesen, E.; Jensen, J. P.; Kristensen, P. 1996. Modelling of seasonal variation in nitrogen retention and in-lake concentration: a four year mass balance study in 16 shallow Danish lakes, *Biogeochemistry* 33: 25–44. http://dx.doi.org/10.1007/BF00000968

Vitousek, P.; Moony, H.; Lubchenco, J.; Melillo, J. 2004. Human domination of Earth's ecosystems, *Science* 277: 494–499. http://dx.doi.org/10.1126/science.277.5325.494

Vuorenmaa, S.; Rekolainen, S.; Lepisto, A.; Kenttamies, K.; Kauppila, P. 2002. Losses of nitrogen and phosphorus from agricultural and forest areas in Finland during the 1980s and 1990s, *Environmental Monitoring and Assessment* 76: 213–248. http://dx.doi.org/10.1023/A:1015584014417

Vuoristo, H. 1998. Water quality classification of Finnish inland waters, *Eur. Water. Manag.* 6: 35–41.

Wulf, F.; Rahm, L. 1988. Long-term, seasonal and spatial variations of nitrogen, phosphorus and silicate in the Baltic: an overview, *Marine Environmental Research* 26: 19–37. http://dx.doi.org/10.1016/0141-1136(88)90032-3

Antanas Sigitas ŠILEIKA. Dr, chief researcher, Water Research Institute of the Aleksandras Stulginskis University. Research interest: Surface water quality modelling, assessment and search of measures for pollution reduction.

Mats WALLIN. PhD, director, Department of Aquatic Science and Assessment, Swedish University of Agricultural Sciences Research interest: Catchment scale water quality modelling and assessment of the impact of point source and non-point source pollution on rivers and lakes.

Kazimieras GAIGALIS. Dr, senior researcher, Water Research Institute of the Aleksandras Stulginskis University. Research interest: Hydrology, water flow and quality monitoring.

3

THE INVESTIGATION OF TOBACCO SMOKE INFLUENCE ON THE CHANGES OF INDOOR RADON AND ITS SHORT-LIVED DECAY PRODUCTS VOLUMETRIC ACTIVITIES

Dainius Jasaitis[1], Aloyzas Girgždys[2]

Department of Physics, Vilnius Gediminas Technical University, Saulėtekio al. 11, LT-10223 Vilnius, Lithuania

E-mail: [1]dainius.jasaitis@vgtu.lt (corresponding author); [2]aloyzas.girgzdys@vgtu.lt

Abstract. The changes of radon and its short-lived decay products were investigated in accommodations under natural living conditions and in tobacco smoke-filled premises. The measured radon and its short-lived decay products volumetric activities, aerosol particle concentration in the air, radioactive equilibrium factor, unattached fraction factor values are presented. It was identified that the increase of aerosol particles concentration in the air (in smoke-filled premises) determines the increase of the unattached nuclides of radon short-lived decay products attached to aerosol particles (the average values of radioactive equilibrium factor varied from 0.35 to 0.72). In this case, larger volumetric activity of the alpha particles is registered. Therefore larger amount of radon progeny is inhaled in smoke-filled premises and there is an increased possibility of damaging the organism. Positive correlation ($r = 0.9$) between the radioactive equilibrium factor and aerosol particle concentration in the air of accommodation, as well as negative correlation ($r = -0.64$) between the radioactive equilibrium factor and unattached fraction factor have been determined. Seasonal changes of the radioactive equilibrium factor are presented.

Keywords: radon, short-term decay products, volumetric activity, aerosol particle, tobacco smoke, radioactive equilibrium factor, unattached fraction factor.

Introduction

The ^{222}Rn is radioactive scentless and colorless inert gas. It is formed in the result of decaying of ^{226}Ra, which is the member of decaying chain of ^{238}U isotope. Radon and its decay products compose most of ionizing radiation (Girgždys, Ulevičius 1985; Gesell 1983). The average annual irradiance dose of Lithuanian inhabitants, aroused by natural sources, is about 2.2 mSv, about 1 mSv is conditioned by radon and its decay products‘ ionizing radiation (UNSCEAR 2000).

Radon is inert gas (half-life duration 3.824 days), therefore humans‘ health is more effected by its non-gas short-lived decay products: ^{218}Po (half-life 3 minutes), ^{214}Pb (27 minutes), ^{214}Bi (20 minutes), ^{214}Po (164 μs).

Most of radon radiation and its progeny is easily absorbed by human‘s skin surface, therefore human‘s inner irradiance is the most important, that is the irradiance due to inhaled radon decay products in the organism (Hess *et al.* 1983; Mohammed 1999). Having inhaled the air filled with radon short-lived decay products, they settle and decay on breathing organs. Then exuded alpha particles penetrate and damage the tissue. Due to that lungs or breathing organs cancer can develop (Lubin 2003; Ishikawa *et al.* 2003) (Fig. 1).

In Lithuania as in many other countries the largest source of radon is soil (Clavensjo *et al.* 1999; Mastauskas, Morkūnas 1996). The radon inert gas easily enters the premises through the gaps from the soil, which the house is built on, as well it is exuded from construction materials or even from the water. Thus, the big amount of radon is stored in the premises and after its decaying the hazardous short-lived products volumetric activities are formed, especially if the premises are not winded (Morkūnas 2000; El-Hussein *et al.* 1999; Andersen *et al.* 1997).

The larger the amount of aerosol particles in the air, the larger part of radon progeny settles on these particles but not on furniture, walls, or curtains. Thus, there are more radon short-lived decay products in a smoke-filled room than in a “clean” room with the same concentration of radon in the air (Dwiwedi *et al.* 2001; Mohamed 2005).

The influence of radon progeny on health depends on its behavior in the accommodations. There are two types of decay products: “unattached to aerosol particles” (the diameter of their diffusion equivalent is from 0.5 nm to 5 nm), and “attached to aerosol particles” (the diameter of particle is from 5 to 3 000 nm) (Hopke *et al.* 1992; Reineking *et al.* 1994).

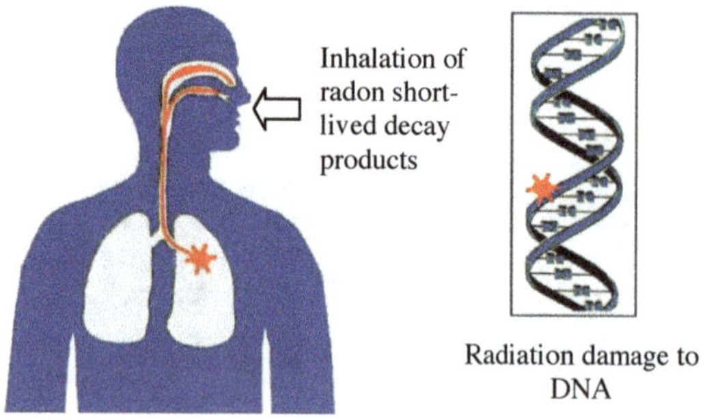

Fig. 1. The influence of radon short-lived decay products on human's organism

The nuclides of attached and unattached to aerosol particles are not separated while carrying different research on radon progeny out. It is important to evaluate the amount of attached and free nuclides in premises because due to different diffusion of molecules the short-lived decay products settle in different places of the airway and in different amounts (Castren *et al.* 1985; Swedjemark 1990).

It is proven by research that smoking strengthens the negative effect of radon. A smoker, who is together with a non-smoker in premises with the same amount of radon, risks to get a cancer even more than a non-smoker because radon progeny settle on tiny particles of tobacco smoke (i.e. aerosol), and newly appeared particles affect the airways much more. Though general effect of smoking and radon is still under investigation, there are some data showing that if the amount of radon in the premises increases 4 times, the possibility for a smoker to get lungs cancer is about 9 times higher than for a non-smoker (Radiation Protection Centre newsletter 2008).

The concentration of radon progeny and their location in the environment depend on the "behavior" of the first radon decay product ^{218}Po (Ho *et al.* 1982; Leung *et al.* 1998). This "behavior" is conditioned by such processes as cluster formation, neutralization of electric charge, attachment to the aerosol particles, and deposition on the surfaces. The electric charge of particles with ^{218}Po nuclides has a big impact on these processes and on the spectrum of radon short-lived decay products bearers (i.e. aerosol particles) (Porstendorfer 1994).

About 90% of ^{218}Po ions appeared after the decay of radon are positively charged. Part of ^{218}Po ions is neutralized; therefore another part of ions is neutral. ^{218}Po attaches to the aerosol particles, which are always in the air in natural environment, or clusters. Some of ^{218}Po ions, attached to clusters or aerosol particles, deposit on the surface. These processes of attachment to aerosol particles and surfaces depend on clusters and aerosol diffusions, and electric charges (Porstendorfer *et al.* 2005; Li, Hopke 1992).

The amount ratio of radon to its decay products is defined by radioactive equilibrium factor, usually named as F factor (Mohamed 2005; Baixeras *et al.* 1999):

$$F = \frac{(0,105C_1 + 0,516C_2 + 0,379C_3)}{C_0}, \quad (1)$$

where C_0 is radon volumetric activity; C_1, C_2, C_3 are ^{218}Po, ^{214}Pb and ^{214}Bi/^{214}Po volumetric activities in the air (Bq·m^{-3}).

This factor shows radioactive balance between radon and its progeny. The radioactive equilibrium factor depends mostly on winding intensity and concentration of aerosol particles in the air (Clavensjo, Akerblom 1994; Ramola *et al.* 2003).

Radioactive equilibrium factor is used as a factor to proceed from radon volumetric activity to its decay products volumetric activity in the premises, or vice versa. Radioactive equilibrium factor makes it possible to calculate the conditional effecting dose of radon in the premises (Baeza *et al.* 2003).

Another important parameter, which characterizes radon progeny indoors, is unattached fraction factor, which shows which part of radon decay products is not settled on aerosol particles in accommodations (Yu *et al.* 1996):

$$f_i = \frac{C^i_{eq}}{C_i}, \quad (2)$$

where C^i_{eq} is balanced equal activity. C_i is volumetric activity of itch radon decay product unattached nuclides.

Equivalent equilibrium concentration is expressed (Mohamed 2005; Yu *et al.* 1996):

$$C_{eq} = 0.105C_1 + 0.516C_2 + 0.379C_3, \quad (3)$$

where C_1, C_2, C_3 are radon short - term decay products (^{218}Po, ^{214}Pb, ^{214}Bi) volumetric activities in the air.

The aim of this work is to investigate the influence of tobacco smoke on changes of indoor radon and its short-lived decay products volumetric activities, evaluate the dependence of unattached fraction factor and radioactive equilibrium factor on concentration of aerosol particles in the air.

1. Materials and methods

Measurements were carried out applying invented equipment and method for constant measurements of radon short-lived decay products volumetric activities in the air (Jasaitis, Girgždys 2007).

The measuring equipment consists of optical aerosol particles counter, filtration device with a radiometer and air volume meter with a pump (Fig. 2).

The metal frame of filtration device has inlet and outlet. Inlet is used for air suction. A pump is attached to the outlet. The diameter of the air inlet is 20 mm.

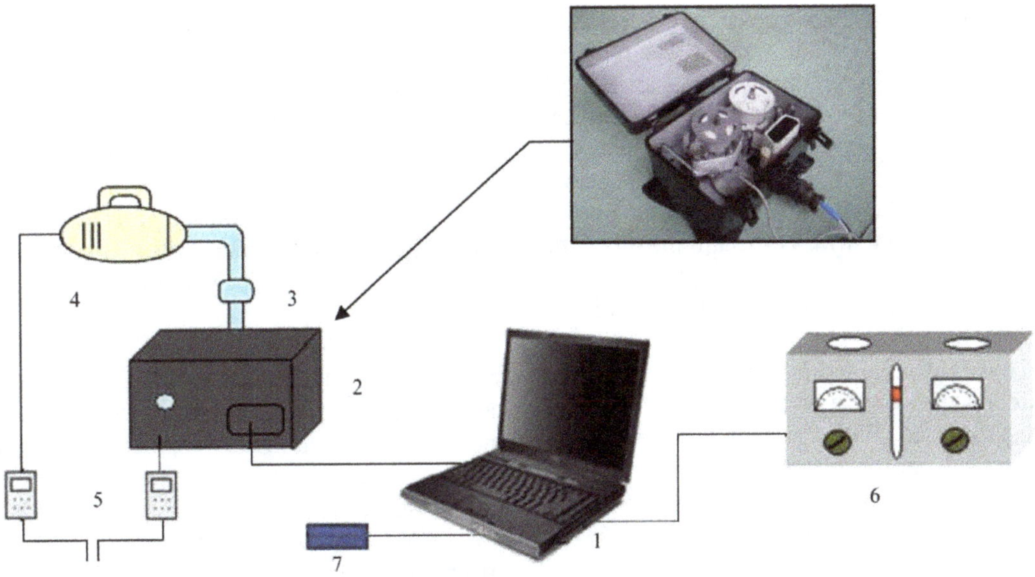

Fig. 2. Measurement scheme: 1 – computer; 2 – filtration device; 3 – air volume meter; 4 – air pump; 5 – stopwatches; 6 – optical aerosol particle counter; 7– multifunctional data logger

The air suction rate is 40 l/min. and it is controlled by air flow meter. The air in the device frame changes within 25 seconds. The 800 W powered air pump sucks the air through filter band "Fiberglaz" which diameter is 50 mm. The air is sucked through the part of filter band which is between a suction channel and a radiometer. The band is stopped during the measurement period.

When the measurement is over, the band is overwound so that the "clean" (i.e. without radon progenies) part of the filter appears above the radiometer and the air suction channel of the pump. The radiation of the particles, settled on the filter, is measured with the radiometer GM-45.

The radiometer GM-45 is small, light and extremely sensitive detector of the ionizing radiation. It contains a Geiger - Müller counter, which is sensitive to alpha, beta and gamma radiation (alpha above 3 MeV, beta above 50 keV, gamma above 7 keV). The diameter of isinglass is 42 mm. The RAD (Radiation Acquisition and Display) programme is included in this device. The usage of RAD programme enabled data to be stored and transferred into MS Excel programme, where they could be processed. During the measurement data were automatically recorded and saved.

The radiometer is fixed so that the filter band through which the air is sucked moves beside the radiometer measurement window (Fig. 3).

Electronic stopwatches TS-ED1 are set, and the time of suction and when the filter band is turned are programmed. Having set the hours when the measurements are automatically carried out, the device can operate permanently.

The concentration of particles attached to aerosol and "free" nuclides of radon progeny and their small clusters in the air are assessed by plugging in and then unplugging the collector of "free" nuclides and clusters to the entrance of the air flow in the control device. This collector is made of 10 layers of a wire (the diameter is 0.2 mm) net bundle. The net bundle is connected to soil so that the charged nuclides can deposit on the surface of the wire more efficiently. Free positively charged nuclides of ^{218}Po and other short-lived decay products and small clusters deposit on the surface of the net wire due to high mobility. The nuclides attached to aerosol particles pass the net bundle; they are collected by the filter band "Fiberglaz" where after the radioactive decay they are registered with the radiometer.

The aerosol particle concentration is measured with an optical aerosol particle counter AZ-5. The aerosol counter measures the light scatter of separate

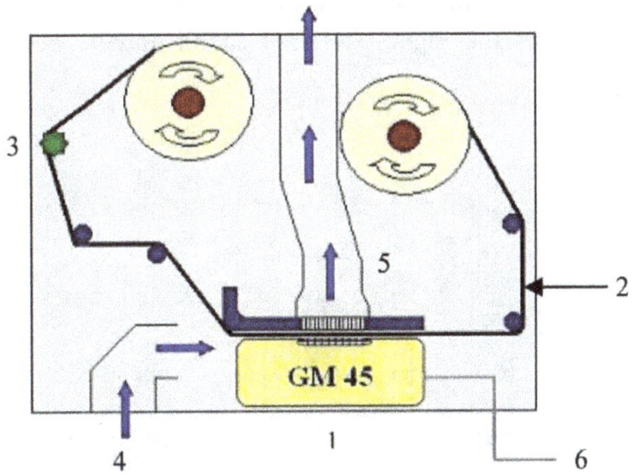

Fig. 3. Filtration device: 1 – radiometer; 2 – filter band; 3 – filter band speed regulator; 4 – air flow pumped from outside; 5 – air flow pumped by pump; 6 – data sent to the computer

aerosol particles and can rate aerosol particles sized from 0.4 to 10 μm.

The aerosol counter AZ-5 consists of an optical sensor, a pump and an electrical part. The optical sensor creates an electric pulse for each particle. The amplitude depends on the size of particles. The pump sucks the air through the measuring zone of the optical sensor (capacity 1.2 l/min). AZ-5 can measure the concentration of aerosol particles in the air from 0 to 310^6 particles per liter. Possible systematic bias is ±20%.

In this measuring method the aerosol particle counter is connected to the computer, to which data are sent and recorded. With the help of a data logger ADC-16 the data of continuous measurements are accumulated in the computer. The concentration of aerosol particles and its alterations are constantly controlled.

The bias of radon short-lived decay products measuring method was evaluated. It depends on the efficiency changes of the filter band, the speed of air flow through it, exposition time, and the concentration of progeny in the air. The established bias was not higher than 6%.

2. The Results of Measurement

The research was carried out in the rooms of an individual house under natural living conditions and having increased the concentration of aerosol particles in the air, i.e. after having smoked in the room.

The average radon short-lived decay products volumetric activities were measured under natural living conditions on all the floors. The results of the measurements are presented in Fig. 4.

Considering the measurement results in the cellar, on the ground and first floors, it can be stated that the main source of radon progeny in the investigated house is the soil under the house, as it is usual in Lithuania and other countries (Clavensjo *et al.* 1999), which conditions the highest level of radon and its decay products volumetric activities in the cellars or on the ground floors (if there is no cellar).

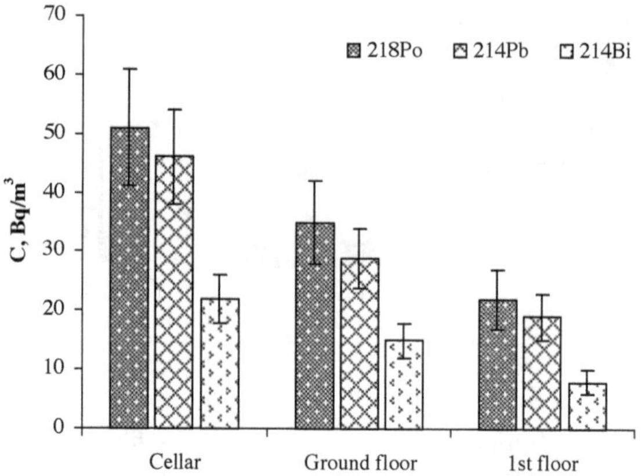

Fig. 4. Average radon short-lived decay products volumetric activities on different floors of the house

Further measurements were carried out on the ground floor of the house. It was chosen because inhabitants of the house usually spend most of their time there, and the first floor is used for bedrooms. The volumetric activity of radon progeny in bedrooms is low (Fig. 4), what is more, these premises are usually winded before sleep. The plan of the house is presented in Fig. 5.

Radon volumetric activities (*C*), unattached fraction factor (f_{Po}), radioactive equilibrium factor (*F*) and aerosol particle concentration (*N*) in the air under natural living conditions were measured in all rooms of the investigated house. The results of measurements are presented in Table 1.

The values of radon volumetric activities altered from 16 to 45 $Bq{\cdot}m^{-3}$. The statistical bias of the measurement reached up till 15%. It was determined that the average radon volumetric activity on the ground floor of the investigated house was 34±4 $Bq{\cdot}m^{-3}$. To compare the average volumetric activity of individual houses in Lithuania is about 56 $Bq{\cdot}m^{-3}$ (Clavensjo *et al.* 1999).

According to valid hygienic norms in Lithuania the volumetric activity of radon should not be higher than 200 $Bq{\cdot}m^{-3}$ in newly built premises, and 400 $Bq{\cdot}m^{-3}$ in already existing premises. As it is seen in Table 1, radon volumetric activities are lower in those premises, which are winded better and which are less hermetic (Fig. 5).

The values of the unattached fraction factor altered from 0.05 to 0.11, calculated average value is 0.08±0.02. It was determined that the increase of aerosol particles concentration in the air of accommodations conditions the decrease of the unattached fraction factor. Also the inverse dependence between unattached fraction factor and radioactive equilibrium factor was determined.

The values of radioactive equilibrium factor altered from 0.33 to 0.65, calculated average value is 0.39±0.03. A tight correlation (correlation coefficient r = 0.85) between the radioactive equilibrium factor and aerosol particles concentration in the air was

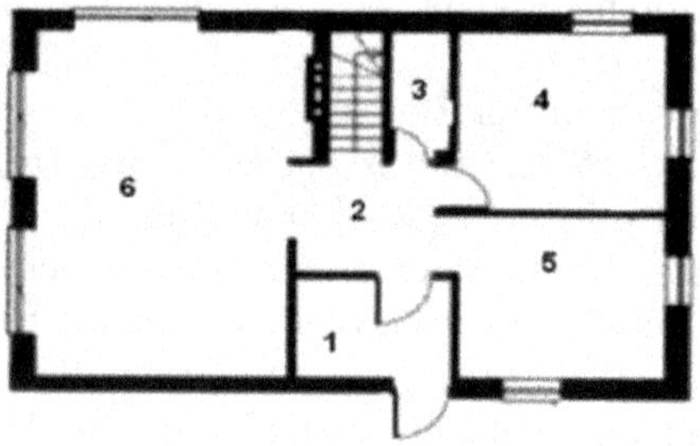

Fig. 5. The plan of the ground floor of the investigated house: 1 – anteroom; 2 – hall; 3 – toilet; 4 – kitchen; 5 – office room; 6 – living-room

Table 1. Measured values of radon volumetric activities (*C*), unattached fraction factor (f_{Po}), radioactive equilibrium factor (*F*) and aerosol concentration (*N*) in the air on the ground floor of the investigated house under natural living conditions

Nr.	C, Bq·m^{-3}	f_{Po}	F	$N \times 10^3$, cm^{-3}
1	16 ± 3	0.11 ± 0.02	0.33 ± 0.03	1.6 ± 0.19
2	29 ± 4	0.08 ± 0.02	0.34 ± 0.03	1.9 ± 0.20
3	32 ± 4	0.08 ± 0.02	0.37 ± 0.03	2.4 ± 0.20
4	45 ± 6	0.05 ± 0.01	0.65 ± 0.05	3.4 ± 0.24
5	40 ± 5	0.07 ± 0.02	0.37 ± 0.03	2.8 ± 0.22
6	42 ± 5	0.07 ± 0.02	0.38 ± 0.03	3.2 ± 0.22

determined. The higher aerosol particle concentration, the higher is the radioactive equilibrium factor, as the process of attachment becomes faster than the deposition of unattached particles on the surface.

The average absolute bias of measured aerosol particles concentration in the air reached 8%.

Further research was carried out to investigate what impact the increased concentration of aerosol particles has on indoor radon short-lived decay products alterations.

The measurements were carried out having smoked in the rooms of an individual house.

During the research the temperature in premises and relative humidity was about 21 °C and 47%, the average aerosol particles concentration reached 1.910^3 cm^{-3}, the average radon volumetric activity was 32 Bq·m^{-3}. The speed of winding in the premises was about 0.2 h^{-1}. Having smoked in the premises the aerosol particles concentration had increased till 95.10^3 cm^{-3}. The results of measurement are presented in Table 2. Statistical averages of the measurements are presented there. In total 45 measurements were carried out.

Under natural conditions indoors the values of unattached fraction factor varied from 0.05 to 0.20,

Table 2. Aerosol concentration in the air (*N*), unattached fraction factors (f_{Po}) and radioactive equilibrium factors (*F*) in smoke-filled and non-smoked premises

	Under natural conditions	*Smoke-filled premises*
$N \times 10^3$, cm^{-3}	1.90 ± 0.20	92.5 ± 6.5
F	0.35 ± 0.03	0.72 ± 0.06
f_{Po}	0.09 ± 0.02	0.03 ± 0.01

and the average value was 0.09±0.02. The values of radioactive equilibrium factor altered from 0.15 to 0.50, and the average value was 0.35±0.03.

Usually the indoor aerosol particles concentration is low, therefore only very small amount of radon short-lived decay products attaches to the particles. It means that most of radon progeny appeared after the decay are free (unattached), and they settle on different surfaces. Facts mentioned above condition the radioactive disbalance between radon and its short-lived decay products. As it was mentioned before with the increase of aerosol particles concentration the radioactive equilibrium factor increases too.

Having smoked in the accommodation the aerosol particles concentration in the air has increased, which conditioned the increase of average value of radioactive equilibrium factor till 0.72±0.06 and the increase of average value of unattached fraction factor till 0.03±0.01.

Figure 6 shows a typical example of changes of radioactive equilibrium factor and unattached fraction factor during day and night period, where the concentration of indoor aerosol particles is changing under natural conditions and in smoke-filled premises.

It was noticed that with the increase of aerosol particles concentration in the air the unattached fraction factor was decreasing – the values of factors were only reaching 0.03–0.05. At the same time the

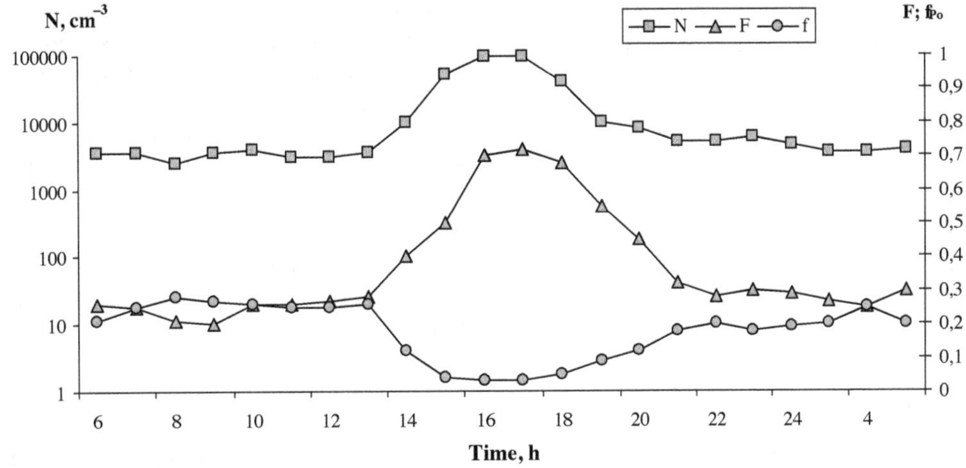

Fig. 6. Changes of unattached fraction factor and radioactive equilibrium factor when the concentration of aerosol particles in the air alters due to the smoke in the premises

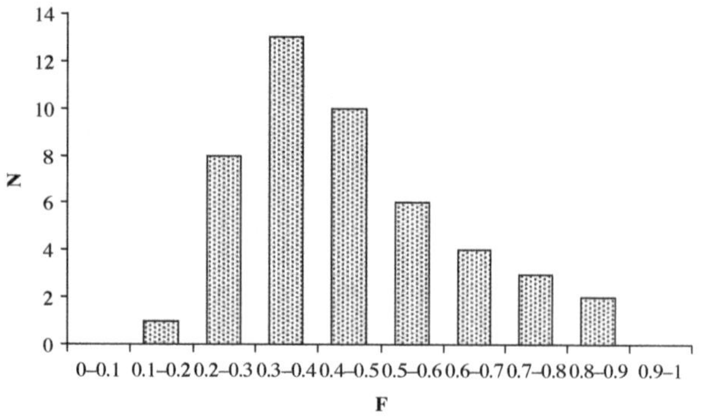

Fig. 7. Repeated frequency of indoor radioactive equilibrium factor

radioactive equilibrium factor was increasing, its values reached 0.6–0.7, and in some cases even about 0.8–0.9.

As it is seen in figure 6, there is an inverse dependence between radioactive equilibrium factor and unattached fraction factor. The negative correlation between the unattached fraction factor and the radioactive equilibrium factor was obtained, which can be expressed as $f = 0.06 \cdot F^{-0.64}$. Such dependence was also determined during other measurements (Politova, Jasaitis 2010). What is more, a clear correlation between radioactive equilibrium factor and indoor aerosol particles concentration was determined, the correlation coefficient $r = 0.9$.

Figure 7 presents the repeated frequency of calculated radioactive equilibrium factors in different investigated accommodations.

Usual repeated value of radioactive equilibrium factor is 0.3–0.4.

The seasonal average values of the radioactive equilibrium factor in the premises are presented in Figure 8.

It was determined that the seasonal values of the radioactive equilibrium factor in the investigated house varied from 0.43 to 0.52 (Jasaitis, Girgždys 2011).

Figure 8 shows, that the values of radioactive equilibrium factor are the highest in winter and the lowest in summer. In winter the accommodations are

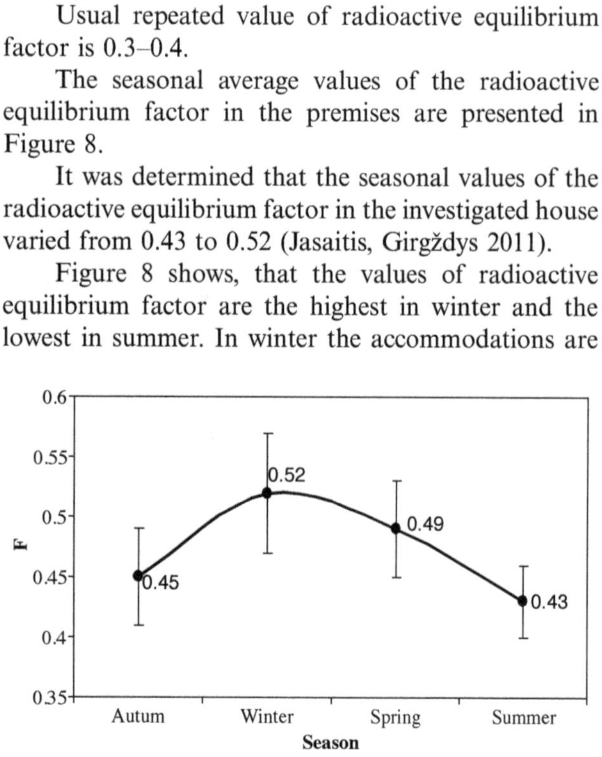

Fig. 8. Seasonal variations of radioactive equilibrium factor in the investigated accommodations

less winded and the house is more hermetic than during other seasons. This allows radon and its progeny to accumulate. Therefore the radioactive equilibrium factor between radon and its progeny increases. In summer accommodations are well ventilated and this lowers the radioactive equilibrium factor between radon and its progeny (Porstendorfer *et al.* 1994).

The obtained measurement values of radioactive equilibrium factor are similar to the values calculated in other European countries. The received values are similar to the measurements in Holland and Germany (Cavallo 2000; Kreuzer *et al.* 2003). It is noticed that the radioactive equilibrium factor is higher in cold countries than in tropical countries. However, having in mind a wide range of radioactive equilibrium factor variations in different places, it is recommended to calculate it separately for specific regions.

Conclusions

1. The measured average radon volumetric activity in the investigated house was 34 ± 4 $Bq \cdot m^{-3}$. It is lower than average volumetric activity in individual houses in Lithuania (about 56 $Bq \cdot m^{-3}$). It was determined that radon volumetric activities were lower in better winded, less hermetic accommodations of the investigated house.

2. Considering the results of volumetric activities measurements in the cellar, on the ground and first floors of the house, it has been determined that the main source of radon progeny in the investigated house is the soil under the house.

3. Under natural conditions the average values of unattached fraction factor and aerosol particles concentration were 0.09 ± 0.02 and $(1.90 \pm 0.20) \cdot 10^3$ cm^{-3}. In smoked premises respectively – 0.03 ± 0.01 and $(92.5 \pm 6.5) \cdot 10^3$ cm^{-3}. Having increased aerosol particles concentration, the amount of attached free nuclides of radon short – lived decay products to the aerosol particles increases, therefore higher volumetric activity of alpha particles is registered.

4. The average value of measured radioactive equilibrium factor under natural conditions was 0.35 ± 0.03, at the same time the value in smoked premises often reached 0.7–0.8, therefore there is a higher possibility to inhale radon progeny and damage organism in the smoked premises.

5. It was determined that with the increase of indoor aerosol particles concentration unattached fraction factor decreased. What is more, the invert dependence between unattached fraction factor and radioactive equilibrium factor was determined.

6. Clear positive correlation between the radioactive equilibrium factor and aerosol particles concentration in the premises was determined, correlation coefficient $r = 0.9$. Correlation between radioactive

equilibrium factor and unattached fraction factor was negative.

7. The most frequently repeated value of the radioactive equilibrium factor is 0.3–0.4. It has been determined that radioactive equilibrium factor is lower in summer than in other seasons. Such seasonal change of the radioactive equilibrium factor due to better ventilated accommodations during warm season.

References

Andersen, C.; Bergsoe, N. C.; Majborn, B.; Ulbak, K. 1997. Radon and ventilation in newer Danish single – family houses, *Indoor Air* 7(4): 167–185. http://dx.doi.org/10.1111/j.1600-0668.1997.00007.x

Baeza, A.; Navarro, E.; Roldan, C.; Ferrero, J. L.; Juanes, D.; Corbacho, J. A.; Guillen, F. J. 2003. Indoor radon levels in buildings in the Autonomous Community of Extremadura (Spain), *Radiation Protection Dosimetry* 103(3): 263–268. http://dx.doi.org/10.1093/oxfordjournals.rpd.a006142

Baixeras, C.; Amgarou, K.; Font, L.; Domingo, C. 1999. Long-term radon levels and equilibrium factor in some Spanish workplaces measured with a passive integrating detector, *Radiation Protection Dosimetry* 85(1–4): 233–236. http://dx.doi.org/10.1093/oxfordjournals.rpd.a032841

Cavallo, A. 2000. The radon equilibrium factor and comparative dosimetry in homes and mines, *Radiation Protection Dosimetry* 92(4): 295–298. http://dx.doi.org/10.1093/oxfordjournals.rpd.a033295

Castren, O.; Voutilainen, A.; Winqvist, K. 1985. Studies of high indoor radon areas in Finland, *Science Total Environment 45*: 311–318. http://dx.doi.org/10.1016/0048-9697(85)90232-3

Clavensjo, B.; Akerblom, G. 1994. *The radon book. Measures against radon.* The Swedish Council for Building Research. Stockholm. 129 p.

Clavensjo, B.; Akerblom, G.; Morkunas, G. 1999. *Radonas patalpose. Jo kiekio mazinimo būdai* [Radon indoors. Its reduction techniques]. Vilnius: Litima. 126 p.

Dwiwedi, K. K.; Mishra, R.; Tripathy, S. P.; Kulshreshtha, A.; Sinha, D.; Srivastava, A.; Deka, P.; Bhattacharjee, B.; Ramachandran, T. V.; Nambi, K. S. V. 2001. Simultaneous determination of radon, thoron and their progeny in dwellings, *Radiation Measurements* 33(1): 7–11. http://dx.doi.org/10.1016/S1350-4487(00)00131-1

El-Hussein, A.; Mohhamed, A.; Ahmed, A. 1999. Radon exhalation and ultrafine fraktion of radon progeny in closed room air, *Atmospheric Environment* 33(2): 183–190. http://dx.doi.org/10.1016/S1352-2310(98)00146-0

Gesell, T. F. 1983. Background atmospheric ^{222}Rn concentrations outdoors and indoors: a review, *Health Physics* 45(2): 289–302. http://dx.doi.org/10.1097/00004032-198308000-00002

Girgždys, A.; Ulevičius, V. 1985. The concentration of radon and its short-term decay products in the air, *Atmospheric Environment* 985: 28–33.

Hess, C. T.; Weiffenback, C. V.; Norton, S. A. 1983. Environmental radon and cancer correlations in Maine, *Health Physics* 45(2): 339–348. http://dx.doi.org/10.1097/00004032-198308000-00006

Ho, W. L.; Hopke, P. K.; Stukel, J. J. 1982. The attachment of RaA (^{218}Po) to monodisperse aerosol, *Atmospheric Environment* 16(4): 825–836. http://dx.doi.org/10.1016/0004-6981(82)90401-2

Hopke, P. K.; Wasiolek, P.; Montassier, N.; Cavallo, A.; Gadsby, K.; Socolow, R. 1992. Measurement of activity-weighted size distributions of radon decay products in a normally occupied home, *Radiation Protection Dosimetry* 45(1): 329–331.

Ishikawa, T.; Yamada, Y.; Fukutsu, K.; Tokonami, S. 2003. Deposition and clearance for radon progeny in the human respiratory tract, *Radiation Protection Dosimetry* 105(1–4): 143–148. http://dx.doi.org/10.1093/oxfordjournals.rpd.a006210

Jasaitis, D.; Girgždys, A. 2007. Hourly measurement method for radon progeny volumetric activity in air, *Journal of Environmental Engineering and Landscape Management* 15(3): 158–165.

Jasaitis, D.; Girgždys, A. 2011. Influence of aerosol particle concentration on volumetric activities of indoor radon progeny, *Lithuanian Journal of Physics* 51(2): 155–161.

Kreuzer, M.; Heinrich, J.; Wölke, G.; Rosario, A.; Gerken, M.; Wellmann, J.; Keller, G.; Kreienbrock, L.; Wichmann, H. 2003. Residential radon and risk of lung cancer in Eastern Germany, *Epidemiology* 14(5): 559–568. http://dx.doi.org/10.1097/01.ede.0000071410.26053.c4

Li, C. S.; Hopke, E. K. 1992. Air filtration and radon decay product mitigation, *Indoor Air* 2(2): 84–100. http://dx.doi.org/10.1111/j.1600-0668.1992.03-22.x

Leung, J. K. C.; Tso, M. Y. W.; Ho, C. V. 1998. Behavior of ^{222}Rn and its progeny in high-rise building, *Health Physics* 75(3): 303–312. http://dx.doi.org/10.1097/00004032-199809000-00010

Lubin, J. H. 2003. Studies of radon and lung cancer in North America and China, *Radiation Protection Dosimetry* 104(4): 315–319. http://dx.doi.org/10.1093/oxfordjournals.rpd.a006194

Mastauskas, A.; Morkūnas, G. 1996. Problem of indoor radon in Lithuania, *Health Physic* 70(6): 581.

Mohammed, A. 1999. Activity size distributions of short-lived radon progeny in indoor air, *Radiation Protection Dosimetry* 86(2): 139–145. http://dx.doi.org/10.1093/oxfordjournals.rpd.a032933

Mohamed, A. 2005. Study on radon and radon progeny in some living rooms, *Radiation Protection Dosimetry* 105(4): 143–148.

Morkūnas, G. 2000. *Assessment of effective dose caused by radon in detached houses*: PhD thesis. 89 p.

Politova, D.; Jasaitis, D. 2010. The influence of aerosol concentration on changes of volumetric activities of indoor radon short-term decay products, *Science – Future of Lithuania: Environment Protect Engineering* 2(5): 81–86.

Porstendorfer, J. 1994. Properties and behaviour of radon and thoron and their decay products in the air, *Journal of*

Aerosol Science 25(2): 219–263. http://dx.doi.org/10.1016/0021-8502(94)90077-9

Porstendorfer, J.; Butterweck, G.; Reineking, A. 1994. Daily variation of indoor radon concentration indoors and outdoors and the influence of meteorological parameters, *Health Physics* 67(3): 283–287. http://dx.doi.org/10.1097/00004032-199409000-00011

Porstendorfer, J.; Pagelkopf, P.; Grundel, M. 2005. Fraction of the positive ^{218}Po and ^{214}Pb clusters in indoor air, *Radiation Protection Dosimetry* 113(3): 342–351. http://dx.doi.org/10.1093/rpd/nch465

Radiation Protection Centre newsletter 2008. *2*(6).

Ramola, R. C.; Negi, M. S.; Choubey, V. M. 2003. Measurement of equilibrium factor "F" between radon and its progeny and thoron and its progeny in the indoor atmosphere using nuclear track detectors, *Indoor and Built Environment* 12: 351–355. http://dx.doi.org/10.1177/142032603035368

Reineking, A.; Knutson, E. A.; Geoge, A. C.; Solomon, S. B.; Kesten, J.; Butterweck, G.; Porstendorfer, J. 1994. Size distribution of unattached and aerosol-attached short-lived radon decay products: some results of intercomparison measurements, *Radiation Protection Dosimetry* 56(1–4): 113–118.

Swedjemark, G. A. 1990. Recent Swedish experiences in radon control, *Health Physics* 58: 453–456. http://dx.doi.org/10.1097/00004032-199004000-00007

UNSCEAR 2000. *Report to the General Assembly. Annex B: Exposures from natural radiation sources.* United Nations, New York. 89–92.

Yu, K. N.; Young E. C. M.; Li, K. C. 1996. A study of factors affecting indoor radon properties, *Health Physics* 71(2): 179–184. http://dx.doi.org/10.1097/00004032-199608000-00008

Dainius JASAITIS. Dr, Assoc. Prof., Department of Physics, Vilnius Gediminas Technical University (VGTU), Saulėtekio al. 11, LT–10223, Vilnius, Lithuania.
Doctor of Technological Sciences (Environmental Engineering and Landscape Management), VGTU, 2007. Master of Science (Ecology and Environment), VGTU, 2003. Publications: co-author of more than 15 research papers. Research interests: environmental physics, natural radioactivity, ionizing radiation.

Aloyzas GIRGŽDYS. Dr, Prof., Department of Physics, Vilnius Gediminas Technical University (VGTU), Saulėtekio al. 11, LT–10223, Vilnius, Lithuania.
Head of Laboratory of Nuclear Hydrophysics (VGTU).
Doctor of Science (environmental physics), Moscow Institute of Atmospheric Physics, 1985. First degree in Physics, Vilnius University (VU), 1970. Publications: author of 1 monograph, over 180 scientific articles. Research interests: environmental physics, aerosol physics.

4

THE INFLUENCE OF TREATED OLEO-CHEMICAL WASTEWATER APPLICATIONS ON THE METAL SPECIATION FORMS IN SOILS

Sławomir Żak[a], Terese Rauckyte-Żak[a], Alfredas Laurinavičius[b]

[a]*Faculty of Chemical Technology and Engineering, University of Technology and Life Sciences, 3 Seminaryjna Street, 85-326 Bydgoszcz, Poland*

[b]*Faculty of Environmental Engineering, Vilnius Gediminas Technical University, Sauletekio al. 11, LT-10223 Vilnius, Lithuania*

Abstract. This paper presents a developed method of pretreating wastewaters coming from small-tonnage oleo-chemical production and also the research results on the variability of concentrations of heavy metals (Cd, Cr, Cu, Ni, Pb, Zn) in the soil irrigated and fertilized with wastewater from this production. The processing wastewater pretreatment was carried out at two stages. The first stage eliminated fatty substances by using dispersed air flotation aided with hydrogen peroxide as an oxidizer. The second stage included coagulation and chemical precipitation with milk of lime, flocculation, final dissolved air flotation (DAF), and also oxidation with hydrogen peroxide. Field experiments, the aim of which was to compare variability levels for speciation fractions of heavy metals, were carried out according to Tessier's method for four years. The study revealed that the average concentrations of the given metals fractions in soil samples where raw wastewaters were directed did not significantly change during the experiment, which clearly justifies the need for their physical–chemical pretreatment and further agro-utilization without the risk of soil environment degradation. The successive decrease of exchangeable fraction, the increase in the content of the fraction bound with carbonate forms, and stability of the fraction bound with hydrated iron and manganese oxides as well as organic and residual fractions were found in soil samples treated with pretreated wastewaters.

Keywords: irrigation and fertilization with oleo-chemical plant processing wastewater; heavy metal fractional content; flotation.

Reference to this paper should be made as follows: Żak, S.; Rauckyte-Żak, T.; Laurinavičius, A. 2013. The influence of treated oleo-chemical wastewater applications on the metal speciation forms in soils, *Journal of Environmental Engineering and Landscape Management* 21(2): 85–95. http://dx.doi.org/10.3846/16486897.2013.773259

Introduction

Small oleo-chemical plants generally deal with manufacturing fatty products such as vegetable and animal technical greases and technical fatty acids. The plants are located mainly in rural areas and their products are designed for biodiesel production, surface-active agents (SAA) synthesis, feed enrichment, and cosmetic industry (Meher *et al.* 2006; Tashtoush *et al.* 2004; Ullmanns Encyclopedia of Industrial Chemistry 1992; Haas *et al.* 2000; Angenent *et al.* 2004; Baltrenas, Kvasauskas 2008). Basic raw materials used for oleo-chemical production mostly come from the postproduction of animal fats (raw materials of the third class) from meat factories and include pork, beef, and poultry fats (e.g. tallow, inguinal fat, intestinal fat, etc.) (Commission Regulation (EC) No 92/2005; Żak 2005; Żak, Pawlak 2006a). As regards fats of plant origin, they are post-refining wastes, soapstock wastes, wastes from the process of degumming, fats from gravitational separators, post-flotation fats, fats recovered from bleaching soils, etc. (Żak 2005; Żak, Pawlak 2006b). The production processes regarding fats of animal or vegetable origin are inextricably linked with the formation of industrial wastewaters which after pretreatment can be utilized in plant farming by irrigation and fertilization because they are biodegradable (Kvasauskas, Baltrenas 2009; Aikaite-Stanaitiene *et al.* 2010; Directive 76/464/EEC). Technological wastewaters from vegetable fat production are rich in phosphorus compounds and poor in nitrogen compounds (Rüffer, Rosenwinkel 1991). The industrial wastewaters from animal fat production are rich in nitrogen compounds and potassium, and poor in forms of phosphorus (Rüffer, Rosenwinkel 1991; Londong, Rosenwinkel 2007). Therefore, it is possible to optimize and regulate the proportions N:P:K in the mixed wastewaters, pretreated

Corresponding author. Terese Rauckyte-Żak
E-mail: terra@mail.utp.edu.pl

or directed for watering and fertilization of soils. Due to the significant contents of fats, suspensions and colloids generally found in form of stable dispersions, it is necessary to use physical–chemical pretreatment (Aktas *et al.* 2001). A basic way of pretreating wastewaters from such production is their chemical treatment by using flocculation-aided coagulation and flotation (Rüffer, Rosenwinkel 1991; Londong, Rosenwinkel 2007; Ng *et al.* 1988; Mittal 2006; Żak, Zabłocki 2005; Rubio *et al.* 2002; Żak 2008b). One of the main parameters limiting the agro-utilization of the tested wastewaters is the content of heavy metals in soils where wastewaters are found (Directive 76/464/EEC). The concentration level of wastewaters, including their volume, dosages, and loads of pollutants (Directive 76/464/EEC; Lepp 1981; Ociepa *et al.* 2008), affects a given type of soil and limits their use in agriculture. A total volume of the post-pretreated substances contains mainly dissolved and colloidal proteins and polyoses which, after being introduced into the soil, undergo complex chemical and biochemical processes (Lepp 1981; Rattan *et al.* 2005). In literature, one can find various aspects of the agricultural use of wastewater from animal and plant fat production (Russell *et al.* 1993), but there is no discussion on issues concerning speciation variability of metals during multiyear agro-utilization. These aspects seem to be crucial, considering the mobility of individual metal ions and transport of some of their forms in a plant system during vegetation (Vaišis, Šuksta 2010). The problem may become important as regards cultivation technology and practical usability of plants cultivated on such irrigated arable grounds (Lepp 1981). The investigated production of technical fats of vegetable and animal origin was carried out by a small oleo-chemical factory in the countryside and the research results presented here are the continuation of the previous study (Żak, Rauckyte 2008b).

The aim of the present study was to compare the level of the content and the changeability of heavy metals (Cd, Cr, Cu, Ni, Pb, Zn) in the surface layer of soil (0.0–20.0 cm) during the four-year agro-utilization of the defatted raw and physico-chemically pretreated mixed wastewaters from oleo-chemical production.

1. Experimental part

1.1. Field plot trial

The soil samples taken from the experiment plots were classified as light soil (with its grain-size distribution at the plough-humic horizon – loamy sand). The investigated soils came from boulder clays and had a sequence of genetic horizons appropriate for Cambisols, the soil profile: Ap-Bw-Ck (IUSS WRB 2006). Before the experiment started, the pH of soil was acidic, pH in 1 M KCl 3.7 $\pm$ 0.2 and the active acidity (in distilled H_2O), measured potentiometrically was pH 6.7 $\pm$ 0.2. The average content of humus in depositional levels was 13.206 $g{\cdot}kg^{-1}$, organic carbon amounted 7.66 $g{\cdot}kg^{-1}$, and the total content of: nitrogen 0.70 $g{\cdot}kg^{-1}$, phosphorus (expressed as P_2O_5) 52 $mg{\cdot}kg^{-1}$, potassium (as K_2O) 27 $mg{\cdot}kg^{-1}$, and magnesium 14 $mg{\cdot}kg^{-1}$.

The doses directed to experimental plots were limited by the total nitrogen concentration in all wastewater batches, not exceeding the limit account of 170 kgN/ha $\times$ year. The following plants were sown on the experimental plots every year: oat, rape, triticale, and mustard, which were introduced into soil in the form of total mass of green manure.

Soil samples (the so-called light soils) from two experimental plots (50.0 ares each), irrigated and fertilized with defatted mixed raw and pretreated wastewaters at autumn every year, were taken from a surface soil layer at 0.0–20.0 cm. Defatted raw and processing wastewaters from oleo-chemical production used for fertilization were free from heavy metals.

The wastewaters used in the experiments originated from the oleo-chemical facility, category III, functioning in accordance with Directive 1774/EEC, and located in the Urszulewo Plain (315.16) of the physico-geographical mezoregion in the eastern part of Chełmno-Dobrzyń Lake District, Poland. The experimental plots were located in the arable areas near the facility.

Data-sets of investigated soils

Soil material for the analysis of the total contents of heavy metals was sampled for four years, in equal intervals – one series every six months (one in April, the second one in October) from the experimental plots. Air-dried soil samples were mineralized and extracted with a mixture of acids (HCl, ρ = 1.1830 g/ml, and HNO_3, ρ = 1.3410 g/ml in volume ratio 3:1), according to the Polish Standard PN-ISO 11047:2001. The concentrations were determined by using the atomic absorption spectrophotometry AAS (apparatus: BUCK Scientific 210 VGP or Spectr AA Varian 220 SS).

1.2. Measurement methods

Soil samples collected from the areas of experimental plots were also sequentially extracted according to Tessier's method (Tessier *et al.* 1979), by using air-dried analytical samples (40.0 g each) which were treated with extracting solutions of increasing extraction power.

After filtering solid parts, the filtrate was analyzed by using AAS.

The soil specimens, washed with water and dried, were used at the next stage of the extraction – at II–IV fractions. The results of the filtrate analyses for the contents of heavy metals in individual speciation fractions during four experimental years are presented in Figures 2–7.

1.3. Experimental technological equipment

Figure 1 presents a simplified flowchart of a physico-chemical treatment system where the pretreatment of mixed wastewaters from the tested oleo-chemical production was carried out (Żak, Rauckyte 2008b).

Post-processing wastewaters from the vegetable (CW1) and animal (CW2) fat plant were directed into a preliminary chamber of wastewater equalization, volume 5.0 m^3, equipped with a bypass (1.1), and after that the averaged wastewater was pumped into an aerated chamber (1.3), volume 50.0 m^3 (the first stage of pretreatment), equipped with fatty flotate skimmer (1.6).

Then, hydrogen peroxide was introduced from the station (12) into the chamber in doses 150.0–175.0 gH_2O_2 per m^3 of wastewater mixed with this agent in a preliminary pipe reactor (2).

The second, chemical stage of pretreatment was conducted in a batch reactor (4) equipped with a vertical gate agitator (4.1) and a circulating mixing set with a pipe reactor (3) where lime milk was introduced from station (11) in dose of 1.25–1.50 kg CaO/m^3 of the defatted mixed wastewaters and finally enhanced by flocculation with 0.3% solution of Praestol 859 BC Stockhausen added from station (10) in dose 50.0 g of pure agent per m^3 of wastewater.

The process was additionally enhanced by dissolved air flotation (DAF) (saturation station (5)) (saturation time: 10.0 min, saturation pressure: 500 kPa, pretreated wastewater recirculation at 20%) aided by oxidation with hydrogen peroxide added from station (12) in dose of 250.0 gH_2O_2/m^3 of wastewaters. After oxidation and separation of phases, a flotate was directed into an additional tank for thickening (7), and then into a frame filter press in order to be dewatered (8), and the obtained filtrate was recirculated periodically to the aerated tank (1.3).

1.4. Wastewater

Wastewater parameters were determined according to the following standards: reaction/acidity (pH) – PN-90/C-04540/01, total nitrogen (TN) – PN-73/C-04576/12, total phosphorus (TP) – PN-EN 1189-2000,

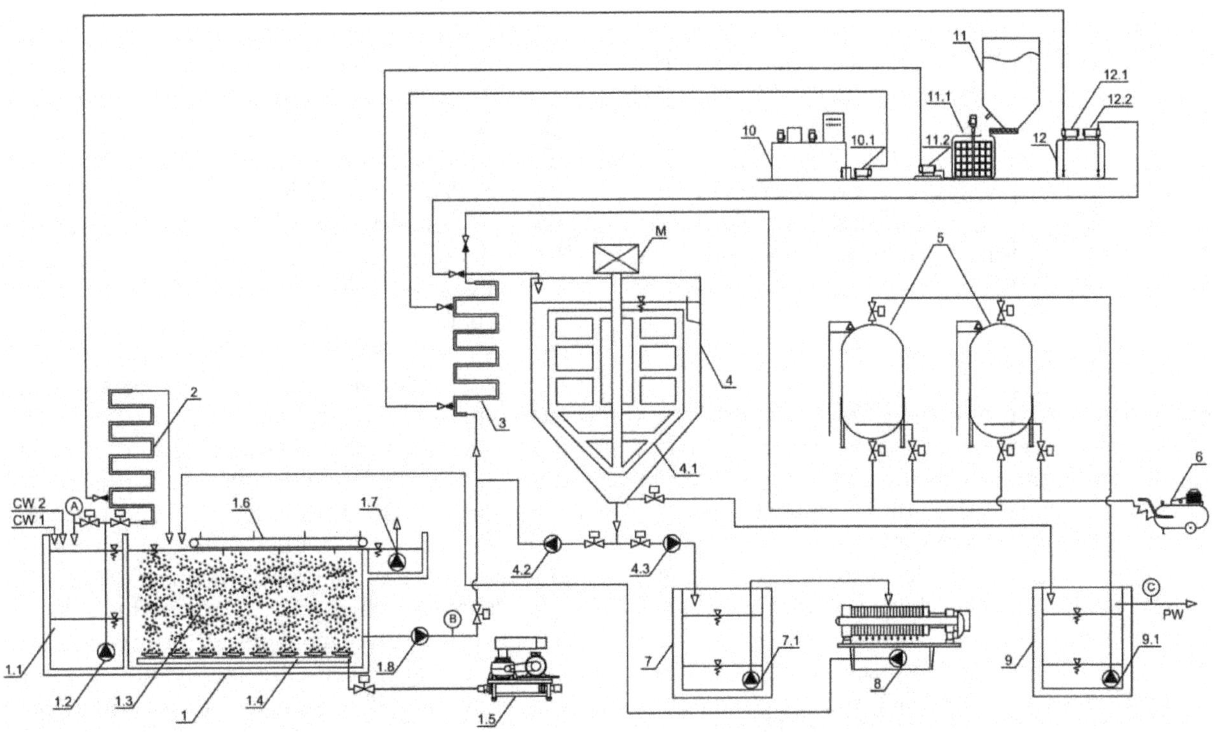

Fig. 1. Flowchart of the experimental installation; CW1 and CW2 – raw wastewaters from vegetable and animal fat production plants; PW – pretreated mixed wastewater; 1 – two-chamber storage-average tank for raw wastewater; 1.1 – preliminary averaging chamber with bypass; 1.2 – bypass pump; 1.3 – aerated preliminary chamber; 1.4 – set of aerating diffusers; 1.5 – compressor; 1.6 – flotate skimmer; 1.7 – pump for thickened fatty flotate; 2 – preliminary pipe reactor; 3 – 2nd stage pretreatment pipe reactor; 4 – batch process reactor; 4.1 – gate agitator; 4.2 – wastewater circulation pump; 4.3 – sludge pump; 5 – saturation station for pretreated wastewater; 6 – compressor; 7 – sludge thickening tank; 8 – frame filter press; 9 – pretreated wastewater tank; 9.1 – wastewater tank; 10 – flocculant preparation and metering unit; 10.1 – flocculant pump; 11 – CaO storage bunker; 11.1 – milk of lime preparation unit; 11.2 – pump for milk of lime; 12 – perhydrol unit; 12.1 and 12.2 – perhydrol pumps; M – mixer drive unit with gear-motor; A, B and C – wastewater sampling points for analyses

potassium (K) by using AAS (equipment: BUCK Scientific 210 VGP and Spectr AA Varian 220 SS) – PN-ISO 9964-1:1994, ether extract (EE) – PN-86/C-04573/01, total suspended solids (TSS) – PN-EN 872:2002, chemical oxygen demand determined by dichromate method (COD) – PN-ISO 15705:2005, biological oxygen demand (BOD_n) – PN-EN 1899-1-2002. The COD parameter concerning the pretreated wastewaters was corrected by deducting the amount connected with hydrogen peroxide, according to the relationship $COD_r = COD_p - f \times d$ (COD_r – corrected by subtracting the fraction of hydrogen peroxide, the concentration of which (d) was determined iodometrically, f = 0.25 – literature correction factor (Talinli, Anderson 1992; Żak 2008a), COD_p – measured).

During the experiment, basic parameters for mixed raw wastewaters ranged as follows (Żak, Rauckyte 2008b): reaction (pH) 3.9–5.0; total nitrogen (TN) 175.3–720.9 mg/l; total phosphorus (TP) 60.7–220.8 mg/l; potassium (K) 149.4–606.2 mg/l; ether extract (EE) 703.3–5999.0 mg/l; total suspended solids (TSS) 477.3–2530.0 mg/l; chemical oxygen demand (COD) 3348.9–10201.6 mgO_2/l; and biochemical oxygen demand (BOD_5) 905.8–4593.9 mgO_2/l.

During the experiment, after the second physicochemical stage, the pretreated wastewaters, achieved the following values of basic parameters (Żak, Rauckyte 2008b): reaction (pH) 8.4–9.3; total nitrogen (TN) 88.8–170.7 mg/l; total phosphorus (TP) 11.3–40.8 mg/l; potassium (K) 66.9–483.7 mg/l; ether extract (EE) 3.7–27.3 mg/l; total suspended solids (TSS) 18.9–42.7 mg/l; chemical oxygen demand (corrected COD_r, for f = 0.25 (Talinli, Anderson 1992; Żak 2008a)) 809.6–4405.7 mgO_2/l, biochemical oxygen demand (BOD_5) 467.9–2067.8 mgO_2/l.

2. Results and discussion

To compare analytical results concerning the heavy metals content, the area of the experiment plots was irrigated and fertilized separately by mixed raw wastewaters after their defatting and by pretreated ones. In the literature, numerous reports can be found on agro-utilization of wastewaters from animal or vegetable fats production, including changes in the total content of heavy metals induced by agro-utilization (Lepp 1981; Rattan *et al.* 2005; Russell *et al.* 1993; Żak, Rauckyte 2008a, b; Lassat 2002; Rauckyte *et al.* 2008; Manz *et al.* 1999; Paredes *et al.* 1986, 1987; Tardioli *et al.* 1997).

However, there are no reports concerning changeability in the individual fractional fields during the multi-year research on fertilization and irrigation of soils with such wastewaters. This problem seems to be particularly interesting because the total content of heavy metals does not show transformations which take place with their participation or the components of wastewater load (or possible products of their metabolism in soil) during a multi-year supply of those substances into soil. In that case, mean daily volumes of wastewaters from individual processing lines were as follows: from the production of vegetable originated fatty acids, 5.5–10.0 m^3/day, and from the production of technical greases of animal origin by using the "wet method," 7.5–12.0 m^3/day (Rauckyte *et al.* 2008; Żak, Rauckyte 2008a). Collecting raw wastewaters in the storage-equalizing tank aimed at stabilizing the contents (1), regulating the ratio of N:P:K, cooling (the temperature of raw wastewaters directly from the production was not less than 80 °C) and defattening by induced air flotation (IAF).

Hydrogen peroxide added at this stage significantly stopped microbiological reduction processes, which substantially stabilized the process of fat separation. The oxidizer introduced in dosage 150.0–175.0 gH_2O_2/m^3 ceased the processes revealing characteristic thickening of waste volume, which was a result of biochemical transformations leading to evolving gaseous products, CO_2 in particular (Krause 1982). Hydrogen peroxide is characterized by well-known bactericidal and fungicidal properties, considerably limiting or eliminating undesirable biological processes and increasing biological stability of the pretreated wastewaters (Krause 1982).

An advantageous effect of hydrogen peroxide on the process of the destruction of protein-fats stable colloids was stated and, at the same time, elimination of fats by their flotating separation efficiently increased (efficiency 97–99%). Additionally, oxygen liberating during hydrogen peroxide auto-decomposition ($H_2O_2 \rightarrow O_2 + 2H^+ + 2e^-$) considerably improved the process of impurities flotation by forming gaseous micro-bubbles that intensified the process of floating fatty contaminations.

The second-stage installation worked at the maximum output of 20.0 m^3/day. The analyses of soil specimens showed that concentrations of individual metals were at the level of natural background of the tested soils. Slight fluctuations observed in metal concentrations in soil surfaces of the experimental plots were rather a result of vegetation processes of the sown plants. No effect of wastewater dosage on the level of the observed contents of heavy metals was found. The total content of metals did not vary during the experiment and was in accordance with a multinomial equation type $C_M(t) = at^2 + bt + c$ (where C_M – total concentration of analyzed metal, determined by the AAS method, and the parameter t means time of the experiment in monthly units) (Żak, Rauckyte 2008b).

The registered concentration of cadmium did not exceed 1.0 mg/kg d.m., ranging within the scope predicted for light soils, 1.0 mg/kg d.m. (Directive 76/464/EEC). Contents of total chromium, nickel, copper, lead, and zinc in the analyzed soil specimens varied respectively: 13.0 ± 0.8, 9.2 ± 1.3, 11.6 ± 1.9, 10.9 ± 2.1, and 49.7 ± 2.7 mg/kg d.m., not exceeding the threshold values predicted for this type of soils (Directive 76/464/EEC).

Despite using pretreated wastewaters of pH = 7.8–8.8 in the four-year field experiment, no significant change was found in the experimental plots because the level of this parameter was within pH ≈ 6.7 ± 0.2 (pH_{KCl} = 3.7 ± 0.2). Thus, this is the reaction level at which ionic forms of metals (M), such as chromium, zinc, copper, lead, show natural preference to form sparingly soluble hydroxide forms of type $M(OH)_{n(s)}$. As a result, the content of fraction I for all metals registered during the experiment gradually decreased as far as specimens of soils irrigated with the pretreated wastewaters were concerned (Figs 2–7). Other ingredients (average concentrations of: humus, organic carbon, P_2O_5, K_2O, and magnesium) found in soil samples in that period showed no significant quantitative changes with reference to the values before starting the experiment. However, specimens of soil treated with defatted raw wastewater were acidified from the initial level of pH_0 ≈ 6.7 ± 0.3 (pH_{KCl} ≈3.7 ± 0.3) up to the final value pH_K ≈5.9 ± 0.4 (pH_{KCl} ≈ 3.1 ± 0.4) (the plot was not limed, and the raw wastewaters, after their defatting, had the reaction pH ≈ 3.9–5.0). Such soil components as P_2O_5, K_2O, and Mg did not reveal any significant quantitative changes during the experiment with reference to the values before the experiment. But in the case of humus, after the second year of the experiment a decreasing trend was observed (on average by 3.1% after the second year, 4.3% after the third year, and 4.9% after the fourth year). In the case of nitrogen, the increasing trend was noticed after one year in soil samples from the starting level ratio C:N – 10.9 to 9.8 after four years of the experiment.

It induced an increase in the content of an exchangeable fraction for each metal (Figs 2–4, 6 and 7), excluding copper (Fig. 5), for which no quantitative amounts were registered during the experiment (Fig. 5).

This may be explained by appearance of such conditions that enabled metals to release soluble forms, probably aquacomplexes: $[M(H_2O)]^{+m}$ as well as organic products of soil transformations, especially low-molecular weight acids during vegetation periods (e.g. from micro- and myco-logical soil transformations of poly- or oligosaccharides, e.g. glycogen supplied with wastewaters). This group may include both complex systems of type $[M(R)]^{+m+n}$ (where R – organic ligand, e.g. products of hydrolysis of polysaccharides: monoses or dextrins; products of soil metabolism: low-molecular weight acids, polyacids, low-molecular weight products of condensation, etc., both m and n – loads of metal, its organic ligand) and

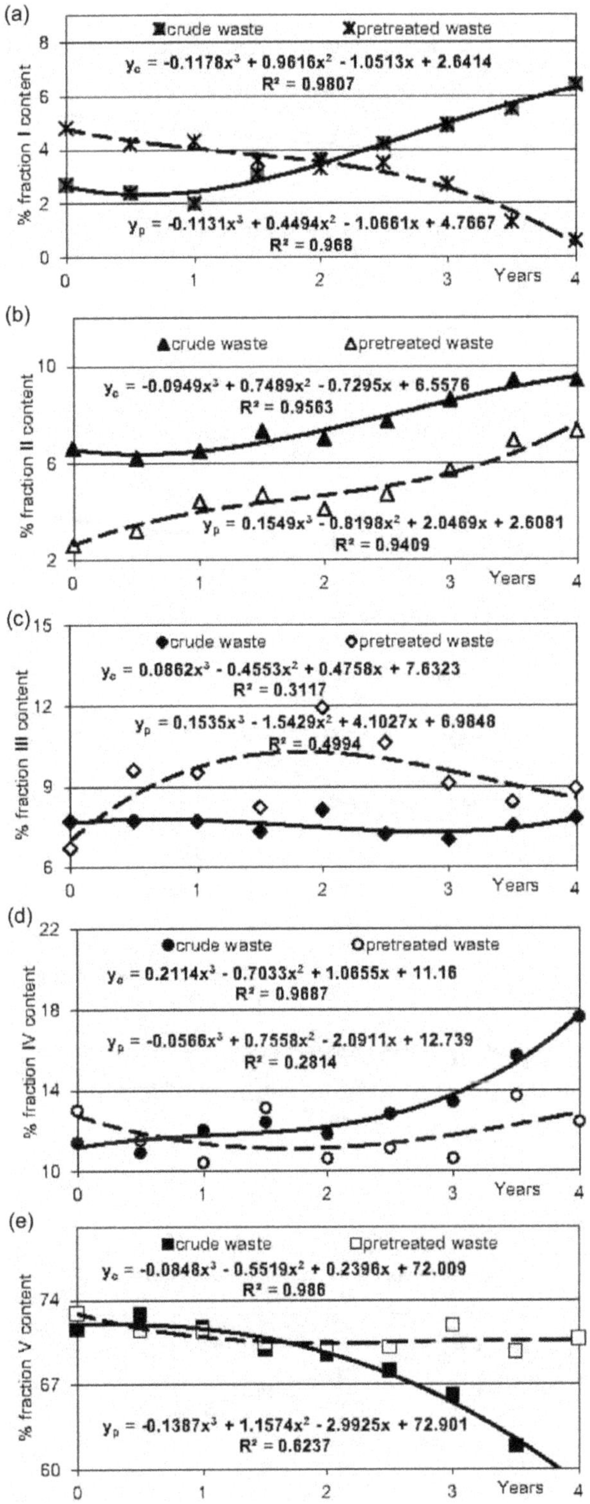

Fig. 2. Changes in the fractional content of cadmium (Cd) in the surface layer of soil from the experimental plots irrigated and fertilized with mixed wastewaters from an oleo-chemical factory – defatted raw (solid line of the trend – y_c) and pretreated (broken line of the trend – y_p) ($C_{0(Cd)}$ – 0.6 ± 0.2 mg/kg d.m.). Individual fractions are given in diagrams: (a) fraction I – exchanged, (b) fraction II – carbonate, (c) fraction III – connected with hydrated oxides of manganese and iron, (d) fraction IV – organic,e) fraction V – the residual

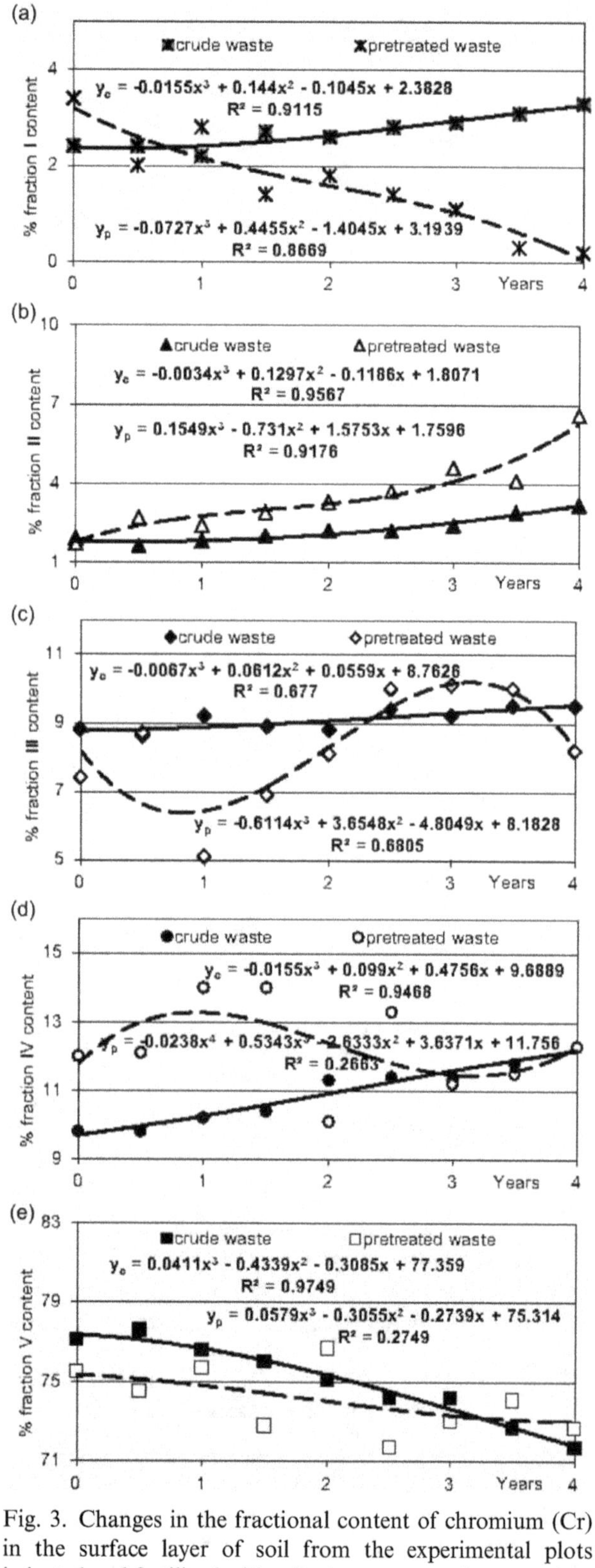

Fig. 3. Changes in the fractional content of chromium (Cr) in the surface layer of soil from the experimental plots irrigated and fertilized with mixed wastewaters from an oleo-chemical factory – defatted raw (solid line of the trend – y_c) and pretreated (broken line of the trend – y_p) ($C_{0(Cr)}$ – 13.0 ± 0.6 mg/kg d.m.). Individual fractions are given in diagrams: (a) fraction I – exchanged, (b) fraction II – carbonate, (c) fraction III – connected with hydrated oxides of manganese and iron, (d) fraction IV – organic, (e) fraction V – the residual

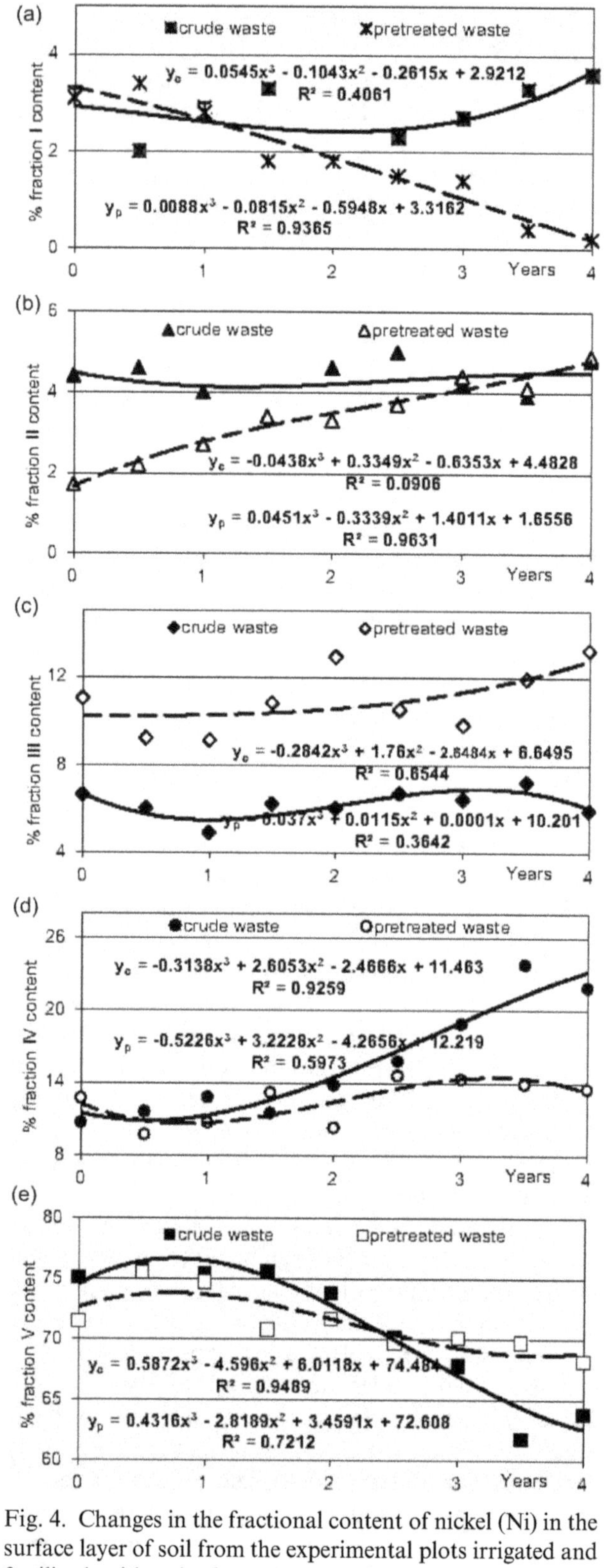

Fig. 4. Changes in the fractional content of nickel (Ni) in the surface layer of soil from the experimental plots irrigated and fertilized with mixed wastewaters from an oleo-chemical factory – defatted raw (solid line of the trend – y_c) and pretreated (broken line of the trend – y_p) ($C_{0(Ni)}$ – 9.3 ± 0.6 mg/kg d.m.). Individual fractions are given in diagrams: (a) fraction I – exchanged, (b) fraction II – carbonate, (c) fraction III – connected with hydrated oxides of manganese and iron, (d) fraction IV – organic, (e) fraction V – the residual

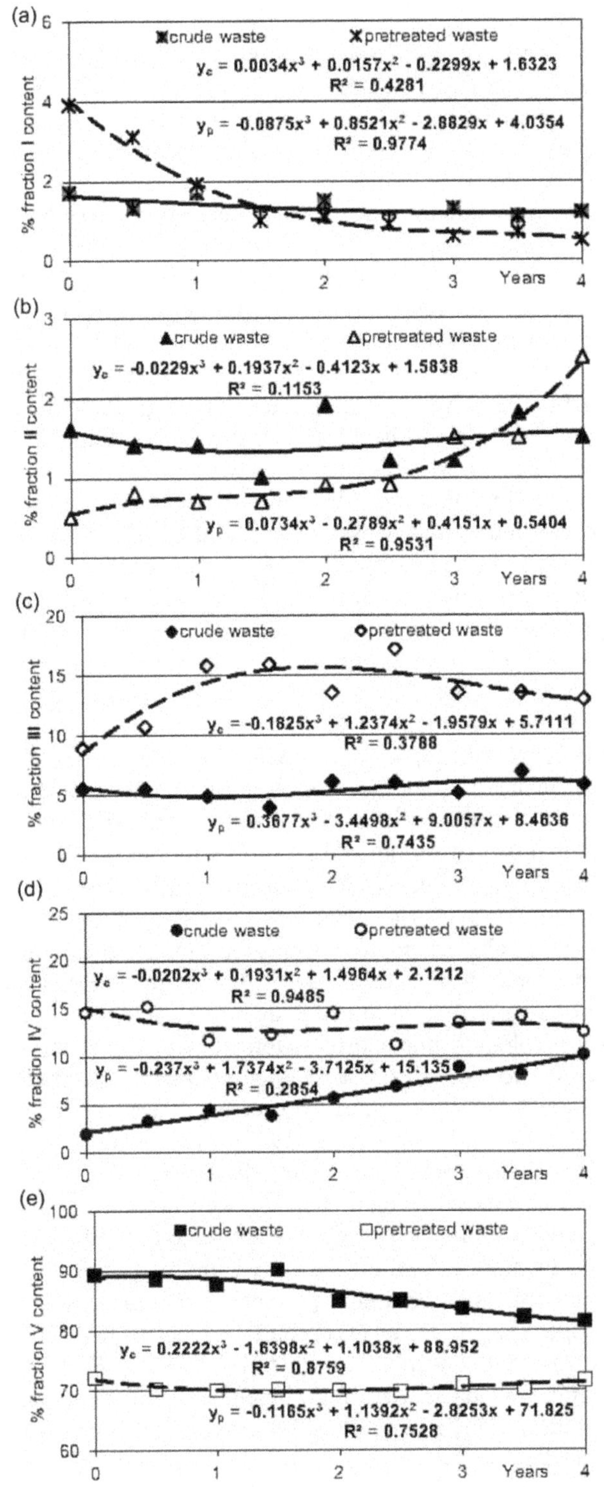

Fig. 5. Changes in the fractional content of copper (Cu) in the surface layer of soil from the experimental plots irrigated and fertilized with mixed wastewaters from an oleo-chemical factory – defatted raw (solid line of the trend – y_c) and pretreated (broken line of the trend – y_p) ($C_{0(Cu)}$ – 11.6 ± 0.8 mg/kg d.m.). Individual fractions are given in diagrams: (a) fraction I – exchanged, (b) fraction II – carbonate, (c) fraction III – connected with hydrated oxides of manganese and iron, (d) fraction IV – organic, (e) fraction V – the residual

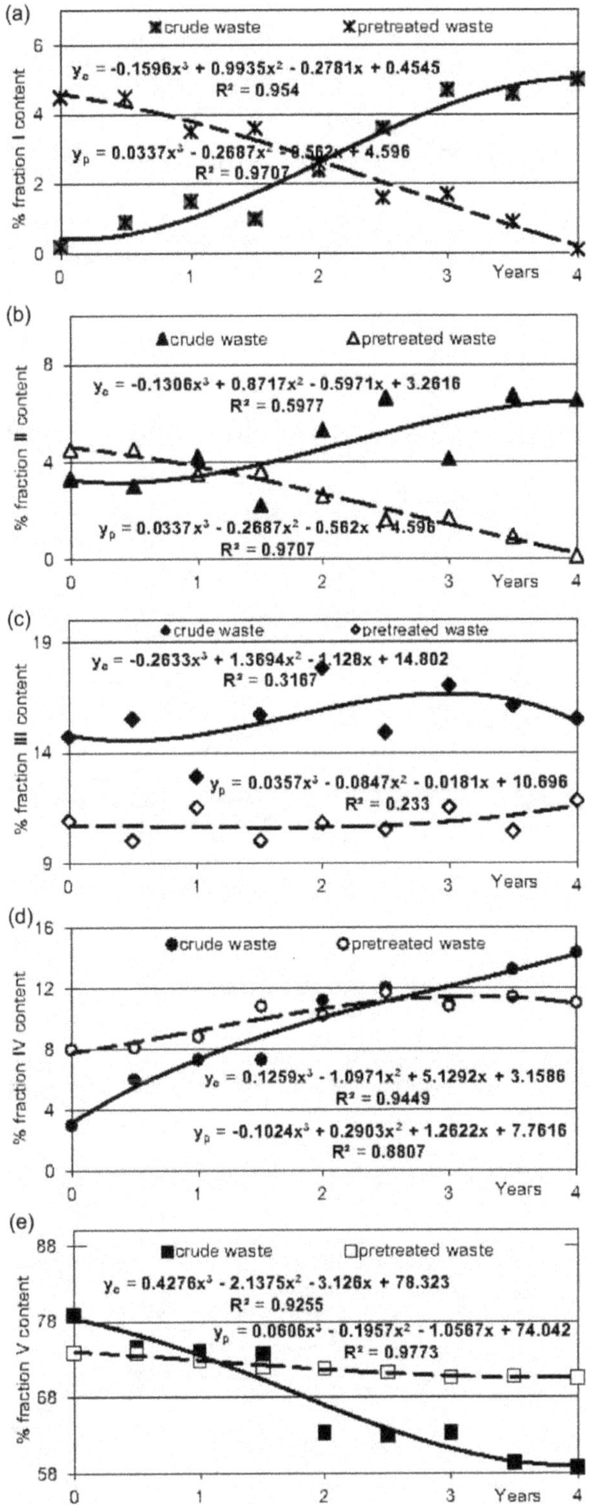

Fig. 6. Changes in the fractional content of lead (Pb) in the surface layer of soil from the experimental plots irrigated and fertilized with mixed wastewaters from an oleo-chemical factory – defatted raw (solid line of the trend – y_c) and pretreated (broken line of the trend – y_p) ($C_{0(Pb)}$ – 10.9 ± 1.2 mg/kg d.m.). Individual fractions are given in diagrams: (a) fraction I – exchanged, (b) fraction II – carbonate, (c) fraction III – connected with hydrated oxides of manganese and iron, (d) fraction IV – organic, (e) fraction V – the residual

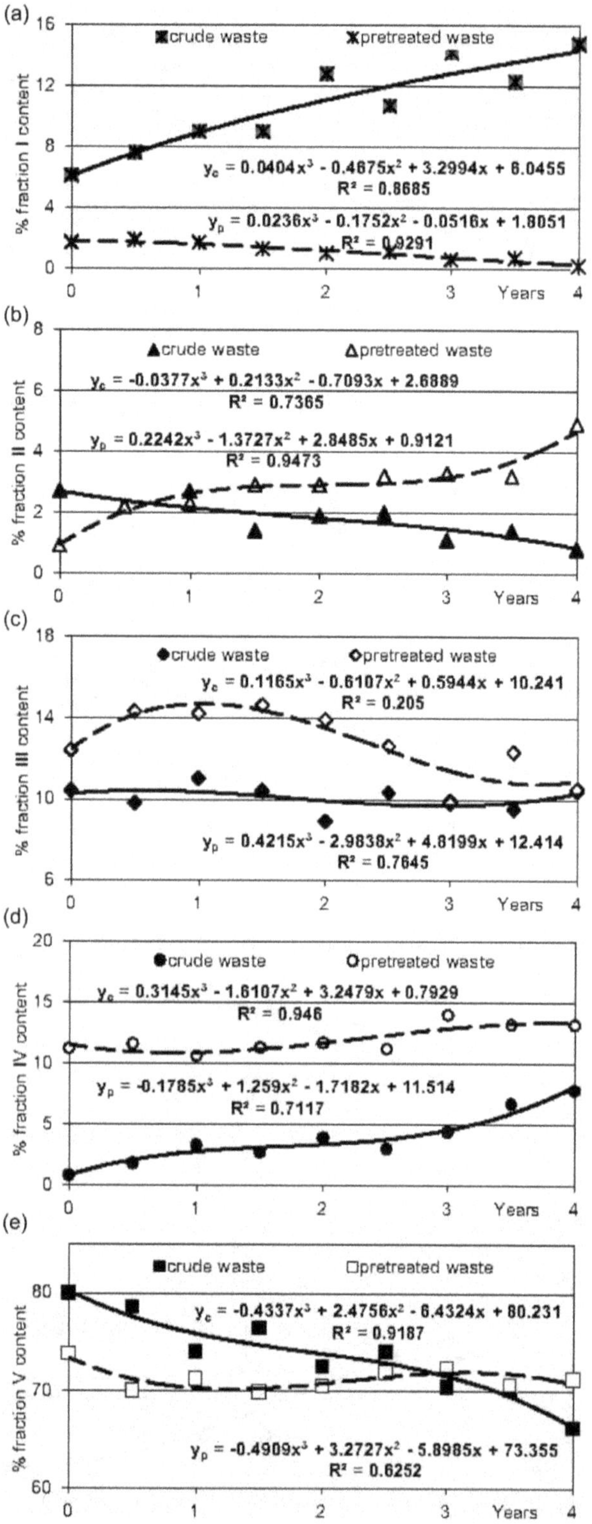

Fig. 7. Changes in the fractional content of zinc (Zn) in the surface layer of soil from the experimental plots irrigated and fertilized with mixed wastewaters from an oleo-chemical factory – defatted raw (solid line of the trend – y_c) and pretreated (broken line of the trend – y_p) ($C_{0(Zn)}$ – 49.7 ± 1.6 mg/kg d.m.). Individual fractions are given in diagrams: (a) fraction I – exchanged, (b) fraction II – carbonate, (c) fraction III – connected with hydrated oxides of manganese and iron, (d) fraction IV – organic, (e) fraction V – the residual

other organic, noncoordination soluble forms. The formation of metallo-organic bonds is justified by a permanent supply of organic substances to soil, inducing increased transformations of these substances and their products of decay, especially during vegetation. Thus, for soil treated with defatted raw wastewaters but not limed, an increase in cation-exchange capacity is observed.

This parameter and its unsteadiness strictly depend on the reaction value (pH) of the tested soil environment (Lepp 1981; Rattan *et al.* 2005; Lassat 2002). In the case of fraction II, connected with carbonate forms of metal compounds, an increase in concentrations of these forms of metals in specimens of soils treated with defatted raw wastewaters for cadmium, chromium, and lead was found (Figs 2, 3, and 6). For nickel and copper (Figs 4 and 5), no significant changes were observed and the zinc concentration decreased (Fig. 7). But for specimens irrigated with pretreated wastewaters, an obvious increase in the content of this fraction for metals was found (Figs 2–5 and 7), with the exception of lead (Fig. 6). The solubility products for carbonates of the tested heavy metals ($MCO_{3(s)}$) were at low level (below $1.0 \cdot 10^{-10}$), which means that such forms are naturally preferred under circumstances of the registered minimal variations in soil reaction (pH). With many diverse factors that affect the properties of a given type of soil, the introduction of the alien substances must result in essential changes in chemical, biochemical, and biological mechanisms (Lepp 1981; Manz *et al.* 1999). This also happens while introducing wastewaters from the tested oleo-chemical production or with a real structural probability of a new substrate to intensify established soil-forming processes (Paredes *et al.* 1986, 1987; Tardioli *et al.* 1997). Wastewaters, especially the pretreated ones, contain additionally mineral salts – ions of calcium (e.g. in a form of $Ca^{2+} \cdot aq$, $CaOH^+$ or pairs of ions: $Ca^{2+} \cdot CO_3^{2-}$ and $Ca^{2+} \cdot HCO_3^-$) which show a definite reactivity in relation to soil components by changing to some degree macroscopic properties and decide about buffer capacity (Tashtoush *et al.* 2004; Rüffer, Rosenwinkel 1991). With the significant participation of chemical and biological reactions which mineralise some organic wastes introduced to soil (where CO_2 is a final product of those transformations), the stability of soil reaction (pH) or its lack may also be a result of buffer capacity of the carbonate–hydrogencarbonate system. A system of equilibrium reactions: $CO_{2(g)} \leftrightarrow CO_{2(aq)} + H_2O \leftrightarrow H_2CO_{3(aq)} \leftrightarrow H^+_{(aq)} + HCO^-_{3\,(aq)} \leftrightarrow H^+_{(aq)} + CO^{2-}_{3(aq)}$ essentially stabilizes and prefers the type of transformations defined by energy (Lassat 2002).

In the case of fraction III (related to hydrated forms of manganese and iron oxides), a high stability was stated for soil specimens irrigated with the defatted raw wastewaters. It is understood as a permanent trend to keep the content similar to the value of natural background for all metals tested during the experiment.

That means that the content of the substances in the wastewaters does not affect total changes in concentrations of this fractional group. As regards soil specimens irrigated with pretreated wastewaters, no statistical trends suggesting definite transformations were registered. It shows high contents instability of this fraction. It may be a result of content variability or the reaction courses of diverse vegetation processes of plants cultivated on the experimental plots. The participation of speciation forms of metals in soil as a multi-phase poly-dispersion system depends on several factors such as minerals forming a solid phase, the content and composition of organic matter, a volume of soil water, contents of soluble substances in soil water, a form and flow of soil water, local climate, diversity of microorganisms, the level of energy processes, etc. (Lepp 1981; Rattan *et al.* 2005; Manz *et al.* 1999). Legal environmental rules state that only toxic chemical components present in wastewaters are limited by norms, but the volume of water is not included into any legal restrictions; this factor has a significant effect on the dynamics of transformations and transport in soil, especially for outer substances introduced with pollution load. Introducing a new batch of organic-mineral substances with a considerable amount of water may lead to soil transformations and may change the content of speciation forms of metals, especially when wastewaters are used for a long time in agricultural production (Żak 2005). Defatted raw, pretreated processing wastewaters from oleo-chemical plant generally contain low-weight molecular substances of vegetable origin (e.g. lecithins, glicerols, proteins, aldehydes, ketones, lactones, sterols, tocopherols) or of animal origin (e.g. amino acids, products of protein degradation, mono- and oligosaccharides) (Rüffer, Rosenwinkel 1991; Rack 2007).

The obtained results suggest that the substances have no influence on the content of fraction III, which should be recognized as a desired effect of agro-utilization on soil quality. A clear increase in contents of fraction IV for each tested metal was found in the analyzed specimens of soils irrigated with defatted raw wastewaters.

The growth was registered at the following level for the individual metals: cadmium (Cd) ca 50%, chromium (Cr) ca 20%, nickel (Ni) ca 260%, copper (Cu) ca 500%, lead (Pb) ca 470%, and zinc (Zn) ca 350% (Figs 2–7). The increase in concentration of fraction IV varies for the individual metals and depends on a pure specificity of individual elements, on the properties of organic components introduced into soil with wastewaters (basic–acidic properties, condensation susceptibility, donor–acceptor properties, humification susceptibility, etc.) and also on the products of their bioconversions. Undoubtedly, the dynamics and characteristics of the humification processes with participation of outer organic substances introduced with the wastewaters affected such an essential change with reference to the initial conditions. However, the course of changes in concentration of this fraction for specimens of soil irrigated with the pretreated wastewaters shaped differently. No increase in the content of this fraction for cadmium, nickel, copper, and zinc was observed (Figs 2, 4, 5, and 7). For lead, the concentration of this fraction (IV) increased by 30–40% (Fig. 6) and for chromium the dispersion of the results was so large that it was impossible to interpret them explicitly and reliably (Fig. 3).

The results show good efficiency of physico-chemical pretreatment which eliminates organic matter that can essentially modify the character of metallo-organic transformations, which is undoubtedly a desirable effect. Analyzing fractional contents in specimens of soil irrigated with wastewaters, the definitely highest percent fraction was registered for each metal in the so-called residual fraction (fraction V). For specimens of soil irrigated with pretreated wastewaters, the content of cadmium and lead was at level of ca 70%, but for the rest of metals – more than 85%. For fraction V, a residual of specimens of soil irrigated with defatted raw wastewaters, a decrease in the content of each metal was observed in time (Figs 2–7). The lowest decrease was for chromium, copper, and zinc, approximately 10–15% (Figs 3–5), while for the other metals it shaped at 20–35% (Figs 2, 4, and 6). Fraction V is a group of compounds almost unrecognized both in content and structure, and its properties depend on local soil formation and environmental conditions, which makes a deeper interpretation difficult, basing only on quantitative transformations in time (Lepp 1981).

To sum up, we are able to suggest a thesis that, for specimens of soil irrigated with pretreated wastewaters, the lack of evidence to support a significant change in quantitative relations of the individual fractions resulted from the increase in buffer capacity of the investigated soil for which no significant variation in the reaction was found (excluding periodical minimal pH fluctuation approximately 0.3 of the unit). The parameter of pH is essential for almost all soil chemical and biological transformations. For cadmium, the fact of considerable diversity of the percent participation of fractions in time should also be assessed with respect to the dispersion of the achieved results, taking into account relatively low concentrations of this element in soil (for $Cd_{(0)} = 0.6\ (\pm 0.2)$ mg/kg d.m.), which affects statistical assessment of the obtained results. Basing on the studies carried out, it was found that the method of agro-utilization of the processing wastewaters from low tonnage oleo-chemical production is an effective way of the tested wastes management, not resulting in significant changes in the contents of heavy metals in the irrigated and/or fertilized soils. The research carried out explicitly proved the necessity of pretreating mixed wastewaters from oleo-chemical production by using milk of lime, the application of which regulates advantageously the ratios N:P:K and eliminates poly- and oligostructural substances as well as excessive amounts of soluble forms of phosphorus.

Conclusions

1. The content of mixed effluents from oleo-chemical industry, both raw and defatted as well as the pretreated physico-chemically, used in the process of their agro-utilization in light soils, does not affect the total content of heavy metals (Cd, Cr, Cu, Ni, Pb, Zn) in the four-year period of irrigation and spreading with these wastewaters.
2. Mixed wastewaters from the oleo-chemical industry, pretreated in a two-stage physico-chemical process (the dispersed air flotation (DAF) and the coagulation combined with the precipitation with lime milk and final flotation enhanced additionally by the oxidation with hydrogen peroxide), disposed to light brown soils, do not induce any significant changes in basic chemical composition of the soils, however they affect the quantitative fluctuations in both elutable and carbonate fractions.
3. The mixed wastewaters from oleo-chemical industry, pretreated only by flotation defatting and directed to their agro-utilization in light Cambisols, induced small changes in chemical composition of these soils during the period of four-year agro-utilization, however, the significant quantitative changes were observed in fractional pools of metals, especially in interchangeable organic, carbonate fractions, and the so-called residuals.

References

Aikaitė-Stanaitienė, J.; Grigiškis, S.; Levišauskas, D.; Čipinyte, V.; Baškys, E.; Kačkytė, V. 2010. Development of fatty waste composting technology using bacterial preparation with lipolytic activity, *Journal of Environmental Engineering and Landscape Management* 18(4): 296–305. http://dx.doi.org/10.3846/jeelm.2010.34

Aktas, E. S.; Imre, S.; Ersoy, L. 2001. Characterization and lime treatment of olive mill wastewater, *Water Research* 35(9): 2336–2340. http://dx.doi.org/10.1016/S0043-1354(00)00490-5

Angenent, L. T.; Karim, K.; Al-Dahhan, M. H.; Wrenn, B. A.; Domíguez-Espinosa, R. 2004. Production of bioenergy and biochemicals from industrial and agricultural wastewater, *Trends in Biotechnology* 22(9): 477–485. http://dx.doi.org/10.1016/j.tibtech.2004.07.001

Baltrenas, P.; Kvasauskas, M. 2008. Experimental investigation of biogas production using fatty waste, *Journal of Environmental Engineering and Landscape Management* 16(4): 178–187. http://dx.doi.org/10.3846/1648-6897.2008.16.178-187

Commission Regulation (EC) No 92/2005 of 19 January 2005 implementing Regulation (EC) No 1774/2002 of the European Parliament and of the Council as regards means of disposal or uses of animal by-products and amending its Annex VI as regards biogas transformation and processing of rendered fats.

Directive 76/464/EEC Water pollution by discharges of certain dangerous substances.

Haas, M. J.; Bloomer, S.; Scott, K. 2000. Simple, high-efficiency synthesis of fatty acid methyl esters from soapstock, *Journal of American Oil Chemists' Society* 77(4): 373–379.

IUSS Working Group WRB. 2006. *World reference base for soil resources 2006*. World Soil Resources Reports No. 103. Rome: FAO.

Krause, A. 1982. Vegetable and animal fats – their influence on microorganisms in waste water treatment plants, *European Journal of Lipid Science and Technology* 84(S1): 536–543. http://dx.doi.org/10.1002/lipi.19820841310

Kvasauskas, M.; Baltrenas, P. 2009. Research on anaerobically treated organic waste suitability for soil fertilization, *Journal of Environmental Engineering and Landscape Management* 17(4): 205–211. http://dx.doi.org/10.3846/1648-6897.2009.17.205-211

Lassat, M. M. 2002. Phytoextraction of toxic metals: a review of biological mechanisms, *Journal of Environmental Quality* 31(1): 109–120.

Lepp, N. W. 1981. Effects of heavy metals pollution in plants, in Lepp, N. W. (Ed.). *Effects of heavy metals on plant function*. London: Applied Science Publishers, 1. 111–143.

Londong, J.; Rosenwinkel, K. H. (Eds.) 2007. *Industrial wastewater treatment*. 1st ed. Weimer: Bahaus Weimar University Press, 2007; 140–143.

Manz, M.; Weissflog, L.; Kühne, R.; Schürmann, G. 1999. Ecotoxicological hazard and risk assessment of heavy metal contents in agricultural of central Germany, *Ecotoxicology and Environmental Safety* 42(2): 191–201. http://dx.doi.org/10.1006/eesa.1998.1741

Meher, L. C.; Sagar, D. V.; Naik, S. N. 2006. Technical aspects of biodiesel production by transesterification – a review, *Renewable and Sustainable Energy Reviews* 10(3): 248–268. http://dx.doi.org/10.1016/j.rser.2004.09.002

Mittal, G. S. 2006. Treatment of wastewater from abattoirs before land application – a review, *Bioresource Technology* 97(9): 1119–1135. http://dx.doi.org/10.1016/j.biortech.2004.11.021

Ng, W. J.; Goh, A. C. C.; Tay, J. H. 1988. Palm oil mill effluent treatment – liquid-solid separation with dissolved air flotation, *Biological Wastes* 25(4): 257–268. http://dx.doi.org/10.1016/0269-7483(88)90087-0

Ociepa, A.; Pruszek, K.; Lach, J.; Ociepa, E. 2008. Influence of long-term cultivation of soils by means of manure and sludge on the increase of heavy metals content in soils, *Ecological Chemistry and Engineering S* 15(1): 103–109.

Paredes, M. J.; Monteoliva-Sanchez, M.; Moreno, E.; Perez, J.; Ramos-Cormenzana, A.; Martinez, J. 1986. Effect of waste waters from olive oil extraction plants on the bacterial population of soil, *Chemosphere* 15(5): 659–664. http://dx.doi.org/10.1016/0045-6535(86)90015-9

Paredes, M. J.; Moreno, E.; Ramos-Cormenzana, A.; Martinez, J. 1987. Characteristics of soil after pollution with waste waters from olive oil extraction plants, *Chemosphere* 16(7): 1557–1567. http://dx.doi.org/10.1016/0045-6535(87)90096-8

Rack, M. 2007. *Oilseed processing, edible fat – and edible oil refining – Details and explanations of Annex 4 of the Waste Water Ordinance.* Federal Gazette Supplement No. 3a/2007. Frankfurt.

Rattan, R. K.; Datta, S. P.; Chhonkar, P. K.; Suribabu, K.; Singh, A. K. 2005. Long-term impact of irrigation with sewage effluents on heavy metal content in soils, crops and groundwater – a case study, *Agriculture, Ecosystems & Environment* 109(3–4): 310–322. http://dx.doi.org/10.1016/j.agee.2005.02.025

Rauckyte, T.; Żak, S.; Pawlak, Z. 2008. Research on heavy metal contents in soils used for utilization of wastewater from vegetable fats production, *Ecological Chemistry and Engineering A* 15(4–5): 405–413.

Rubio, J.; Souza, M. L.; Smith, R. W. 2002. Overview of flotation as a wastewater treatment technique, *Minerals Engineering* 15(3): 139–155. http://dx.doi.org/10.1016/S0892-6875(01)00216-3

Rüffer, H.; Rosenwinkel, K. H. 1991. *Handbook of industrial wastewater treatment*. Oldenbourg Verlag: Munich, Vienna, 115–134.

Russell, J. M.; Cooper, R. N.; Lindsey, S. B. 1993. Soil denitrification rates at wastewater irrigation sites receiving primary-treated and anaerobically treated meat-processing effluent, *Bioresource Technology* 43(1): 41–46. http://dx.doi.org/10.1016/0960-8524(93)90080-U

Talinli, I.; Anderson, G. K. 1992. Interference of hydrogen peroxide on the standard COD test, *Water Research* 26(1): 107–110. http://dx.doi.org/10.1016/0043-1354(92)90118-N

Tardioli, S.; Bànnè, E. T. G.; Santori, F. 1997. Species-specific selection on soil fungal population after olive mill wastewater treatment, *Chemosphere* 34(11): 2329–2336. http://dx.doi.org/10.1016/S0045-6535(97)00044-1

Tashtoush, G.; Al-Widyan, M.; Al-Jarrah, M. 2004. Experimental study on evaluation and optimization of conversion of waste animal fat into biodiesel, *Energy Conversion and Management* 45(17): 2697–2711. http://dx.doi.org/10.1016/j.enconman.2003.12.009

Tessier, A.; Campbell, P. G. C.; Bisson, M. 1979. Sequential extraction procedure for the speciation of particulate trace metals, *Analytical Chemistry* 51(7): 844–851. http://dx.doi.org/10.1021/ac50043a017

Ullmanns Encyclopedia of Industrial Chemistry. 1992. *Fats and fatty oils*, vol. A 10. New York: VCH, 173–244.

Vaišis, V.; Šuksta, L. 2010. Experimental analysis of iron impact on heavy metal (copper) migration in the soil, *Journal of Environmental Engineering and Landscape Management* 18(4): 306–311. http://dx.doi.org/10.3846/jeelm.2010.35

Żak, S. 2005. Processing of organic waste from vegetable and animal fat production, *Polish Journal of Chemical Technology* 7(1): 74–79.

Żak, S. 2008a. Problem of correction of the chemical oxygen demand values determined in wastewaters treated by methods with hydrogen peroxide, *Proceedings of ECOpole* 2(2): 409–414.

Żak, S. 2008b. Separation of fats from wastewaters and processing wastes using flotation for the needs of efficient sludge management, in Pawłowska, M.; Pawłowski, L. (Eds.). *Management of pollutant emission from landfills and sludge.* London: Taylor & Francis Group, 265–271.

Żak, S.; Pawlak, Z. 2006a. The use of flotation in fat recovery and the pretreatment of wastewaters from animal fat production, *Polish Journal of Environmental Studies* 15(3): 507–515.

Żak, S.; Pawlak, Z. 2006b. Investigations on recovering fats from industrial wastes by flotating separation, *Polish Journal of Environmental Studies* 15(5): 819–826.

Żak, S.; Rauckyte, T. 2008a. Analysis of heavy metal contents in soil irrigated and fertilized with wastewaters from animal fats production, *Ecological Chemistry and Engineering A* 15(4–5): 451–459.

Żak, S.; Rauckyte, T. 2008b. Research on the contents of heavy metals in soils subjected to many-year agroutilization of wastewaters from low tonnage oleochemical production, *Ecological Chemistry and Engineering A* 15(9): 997–1003.

Żak, S.; Zabłocki, L. 2005. The use of flotation to pretreat wastewater from animal fat rendering, *Polish Journal of Chemical Technology* 7(2): 94–99.

Sławomir ŻAK. Dr, Department of Technology and Chemical Engineering, University of Technology and Life Sciences, Bydgoszcz, Poland. His research interests are environmental technology and analysis of environmental pollution.

Terese RAUCKYTE-ŻAK. Dr, Department of Technology and Chemical Engineering, University of Technology and Life Sciences, Bydgoszcz, Poland. Her research interests are heavy metal ions in used engine oils, wastes, soil and environmental pollution.

Alfredas LAURINAVIČIUS. Professor, Dr, Head of Department of Road, Vilnius Gediminas Technical University, Lithuania. He has published about 160 scientific papers, 18 of them are included in Thomson Masters ISI Web of Science database. His research interests are road construction technologies and materials, Road Climatology, road safety.

5

POSSIBILITIES OF USING SULPHUR AND PLANT ORIGIN WASTE FOR LIGHT-WEIGHT AND THERMAL INSULATION CONSTRUCTION COMPOSITES

Gediminas Marčiukaitis, Gintaras Jurkėnas

Department of Reinforced Concrete and Masonry Structures, Vilnius Gediminas Technical University, Sauletekio al. 11, 10223 Vilnius, Lithuania

Abstract. The article notes that not all types of plant origin waste are appropriate for being used as fuel or left for decay. The environment is polluted by noxious gas that is generated in the process of heating or decaying. Sulphur waste is produced at chemical and oil industry plants as well as during the treatment of emissions resulting from fuel combustion. Thermal energy is produced in the process of combusting plant origin waste; however, the emitted fumes are pollutants. A much more effective way for the use of this waste is proposed, i.e. to produce and use a new type of thermal insulation construction composites that would allow long-term heat saving buildings. The performed tests covered the properties of components and the influence of plasticisers on the plasticity and compressive strength of hardened sulphur. The common behaviour of sulphur and plant origin waste – sawdust, boon, straw – in composite was also analysed. All the properties necessary for thermal insulation materials were identified – compressibility, compression and bending strength, thermal conductivity. They have the same value as other classical thermal insulation materials.

Keywords: solid waste; physical pollution; environmental sustainability; sulphur and bio-waste; thermal insulation composite; properties.

Reference to this paper should be made as follows: Marčiukaitis, G.; Jurkėnas, G. 2013. Possibilities of using sulphur and plant origin waste for light-weight and thermal insulation construction composites, *Journal of Environmental Engineering and Landscape Management* 21(2): 106–113. http://dx.doi.org/10.3846/16486897.2013.767805

Introduction

With the development of civilisation, the range of application of natural resources in human life and activities was expanding. Upon reaching the present level, it has become necessary to utilise vegetal waste. Large amounts of waste are produced in the construction industry and agricultural production sectors – wood sawdust, chips, boon and straw. Large quantities of vegetation found in water bodies, such as reeds and bulrushes, remain unused.

In recent years, a large share of all cellulose raw materials and related wastes has been used as a source of energy produced by combustion.

A more versatile and effective use of such wastes for the creation of thermal isolation products would enable the reduction of thermal energy inputs. Major energy inputs go to the construction and heating of premises, and therefore, as investigations have revealed (Nanazashvili 1990; Kazragis *et al.* 2004; Kazragis, Gailius 2006; Zavadskas *et al.* 2009), the aforementioned waste can be used for the production of construction composite products with enhanced thermal insulation properties. However, this requires an appropriate binding agent. For the purpose of combining cellulose materials, i.e. composite inserts, cement slurry as a composite matrix poorly interacts with cement or requires special additives.

Cement is a material of high energy intensity; its production results in large amount of CO_2 emissions and, in addition, it is a zero waste material. Besides, according to Mechta (2009), Portland cement contains about 85% clinker and the manufacturing process releases 0.9 tonne of CO_2 per tonne of clinker. These numbers show that cement manufacturing is one among such leaders producing large amounts of CO_2 as transportation, power generation, treatment of oil and gas and steel manufacturing. This is a problem to reducing the production of the main component of cement clinker. According to CEMBUREAU, it is necessary to reduce the portion of clinker up to 60% and to replace it by other materials with binder properties, such as waste, various ashes, slag, and silicium microdust or to replace all cement by another binder material. At the same time, concrete with new

Corresponding author. Gediminas Marčiukaitis
E-mail: gelz@vgtu.lt

binder materials must have more durability than with cement. Such materials are sulphur or its waste. Generally, the use of sulphur, instead of cement, for the production of construction products was started nearly 50 years ago (Vroom 1992). The main reason for this replacement was quick setting, water-proof, chemical resistance and mechanical strength (Vroom 1992; Weber *et al.* 1992).

Nature is rich in highly sulphurous ore deposits reaching even the surface. However, only around 20–25% of pure sulphur is obtained from this ore and large amounts are left non-segregated from the impurities of various countries. Currently, plants of many countries, such as those of mineral fertilisers producing sulphuric acid, oil, gas, and other chemical industry plants, generate large amounts of sulphuric waste containing 2–60% of elementary sulphur (Michailov *et al.* 1989; Orlowsky 1988, 1990; Sulphur and Sulfur Containing Wastes 1990; Komnitsas *et al.* 2001; Gracia *et al.* 2004; Krishnan, Freeman 2008). The same amount of elementary sulphur can contain sulphur waste of proper chemical plants and energy powers as in Lithuania.

During solid and liquid fuel combustion, sulphur dioxide (SO_2) and trioxide (SO_3) emit into air together with small particles and various compounds that are harmful to the human organism form on the ground (Peavy *et al.* 1985).

The first in-depth research into this new material, sulphur concrete, was started in the USA and Canada. In Lithuania, research on the use of sulphur for fine concrete was launched in 1988 (Gauronskis *et al.* 1990; Marčiukaitis 1994).

As investigations carried out by a number of authors (Paturoev *et al.* 1985; Czarnecki, Gillot 1989a, b; Michailov *et al.* 1989; ACI Committee 548 1993; Mohamed, El Gamal 2010) show, the use of sulphur, as a binding agent for concrete production, ensures such main mechanical and physical properties of concrete which nearly do not differ from those of cement concrete, and some properties, such as resistance to corrosion and water-proof capacity, are even much better (Table 1). However, its wider application was impeded by higher energy inputs as a temperature of 140 ± 5 °C is required for concrete components within the entire process of concrete production. A large amount of sulphur has to be melted. The melting temperature is 130 ± 5 °C. Specially heated stirrers are required in the production of heavy sulphur concrete. If industrial sulphur is used, fine aggregate particles are necessary (Czarnecki, Gillot 1989b; Mohamed, El Gamal 2009, 2010).

However, little research has been done into sulphur waste by using the plant origin waste as a binding agent in the production of construction thermal insulation composites. The advantage of sulphur in all areas of its use, when compared to other binding agents of construction materials, is its quick setting in a normal environment (12–20 h).

Table 1. Comparison of the properties of sulfur concrete and cement concrete

Properties	Parameters of concrete: Sulphur concrete	Cement concrete
Mass (kg/m^3)	2300–2400	The same
Short-term strength (N/mm^2)		
Compression	45–70	The same
Bending	10–12	Lower
Tension	5–7	Considerably lower
Compressive modulus of elasticity (N/mm^2)	$(3.5–5.0) \times 10^4$	Nearly the same
Poisson's ratio	0.20–0.25	Nearly the same
Frost resistance	Cycles	Considerably lower
Resistance to aggressive acids		Considerably lower

In order to improve the inefficient use of this waste and minimise the nature pollution, it is appropriate to develop the ways of using it in production and construction of construction products with reduced thermal conductivity by reducing energy inputs (Kazragis, Gailius 2006; Gravitis *et al.* 2010). Therefore, it is advisable to analyse the combining of these types of waste with different properties into a common effective construction composite, thus identifying one of the ways of a more effective use of nature- and air-polluting waste.

1. Components and methods of research into their properties

To produce binding agents containing sulphur, it is possible to use pure sulphur by adding fine particles of another type, also sulphur waste.

Currently, considering the need to reduce pollutant emissions and preserve nature, thousands of tonnes of mixes of various particles containing sulphur are caught in equipment cleaning gas emitted during fuel combustion. Sulphur waste is very diverse and, as data show, can be divided into three types:

(1) Sulphur with a large content of solid mineral impurities of different sizes which are similar to those of aggregates used for concrete ($12 \div 35$ mm).
(2) Sulphur with fine impurities, such as dusty sand ($0.12 \div 1.25$ mm).
(3) Sulphur with minor mineral particles, such as flour ($0.12 \div 0.32$ mm).

Sulphurous waste of a similar type is produced by some enterprises of our country, i.e. a crude-oil refinery in Mažeikiai, fertiliser plants in Kėdainiai and others which are related to the production and use of sulphuric acid. Large amounts of it can be collected in plants consuming much fuel and cleaning emitted fumes and steam. The properties of sulphur are not exactly identical; they depend on various factors (Lvov Ukraina 1990; Abdel-Jawad, Al-Qudah 1994; Mohamed, El Gamal 2010; Vlahovič 2011). A major impact is made by the concentration of H_2S. With its content increasing, the strength of sulphur is decreasing. Therefore, sulphur obtained from different sources or its waste should satisfy the following general criteria:

- Density – 2 100 kg/m^3;
- Strength (N/mm^2): compression 18–20; bending 5–7;
- Coefficient of temperature deformations 5.7×10^{-6} °C;
- Thermal conductivity (W/(m °C)) 0.27–0.28.

Our investigations used sulphur from Mažeikiai Crude Oil Refinery, i.e. grain sulphur, 99.01%; ash of different compositions, 0.4%; various acids, 0.01%; organic admixtures, 0.4%; humidity, 0.15% and other chemical elements, 0.03%. Sulphur was melted in a special electrically heated tank at a temperature of (140±5) °C. In order to determine the physical and mechanical properties of hardened sulphur, specimens were produced and tested in accordance with the standard requirements.

The following plant origin wastes were used as inserts for a composite with sulphur matrix:

(1) Coniferous timber sawdust and chips. Length of the particles, 1–30 mm; natural humidity, 10–18%; bulk volume mass, 110–150 kg/m^3; and coefficient of thermal conductivity, 0.07–0.11 W/(m °C).
(2) Flax boon – dimensions of the particles up to $50 \times 3 \times 0.3$ mm; bulk volume mass, 110–120 kg/m^3; average natural humidity, 15–20%; water adsorption capacity, 220–240% (according to mass) and thermal conductivity of dried boon, 0.04–0.1 W/(m °C).
(3) Barley and wheat straw cut in 30–50 mm long pieces; humidity, 10–15%; bulk volume mass 35–50 kg/m^3 and water absorption capacity, 50–80%.

Plasticising agents of three known types (paraffin, thiokol and dicyclopentadiene) were used. The best plasticising agent (no. 1) out of these types and their mixtures was suggested. It also mainly consists of cheap production waste. Generally, the impact of various plasticising agents on the fragility (plasticity) of sulphur has been widely researched (Jordan *et al.* 1978; McBee *et al.* 1981; Darnell 1991). The impact of the accepted plasticising agent (no. 1) on the mechanical properties of sulphur was analysed in accordance with the adopted methodology.

The investigations of composite products were carried out using mixtures of different compositions with wood sawdust, flax boon and cut straw as inserts. Specimen formation technology was the following: dried inserts were placed in a special holey form and pressed up to the desired density by putting a mesh. The product was submerged into melted sulphur for 15–30 seconds, then left for 20–30 seconds for it to trickle and was placed in a setting place. Melted sulphur fully sets after 3–5 h. Plasticising agent no. 1 was used as the main plasticising agent 1, as preliminary investigations with other plasticising agents produced much worse results.

The main mechanical properties of nearly all construction materials are compression and bending strength, for thermal insulation materials also compressibility. Compression strength is closely related to tensile strength. Four cubes, $100 \times 100 \times 100$ mm, and four prisms, $40 \times 40 \times 200$ mm, were tested for the determination of each of the indicators. Prior to testing, their dimensions were specified and density of the specimen was identified, as presented in Table 2. Compressibility was determined by two methods: assuming that a load is equal to 2 kPa and that the

Table 2. Composition of mixtures and density of products

Composition no.	Product density (kg/m^3)	Component quantities (kg/m^3)				
		Sulphur	Plasticising agent no. 1	Boon	Straw	Sawdust
1	350	220	1.1	150	–	–
2	400	250	1.25	150	–	–
3	480	310	1.5	170	–	–
4	200	140	0.70	–	60	–
5	250	190	0.95	–	60	–
6	320	240	1.20	–	80	–
7	450	240	0.60	–	–	150
8	520	340	0.60	–	–	180

load accounts for 30% of failure load F. In accordance with this method, compressibility S is:

$$S = \frac{\Delta h}{h \cdot \sigma}, \quad (1)$$

where Δh – absolute compression at 0.3 failure load F; h – initial height of the specimen; σ – stresses when $\sigma = 0.3F/A$; A – specimen's area under compression.

The properties of any composite, particularly mechanical, are largely dependent on the cohesion between the matrix and inserts (Marčiukaitis 1998; Blankenhorn *et al.* 2001). The cohesion between sulphur, as a matrix, and plant origin waste, and inserts, was determined in two ways. Using wood inserts, the wood specimens, 100 × 100 × 30 mm, were produced and glued together with the sulphur binder at issue. Prior to gluing, they were dried to a constant weight. The second way – an indirect investigation of cohesion during the testing of composite elements by compression and bending in accordance with the nature of failure by examining the failure area with a microscope with a 150 times magnifying capacity.

For the purpose of determining the thermal conductivity of the products, the specimens, 200 × 250 mm, with a thickness of 66 and 80 mm (boons), 80 mm (straw) and 100 mm (sawdust) were made. They were made in accordance with the composition of components indicated in Table 2. Tests were carried out in accordance with the methodology that meets standard requirements.

2. Research results and their analysis

2.1. The properties of melted sulphur and its cohesion with inserts of plant origin

Sulphur was melted – its melt is highly penetrative and plastic in a liquid state. It, therefore, perfectly fills volumes composed of bulk or fibre thermal insulation materials binding them into a monolith.

During the setting, sulphur converts from amorphous into crystalline, therefore, its density and strength rapidly increase. Quick setting, either in contact with another material or in its own volume, results in the occurrence of internal stresses. As investigations carried out by other authors and by us show, sulphur is a fragile substance (Mohamed, El Gamal 2009, 2010). However, this property is enhanced either by plasticising a melted sulphur or adding the stabilisers of its structure (McBee *et al.* 1981; Darnell 1991; Gracia *et al.* 2004). However, the use of plasticising agents reduces the strength of sulphur. As the charts presented in Figure 1 show, the strength of plasticised sulphur depends on the content of the plasticising agent. With the amount of plasticising agent increasing, the strength is decreasing.

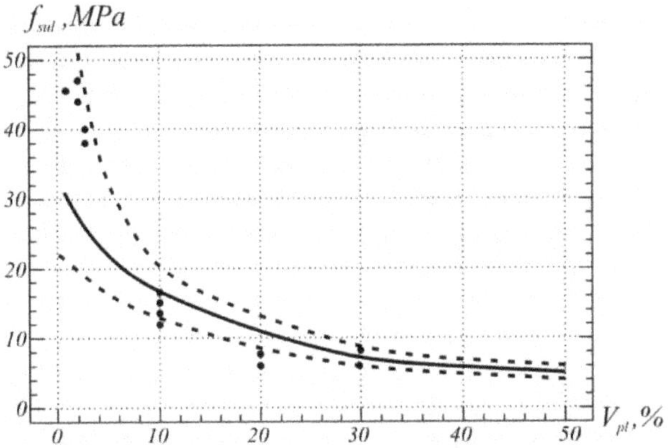

Fig. 1. Influence of plasticiser no. 1 on the compressive strength of sulphur

As the strength of thermal insulation materials is relatively low and the volume ratio of a binding component is small, a lower strength of a matrix does not have a significant impact. On the other hand, as investigations have shown, a plasticising agent proportion of 0.065–0.60% is sufficient to enhance the plasticity of sulphur. This also depends on the type of a plasticising agent. The quantity of plasticiser used for the production of our specimens is presented in Table 2. However, a plasticiser has a significant impact on the plasticity of hardened sulphur. This is proved by the investigations carried out by us (Fig. 2) and other authors (Jordan *et al.* 1978; McBee *et al.* 1981).

An important factor of the properties of composite materials is cohesion between the matrix (sulphur) and inserts (plant origin waste).

The theory of construction materials, including composites, covers not only a widely applied law of mixture but also a dependence approximating the dependence of their strength on a number of factors according to Nanazashvili (1990):

$$f_c = \frac{f_m}{\left(t_m / t_0\right)^n}, \quad (2)$$

where f_c – composite strength; f_m – binder strength; t_m – average thickness of binder's film between inserts;

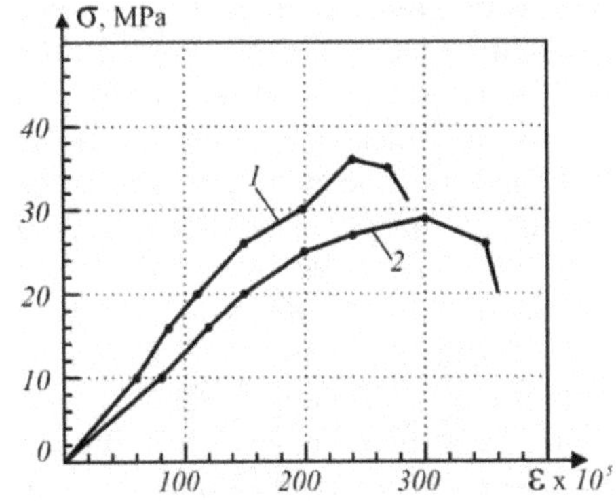

Fig. 2. Deformation character of hardened sulphur during compression: (1) non-plasticised, (2) plasticised

t_0 – thickness of a binder film of inserts of the optimum composition; n – indicator dependent on the physical and mechanical properties of inserts, their form, cohesion strength, film thickness, etc.

As the physical and mechanical properties of the aforementioned inserts do not differ much and have a similar geometrical form, only an impact of a film (layer thickness) on cohesion between inserts was investigated.

As the analysis of the curves in Figures 2 and 3 shows, when a thickness of the binder film is smaller, cohesion strength is bigger. This can be explained by the inter-molecular interaction of components in the pre-contact zone. The thinner a sulphur layer is, the relatively bigger its depth covered by this interaction. In addition, when a sulphur layer is thicker, greater internal stresses, and even micro cracks, occur when sulphur shrinks during the setting. This weakens relationships between both the components.

In all the specimens, failure occurred within the limits of the hardened layer of sulphur without affecting wood whose surface was polished. It should, therefore, be considered that such a surface corresponds to the smoothness of all straw and boon.

Cohesion among the components was analysed in an indirect way in accordance with the nature of surface that has disintegrated during testing.

As the analysis of specimen disintegration shows, cohesion between components depends on the surface of inserts: the surface of wood and boon is rougher than that of straw and, therefore, nearly no destruction takes place via the surface of component contact, while the surface of straw is much smoother and, therefore, disintegration most frequently takes place via the surface of contact.

2.2. Physical and mechanical properties of composite

The compressibility of light-weight and thermal insulation materials is closely related to strength as well as density.

As the obtained data about compressibility show (Table 3), the compressibility of the articles whose compositions are presented in Table 1 depends on the properties and quantities of components. The compressibility of composites with straw inserts exceeds 1%. However, it is nearly by three times below the compressibility of some mineral wool articles. The compressibility S of products with inserts of other types of plant origin is yet lower. For comparison, Table 3 shows the compressibility when calculating the value S (Eqn 1). As the comparison of compressibility results obtained by both methods shows, in both cases, articles with straw inserts have higher compressibility. This can be explained by a higher void content of straw structure and at the same time its lower density (kg/m^3).

Table 3. Compressibility of composites with components of sulphur and plant origin waste

Composition no.	Insert type	Compressibility: Percent	Compressibility: S (MPa^{-1})
1	Boon	0.05	0.25
2		0.04	0.18
3		0.02	0.11
4	Straw	2.05	12.05
5		1.80	9.00
6		1.60	8.00
7	Sawdust	0.02	0.095
8		0.17	0.083

This testifies to the fact that it is possible to obtain a thermal insulation building material which can resist higher pressure loads than a number of other thermal insulation materials used without infringing structural and thermal insulation requirements.

When testing compressibility up to 10% of deformation and continuing to apply a load, there was an identified compressive strength before destruction and, on the basis of bend specimen results, the bending strength and character of destruction were determined. The obtained results of strength identification are shown in Table 4.

As the data presented in Table 4 show, strength depends on the material density of products, as in the case of many other materials of low density.

As the analysis of the disintegrating nature of bending elements and surface shows, the failure nature of composite products with the matrix of sulphuric waste and inserts of plant origin waste is typical of the composites of other components with dispersive inserts (Marčiukaitis 1998). The performed investigations make it possible to point out three different features of failure which are also characteristic of other composites with dispersive inserts: failure via (1) matrix (sulphur); (2) a surface of contact between

Table 4. Compressive and flexural strength of composites made of sulphur and plant origin waste

Composition no.	Compressive strength (MPa): Deformation (10%)	Compressive strength (MPa): Failure strength	Flexural strength (MPa)
1	0.234	–	0.203
2	0.308	–	0.231
3	0.394	–	0.317
4	0.019	0.025	0.023
5	0.023	0.027	0.048
6	0.041	0.050	0.075
7	–	0.424	0.295
8	–	0.504	0.340

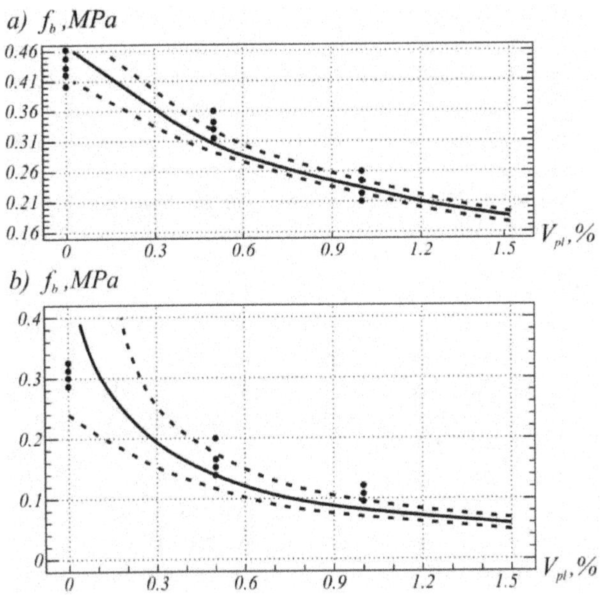

Fig. 3. Influence of plasticiser's quantity on the strength of binder's film during splitting, when film thickness is: (a) 0.5 mm, (b) 5 mm

the matrix and inserts and (c) inserts and matrix (Fig. 4).

As the thermal conductivity of sulphur λ (0.27–0.28 W/(m °C)) is lower than that of cement slurry, it was necessary to carry out an analysis of the thermal properties of composite with sulphur matrix and interests of plant origin waste. This is also predetermined by the thermal conductivity λ_c of obtained composite material. There is a direct relationship between this characteristic and heat resistance R (m°C/W). During experimental investigation, the thermal conductivity λ was obtained according to the experimental results of heat resistances of specimen R_{sp} and of contact between specimen and source of heat R_k. Then $R = R_{sp} - R_k$, as it is known $R_{sp} = \Delta T / q_{vid}$ and $\lambda = \delta / R_{sp}$. According to this data and other known suggestions (Fokin 2006), formula for the calculation of thermal conductivity of investigated composite material is:

$$\lambda_c = \frac{\delta}{\Delta T \big/ q_{vid} - nR_k}, \tag{3}$$

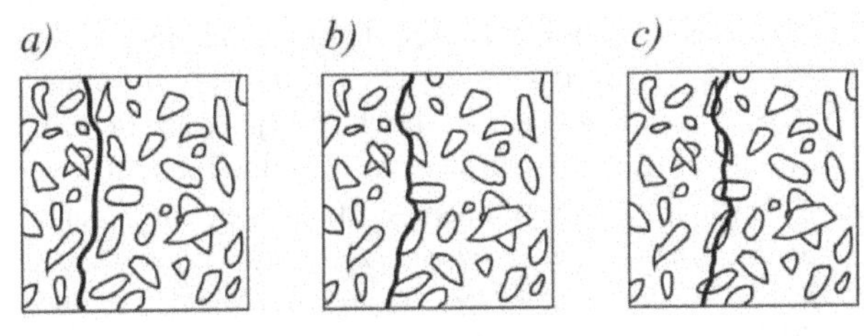

Fig. 4. Types of destruction of specimens made of sulphur and plant origin inserts: (a) typical of boon and sawdust inserts, (b) straw, (c) sawdust

where δ – specimen thickness (m); ΔT – difference in surface temperatures (°C); q_{vid} – density of the average heat flow passing through the specimen (W/m^2); R_k – thermal resistance of contact between the specimen and the source of heat (m^2 °C/W); n – the number of contacts.

As proved by investigations, the thermal conductivity of a composite with chaotically arranged inserts can also be determined analytically with sufficient accuracy. One of the best of such analytical equations according to G. Dulnev is the following (Marčiukaitis 1998):

$$\lambda_c = \lambda_m \left[1 - \frac{3\,V_{inc}\left(1 - \frac{\lambda_{inc}}{\lambda_m}\right)}{2 + V_{inc} - \lambda_{inc}\left(1 - \frac{\lambda_{inc}}{\lambda_m}\right)} \right], \tag{4}$$

where λ_c – thermal conductivity of composite; λ_m – thermal conductivity of matrix (sulphur); λ_{inc} – thermal conductivity of inserts (boon etc.); V_{inc} – relative volume of inserts.

As the comparison of experimental and calculated values of thermal conductivities λ_c obtained using Eqn (4) (Table 5) shows, the analysed composites made of waste have better properties than solid mineral wool boards whose production requires big thermal energy inputs. As the comparison of data obtained by other authors (Nanazashvili 1990; Kazragis, Gailius 2006) when using cement slurry as a binding agent shows, the thermal conductivity of the composite in question is nearly twofold lower. In addition, the proposed material requires less deficit binding agent and utilises production waste which pollutes the environment (Fig. 5).

Naturally, the authors' investigations did not cover various technological factors which have an influence on the physical and mechanical properties of the composite material of production waste in question. However, as proved by these investigations, production waste can be used in the areas causing lesser environmental pollution and producing a higher energy effect.

Table 5. Thermal conductivity of the composite material made of waste

Waste (insert) type	Density (kg/m^3)	Conductivity with sulphur matrix	
		λ_{teor} (W/(m °C))	λ_{eks} (W/(m °C))
Boon	350	0.051	0.043
	400	0.055	0.040
Straw	200	0.042	0.035
Sawdust	450	0.075	0.076

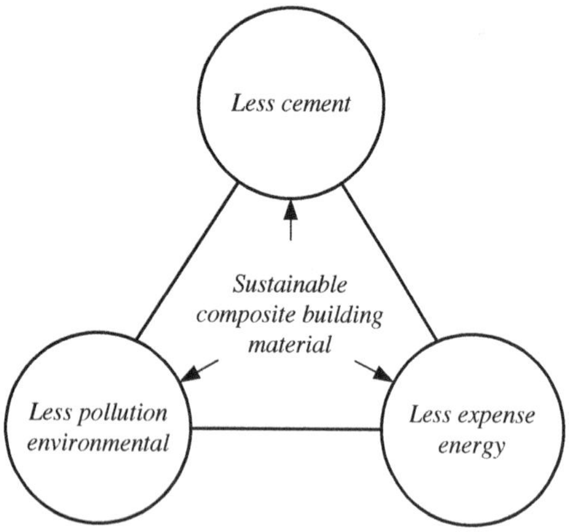

Fig. 5. Model of sustainable composite material using sulphur and plant origin waste

Conclusions

As the investigations carried out with sulphur and plant origin waste polluting the environment show, the production of composite construction of light-weight and thermal insulation materials and the products of this waste are more effective alternatives to combustion (fuel). Proper selection of composites' compositions makes it possible to obtain zero waste, energy-saving products, which fully utilise sulphur and plant origin waste and does not pollute the environment, which happens when it is combusted or left to decay.

The composite material obtained from the analysed waste polluting the environment is, by a number of its properties, equal to the currently widely used thermal insulation materials whose production requires larger energy inputs. Such important parameters of thermal insulation materials as compressibility and thermal conductivity have the same value as other classical thermal insulation materials used for building structures.

References

Abdel-Jawad, Y.; Al-Qudah, M. 1994. The combined effect of water and temperature on the strength of sulfur concrete, *Cement and Concrete Research* 24: 165–175. http://dx.doi.org/10.1016/0008-8846(94)90098-1

ACI Committee 548. 1993. *Guide for mixing and placing sulfur concrete in construction*, ACI548-2R-93. MI: American Concrete Institute. 28 p.

Blankenhorn, P. R.; Blankerhorn, B. D.; Silsbee, M. R.; DiCola, M. 2001. Effects of fiber surface treatments on mechanical properties of wood fiber–cement composites, *Cement and Concrete Research* 31(7): 1049–1155. http://dx.doi.org/10.1016/S0008-8846(01)00528-2

Czarnecki, B.; Gillot, I. E. 1989a. Effect of different admixtures on strength of sulfure concrete, *Cement, Concrete and Aggregates* 11: 2–10.

Czarnecki, B.; Gillot, I. E. 1989b. Stress–strain behaviour of sulfur concrete made with different aggregate and mixture, *Engineering Geology and Hydrogeology* 22(3): 195–206.

Darnell, G. R. 1991. Sulfur polymer cement. A new stabilization agent for mixed and low-level radio-active waste, in *Proceedings of the First International Symposium on Mixed Waste*, 1991 Baltimore, MD, USA, 7.1.1 – 7.1.11.

Fokin, K. F. 2006. *Thermal Engineering enclosing parts of the buildings*. Moscow: AVOK-PRESS. 272 p. (in Russian).

Gauronskis, V.; Jurkėnas, G.; Marčiukaitis, G. 1990. Betonas su Kėdainių chemijos gamyklos sieros atliekomis [Sulphur concrete with sulphur waste of Kėdainiai chemical factory], in *Proceedings of the Republic Scientifical and Technical Conference "Achievement of Building Sciences and Application Their Results"*, Vilnius: VISI, 8–9.

Gracia, V.; Vazquez, E.; Camona, S. 2004. Utilization by-produced sulfur for manufacture of unmodified sulfur concrete, in *International RILEM Conference Recycle Material for Building Structure* [online], 1054–1063. [Cited 11 March 2011]. Available from Internet: http://congresscimne.ups.es/rilem04

Gravitis, J.; Abolinš, J.; Tupčiauskas, R.; Veveris, A. 2010. Lignin from stem-exploded wood as binder in wood composites, *Journal of Environmental Engineering and Landscape Management* 18(2): 75–84. http://dx.doi.org/10.3846/jeelm.2010.09

Jordan, I. J.; Gillott, J. E.; Loo, V. R. E.; Shrive, N. G. 1978. Improved ductility of sulfur concretes and its relation to strength, in *Proceedings of the International Conference on Sulfur in Construction*, 1978, Canada, Centre for Mineral and Energy Technology in Ottawa, 475–488.

Kazragis, A.; Gailius, A. 2006. *Kompozitinės medžiagos ir dirbiniai su gamtiniais organiniais užpildais* [Composites materials and product with natural organic fillers]. Vilnius: Technika. 255 p.

Kazragis, A.; Juknevičiūtė, A.; Gailius, A.; Zaleckienė, E. 2004. The utilization use boon and chaff for manufacturing lightweight walling materials, *Journal of Environmental Engineering and Landscape Management* 12(1): 12–21.

Komnitsas, K.; Stouraiti, C.; Paspaliaris, I. 2001. Sulfur removal from coal beneficiation Wastes. *Paper Presented at the 7th Conference on Environmental Science and Technology* [online]. Greese, Syros, [cited 21 March 2011]. Available from Internet: http://www.labmet.ntua.gr/enviman/Papers/7TH_CEST_Conference_2001.pdf

Krishnan, P. K.; Freeman, B. 2008. Utilization of sulfur wastes from sour gas and crude oil production, in *Gas Field Development and Processing Symposium* [online], 2008, Kuwait, [cited 21 March 2011]. Available from Internet: http://www.iestech.net/download/d9.pdf.

Marčiukaitis, G. 1994. Properties of concrete from sulfur production waste, in *12 International Conference of Building Materials and Silicates (Ibausil)*, Reports. Weimar, 183–189 (in German).

Marčiukaitis, G. 1998. *Statybinių kompozitų kūrimo ir savybių prognozavimo principai* [Principles of creation and prognosis properties of building composites]. Vilnius: Technika. 133 p.

McBee, W. C.; Sullivan, T. A.; Jong, B. W. 1981. Modified sulfur concrete technology, in *Proceedings of the SULFUR-81 International Conference on Sulfur*, 1981, Calgary: Sulphur Development Institute of Canada, 367–388.

Mechta, P. K. 2009. Global concrete industry sustainable. Tools for moving forward to cut carbon emissions, *Concrete International*, Society of National Association Publications (SNAP), 45–48.

Michailov, K. V.; Paturoev, V. V.; Krais, R. 1989. *Polymers and structures with polymers.* Moscow: Strojizdat. 302 p.

Mohamed, A. M. O.; El Gamal, M. M. 2009. Hydro-mechanical behavior of a newly developed sulfur polymer concrete, *Cement and Concrete Composites* 31: 186–194. http://dx.doi.org/10.1016/j.cemconcomp.2008.12.006

Mohamed, A. M. O.; El Gamal, M. M. 2010. *Sulfur concrete for the construction industry. A sustainable development approach.* Fort Laudendale: J. Ross Publishing. 425 p.

Nanazashvili, I. H. 1990. *Build materials from wood–cement composites.* Scientific edition. Moscow: Strojizdat. 415 p. (in Russian).

Orlowsky, I. 1988. *Sulfur concrete and cement concrete modified with sulfur*, vol 1. Ukraina, Lvov: Politechnical institute, 48 p. (in Russian).

Orlowsky, I. 1990. Concrete and goods on the sulfur containing wastes foundation, *Concrete and Reinforced Concrete* 1: 24–26 (in Russian).

Paturoev, V. V.; Volgushew, A. N.; Yu, I. 1985. Peculiarities and prospects of sulfur concrete utilization, *Concrete and Reinforced Concrete* 5: 16–17.

Peavy, H. S.; Rowe, D. R.; Tchobanoglous, G. 1985. *Environmental engineering.* New York: McGraw-Hill Book Company. 699 p.

Sulphur and Sulfur Containing Wastes. 1990. Utilization in Building Industry, in Orlowsky, I.; Paturoev, V. (Eds.). *Thesis of Scientific Reports. First All-union Conference,* Ukraina, Lvov, Engineering center "Builder". 95 p.

Vlahovič, M. M.; Martinovič, S. P.; Boljanac, T. D.; Janovic, P. B.; Volkov-Husovic, T. D. 2011. Durability of sulfur concrete in various aggressive environments, *Construction and Building Materials* 25(10): 3926–3934. http://dx.doi.org/10.1016/j.conbuildmat.2011.04.024

Vroom, A. H. 1992. Sulfur concrete and its application. Polymers in concrete, in *VII International Congress on Polymers in Concrete.* Moscow: "BETECOM", 606–621.

Weber, H.; Apsleff, L.; Kogler, M. 1992. Precast techniques and applications for sulfur concrete, in *VII International Congress on Polymers in Concrete*, 1992. Moscow: "BETECOM", 627–634.

Zavadskas, E. K.; Kaklauskas, A.; Turskis, Z.; Kalibotas, D. 2009. An approach to multi-attribute assessment of indoor environment before and after refurbishment of dwellings, *Journal of Environmental Engineering and Landscape Management* 17(1): 5–11. http://dx.doi.org/10.3846/1648-6897.2009.17.5-11

Gediminas MARČIUKAITIS. Prof., Dr Habil, the Department of Reinforced Concrete and Masonry Structures, Vilnius Gediminas Technical University (VGTU). He had received a PhD from the Kaunas Polytechnic Institute in 1963. He had made a research visit to the University of Illinois (1969) and was a habilitated doctor from the Moscow Civil Engineering University in 1980 and was a professor in 1982. He is the author and co-author of five monographs, nine textbooks, six course books and more than 300 scientific articles. His research interests include new composite materials, mechanics of reinforced concrete, masonry and layered structures and investigation and renovation of buildings.

Gintaras JURKĖNAS. Engineer and head of laboratory at the Department of Reinforced Concrete and Masonry Structures, Vilnius Gediminas Technical University (VGTU). The author and co-author of six scientific articles, more than 100 expertises of properties, various types of materials and structures of buildings. Research interests include structural properties of construction materials and investigation of engineering structures and buildings.

6

A COMPARISON STUDY OF REMOVAL OF METHYLENE BLUE DYE BY ADSORPTION ON NEEM LEAF POWDER (NLP) AND ACTIVATED NLP

Himanshu Patel[1], R. T. Vashi[2]

Department of Chemistry, Navyug Science College, Rander Road, Surat 395009, Gujarat, India
E-mails: [1]hjpatel123@yahoo.co.in (corresponding author); [2]vashirajendra@yahoo.co.in

Abstract. The adsorption of basic dye, Methylene Blue Dye (MBD) onto neem (*Azadirachta indica*) leaf powder (NLP) and activated NLP with sulphuric acid was compared in this investigation. The surface analyses of NLP and activated NLP (particle size distribution, porosity, pore volume, pore diameter and BET surface area) were conducted using sophisticated instrument. The process parameters like adsorbent dose, pH, temperature and contact duration have been studied. The experimental data were analyzed using Langmuir and Freundlich isotherm, in which adsorption capacities and intensities were evaluated. Three kinetic models, pseudo first-, second-order and Elovic equations were applied to describe the adsorption process, in which correlation coefficient for each kinetic equation were calculated and evaluated.

Keywords: Adsorption, methylene blue dye, neem leaf powder, isotherm models, kinetic equation.

Introduction

Synthetic dyes have a wide application in the food, pharmaceutical, textile, leather, cosmetics and paper industries due to their ease of production, fastness, and variety in color compared to natural dyes (Adedayo *et al.* 2004). Dyes usually have a synthetic origin and complex aromatic molecular structures which make them more stable and more difficult to biodegrade. Degradation of dyes is typically a slow process (Carlos *et al.* 2009). The removal of color is needed to be considered in the disposal of textile wastewater due to aesthetic deterioration as well as the obstruction of penetration of dissolved oxygen and sunlight into water bodies, which seriously affects aquatic life. Besides, the dye precursors and degradation products are proven carcinogenic and mutagenic in nature (Somasiri *et al.* 2006). Consumption of dye-polluted water can cause allergy reactions, dermatitis, skin irritation, cancer and mutation both in babies and matures. (Kalyuzhnyi, Sklyar 2000).

Many techniques have been used to remove harmful dyes from colored wastewater, including chemical coagulation/ flocculation, precipitation, ozonation, adsorption, oxidation, ion exchange and photo degradation. Some of these techniques have shown to be effective, although they have limitations. Among these are: excess amount of chemical usage, or accumulation of concentrated sludge with disposal problems; expensive plant requirements and operational costs; lack of effective color reduction; and sensitivity to a variable wastewater input (Jaikumar, Ramamurthi 2008). One of the powerful and convenient treatment processes is adsorption. Some natural adsorbents such as fly ash (Lin *et al.* 2008), zeolite (Alpat *et al.* 2008; Mazeikiene *et al.* 2010), gulmohar (*Delonix regia*) plant leaf powder (Ponnusami *et al.* 2009), orange peel (Khaled *et al.* 2009), walnut shell (Nazari-Moghaddam *et al.* 2010) etc. are exploited for removal of various types of dyes. Raising Neem Juss seedling (Singh *et al.* 2009) and Neem leaf powder (Bhattacharyya, Sharma 2005) was also utilized for treatment of textile industrial effluent. Some acids were also utilized for activation of adsorbents (Rozic *et al.* 2008, Bhattacharyya, Gupta 2008).

This investigation is indented to prepare NLP and activated NLP using sulphuric acid and compare for the adsorption of MBD. The process parameters like adsorption dose, pH, temperature and contact duration were explored. The data were analyzed using adsorption isotherms (Freundlich and Langmuir isotherm) and kinetic models (pseudo first-, second-order and Elovic equation) models.

1. Material and methods

1.1. Adsorbent

The neem (scientific name: *Azadirachta indica*) belongs to the meliaceae family and is native to Indian subcontinent. Its seeds and leaves have been in use since ancient times to treat a number of human ailments and also as a household pesticide. The structure and ingredient presented in *Azadirachta indica* was mentioned by Sethu (Sethu, Andersen 2009). The mature leaf of plant was washed thrice with de-ionized water to remove dust and water soluble impurities and was dried until the leaves become crisp. The dried leaves were crushed and powdered and further washed with de-ionized water till the washings were free of color and turbidity. Then this powder was dried in an oven at 60 ± 2 °C and placed in desiccator for the adsorption studies, thus NLP prepared.

For activated NLP, it was stirred with 0.1 N sulphuric acid for 30 min. Then after, it washed thrice with 1 liter de-ionized water to remove untreated acid and dried in an oven at 60 ± 2 °C. The surface characteristics of NLP and activated NLP, such as particle size distribution was investigated using Laser Particle Size Analyzer, Sympatec, Germany (Model No.: Helos-BF). The porosity, pore diameter and pore volume of adsorbents were determined by Mercury Porosimeter, Thermo Quest (Model No.: PASCAL – 140). The surface area was calculated using the multipoint Brunauer-Emmett-Teller (BET) model.

1.2. Adsorbate

Methylene blue (CI No. 52015) is a heterocyclic aromatic chemical compound with molecular weight 319.851. The dye was purchased from Sigma Aldrich, India and structure of same was mentioned in Fig. 1. It has many uses in a range of different fields, such as biology, chemistry, medicine, etc. The concentration of MB in each aqueous solution was measured on an UV–visible spectrophotometer (ELICO SL 164 Double Beam UV-VIS Spectrophotometer) at λ_{max} = 665 nm (Wang *et al.* 2005).

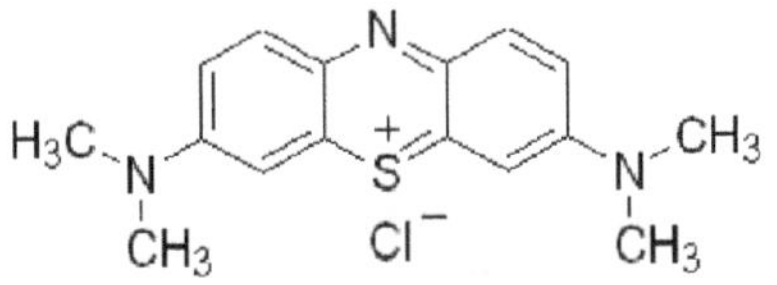

Fig. 1. Structure of Methylene Blue dye

1.3. Experimental details

The experiments were carried out as shown in Table 1 and maintaining process parameters of dye initial concentration of 200 mg/L and agitation speed of 200 rpm. The pH of system was maintained by 1.0 N HCl or 1.0 N NaOH during experiment. All other chemicals used were of analytical reagent grade.

Table 1. Experimental details for comparison of NLP and activated NLP for adsorption of MBD

Effect of System	Adsorption Dose (g/L)	pH	Temperature (K)	Contact Duration (min)
Effect of Adsorption dose	0.5, 1.0, 1.5, 2.0, 2.5 and 3.0	7	300	60
Effect of pH	1.0	1, 3, 5, 7, 9 and 11	300	60
Effect of Temperature	1.0	7	310, 320, 330, 340, 350 and 360	60
Effect of Contact duration	1.0	7	300	30, 45, 60, 75, 90 and 105

The percentage of removal and quantity of dye adsorbed, q_e (mg/g) was calculated using the following formula:

$$\% \text{ Removal} = (C_o - C_e) \text{ X } 100 / C_o, \quad (1)$$

$$q_e = (C_o - C_e) \text{ X } V / W, \quad (2)$$

were, C_o and C_e are initial and equilibrium concentration of dye respectively. V the volume of the solution and W the weight of the adsorbent used.

1.4. Equilibrium isotherm

The Freundlich isotherm is an empirical equation employed to describe heterogeneous systems and is expressed by the following equation.

$$q_e = K_F C_e^{1/n} \text{ OR } \log q_e = \log K_F + 1/n \log C_e, \quad (3)$$

where K_F (L/g) and n are the Freundlich constants reflecting the adsorption capacity and intensity respectively which are calculated from the intercept and slope of the plots of log C_e *vs.* log q_e.

The Langmuir equation is widely used for adsorption equilibria because of its thermodynamic basis. The Langmuir isotherm model assumes monolayer coverage of adsorbate over a homogeneous adsorbent surface, and at equilibrium, a saturation point is reached where no further adsorption can occur. The Langmuir isotherm model is expressed as

$$1/q_e = 1/q_{max} + (1/q_{max} K_L)(1/C_e), \quad (4)$$

where C_e is the equilibrium concentration (mg/L); q_e is the quantity of dye adsorbed onto adsorbents (mg/g); q_{max} is q_e for a complete monolayer (mg/g), a constant related to sorption capacity; and K_L is the Langmuir constant related to the affinity of the binding sites and energy of adsorption (L/mg) (Venkateswarlu *et al.* 2007).

1.5. Kinetic isotherm

There are different variables that depend upon rate of sorption such as surface area, porosity, particle size, etc. of adsorbent and properties of adsorbate. In order to investigate the mechanism of adsorption and potential rate controlling steps such as chemical reaction, diffusion control and mass transport processes, kinetic models have been used to test experimental data. These kinetic models included the pseudo first-order equation, the pseudo second-order equation and the Elovich equation.

The pseudo first-order equation of Lagergren is generally expressed as follows

$$\ln (q_e - q_t) = \ln q_e - k_1 t, \quad (5)$$

where q_t and q_e are the amounts of MBD adsorbed at time t and equilibrium (mg/g), respectively, and k_1 is the pseudo first-order rate constant for the adsorption process (1/min). The linear graph of ln (q_e − q_t) *vs.* t shows the applicability of first order kinetic. Also, the pseudo second-order chemisorption kinetic rate equation is expressed

$$t / q_t = 1/ k_2 q_e^2 + (1/q_e)t, \quad (6)$$

where k_2 is the equilibrium rate constant of pseudo second order equation (g/mg min). The linearity of t / q_t *vs* t suggests the best fit with pseudo second order kinetic (Paliulis 2006).

The Elovic equation was firstly used in the kinetics of chemisorption of gases on solids, it has been successfully applied for the adsorption of solutes from a liquid solution. The Elovic equation is given as follows:

$$q_t = 1/\beta \ln(\alpha\beta) + 1/\beta \ln t, \quad (7)$$

where α is the initial sorption rate (mg/g min) and the parameter β is related to the extent of surface coverage and activation energy for chemisorption (g/mg). The linear graph of q_t *vs.* ln t shows the applicability of Elovic kinetic (Bulut, Tez 2007).

2. Results and discussion

2.1. Surface analysis of adsorbents

Table 2 compares the surface analysis of NLP and activated NLP using H_2SO_4, in which particle size, porosity, pore diameter, pore volume and surface area of activated NLP were moderately increased than those of NLP. While in comparison to NLP, pore diameter of activated NLP was slightly increased. This shows that H_2SO_4 was effective in creating well-developed pores on the surface of NLP with large surface area and porous structure.

Table 2. Surface analysis of Adsorbents

Analysis	NLP	Activated NLP
Particle Size (mesh)	122	184
Porosity (%)	24	37
Pore Volume (cm^3/g)	0.052	0.087
Pore Diameter (nm)	8.7–9.5	9.0–9.7
BET surface Area (m^2/g)	412	524

2.2. Effect of adsorbent dose

Fig. 2 depicted effect of adsorption doses of activated NLP and NLP (0.5 to 3.0 gm/L) for adsorption MBD at temperature of 300 K and contact duration of 60 min at neutral pH, in which constant increment in removal of MBD was found. The value of percentage removals were found to be 41.1 to 82.9% using activated NLP and also, 37.1 to 72.2% of removal using NLP at adsorbent dosage of 0.5 to 3.0 g/L respectively. This may be due to the increase in availability of surface active sites resulting from the increased dose and conglomeration of the adsorbents (D'Ilario *et al.* 2008). Also, the removal of MBD by activated NLP was found higher than NLP. The pore sizes after acid treatment of NLP are bigger and also, possibility that the small increase in the pore sizes of activated NLP as compared to normal NLP was due to formation of small new pores during acid activation at the higher acid concentration (Brezovska *et al.* 2005).

2.3. Effect of pH

The pH exploits an important part on the quantity of the dye fixed. The choice of the pH values was carried out in order to examine the evolution of the adsorption of the dye associated with the various chemical forms present. The influence of pH for percentage adsorption of MBD onto activated NLP and NLP was represented in Fig. 3, in which maximum MBD removal was found to be 72.3 and 79.9% using NLP and activated NLP respectively. The adsorption of MBD on activated NLP and NLP shows that the best adsorption may be obtained at basic pH which shows there is protons release from the surface of the carbon due to cation exchange between the sorbent and the dye solution. This reveals that the increase of pH deprotonate the acidic groups on the surface of the carbon and provides more negative sites for the sorption of cationic form of the dye molecules (Malarvizhi, Sulochana 2008). It is observed that percentage removal of activated NLP is higher than that of NLP.

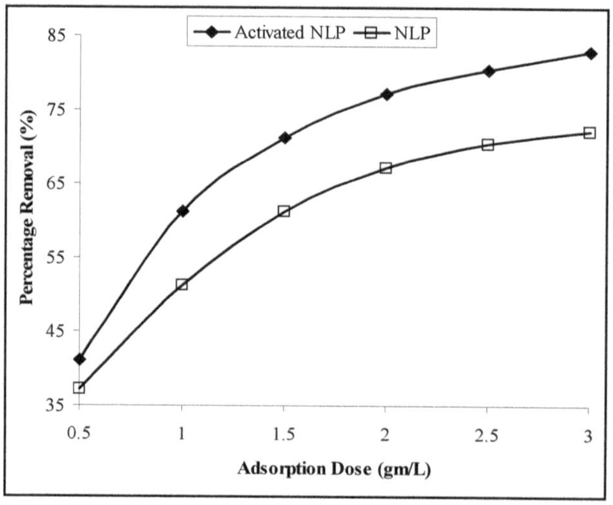

Fig. 2. Effect of adsorption dose on adsorption of MBD by activated NLP and NLP

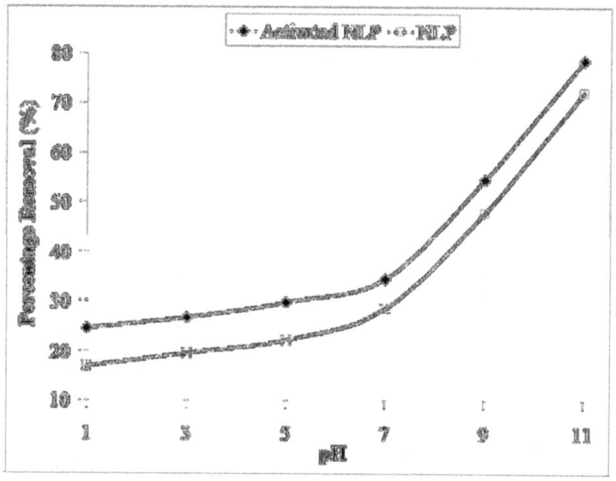

Fig. 3. Influence of pH on adsorption of MBD by activated NLP and NLP respectively at pH 11

2.4. Effect of temperature

To study the effect of temperature on the color removal, experiment was performed at temperature of 310, 320, 330, 340, 350 and 360 K (Fig. 4). The nature of curves of graph is considerable increase as increasing temperature and highest removal was found to be 44.8 and 49.8% using NLP and activated NLP respectively at temperature of 360 K. The adsorbed amounts of dyes decrease with the rise in temperature, which indicate that the adsorption process is of exothermic nature. This may be due to a tendency for the dye molecules to escape from the solid phase to be bulk phase with the increasing temperature of the solution (Ozacar, Sengil 2004). It is verified that removal of MBD by activated NLP was higher than the NLP.

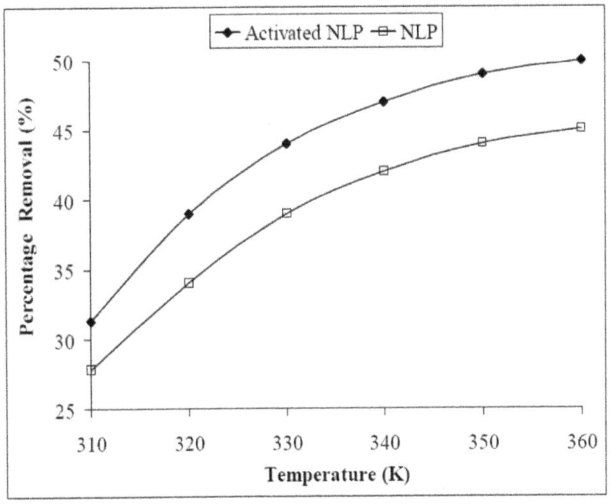

Fig. 4. Effect of temperature on adsorption of MBD by activated NLP and NLP

2.5. Effect of contact duration

The influence of contact duration on adsorption of MBD was presented in Fig. 5, in which amount of adsorption of dye are continuously increasing with increasing contact durations and The value of percentage removals were found to be 37.2 to 63.7% using activated NLP and also, 32.1 to 56.2% of removal using NLP at contact duration of 30 to 105 min respectively. That is probably due to the larger surface area of natural materials at the beginning for the adsorption of color. As the surface adsorption sites become exhausted, the uptake rate is controlled by the rate at which the adsorbate is transported from the exterior to the interior sites of the adsorbent particles (Bulut, Tez 2007). Also, it is confirm that adsorption of MBD on activated NLP is higher than NLP.

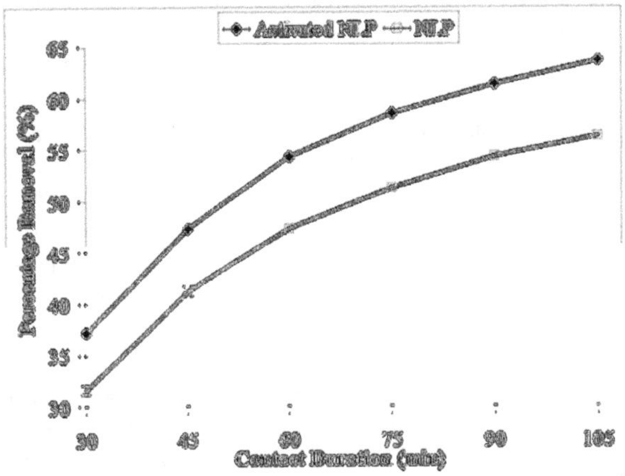

Fig. 5. Influence of contact duration on adsorption of MBD by activated NLP and NLP

2.6. Adsorption isotherm:

The isotherms parameters and correlation coefficient values were mentioned in Table 3, in which maximum adsorption capacity (Q_{max}) of activated NLP (401.6 mg/g) was found higher than NLP (352.6 mg/g). So, the activated NLP is more suitable adsorbent than NLP. Many studies are conducted for adsorption of Methylene Blue removal using adsorbent, rice husk, cotton waste, hair, coal, GLP and MLP and their Q_{max} values are 312, 277, 158, 250, 315.6 and 304.6 mg/g respectively (McKay *et al.* 1999; Patel, Vashi 2009). The coefficient values (r^2) of Freundlich and Langmuir models represented that the data were fitted to both isotherm. Also, we observed that r^2 values of Freundlich isotherm were found higher than Langmuir isotherm for both adsorbents, so, adsorption process is well fitted to Freundlich isotherm, confirming multilayer coverage of adsorbate having heterogenic adsorbent surface. The same pattern was observed for absorption of various types of dye onto adsorbents (Malarvizhi, Sulochana 2008; Inbaraj, Sulochana 2002; Prahas *et al.* 2008).

Table 3. Isothermal and kinetic parameters for adsorption of MBD using activated NLP and NLP

Model	Parameters	Adsorbent	
		Activated NLP	NLP
Freundlich	K_F (L/g)	4.99	4.84
	n	1.43	1.46
	r^2	0.994	0.997
Langmuir	K_L (L/g)	0.032	0.039
	Q_{max} (mg/g)	402	353
	r^2	0.998	0.993
Pseudo first-order	$K_1 X 10^{-2}$ (1/min)	1.97	1.73
	r^2	0.985	0.954
Pseudo second-order	$K_2 X 10^{-2}$ (g/mg min)	1.42	1.13
	r^2	0.996	0.991
Elovich equation	α (mg/g min)	30.1	25.5
	β (g/mg)	0.12	0.11
	r^2	0.995	0.967

2.7. Kinetic studies

Values of kinetic parameters (derived from slope and intercept) and the coefficients for these equations determined by non-linear regression for all kinetic models, Pseudo first-, Pseudo-second order and Elovich equation for adsorption of MBD onto activated NLP and NLP are listed in Table 3. High r^2 (> 0.99) implies that the correlations fit well with the experimental data. It can be observed that correlation coefficient values, 0.9960 and 0.9909 of Pseudo-second order equation, for activated NLP and NLP respectively, were higher than other kinetic isotherms. So, Pseudo-second order equation is most fitted than other kinetic isotherm. Similar views have been expressed by earlier scientists (Lin *et al.* 2008; Alpat *et al.* 2008; Nagda, Ghole 2008).

Conclusions

1. According to the surface analysis of activated NLP and NLP, the surface of activated NLP is more porous than NLP.

2. The comparison study of adsorption of MBD by activated NLP and NLP was conducted, in which effect of adsorption dose, pH, temperature and contact duration was exploited, and concluded that percentage removal of MBD by activated NLP was found higher than NLP. Also, parameter of adsorption dose is found to be more effective for removal of MBD onto activated NLP and NLP. The highest removal of MBD (82.9%) was achieved using 3.0 gm/L of activated charcoal at pH 7 and contact duration of 60 min.

3. The Freundlich and Langmuir isotherms were utilized, in which maximum adsorption capacity (Q_{max}) of activated NLP and NLP was found to be 401.6 and 352.6 mg/g respectively, indicating activated NLP is more efficient than NLP. Also, Freundlich isotherm was more application than Langmuir isotherm for MBD adsorption.

4. The data were analyzed using kinetic models, pseudo first- and second-order and Elovich equation, in which pseudo second-order model is more applicable among others investigated models, derived from correlation coefficient.

References

Adedayo, O.; Javadpour, S.; Taylor, C.; Anderson, W. A.; Moo-Young, M. 2004. Decolourization and detoxification of Methyl Red by aerobic bacteria from a wastewater treatment plant, *World Journal of Microbiology and Biotechnology* 20: 545–550. http://dx.doi.org/10.1023/B:WIBI.0000043150.37318.5f

Alpat, S. K.; Ozbayrak, O.; Alpat, S.; Akcay, H. 2008. The adsorption kinetics and removal of cationic dye, Toluidine Blue O, from aqueous solution with Turkish Zeolite, *Journal of Hazardous Materials* 151: 213–220. http://dx.doi.org/10.1016/j.jhazmat.2007.05.071

Bhattacharyya, K. G.; Sharma, A. 2005. Kinetics and thermodynamics of Methylene Blue adsorption on Neem (*Azadirachta indica*) leaf powder, *Dyes and Pigment* 65(1): 51–59. http://dx.doi.org/10.1016/j.dyepig.2004.06.016

Bhattacharyya, K. G.; Gupta, S. S. 2008. Influence of acid activation on adsorption of Ni(II) and Cu(II) on kaolinite and montmorillonite: kinetic and thermodynamic study, *Chemical Engineering Journal* 136(1): 1–13. http://dx.doi.org/10.1016/j.cej.2007.03.005

Brezovska, S.; Marina, B.; Napanova, B.; Burevski, D.; Bosevska, V.; Stoianovska, L. 2005. Adsorption properties and porous structure of sulfuric acid treated bentonites determined by the adsorption isotherms of benzene vapor, *Journal of the Serbian Chemical Society* 70: 33–40. http://dx.doi.org/10.2298/JSC0501033B

Bulut, Y.; Tez, Z. 2007. Adsorption studies on ground shells of Hazelnut and almond, *Journal of Hazardous Materials* 149: 35–41. http://dx.doi.org/10.1016/j.jhazmat.2007.03.044

Carlos, B.; Ivonne, L.; Gabriela, R.; Bryan, B.; Patricia, B. 2009. Removal of biorefractory compounds in industrial wastewater by chemical and electrochemical pretreatments, *Industrial & Engineering Chemistry Research* 48: 1253–1258. http://dx.doi.org/10.1021/ie800560n

D'Ilario, L.; Francolini, I.; Martinelli, A.; Piozzi, A. 2008. Insight into the Heparin–Toluidine Blue (C.I. Basic Blue 17) interaction, *Dyes and Pigments* 80(3): 343–348. http://dx.doi.org/10.1016/j.dyepig.2008.07.015

Inbaraj, B. S.; Sulochana, N. 2002. Basic dye adsorption on a low cost carbonaceous sorbent – kinetic and equilibrium studies, *Indian Journal of Chemical Technology* 9: 201–208. http://dx.doi.org/10.1016/j.dyepig.2008.07.015

Jaikumar, V.; Ramamurthi, V. 2008. Optimization of batch process parameters by response surface methodology for colour removal using spent brewery grains, in *International Conference on Environmental Research and Technology (ICERT 2008),* 28–30 May, 2008, Penang, Malaysia, 840–489.

Kalyuzhnyi, S.; Sklyar, V. 2000. Biomineralisation of azo dyes and their breakdown products in anaerobic-aerobic hybrid reactor, *Water Science and Technology* 41(12): 23–30.

Khaled, A.; El Nemr, A.; El-Sikaily, A.; Abdelwahab, O. 2009. Removal of Direct N Blue-106 from artificial textile dye effluent using activated carbon from orange peel: Adsorption isotherm and kinetic studies, *Journal of Hazardous Materials* 165: 100–110. http://dx.doi.org/10.1016/j.jhazmat.2008.09.122

Lin, J. X.; Zhan, S. L.; Fang, M. H.; Qian, X. Q.; Yang, H. 2008. Adsorption of basic dye from aqueous solution onto fly ash, *Journal of Environmental Management* 87: 193–200. http://dx.doi.org/10.1016/j.jenvman.2007.01.001

Malarvizhi, R.; Sulochana, N. 2008. Sorption isotherm and kinetic studies of methylene blue uptake onto activated carbon prepared from wood apple shell, *Journal of Environmental Protection Science* 2: 40–46.

Mazeikiene, D.; Valentukeviciene, M.; Cyril, J. 2010. Laboratory study of ammonium ion removal by using zeolite (clinoptilolite) to treat drinking water, *Journal of Environmental Engineering and Landscape Management* 18(1): 54–61. http://dx.doi.org/10.3846/jeelm.2010.07

McKay, G.; Porter, J. F.; Prasad, G. R. 1999. The removal of dye colours from aqueous solutions by adsorption on low-cost materials, *Water Air and Soil Pollution* 114(3–4): 423–438. http://dx.doi.org/10.1023/A:1005197308228

Nagda, G. K.; Ghole, V. S. 2008. Utilization of lignocellulosic waste from bidi industry for removal of dye from aqueous solution, *International Journal of Environmental Research* 2(4): 385–390.

Nazari-Moghaddam, A. A.; Najafpor, G. D.; Ghoeyshi, A. A. 2010. Removal of methylene blue from aqueous phase by pretreated walnut shell in a packed column, *Iranica Journal of Energy and Environment* 1(2): 137–143.

Ozacar, M.; Sengil, I. A. 2004. Equilibrium data and process design for adsorption of disperse dyes onto Alunite, *Environmental Geology* 45: 762–768. http://dx.doi.org/10.1007/s00254-003-0936-5

Paliulis, D. 2006. Numerical modeling of kinetics of heavy metal sorption from pollutes water, *Journal of Environmental Engineering and Landscape Management* 14(1): 10–15. http://dx.doi.org/10.1080/16486897.2006.9636873

Patel, H. J.; Vashi, R. T. 2009. Comparative study of methylene blue dye removal from its aqueous solution using guava and mango leaf powder as low cost adsorbent, *Interna-*

tional Journal of Human Geography and Environmental Studies 1(1): 58–62.

Ponnusami, V.; Gunasekar, V.; Srivastava, S. N. 2009. Kinetics of methylene blue removal from aqueous solution using gulmohar (*Delonix regia*) plant leaf powder: multivariate regression analysis, *Journal of Hazardous Materials* 169: 119–127. http://dx.doi.org/10.1016/j.jhazmat.2009.03.066

Prahas, D.; Kartika, Y.; Indraswati, N.; Ismadji, S. 2008. The use of activated carbon prepared from jackfruit (*artocarpus heterophyllus*) peel waste for methylene blue removal, *Journal of Environmental Protection Science* 2: 1–10.

Rozic, L.; Novakovic, T.; Petrovic, S.; Vukovic, Z.; Cupic, Z. 2008. Fractal analysis of physical adsorption on surfaces of acid activated bentonites from Serbia, *Chemical Industry & Chemical Engineering Quarterly* 14(4): 227–229. http://dx.doi.org/10.2298/CICEQ0804227R

Sethu, V. S.; Andersen, J. M.; Cloke, M. 2009. Biosorption of Cu(II) from wastewater using *Azadirachta indica* (Neem leaves), *Journal of Environmental Research and Development* 3(3): 614–620.

Singh, G.; Bala, N.; Rathod, T. R. 2009. Utilization of textile industrial effluent from raising *Azadirachta Indica* A. juss seedling in indian desert, *Journal of Environmental Engineering and Landscape Management* 17(3): 171–180. http://dx.doi.org/10.3846/1648-6897.2009.17.171-180

Somasiri, W.; Ruan, W.; Xiufen, L.; Jian, C. 2006. Decolourization of textile wastewater containing acid dyes in UASB reactor system under mixed anaerobic granular sludge, *Electronic Journal of Environmental, Agricultural and Food Chemistry* 5(1): 1224–1234.

Venkateswarlu, P.; Venkata, R. M.; Rao, D. S.; Rao, M. V. 2007. Removal of chromium from an aqueous solution using *Azadirachta indica* (neem) leaf powder as an adsorbent, *International Journal of Physical Sciences* 2(8): 188–195.

Wang, S.; Boyjoo, Y.; Choueib, A. 2005. A comparative study of dye removal using fly ash treated by different methods, *Chemosphere* 60: 1401–1407. http://dx.doi.org/10.1016/j.chemosphere.2005.01.091

Himanshu PATEL. Dr Himanshu Patel is a Research fellow of Navyug Science College, Gujarat, India. My research area are characterization of textile wastewater and its treatment using various adsorbents and coagulants prepared from natural plant materials like leaf, root, seed, etc. Also, adsorption of various types of dye from its aqueous solution onto natural plant materials. I am a member of Scientific Committee and Editorial Review Board on Medical and Biological Sciences in "World Academy of Science, Engineering and Technology" and volunteer editor in the "E-International Scientific research Journal of Consortium (E-ISRJC)'.

R. T. VASHI. Dr R. T. Vashi is lecturer and research guide at Navyug Science College, Gujarat, India. His research area is environmental and applied chemistry. He is a member of Editorial Advisory Board of Journal of Environmental Research and Development (JEDAR).

7

SIMULATING THE INFLUENCE OF THE HEIGHT OF HPP DAMS ON THE ENVIRONMENTAL CONDITIONS IN SMALL RIVER VIRVYTE

Alfonsas Rimkus[1], Saulius Vaikasas[2], Boriss Gjunsburgs[3]

[1, 2]Water Research Institute, Aleksandras Stulginskis University, Parko g. 6, LT-58102 Vilainiai, Kėdainiai, Lithuania

[3]Department of Water Engineering and Technology, Riga Technical University, Azenes str. 16/20, LV-1048 Riga, Latvia

Abstract. In natural floodplains, the grassed areas usually entrap fine suspended sediments (wash products rich in nutrients) of the floodwater, thus contributing to the process of self-purification. However, hydroelectric power plants (HPPs), built with the purpose of employing the renewable energy, can worsen this process, it occurs if they pond the floodplains and decrease these grassed areas. In the present paper, the sediment deposition in the floodplains with existing weir heights is investigated. The mathematical-hydraulic model is used for it. The 12 km length interval of the river Virvyte, having three ponds was investigated. The average volume of sedimentation in this region is 500–2000 t/year, i.e. quite large. In the investigated region of the river Virvyte, the sediment deposition decreased by about 50%, due to the impact of a too high dam (H = 5.5 m) of the Kapėnai HPP.

Keywords: hydraulic modelling, HPP ponds, water pollution, sediment deposition.

Reference to this paper should be made as follows: Rimkus, A.; Vaikasas, S.; Gjunsburgs, B. 2013. Simulating the influence of the height of HPP dams on the environmental conditions in small river Virvyte, *Journal of Environmental Engineering and Landscape Management* 21(3): 163–170. http://dx.doi.org/10.3846/16486897.2012.721369

Introduction

To utilize the renewable energy, hydroelectric power plants (HPPs) are built in rivers. However, their weirs and ponds affect the natural conditions of the rivers, thus deteriorating the life conditions of water fauna and flora. Ponds have major impact on the river hydrology (Wang *et al.* 2011; Marčiulionienė *et al.* 2011). Weirs also disrupt the natural flow of sediments and organic materials. The organic silt is mostly retained in reservoirs, instead of fertilizing the downstream floodplains. In the reservoirs, anaerobic processes and algal populations tend to dominate, and eutrophication may occur if there is an excess of nutrients in the water and sediments (Povilaitis 2008). Thus, ecosystems can be affected to a large extent when the exchange of water and sediments becomes difficult or impossible and the water quality may get worse (Fraley *et al.* 2007; Hohensinner *et al.* 2004; Vaikasas, Dumbrauskas 2010; Rimkus, Vaikasas 2010). Therefore, it is necessary to analyse these processes and reduce their negative influence by choosing optimal methods of hydro-energy employment. This was the main aim of the present investigation.

The arrangement of ponds impacts such important factors, as the quality of water and the dynamics of sediment transport (Olli 2008). These complex processes depend on a number of factors. This paper presents the results of mathematical-hydraulic modelling of the influence of HPP ponds with various dam heights on the sediment deposition in valleys overflowed during the floods. The present investigation is based on the model of a 12-km interval of the small river Virvytė, where a HPP cascade includes three HPPs: Skleipiai, Kapėnai and Kairiškiai (Fig. 1).

When the dams are not too high and the riverbed volume is sufficient for arrangement of ponds, the valleys are inundated only during the floods. In the flooded meadows, the sediments washed from the fields settle steadily (the process here is much more intensive than in the river beds). Consequently, the quality of river water is improving greatly because the settled particles, brought from the adjacent agricultural lands and deposited here, are rich in adsorbed nutrients (Bakel 2006; Rijn 2007). On the contrary, when the dams are too high, some part of the valley area is always dammed and thus not grassed. In this case, the valley area, useful for sedimentation, decreases. Consequently, the water self-purification process decreases also, as the more significant part of sediments, containing pollutant materials, returns from the valley to the river flow. Therefore, the exceedingly high dams

Corresponding author: Saulius Vaikasas
E-mail: s.vaikasas@delfi.lt

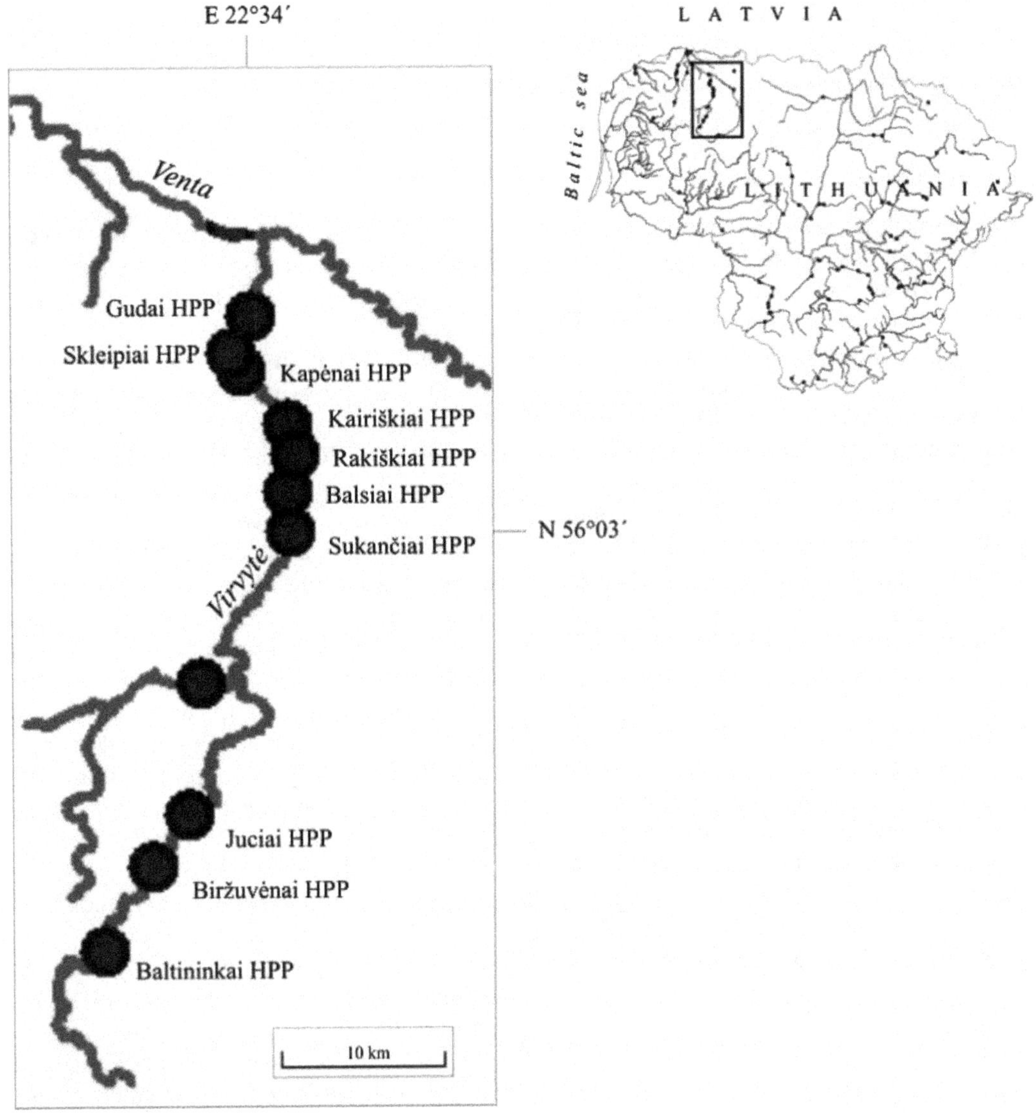

Fig. 1. HES cascades in the river Virvytė

worsen the water quality in the downward river reaches (Vastila *et al.* 2010).

We had to overcome some difficulties in modelling the sediment deposition in periodically flooded grass-covered valleys, since this process has not yet been sufficiently investigated. The necessary field investigations were performed during the floods in the river Nemunas delta and in flooded areas of the Nevėžis valley. A method for calculating sediment deposition in such areas was created (Rimkus *et al.* 2007).

In deep ponds, there are some factors that are unfavourable for the formation of water quality. The low stream velocities in large ponds create favourable conditions for algae and other small vegetation to grow. The decayed fine vegetation pollutes the water. Again, silt sinks to the bottom, although elevated flow velocities could lift it. The velocities increase during the daylong power regulation. With increased turbine discharge, the washed organic silt mixes with water and passes down. The oxidation of these organic materials decreases the amount of dissolved oxygen over a long interval of the river. This process in the Lithuanian rivers has already been discussed earlier (Vaidelienė, Michailov 2008; Zdankus *et al.* 2008).

Most ponds of the river Virvytė fill up only the riverbed, and only few of them overflow into the valley. In the modelled river interval, the pond of Kapėnai HES occupies 30 ha of a dammed valley. This decreases the sediment deposition in the valley. The main amount of sediments settled in the valley deposits during the frequent, although not large floods. In a modelled 50-year interval, an area of 60–70 ha was flooded most frequently. Therefore, it is natural that, in such a long period of time, the sedimentation decreased by half. The average sediment deposition in the valley strips near the riverbed makes 2–3 t/ha/year. The similar quantities were found during the field investigation in the flooded valley of the river Nevėžis (Vaikasas 2010).

This paper presents and discusses results of the above-mentioned mathematical hydraulic modelling of sediment retention in the floodplains and ponds with various weir heights.

1. Investigation method

The method of mathematical hydraulic modelling was used in this study to investigate the sedimentation process in the riverbeds and their valleys. The field investigations of sedimentation process and water quality are discussed in other papers (Rimkus, Vaikasas 1999, 2010; Rimkus *et al.* 2011). The known commercial models, for example, MIKE 21, were created to calculate the sediment movement on the sandy river bottoms; therefore, they are not suitable for grass-covered valleys. A numerical hydrodynamic model, elaborated and verified by the present authors, was employed for investigating the sediment deposition in the ponds and floodplains of the river Virvytė (Rimkus *et al.* 2007; Rimkus, Vaikasas 1999). This model was confirmed in investigations of natural sedimentation in the Nemunas delta valley. The actual sediment deposition in the meadows was found to be several times greater than the calculated one according the method used in commercial models. Therefore, special investigations of the sediment deposition were performed during floods in the Nemunas delta and in the flooded meadows of the river Nevėžis valley. Consequently, the following formulas for sedimentation were created (Rimkus, Vaikasas 2010; Rimkus *et al.* 2007):

$$D = \frac{k_{kor} w \bar{C}}{F}; \tag{1}$$

$$F = \frac{\bar{C}}{C_a} = \left(\frac{a}{h-a}\right)^z \left[\int_a^{h/2} \left(\frac{h-y}{y}\right)^z v_y dy + \int_{h/2}^{h} \exp(-4z(y/h-0.5)) v_y dy\right] \frac{1}{\int_0^h v_y dy}; \tag{2}$$

$$z = \frac{w}{\beta k u_*}, \tag{3}$$

where: D – sediment deposition rate per unit bottom area; k_{cor} – correction coefficient estimating the state of grasses; w – fall velocity of sediment particles in clear fluid; $\bar{C}$ – depth-averaged suspended sediment concentration; C_a – sediment concentration on the surface of grass layer; $a = 0.3h_{gr}$, – thickness of grass layer; h – water depth; v_y – flow velocity at a distance y from the bottom; $\beta = 0.6$, and $k = 0.4$ – von Karman number.

Some difficulties occurred also in choosing a formula for calculating the sediment deposition in the main canal of the river Virvytė. Most of the known formulas were constructed on the basis of laboratory investigations of sandy sediments. They are not valid for fine clay and silt particles (Rijn 2007). Therefore, we used the classical Zamarin formula, created for both fine and coarse particles:

$$C_{tr} = 0.022 \frac{v}{w_0} \sqrt{\frac{Riv}{w}}, \tag{4}$$

where: C_{tr} – transportable sediment concentration; v – average flow velocity; R – hydraulic radius; i – kinetic energy slope; $w_0 = 0.002$ if $w = <0.002$ m/s (for silt and clay) and $w_0 = w$ if $w \geq 0.002$ m/s (for sand).

The calculation of suspended sediment deposition within the regions investigated requires estimation of the stream velocities, which is most often done by using one-dimensional calculation methods. However, these methods determine only an average flow velocity and the sediment discharge in the floodplain as a whole. According to our model, created for estimating the sediment distribution across the valley, the river flow is divided into several strips with equal water discharges, and one-dimensional equations are employed in calculations. Consequently, the model turns to a quasi-two-dimensional one and, therefore, it can give more exact results. The application of real 2D models is very difficult, since, due to a complicated valley relief, a large number (several millions) of net points must be chosen, and the calculations take too much time. The work with such models is not efficient (particularly, when the flow and sediments discharges are variable). Therefore, our quasi-2D model was used. This enabled us to calculate the sediment deposition in a many-year period and to estimate thoroughly the influence of HPP ponds and weir heights on the quality of river water.

The data on suspended sediment concentrations during the floods were taken from the measurements at hydrometric posts (Vaikasas 2010). The investigations were performed with main four sediment fractions and particle diameters 0.001, 0.002, 0.005, and 0.01 mm.

2. Results and discussion

As mentioned above, the sedimentation in the inundated valley meadows reduces contamination of the river. When the weir of HPP dam is not too high, the valleys are overflowed only during sufficiently high floods. If a pond overflows in its lower part into the valley, the useful sediment retention area of the valley decreases, and thus the river self-cleaning capacity also reduces. For the ponds in Skleipiai and Kairiškiai, the riverbed is sufficient; however, in the Kapėnai pond, some part of the floodplain overflows in the valley. The calculation results show a considerable decrease of the sediment deposition in the valley in the lower region of this pond, as illustrated in Fig. 2 by the relations between the sediment deposition and water discharges.

Three groups of curves are plotted in Fig. 2. Thin lines depict the case where the HPP ponds are absent and the sediment deposition is great. Thick lines

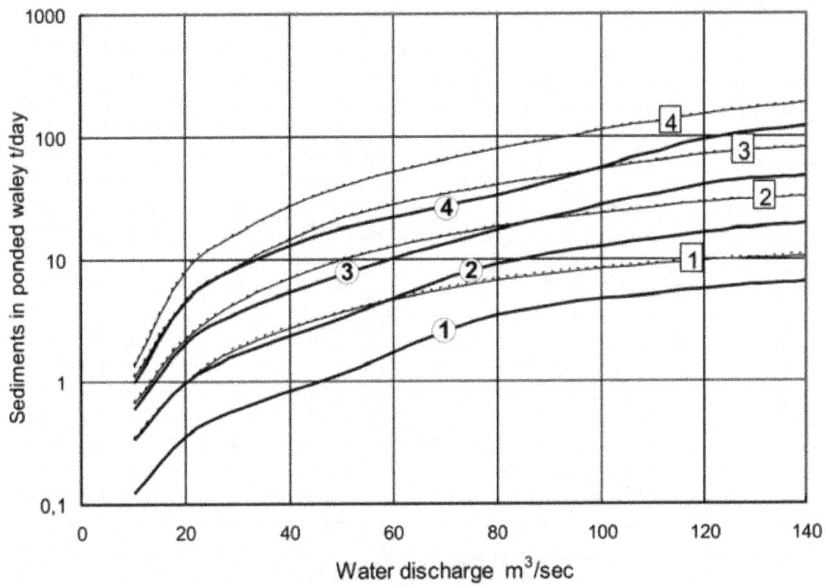

Fig. 2. Amount of deposited sediments in the investigated interval of the river Virvytė valley as a function of flood discharge: thin lines – dams are absent, thick lines – with all 3 HPP dams, dotted lines – only with a dam of Kairiškiai HPP. Particle diameters in the sediment fractions are 0.001 (1), 0.002 (2), 0.005 (3), and 0.01 (4) mm

indicate the case where all ponds are present in the investigated interval and the sediment deposition decreases intensively. During the high floods, the sediment deposition decreases almost by half and, during the small ones, even several times. The dotted lines show the case where only the Kairiškiai pond is equipped. Here, the sediment deposition is practically the same or even somewhat greater compared with the case without HPP dams. Thus, such a HPP has some positive influence, since even low dams slightly pond up the flood water levels.

The sediment deposition and retention in the inundated valley increase quickly with rising water discharges, since in this case the sediment discharge also increases, and the decrease of sediment concentration because of their deposition is then compensated in the main channel. In addition, the area of inundated valley increases with the flood. The floodplain area in the region of these three ponds during the flood of a 1% probability covers 400 ha. The deposition of coarse particles increases with discharge growing much more intensively, because the fall velocity of the particles is also much higher.

In Fig. 3, the deposited sediments are expressed as part of the total amount of sediments brought to the river. The deposition is quite intensive. When the large discharge flows, the deposition of fine sediments increases less intensively than in the case of sediments brought by the river. Therefore, the relative deposition of these sediments starts decreasing a little.

The arrangement of the Kapėnai pond decreases the sediment deposition in the valley more intensively during the relatively low floods, since this pond overflows the part of the valley which would be flooded by the low, but more frequent floods.

Fig. 4 depicts the longitudinal profile of the investigated interval of the river Virvytė during the floods with a water discharge of 20, 50, 100, and 150 m^3/s. These floods pond the water level below the HES and somewhat decrease the power of the turbines. In the Kapėnai pond, there is a greater area of cross-sections near the dam, and thus the water level along the river increases less than in the other two ponds. Therefore, the stream velocities in the ponds of Skleipiai and Kairiškiai are higher, and their water levels increase along the flow more intensively. This leads to the increased inundated area of the valley and increased sediment deposition. As a result, the floodplain meadows are more fertilized and the water quality of the Virvyte River is improved.

During high floods, the sediment deposition in the meadows of Virvyte is more intensive; however, such floods are rare. The lower flood discharges occur more frequently, and the significant part of sediments is deposited then. To estimate the influence of different floods, the calculations were performed for a long-term period. The results are shown in Fig. 5.

The amount of sediments deposited in a one-year period depends on the size of the flood. The most intensive deposition was observed during the large flood with a 1% probability in 1958. Some years, the floods were low and did not overflow in the valley. In this case, the processes of sediment deposition and retention do not proceed. When all three ponds are involved, the sediment retention decreases by about 50% due to the exceedingly high dam of Kapėnai HES. Thus, the volume of sediments deposited in the

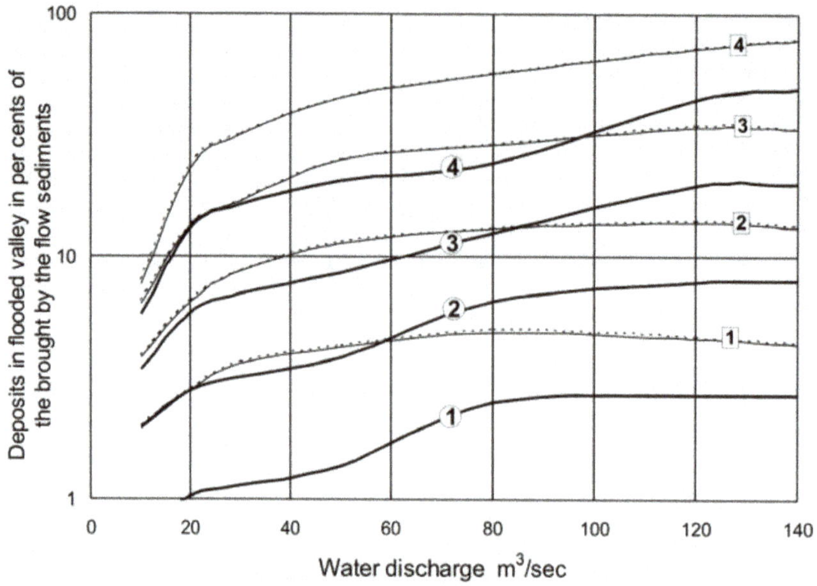

Fig. 3. Intensity of sediment deposition in the inundated valley as part of sediments brought by the river: thin lines – dams are absent, thick lines – with all 3 HPP dams, dotted lines – only with a dam in Kairiškiai HPP. Particle diameters in the sediment fractions were 0.001 (1), 0.002 (2), 0.005 (3), and 0.01 (4) mm

floodplain is quite large. Such amounts of sediments per unit floodplain area were found in field investigations (Rimkus, Vaikasas 2010; Rimkus *et al.* 2007). In periodically flooded meadows, the grasses actually entrap the sediments and favour self-cleaning of the rivers. Therefore, a decrease in the grassed areas worsens this natural process. This fact was also confirmed by other researchers (Jankowska-Huflejet 2006; Deng *et al.* 2007; Vaikasas, Dumbrauskas 2010; Lukianas *et al.* 2006). As one can see, in small rivers, high dams are not desirable from the ecological point of view; they can be important only for energetic purposes, for example, for the daylong regulation of power or for water energy accumulation. One rather large pond at the upper station of a cascade would be enough to successfully regulate the water discharge for the plants in the lower region of the river. Such a pond would better supply the local exchange of power demands, and all the necessary pike energy could not be transported from the system. This would reduce the energy losses in the electricity supply network.

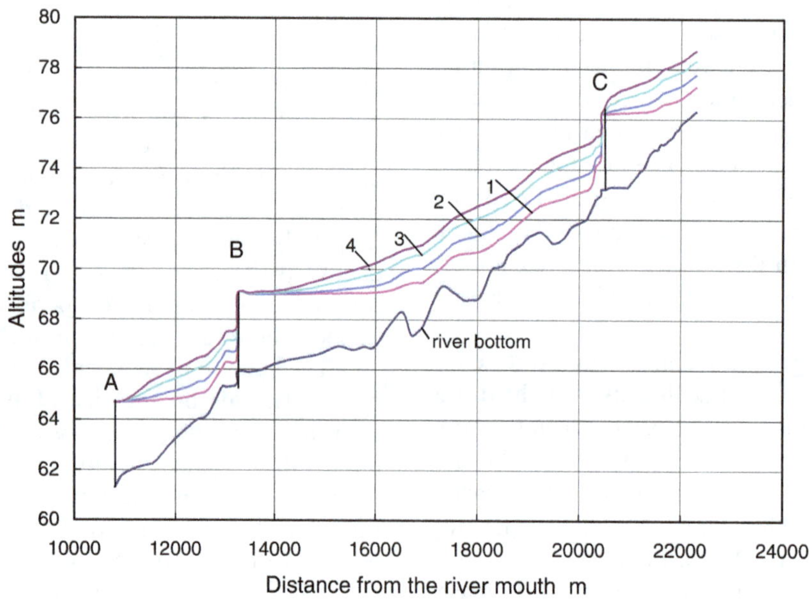

Fig. 4. Longitudinal profile of the investigated interval of the river Virvytė. Water levels of floods with water discharges of 20 (1), 50 (2), 100 (3), and 150 (4) m^3/sec. HES dams: A – Skleipiai, B – Kapėnai, and C – Kairiškiai

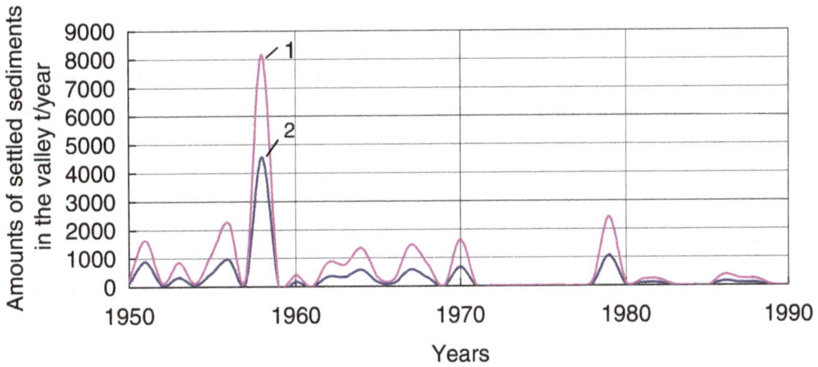

Fig. 5. Amounts of the sediments settled during a many-year period in the interval of Virvytė with the HPP ponds of Skleipiai, Kapėnai, and Kairiškiai: 1 – ponds are absent and 2 – all three ponds are involved

Ten hydropower stations built on the river Virvytė are grouped in two cascades. In the lower one, the upper station (Sukančiai) has a pond sufficient for a daylong power regulation. However, such a regulation and optimal energy production are impossible now, since these stations are equipped with only one or two large propeller-type turbines (for economic reasons). Usually, they operate at their maximum power, which is much higher than the one ensured by the river; otherwise, their efficiency coefficient would be too low. Having worked down the water level in the pond to a permissible limit, the power units are stopped, and only a sanitary discharge is allowed to pass until the pond is filled again. This situation was discussed by Zdankus *et al.* (2008), too. According to the investigations performed in the river Virvytė, such an intensive fluctuation regime is very unfavourable for the environment. The living fish is decreasing in number. Such unfavourable conditions were also found in other rivers (Fraley *et al.* 2007).

To improve this situation, at least one Kaplan-type turbine with a wide power regulation should be installed anew or replace the old one at each station. When designing the power stations on the river Virvytė, the necessity to install better turbines has not been considered, because the total power of all river stations was small compared with the power of the whole system. Therefore, the regulation of small HE power stations was not considered to be of significant importance. Moreover, no attention was paid to the ever-increasing hazard of ecological damage.

The installation of Kaplan-type turbines would be compensated economically in some period of time, since it would be not necessary to pass a sanitary discharge uselessly, when the utilized volume of pond is refilled. The turbines with a power regulation allow one to utilize almost the whole water discharge of the river, except the surplus flowing during the floods. In addition, this makes it possible to produce higher-value energy adapted to the usage exchange. In this way, the ecological and energetic demands will be coordinated. Therefore, it is even supposed that the erection of HPPs in small rivers can be also possible and useful (Hohensinner *et al.* 2004). However, a certain amount of rivers in each region must be left untouched, for preservation of natural environment.

Since the dams of the river Virvytė are not high, the stream velocities in its ponds during the floods are sufficient for transportation of fine sediments. Therefore, they are not silted by the deposition of silt and clay particles. Only coarse particles brought from the fields settle there. This fact has been proved by analysing the ground samples taken from the bottom of all ponds of HPPs.

Conclusions and recommendations

1. The ecological conditions in small rivers with the arranged HPPs are better, when their dams are not too high. Then the water in the ponds does not flood the valley permanently, and the periodically flooded grassed areas, which are of importance for self-cleaning of the river, do not decrease. The suspended sediments, brought by the water and containing the pollutant materials, are intensively settled there. However, from the energetic point of view, the existence of at least one rather large pond in the upper region of a HPP cascade is reasonable for regulating the energy production.

2. Concerning the HPPs in small rivers, the majority of propeller-type turbines without power regulation have been installed earlier. They operate periodically at their maximum power, which is the reason for large fluctuations of stream velocities and water depth. It is very harmful for the water fauna; therefore, it is necessary to modify them or to install additionally at least one Kaplan-type turbine. More perfect small turbines enable one to improve the employment of the river water flow. They recoup themselves due to the increased production of energy.

3. In deep ponds formed by high dams, the unfavourable conditions for algae and small water

vegetation are created. Therefore, high dams can be useful only in large rivers in the case of an unavoidable energetic need (for the flow energy regulation and accumulation). As an example of such an approach, the Kaunas HPP and the high-power hydro-accumulation station built in Lithuania can be considered. The accumulation of water energy will be particularly useful at a wider employment of unstable wind energy. However, the energetic and ecological needs should be coordinated. In most cases, a complex of problems has to be solved, which requires thorough scientific investigations and experimental projecting work in the future.

References

Bakel, J. 2006. Impact of the WFD on agriculture in the Netherlands and possible effect – specific hydrological measures: the Dutch approach, *Journal of Water and Land Development* 10: 45–53. http://dx.doi.org/10.2478/v10025-007-0004-8

Deng, Z.; Guensh, G. R.; Richmond, M. C.; Weiland, M. A.; Carlson, T. J. 2007. Prototype measurements of pressure fluctuations in The Dalles Dam stilling basin, *Journal of Hydraulic Research* 45(5): 674–678. http://dx.doi.org/10.1080/00221686.2007.9521803

Fraley, J.; Marotz, B.; Decker-Hess, J.; Beattie, W.; Zubik, R. 2007. Mitigation, compensation, and future protection for fish populations affected by hydropower development in the upper Columbia system, Montana, U.S.A., *Regulated Rivers: Research and Management* 3(1): 3–18. http://dx.doi.org/10.1002/rrr.3450030103

Hohensinner, S.; Habersack, H.; Jungwirth, M.; Zauner, G. 2004. Reconstruction of the characteristics of a natural alluvial river-floodplain system and hydro morphological changes following human modifications: the Danube River (1812–1991), *River Research and Applications* 20(1): 25–41. http://dx.doi.org/10.1002/rra.719

Jankowska-Huflejet, H. 2006. The function of permanent grasslands in water resources protection, *Journal of Water and Land Development* 10: 55–65. http://dx.doi.org/10.2478/v10025-007-0005-7

Lukianas, A.; Vaikasas; S.; Mališauskas, A. 2006. Water management tasks in the summer polders of the Nemunas lowland, *Irrigation and Drainage* 55(2): 145–156. http://dx.doi.org/10.1002/ird.230

Marčiulionienė, D.; Montvydienė, D.; Kazlauskienė, N.; Kesminas, V. 2011. Changes in macrophytes and fish communities in the cooler of Ignalina power plant (1988–2008), *Journal of Environmental Engineering and Landscape Management* 19(1): 21–33. http://dx.doi.org/10.3846/16486897.2011.557273

Olli, G. 2008. Historic sediment accumulation rates in Karlskä rsviken, a bay of Lake Mälaren, Sweden, *Hydrology Research* 39(2): 123–132. http://dx.doi.org/10.2166/nh.2008.032

Povilaitis, A. 2008. Source apportionment and retention of nutrients and organic matter in the Merkys river basin in southern Lithuania, *Journal of Environmental Engineering and Landscape Management* 16(4): 195–204. http://dx.doi.org/10.3846/1648-6897.2008.16.195-204

Rijn, L. C. van. 2007. Unified view of sediment transport by currents and waves II: suspended transport, *Journal of Hydraulic Engineering* ASCE June: 668–689. http://dx.doi.org/10.1061/(ASCE)0733-9429(2007)133:6(668)

Rimkus, A.; Vaikasas, S. 1999. Calculation of suspended sediment deposition in flooded, overgrown with grass valleys of rivers, in *Proc. of 28th IAHR congress.* 22–27 August, Graz, Austria (CDROM).

Rimkus, A.; Vaikasas, S.; Pukštas, R. 2007. Calculation of suspended sediment deposition intensity in grass-covered floodplains, *Nordic Hydrology* 38(2): 151–163. http://dx.doi.org/10.2166/nh.2007.004

Rimkus, A.; Vaikasas, S. 2010. Possible ways to improve sediment deposition in the Nemunas delta, *Hydrology Research* 41(3–4): 346–354. http://dx.doi.org/10.2166/nh.2010.118

Rimkus, A.; Vaikasas, S.; Gjunsburgs, B. 2011. Bias errors in studies on sediment-laden flow in short experimental channels, *Journal of Environmental Engineering and Landscape Management* 19(3): 198–207. http://dx.doi.org/10.3846/16486897.2011.613981

Vaidelienė, A.; Michailov, N. 2008. Dam influence on the river self-purificatio, in *Proceedings of the 7th International Conference "Environmental Engineering"*, 22–23 May, 2008, Vilnius, Vilnius Gediminas Technical University, 247–250.

Vaikasas, S. 2010. Mathematical modelling of sediment dynamics and their deposition in Lithuanian rivers and their deltas, *Journal of Environmental Engineering and Landscape Management* 18(3): 207–216. http://dx.doi.org/10.3846/jeelm.2010.24

Vaikasas, S.; Dumbrauskas, A. 2010. Self-purification process and retention of nitrogen in floodplains of river Nemunas, *Hydrology Research* 41: 338–345. http://dx.doi.org/10.2166/nh.2010.109

Vastila, K.; Kummu, M.; Sangmanee, C.; Chinvanno, S. 2010. Modelling climate change impacts on the flood pulse in the Lower Mekong floodplains, *Journal of Water and Climate Change* 1(1): 67–86. http://dx.doi.org/10.2166/wcc.2010.008

Wang, W.; Wang, X. G.; Zhou, X. 2011. Impacts of Californian dams on flow regime and maximum/minimum flow probability distribution, *Hydrology Research* 42(4): 275–289. http://dx.doi.org/10.2166/nh.2011.137

Zdankus, N.; Vaikasas, S.; Sabas, G. 2008. Impact of a hydropower plant on the downstream reach of a river, *Journal of Environmental Engineering and Landscape Management* 16(3): 128–134. http://dx.doi.org/10.3846/1648-6897.2008.16.128-134

Alfonsas RIMKUS. Dr Habil of Science, Water Research Institute, Lithuanian University of Agriculture (LUA), 1993. Doctor of Science LUA, 1963. Publications: author of 4 educational books, over 200 scientific publications. Research interests: hydraulics, riverbed processes.

Saulius VAIKASAS. Prof., Dr Habil, Department of Hydraulics, Vilnius Gediminas Technical University (VGTU). Principle researcher of Water Research Institute of ASU, Dr Habil of Science at Lithuanian University of Agriculture and Kaunas University of Techology (environmental engineering and landscape management), Doctor of Science at WMI LUA, 1993. Publications: author of 1 monograph, 2 educational books, over 250 research papers. Research interests: hydraulic modeling, floods, sedimentation and riverbed processes, ecohydraulic.

Boriss GJUNSBURGS. Prof., Dr sc. ing. Head of Department of Water Engineering and Technology, Riga Technical University. Research interests: hydraulic of open flow, sedimentation, flow modeling, scour.

8

FOREST DIVERSITY AND COMMUNITY DYNAMICS ALONG AN ALTITUDINAL GRADIENT OF AILAOSHAN MOUNTAIN (YUANYANG, YUNNAN PROVINCE, CHINA)

Yuan Li[a], Yanqun Zu[a], Bozhi Wu[b], Guoqing Gao[a], Fenggen Guo[b]

[a]*College of Resources and Environment, Yunnan Agricultural University, Kunming, China*
[b]*College of Agronomy and Biotechnology, Yunnan Agricultural University, Kunming, China*

Abstract. In order to evaluate forest community diversity, the quantity of forest community succession along an altitudinal gradient from 1690 to 2016 m above sea level, the understanding of relationships between diversity and community dynamics, and data from sampling plots of forest communities were studied. Forest communities were mid-mountain moist evergreen broad-leaved forests on the southern slope of the Ailaoshan Mountain in Yuanyang County (near the Hani terraces) in Yunnan Province, China. The results showed that (1) the mean species richness, ecological dominance, and evenness (*E*) were 9.16, 0.31, and 0.27, respectively. The mean Shannon-Wiener Index and Coverage Weighted Foliage-Height Diversity Index were 0.84 and 0.41, respectively. (2) The mean live aboveground biomass (AGS) was 99.23 Mg/ha and the mean Composition Index (CI) was 182.17. Species diversity (including *E*, richness, and Shannon-Wiener Index), live AGS, and CI could be predicted by the mid-domain effect (MDE), which displayed a unimodal pattern against elevation. (3) Richness, *E*, Shannon-Wiener Index, and Coverage Weighted Foliage-Height Diversity Index increased with CI. Ecological dominance decreased with increased CI, following a quadratic relationship. Whittaker Index decreased with both increased CI and AGS, with a quadratic relationship. Much forest diversity could be explained by the MDE. However, forest dynamics were seriously disturbed by human activities. More attention should focus on increasing forest diversity in order to prevent the degradation of forest ecological functions thus resulting in threats to the sustainable development of the local ecological systems.

Keywords: landscape management, forest diversity, community dynamics, altitudinal gradient, Ailaoshan Mountain.

Reference to this paper should be made as follows: Li, Y.; Zu, Y.; Wu, B.; Gao, G.; Guo, F. 2013. Forest diversity and community dynamics along an altitudinal gradient of Ailaoshan Mountain (Yuanyang, Yunnan province, China), *Journal of Environmental Engineering and Landscape Management* 21(2): 96–105. http://dx.doi.org/10.3846/16486897.2012.695735

Introduction

Biodiversity is an important attribute of ecosystem function. Forest community diversity usually increases with time and succession to a peak and then declines toward a stable climax community (Odum 1969; Niemela 1997). Many studies on succession patterns have focused on plants and organisms (Whittle *et al.* 1997; Wang *et al.* 2006). Forest communities are in specific successional stages. Thus, it is necessary to discover the relationship between community dynamics and diversity, quantifying the succession of forest community. Composition Index (CI) and biomass are used to quantify forest community succession (Curits, McIntosh 1951; Alves *et al.* 2010).

Biodiversity includes genetic, species, ecosystem, and landscape diversity (Franklin 1988; Li *et al.* 2001). The indices α-diversity and β-diversity are useful for evaluating forest community diversity (Whittaker 1960, 1975). The Index α-diversity expresses biodiversity inside the community, including the Species Richness Index, Species Evenness Index, Shannon-Wiener Index, Simpson Index, horizontal structure diversity, and vertical structure diversity (Magurran 1988). The Index β-diversity expresses diversity among communities, including the Whittaker Index (Whittaker 1960), Cody Index (Cody 1975), Wilson-Shmida Index (Wilson, Shmida 1984), and Jaccard Index (Tang *et al.* 2007). The spatial distribution of biodiversity has long been a key issue in ecology and biogeography (Lomolino 2001; Whittaker *et al.* 2001; Niu *et al.* 2008). Many reports focused on diversity, species, and area along latitudinal gradients (Stevens 1989; Kerr 1999). More attention is now paid to diversity along altitudinal gradients, because of the misunderstandings based on perceptions

Corresponding author. Yanqun Zu
E-mail: zuyanqun@yahoo.com.cn

that biodiversity decreased with both latitude and altitude (Rahbek 1997, 2005; Zhao *et al.* 2009).

The mid-domain effect (MDE) is an important mechanism influencing species diversity along altitudinal gradients (Colwell, Lees 2000; Sanders 2002; Bachman *et al.* 2004; Colwell *et al.* 2004; Kattan, Franco 2004). Colwell and Lees 2000 reported on the MDE, in which plant species diversity reached the maximum in middle elevation regions (Lieberman *et al.* 1996; Tang, Ohsawa 1997; Feng *et al.* 2006). Research suggests that diversity is related to the size of the research area, the total altitudinal range of mountains, plant species composition, and species characteristics (Hu *et al.* 2007; Niu *et al.* 2008).

The aims of this study were to (1) evaluate forest community diversity along an altitudinal gradient; (2) evaluate the quantity of forest community succession along an altitudinal gradient; and (3) understand the relationships between diversity and community dynamics.

1. Materials and methods

1.1. Study site and sampling design

The study sites were located on the southern slope of the Ailaoshan Mountain, which is in the transition zone between the Yunnan Plateau and Western Hengduanshan Mountain (an extension of the Qinghai-Tibet Plateau), and the transition between the middle and southern subtropics. The forest is a mid-mountain moist evergreen broad-leaved forest, in which *Fagaceae, Lauraceae, Magnoliaceae* and *Theaceae* are the dominant species with an average height of 9–9.8 m and stem area at breast height 125–138 cm^2 (He *et al.* 2000; Niu *et al.* 2008). The forest community has a complex floristic composition, and three conspicuous layers: arbor layer, shrub layer, and herbage layer. The richness of interlayer plants is very high (Yang *et al.* 2010a).

This study was conducted on a mountainous, steeply sloping circa 1200 ha site near the Hani terraces, Xinjie town in Yuanyang County, Yunnan Province, China (102°43′38.7″–102°46′24.1″E, 23°05′45″–23°07′32.7″N) at 1547–2019 m above sea level. Local soils are the subtropical red soils (Yuan *et al.* 2010). The climate is a subtropical monsoon, with rainy (May–October) and dry (October–May) seasons. Most of the annual precipitation (circa 1398 mm) falls during the rainy season. The mean temperature is 16.4 °C, ranging from 32.4 to −2.6 °C.

This study was carried out during March–November 2009. Forty typical plots (20 × 20 m) were selected, representing 10 major community types along the altitudinal gradient (Fig. 1 and Table 1). Ten tropical communities were dominated by *Choerospondias axillaris* (Roxb.) Burtt. et Hill., *Helicia clivicola* W.W. Smith., *Lindera communis* Hemsl., *Clerodendrum bungei* Steud., *Neolitsea chuii* Merr., *Cunninghamia C. lanceolata* (Lamb.) Hook., *Castanopsis carlesii* (Hemsl.) Hayata, *Pinus yunnanensis* Franch., *Mallotus barbatus* (Wall. ex Baill.) Muell.-Arg., and *Alnus nepalensis* D. Don. (Table 1). Four replicate plots of each community type were at the same altitude. In each plot, the following data were collected: altitude, plant species, plant density, total coverage, tree coverage, shrub coverage, grass coverage, total height, shrub height, grass height, wood diameter at breast height, number of species, and number of individuals for every identified species.

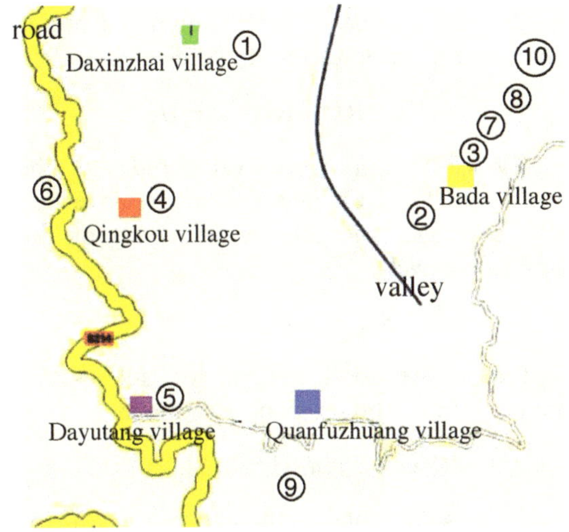

Fig. 1. The map of sampling sites

1.2. Measurement of diversity

1.2.1. α-diversity

Species diversity. Species diversity was calculated using the following indices: Shannon-Wiener Index (H'), importance value (IV_i), species richness (SR), ecological dominance (ED), and evenness (E) (Li *et al.* 2001). The equations for these indicators are:

$$H' = -\Sigma p_i \ln p_i, \tag{1}$$

Table 1. The dominant species and elevation of communities

Sampling No.	Dominant species in community type	Elevation (m)
1	*Choerospondias axillaris*	1690
2	*Helicia Clivicola*	1750
3	*Lindera communis*	1781
4	*Clerodendrum bungei*	1813
5	*Neolitsea chuii*	1876
6	*Cunninghamia lanceolata*	1899
7	*Castanopsis carlesii*	1938
8	*Pinus yunnanensis*	1960
9	*Mallotus barbatus*	1982
10	*Alnus nepalensis*	2016

where H' is the value of the Shannon-Wiener Index and p_i the proportion of the ith species, $p_i = n_i/N$.

$$IV_i = (RD_i + RC_i + RF_i)/3, \quad (2)$$

where IV_i is the importance value Index of the ith species and RD_i, RC_i, and RF_i are relative density, relative coverage, and relative frequency of the ith species, respectively.

$$SR = (s-1)/\lg N, \quad (3)$$

where N is the total number of species in the community and s the total species number.

$$ED = \Sigma n_i(n_i - 1)/N(n_i - 1), \quad (4)$$

where N is the total number of species in the community and n_i the number of the ith species.

$$E = H'/\ln s, \quad (5)$$

where H' is the value of the Shannon-Wiener Index and $\ln s$ the maximum H', s the total species numbers.

Vertical structure diversity. The vertical structure diversity is calculated using the Coverage Weighted Foliage-Height Diversity Index and the equation is (Wang *et al.* 2006):

$$H_c' = -\Sigma(C_i/C \cdot H_i/H)\ln(C_i/C \cdot H_i/H), \quad (6)$$

where H_c' is the Coverage Weighted Foliage-Height Diversity Index, C_i is the coverage of the ith foliage-height, C the total coverage of community, H_i the height of the ith foliage-height, and H is the total height of the community.

1.2.2. β-diversity

$$\text{Whittaker Index } (\beta_\omega) = S/(m_a - 1), \quad (7)$$

where S is the total number of species in the community and m_a is the mean number of species of plots (Li *et al.* 2001).

$$\text{Jaccard Index } (C_j) = j/(a+b-j), \quad (8)$$

where a and b are the total number of species for the two communities, j is the number of common species among the two communities. The Jaccared Index was used to compare each of the different nonclimax plant communities to the climax forest, which was middle-mountain moist evergreen broad-leaved forest in the vertical vegetation zone of a subtropical mountain in Yuanyang County, Yunnan Province, China.

1.3. Quantification of community dynamics

CI and live above-ground biomass (AGS) were used to quantify community dynamics (Curits, McIntosh 1951). CI is the product of the IV_i and the climax adaptation value (CAV_i):

$$CI = \Sigma IV_i \cdot CAV_i. \quad (9)$$

CAV_i values in the southern area of evergreen broad-leaved forest of Yunnan Province are listed in Table 2, as reported by Wang (1987). The values of CAV_i are 1, 3, 5, 7, and 9 for pioneer species, secondary pioneer species, transitional species, secondary climax species, and climax species, respectively.

Table 2. The climate adaptation values (CAV_i) of relatively important trees in the southern area of evergreen broad-leaved forest of Yunnan Province, China

Dominant species	important valuve (IV_i)	Climax adaptation value (CAV_i)
Choerospondias axillaris	21.25	7
Helicia Clivicola	41.19	9
Lindera communis	34.25	9
Clerodendrum bungei	15.06	9
Neolitsea chuii	15.55	9
Cunninghamia lanceolata	65.05	1
Castanopsis carlesii	38.65	9
Pinus yunnanenis	67.81	1
Mallotus barbatus	33.24	5
Alnus nepalensis	71.60	1

Estimation of live AGS

The plot census data (tree stem ≥4.8 cm) were used to estimate stand volume (M) and AGS using an allometric equation, which was a model developed by Zeng (2005) for forest types in Yunnan. The equation is:

$$M = G \cdot H \cdot F, \quad (10)$$

where M is the stand volume, G is the stem area at breast height per hectare, H is the mean tree height, and F is the mean form-factor of the tree species (0.472).

$$AGS = a \cdot M + b, \quad (11)$$

where AGS is the live AGS, M is the stand volume, a and b are constants based on values estimated by Fang *et al.* (2001) and Zeng (2005) (Table 3).

Table 3. The a and b values for calculation of M and AGS

Tree species	a	b
Pinus yunnanenis	0.5101	1.0451
Cunninghamia lanceolata	0.3999	22.5410
Broad-leaved tree	0.4754	30.6034

1.4. Statistical analysis

Effects of altitude were examined by a linear regression, using altitude as the independent variable. Means were used when samples were replicated within a site. Differences between altitudes with multiple comparison were tested using one-way analysis of variance (ANOVA) whenever samples were replicated at each site. Least significant difference (LSD) was used for determining significant differences between properties. SPSS (11.5) and EXCEL (2003) software were used for statistical analysis.

2. Results and discussion

2.1. Community diversity distribution

The total number of forest species was 181, belonging to 73 families and 103 genera in the research area with altitudinal from 1547 to 2019 m above sea level. The typical climax community was *Neolitsea chuii* community, dominated by *Neolitsea*, *Lithocarpus*, *Castanopsis*, and *Schima*. The dominant species of shrubs were *Melastoma normale* D. Don., *Sinocalamus affinis* (Rendle.) McClure and *Camellia pitardii* Coh. stuart, and dominant species of grasses were *Arthraxon hispidus* (Thumb.) Makino, *Gnaphalium affine* D. Don., *Eupatorium coelestrium* L., *Centella asiatica* (L.) Urban, *Selaginella uncinata*, and *Commeline communis* L.

The total number species of every community (N), SR, ED, E, Shannon-Wiener Diversity Index (H'), Coverage Weighted Foliage-Height Diversity Index (H_c'), and β-diversity (β_ω and C_j) are shown in Figure 2. Differences in N, SR, ED, E, H', H_c', β_ω, and C_j were observed in each community.

The data trends for richness and number of species were similar with altitude (Fig. 2a). Mean SR was 9.16 and SR and N at altitudes <1938 m were more than at altitudes >1938 m, except for the community *M. barbatus* at 1982 m. Mean ED and E were 0.31 and 0.27, respectively, with opposing trends. ED (0.08–0.16) was stable for 1750–1876 m, <1690 m (0.34), and <1899 m (0.54), except for the community *C. carlesii* (0.25) (Fig. 2b).

The trends of the Shannon-Wiener Index and Vertical Structure Diversity Index were similar with altitude (Fig. 2c). Mean H' was 0.84, ranging from 0.42 to 1.38. H' was stable (1.13–1.38) from 1750 to 1876 m, and <0.80 at 1690 m and >1899 m. Mean H_c' was 0.41, ranging from 0.17 to 0.68. The trends of β_ω and C_j were similar. Mean β_ω was 10.12 (mainly 5–10), except for the community *C. carlesii* (25.86) (Figs 2d and e). Mean C_j was 0.23 (mainly 0.14–0.21), except for the communities *C. carlesii* (0.04), *Pinus yunnanenis* (0.05), and *Neolitsea chuii* (1.00).

Analysis of forest diversity focused on α-diversity and β-diversity. α-Diversity is intracommunity species diversity and vertical structure diversity (including richness, E, ED, Shannon-Wiener Index, and Vertical Structure Diversity Index). β-Diversity expresses intercommunity diversity and includes β_ω and C_j (Li *et al.* 2001; Wang *et al.* 2001).

From 1750 to 1876 m, species diversity indicators remained relatively stable. Richness, E, and Shannon-Wiener Index from 1750 to 1876 m above sea level were greater than at other elevations, while ED showed the opposite trend. It is suggested that communities at 1750–1876 m would be the dominant and representative community. Ailaoshan Mountain is in the transition region of the middle and southern subtropics (Yuan *et al.* 2008). Typical representative forest in the Ailaoshan Mountain is the middle-mountain moist evergreen broadleaved forest, dominated by *Lithocarpus* and *Castanopsis* (Xie *et al.* 1996; Tang *et al.* 2007; Yang *et al.* 2010b). Dominant species were *Lithocarpus xylocarpus*, *Castanopsis rufes*, *Lithocarpus hancei*, and *Schima noronhas* (He *et al.* 2000; Li *et al.* 2006, 2007; Yuan *et al.* 2008, 2010). In the research area, the dominant species were *H. clivicola*, *L. communis*, *C. bungei*, *Neolitsea chuii*, *Schima superba*, *A. nepalensis*, *Camptotheca acuminata*, and frequently *Lindera Thunb.* The community *N. chuii* was particularly diverse, with 45 plant species, the highest richness, E, Shannon-Wiener Index, and the lowest ED. The community *N. chuii* was considered to be the typical zonal vegetation climax.

Species diversity indicators were less for the community *C. axillaris* at 1690 m than at 1750–1876 m. This might be due to human disturbance, as they are adjacent to the Hani villages, which are located in the mid-mountain area, at 1600–1700 m. Similar disturbances may have affected communities >1899 m. Communities dominated by *Cunninghamia C. lanceolata*, *Pinus yunnanenis*, and *A. nepalensis* were typical secondary communities, with only one dominant species. On the one hand, the invasive species *Eupatorium coelestrium* grew rapidly and restrained other plant species, resulting in restoration difficulty and low species diversity (Ding *et al.* 2007; Guo, Cheng 2008; Liao *et al.* 2008; Chen 2009). On the other hand, *Camellia sinesis* O. Ktze and *Amomum medium* Lour were planted in those communities to increase income, leading to the loss of natural plant species and simpler community structures (Lin 2003; Li *et al.* 2008; Liao *et al.* 2008).

β_ω remained relatively stable, with no significant differences of β_ω between the four communities at 1750–1876 m. The community *C. carlesii* was seriously disturbed in this area, due to fuelwood removal. The β_ω of the *C. carlesii* community was higher than that of other communities, due to the shrub and tree layers being seriously disturbed by human activities.

Community diversity is affected by many factors, including soil type, elevation, temperature, rainfall,

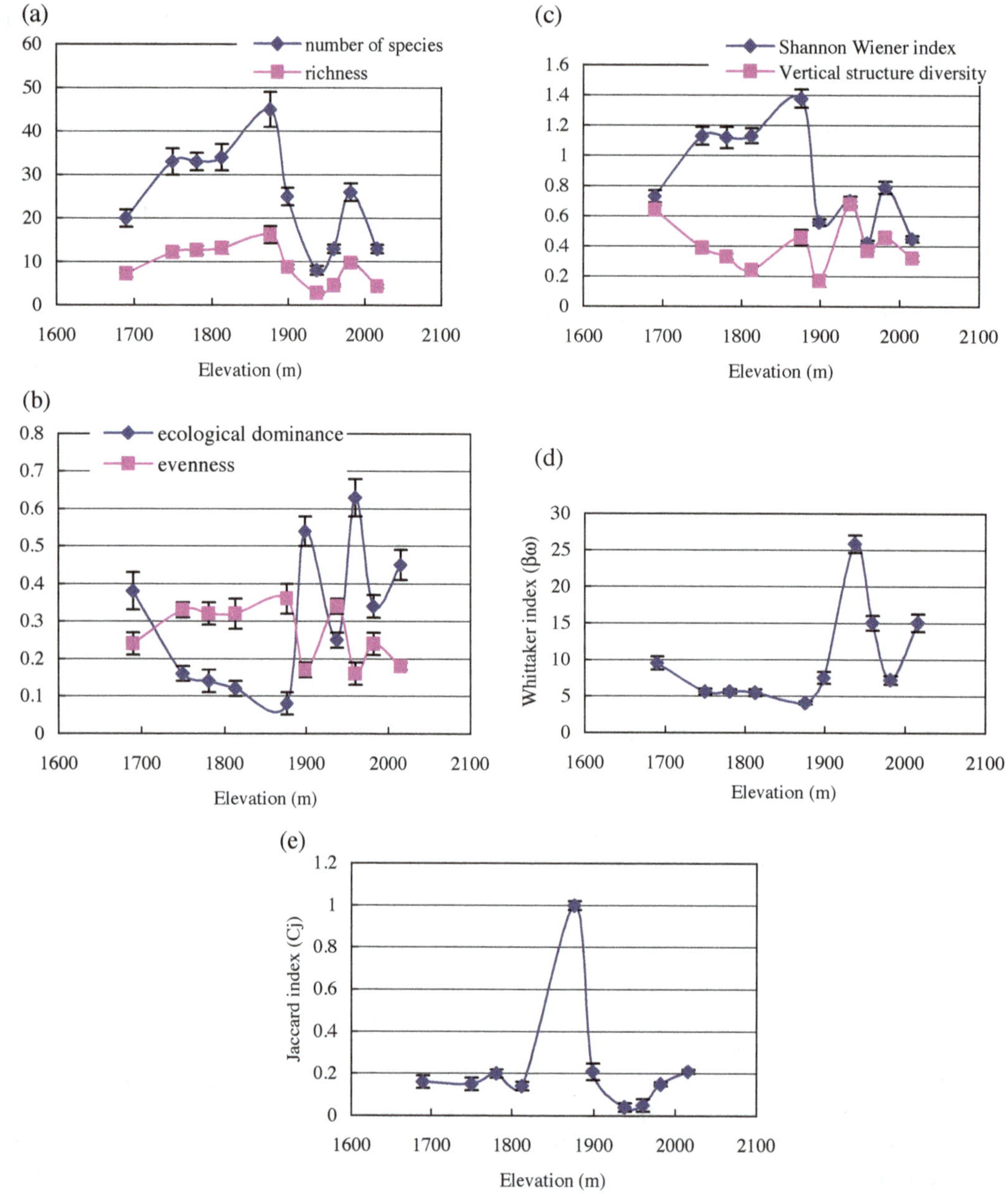

Fig. 2. Values of number species of every community (*N*), species richness (SR), ecological dominance (ED), evenness (*E*), Shannon-Wiener diversity index (*H*'), coverage weighted foliage-height diversity index (H_c') and β-diversity (β_ω and C_j) for each community with altitude

and human disturbance (Kucas 2010). Species diversity (including *E*, richness, and Shannon-Wiener Index) predicted by the MDE first increased and then decreased after reaching their maximum, showing a unimodal pattern against altitude. The MDE showed that species diversity maximized at mid-altitudes. Niu *et al.* (2008) also reported on the distribution of species diversity, in which height and diameter at breast height adopted a single peak pattern along an altitudinal gradient in the Ailaoshan Mountain area. The maximum values of Shannon-Wiener Index and SR occurred at circa 2000 m in semi-moist evergreen broad-leaved forest on the western and eastern slopes of the Ailaoshan Mountain. The distribution of species is successive and distribution area of different species overlap each other, based on the MDE (Colwell, Lees 2000; Zheng *et al.* 2009). The overlapped species are abundant, associated with suitable temperature and rainfall in the mid-altitude region, resulting in high species diversity. Species diversity decreased at both relatively low and high altitudes and was mainly due to few overlapping species, unsuitable temperatures, low soil fertility, and prolonged human disturbance (Tilman *et al.* 1996; O'brien *et al.* 2000).

2.2. Community dynamics variation

Mean AGS was 99.23 Mg/ha (range 34.98–227.15) (Fig. 3). AGS decreased with altitude, except for the

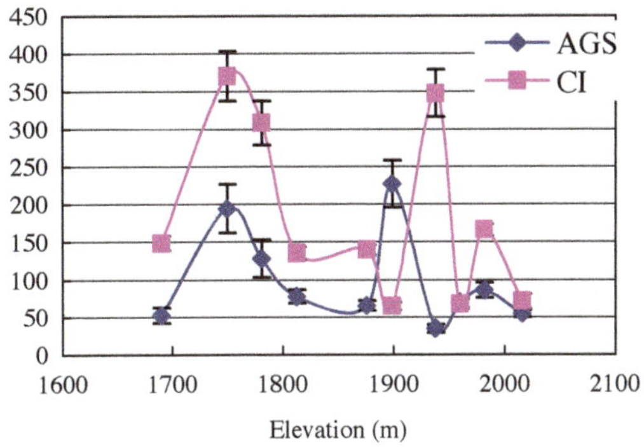

Fig. 3. Composition index (CI) and live aboveground biomass (AGS*)* for each community with altitude

community *C. axillaris* at a relatively low elevation (1690 m) and the community *Cunninghamia C. lanceolata* at 1938 m. Mean CI was 182.17 (range 65.05–370.71). The trend for CI was similar to AGS (Fig. 3). CI decreased with altitude, except for communities at relatively low altitude (1690 m) and the community *C. carlesii* at 1938 m.

The live AGS and CI also followed a unimodal pattern, increasing and then decreasing after reaching their maximum values. This trend is explained by the MDE (Whittaker, Heegaard 2003; Hu *et al.* 2007). AGS and CI reached their maximum at mid-altitude (1750 m). However, the AGS of the community *Cunninghamia C. lanceolata* was an exception, with 227.15 Mg/ha at 1899 m, which was more than that in the community at 1750 m (194.48 Mg/ha). The reason was that most *Cunninghamia C. lanceolata* were planted and managed for well over 10 years by farmers for economic income. Most *Cunninghamia C. lanceolata* were tall and robust, with a mean stem area at breast height of 73.49 m^2/ha.

The CI of the community *C. carlesii* was an exception, with CI 347.82, which was the second highest CI and just less than that in the community at 1750 m. The reason was mainly due to *C. carlesii* being the representative dominant tropical plant species in subtropical evergreen broad-leaved forest with the climax adaptation value of 9. The community *C. carlesii* was located at 1938 m on the southwest slope, in a sunny aspect. The CI of the community *C. carlesii* was high, but succession was considerably restrained by human disturbance.

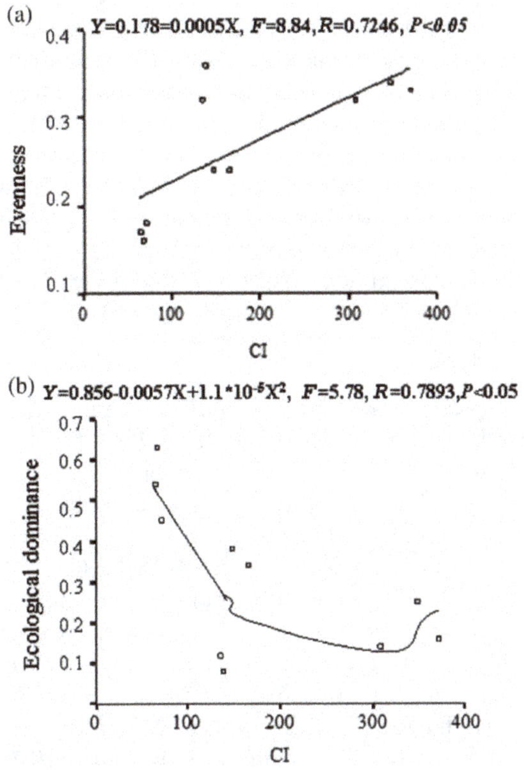

Fig. 4. The relationships between composition index (CI) and evenness (*E*), ecological dominance (ED) ($n = 10$)

2.3. Effect of community diversity and elevation on dynamics variation

A positive linear relationship was observed between *E* and CI (Fig. 4a). *E* increased with CI, which meant *E* had the similar tendency as community dynamics. Ecological dominance decreased with increased CI and increased slightly in the quadratic curve (Fig. 4b). The turning point with ED 0.14 and CI 308.25 was community *L. communis.*

Richness increased with CI, except for one specific point with the lowest richness 2.79 and CI 347.82, which was community *C. carlesii* strongly disturbed due to fuelwood removal (Fig. 5a). The Shannon-Wiener Index was low in the early community succession (Fig. 5b). The *H'* increased with CI and became stable at the stage of nearing the community climax, except for two specific points, one of which was climax community *N. chuii* with the highest *H'* 1.38 and CI 139.92. The other was *C. carlesii* community with *H'* 0.7 strongly disturbed by human activities. It is postulated that the maximum species diversity index occurs before the community climax is achieved. The H_c' increased with CI except for two specific points, which were community *H. clivicola* with H_c' 0.39 and the highest CI 370.71 and community *L. communis* with H_c' 0.33 and CI 308.25 strongly disturbed by human activities and invasion species (Fig. 5c). The vertical structure became abundant as the community approached the climax state.

β_ω (which is β-diversity associated with the intracommunity), decreased with increased CI except for one point, which was community *C. carlesii* with the highest β_ω 25.86 strongly disturbed by human activities (Fig. 6a). β_ω decreased with increased AGS and increased slightly in the quadratic curve (Fig. 6b), which followed a similar trend to ED. It is suggested that species composition among communities become more uniform as communities approach

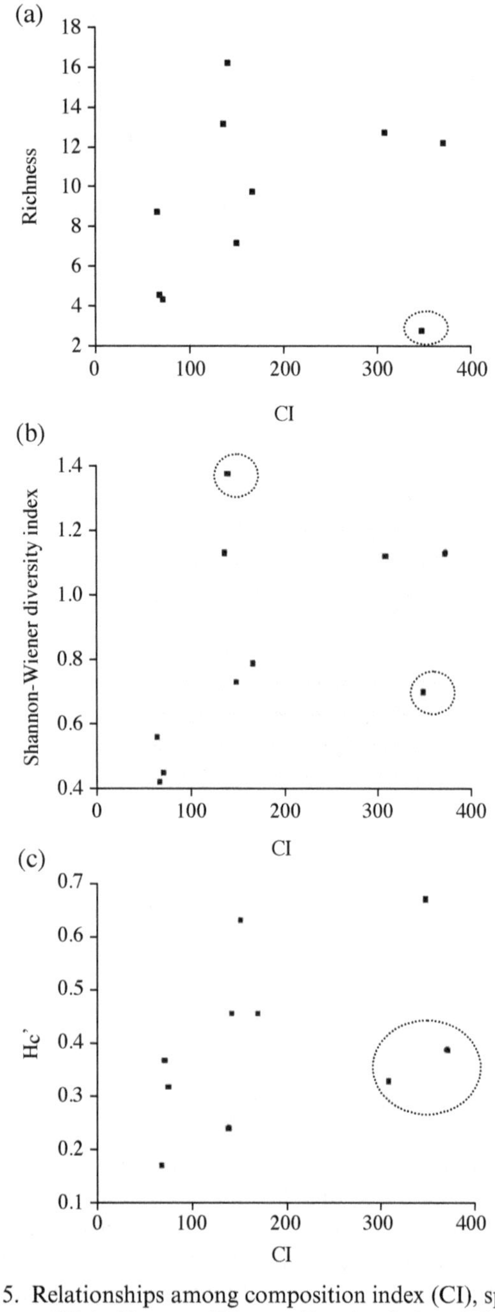

Fig. 5. Relationships among composition index (CI), species richness (SR), Shannon-Wiener diversity index (H'), and coverage weighted foliage-height diversity index (H_c')

the climax state. With the community approaching the climax, every species will be its own niche within the community.

Considering the relationships between forest diversity and dynamic variation, E, richness, Shannon-Wiener index, and vertical structure, the diversity index increased with CI and ED decreased with increased CI. This is because the dominant species become progressively less important as the community approaches the climax state and every species will be its own niche within the community, with the opposite trend to ED and E (Magurran 1988; Wang *et al.* 2006). Meanwhile, in the early stage of community succession, the niches

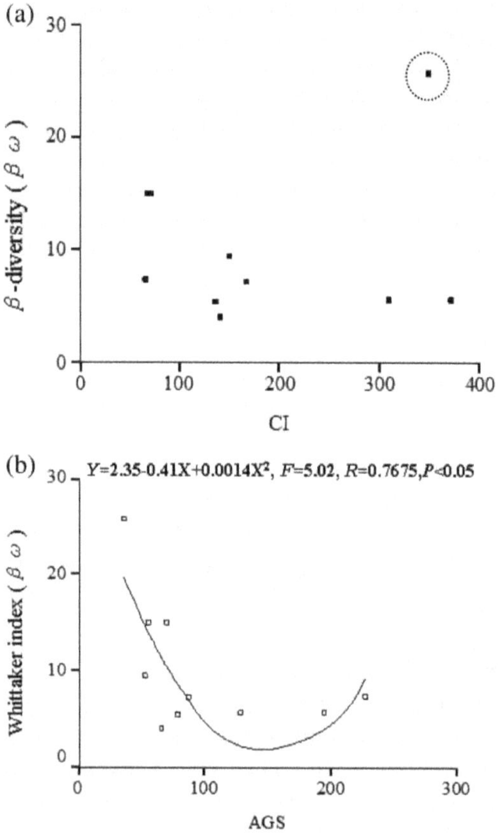

Fig. 6. Relationships between composition index (CI), live aboveground biomass (AGS), and β-diversity (β_ω) ($n = 10$)

of species overlapped, and interspecies and intraspecies competition is very strong. With community succession, every species retains its own niche and vertical diversity increases (Feng *et al.* 2006). The community *N. chuii* was a climax community and more attention should be paid to community species diversity, succession, slope direction, and human disturbance. β-diversity decreased with increased CI and AGS with the quadratic curve. With the community approaching the climax state, every species will be its own niche within the community (Whittaker 1960; Li *et al.* 2001).

Conclusions

The typical climax was *N. chuii* community in the research area with altitudinal from 1547 to 2019 m above sea level. The forest community diversity and succession along an altitudinal gradient and its relationships were as following:

1. Mean SR, ED, and E were 9.16, 0.31, and 0.27, respectively. The mean Shannon-Wiener Index was 0.84, ranging from 0.42 to 1.38. The Mean Coverage Weighted Foliage-Height Diversity Index was 0.41, ranging from 0.17 to 0.68. The mean live AGS was 99.23 Mg/ha, ranging from 34.98 to 227.15 Mg/ha. The mean CI was 182.17, ranging from 65.05 to 370.71.

2. Species diversity (including E, richness, Shannon-Wiener index, live AGS, and CI) could all be

predicted by the MDE, increasing first and then decreasing after reaching their maximum, showing a unimodal pattern with altitude.

3. The richness, *E*, Shannon-Wiener Index, and Coverage Weighted Foliage-Height Diversity Index increased with CI. Species diversity and vertical structure diversity became more abundant as the community approached the climax state.

4. Ecological dominance decreased with increased *CI* in a quadratic relationship. β_ω decreased with increased *CI* and *AGS* in a quadratic relationship. As the community approached the climax state, every species would be its own niche in the community.

Forest diversity, especially species diversity, retained natural characteristics and was influenced by the MDE. However, the forests were seriously disturbed by human activities. It is suggested that decreased forest diversity (including species diversity and vertical structure diversity) would degrade forest ecological functions and threaten the sustainable development of local ecological systems.

Acknowledgements

The study was financed by the National Basic Research Program of China (approval No. 2006CB100203), the National Key Sciences and Technology Program for Water Solution (No.2012ZX07102-003-04), the Research Foundation for Academic Leaders in Yunnan Province, China (approval No. 2006PY01-34) and the project was sponsored by the Scientific Research Foundation for Returned Overseas Chinese Scholars, State Education Ministry (China). We thank Professor Michael A. Fullen for his valuable comments and English writing improvement.

References

Alves, L. F.; Vieira, S. A.; Scaranello, M. A.; Camargo, P. B.; Santos, F. A. M.; Joly, C. A.; Martinelli, L. A. 2010. Forest structure and live above-ground biomass variation along an elevational gradient of tropical Atlantic moist forest (Brazil), *Forest Ecology and Management* 260(5): 679–691. http://dx.doi.org/10.1016/j.foreco.2010.05.023

Bachman, S.; Baker, W. J.; Brummitt, N.; Dransfield, J.; Moat, J. 2004. Elevational gradients, area and tropical island diversity: an example from the palms of New Guinea, *Ecography* 27(3): 299–310. http://dx.doi.org/10.1111/j.0906-7590.2004.03759.x

Chen, Y. X. 2009. Research of mechanism of invasion of *Eupatorium adenophorum*, *Pruataculture & Animal Husbandry* 1: 5–7.

Cody, M. L. 1975. Towards a theory of continental species diversities: bird distributions over Mediterranean habitat gradients, in Cody, M. L.; Diamond, J. M. (Eds.), *Ecology and Evolution of Communities.* Cambridge: Belknap Press, 214–257.

Colwell, R. K.; Lees, D. C. 2000. The mid-domain effect: geometric constraints on the geography of species richness, *Trends in Ecology and Evolution* 15(2): 70–76. http://dx.doi.org/10.1016/S0169-5347(99)01767-X

Colwell, R. K.; Rahbek, C.; Gotelli, N. J. 2004. The mid-domain effect and species richness patterns: what have we learned so far?, *The American Naturalist* 163(3): E1–E23. http://dx.doi.org/10.1086/382056

Curtis, J. T.; McIntosh, R. P. 1951. An upland forest continuum in the prairie-forest region of Wisconsin, *Ecology* 32(3): 476–496. http://dx.doi.org/10.2307/1931725

Ding, H.; Xu, H. G.; Liu, Z. L. 2007. Impacts of invasion of *Eupatorium adenophorum* on vegetation diversity, *Journal of Ecology and Rural Environment* 23(2): 29–32, 75.

Fang, J.; Chen, A.; Peng, C.; Zhao, S.; Ci, L. 2001. Changes in forest biomass carbon storage in China between 1949 and 1998, *Science* 292: 2320–2322. http://dx.doi.org/10.1126/science.1058629

Feng, J. M.; Wang, X. P.; Li, J.; Fang, J. Y. 2006. Effects of area and mid-domain effect on altitudinal pattern of seed plants richness in Lijiang, Yunnan, China, *Biodiversity Science* 14(2): 107–113. http://dx.doi.org/10.1360/biodiv.050123

Franklin, J. F. 1988. Structural and functional diversity in temperate forests, in Wilson, E. O.; Peter, F. M. (Eds.). *Biodiversity.* Washington DC: National Academy Press, 166–175.

Guo, H. M.; Cheng, H. M. 2008. Progress in research on allelopathic potential in an exotic invasive plant *Eupatorium adenophorum*, *Journal of Agricultural Science and Technology* 10(SI): 30–34.

He, Y. T.; Cao, M.; Tang, Y.; Yang, G. P. 2000. A comparative study on tree species diversity of evergreen broad-leaved forest, central Yunnan, *Journal of Mountain Science* 18(4): 322–328.

Hu, Y. K.; Li, K. H.; Wang, X.; Fan, Y. G.; Zhang, X. Y.; Wumaier, W. S. 2007. The biomass of different species communities of Alpine meadow in Bayinbulak Region, *Resources Science* 29(3): 147–151.

Kattan, H. G.; Franco, P. 2004. Bird diversity along elevational gradients in the Andes of Colombia: area and mass effects, *Global Ecology and Biogeography* 13(5): 451–458. http://dx.doi.org/10.1111/j.1466-822X.2004.00117.x

Kerr, J. T. 1999. Weak links: "Rapoport's rule" and large scale species richness patterns, *Global Ecology and Biogeography* 8(1): 47–54. http://dx.doi.org/10.1046/j.1365-2699.1999.00315.x

Kučas, A. 2010. Location prioritization by means of multi-criteria spatial decision-support systems: a case study of forest fragmentation-based ranking of forest administrative areas, *Journal of Environmental Engineering and Landscape Management* 18(4): 312–320. http://dx.doi.org/10.3846/jeelm.2010.36

Li, D. W.; Chen, H. W.; Shi, F. Q.; Yang, B.; Li, J. 2008. Biological and ecological characteristics and geographical distribution of *Alnus nepalensis yunnanensis*, *Forest Inventory and Planning* 33(5): 25–28.

Li, S.; Liu, W. Y.; Wang, L. S.; Yang, G. P.; Li, D. W. 2007. Species diversity and distribution of epiphytic lichens in the primary and secondary forests in Ailao Mountain,

Yunnan, *Biodiversity Science* 15(5): 445–455. http://dx.doi.org/10.1360/biodiv.070084

Li, D.; Tang, J. W.; Luo, C. K.; Li, J. S.; Liu, Z. A. 2006. Analysis on the coecological characteristics of monsoonal evergreen broad-leaved forest communities in Xishuangbanna, *Journal of Mountain Science* 24(3): 257–267.

Li, B.; Yang, C.; Lin, P. 2001. *Ecology.* Beijing: High Education Press, 343–347.

Liao, Y. H.; Chen, C. G.; Chen, H. Y.; Zhang, J.; Wu, Z. Q.; Liu, L. W. 2008. Study on the community characteristics and plant diversity of village forests in Liantang village, Guangzhou, *Ecology and Environment* 17(2): 812–817.

Lieberman, D.; Lieberman, M.; Peralta, R.; Hartshorn, G. S. 1996. Tropical forest structure and composition on a large-scale altitudinal gradient in Costa Rica, *The Journal of Ecology* 84(2): 137–152. http://dx.doi.org/10.2307/2261350

Lin, P. 2003. *Integrated scientific survey at Mangdang Mountains nature reserve, Fujian, China*. Xiamen: Xiamen University Press, 127–135.

Lomolino, M. V. 2001. Elevation gradients of species-density: historical and perspective views, *Global Ecology and Biogeography* 10(1): 3–13. http://dx.doi.org/10.1046/j.1466-822x.2001.00229.x

Magurran, A. E. 1988. *Ecological diversity and its measurement.* Princeton: Princeton University Press, 55–70.

Niemela, J. 1997. Invertebrates and boreal forest management, *Conservation Biology* 11(3): 601–610. http://dx.doi.org/10.1046/j.1523-1739.1997.06008.x

Niu, Y.; Liu, Y.; Zhang, Y. P.; Gao, J. B. 2008. Vertical distribution patterns of plant species diversity in Ailao Mountains, *Journal of West China Forestry Science* 37(4): 27–34.

O'Brien, E. M.; Field, R.; Whittaker, R. J. 2000. Climatic gradients in woody plant (tree and shrub) diversity: water-energy dynamics, residual variation, and topography, *Oikos* 89(3): 588–600. http://dx.doi.org/10.1034/j.1600-0706.2000.890319.x

Odum, E. P. 1969. The strategy of ecosystem development, *Science* 164(3877): 262–270. http://dx.doi.org/10.1126/science.164.3877.262

Rahbek, C. 1997. The relationship among area, elevation, and regional species richness in neotropical birds, *The American Naturalist* 149(5): 875–902. http://dx.doi.org/10.1086/286028

Rahbek, C. 2005. The role of spatial scale and the perception of large-scale species-richness patterns scale and species-richness patterns, *Ecology Letters* 8(2): 224–239. http://dx.doi.org/10.1111/j.1461-0248.2004.00701.x

Sanders, N. J. 2002. Elevational gradients in ant species richness: area, geometry, and Rapoport's rule, *Ecography* 25(1): 25–32. http://dx.doi.org/10.1034/j.1600-0587.2002.250104.x

Stevens, G. C. 1989. The latitudinal gradient in geographical ranges: how so many coexist in the tropics, *The American Naturalist* 133(2): 240–256. http://dx.doi.org/10.1086/284913

Tang, C. Q.; Hou, X.; Gao, K.; Xia, T.; Duan, C.; Fu, D. 2007. Man-made versus natural forests in Mid-Yunnan, Southwestern China: plant diversity and initial data on water and soil conservation, *Mountain Research and Development* 27(3): 242–249. http://dx.doi.org/10.1659/mrd.0732

Tang, C. Q.; Ohsawa, M. 1997. Zonal transition of evergreen, deciduous, and coniferous forests along the altitudinal gradient on a humid subtropical mountain, Mt. Emei, Sichuan, China, *Plant Ecology* 133(1): 63–78. http://dx.doi.org/10.1023/A:1009729027521

Tilman, D.; Wedin, D.; Knops, J. 1996. Productivity and sustainability influenced by biodiversity in grassland ecosystems, *Nature* 379(6567): 718–720. http://dx.doi.org/10.1038/379718a0

Wang, B. S. 1987. *Plant synecology.* Beijing: Higher Education Press.

Wang, D. B.; Ji, S. Y.; Chen, F. P. 2001. A review on the species diversity of plant communities, *Chinese Journal of Ecology* 20(4): 55–60.

Wang, Q.-G.; Xing, Y.-J.; Zhou, X.-F.; Han, S.-J. 2006. Relationship between diversity of forest plant and community dynamics in the eastern mountain area of Heilongjiang Province, China, *Journal of Forestry Research* 17(4): 289–292. http://dx.doi.org/10.1007/s11676-006-0066-5

Whittaker, R. H. 1960. Vegetation of the Siskiyou Mountains, Oregon and California, *Ecological Monographs* 30(3): 279–338. http://dx.doi.org/10.2307/1943563

Whittaker, R. H. 1975. *Communities and ecosystems.* 2nd ed. New York: Macmillan.

Whittaker, R. J.; Heegaard, E. 2003. What is the observed relationship between species richness and productivity: comment, *Ecology* 84(12): 3384–3390. http://dx.doi.org/10.1890/02-3128

Whittaker, R. J.; Willis, K. J.; Field, R. 2001. Scale and species richness: towards a general, hierarchical theory of species diversity, *Journal of Biogeography* 28(4): 453–470. http://dx.doi.org/10.1046/j.1365-2699.2001.00563.x

Whittle, C. A.; Duchesne, L. C.; Needham, T. 1997. The impact of broadcast burning and fire severity on species composition and abundance of surface vegetation in a jack pine (*Pinus banksiana*) clear-cut, *Forest Ecology and Management* 94(1–3): 141–148. http://dx.doi.org/10.1016/S0378-1127(96)03969-2

Wilson, M. V.; Shmida, A. 1984. Measuring beta diversity with presence–absence data, *The Journal of Ecology* 72(3): 1055–1064. http://dx.doi.org/10.2307/2259551

Xie, S. C.; Liu, W. Y.; Li, S. C.; Yang, G. P. 1996. Preliminary studies on the biomass of middle-mountain moist evergreen broadleaved forests in Ailao Mountain, Yunnan, *Acta Phytoecologica Sinica* 20(2): 167–176.

Yang, G. P.; Gong, H. D; Zheng, Z.; Zhang, Y. P.; Liu, Y. H.; Lu, Z. Y. 2010b. Caloric values and ash content of six dominant tree species in an evergreen broadleaf forest of Ailaoshan, Yunnan Province, *Journal of Zhejiang Forestry College* 27(2): 251–258.

Yang, G. P.; Zheng, Z.; Zhang, Y. P.; Liu, Y. H.; Gong, H. D.; Lu, Z. Y. 2010a. Community characteristics of middle-mountain moist evergreen broad-leaved forest in Ailao Mountains, *Journal of Northeast Forestry University* 38(9): 16–19.

Yuan, C. M.; Liu, W. Y.; Yang, G. P. 2008. Species composition and diversity of lianas in forest in Ailao Mountains, Yunnan, China, *Journal of Mountain Science* 26(1): 29–35.

Yuan, C. M.; Liu, W. Y.; Yang, G. P.; Li, X. S. 2010. Liana species diversity and relationships with its host trees in the moist evergreen broad-leaved forest in the Ailao Mountains, Southwest China, *Scientia Silvae Sinicae* 46(1): 15–22.

Zeng, W. S. 2005. Research on forest biomass and productivity in Yunnan, *Central South Forest Inventory and Planning* 24(4): 1–3, 13.

Zhao, J.; Chen, J. W.; Lv, G. 2009. Plant diversity of primary forest vegetation type at Laotudingzi Nature Reserve, *Journal of Liaoning Technical University (Natural Science)* 28(5): 849–852.

Zheng, J. M.; Wang, L. Y.; Li, S. Y.; Zhou, J. X.; Sun, Q. X. 2009. Relationship between community type of wetland plants and site elevation on sandbars of the East Dongting Lake, China, *Forestry Studies in China* 11(1): 44–48.

Yuan LI. Dr, Professor at the College of Resources and Environment, Yunnan Agricultural University. He received his PhD in Ecology in 1998 from Lanzhou University, China. Publications: author/co-authors of ~80 scientific papers. Research interests: community ecology, pollution ecology, and ecological restoration.

Yanqun ZU. Dr, Professor at the College of Resources and Environment, Yunnan Agricultural University. She received her PhD in Agronomy and Biological Engineering in 2008 from Gembloux Agricultural University, Belgium. Publications: author/co-authors of ~60 scientific papers. Research interests: environmental ecology, ecological restoration, and water and soil conservation.

Bozhi WU. Dr, Professor at the Department of Agronomy and Biotechnology, Yunnan Agricultural University. He received his PhD in Water and Soil Conservation in 2000 from Wolverhamton University, UK. Publications: author/co-authors of ~80 scientific papers. Research interests: water and soil conservation and agricultural ecology.

Guoqing GAO. is an MA student at the College of Resources and Environment, Yunnan Agricultural University. He received his Bachelor's degree in Environmental Science in 2009 from Yunnan Agricultural University. Publications: author of one scientific paper. Research interests: water and soil conservation and forest ecology.

Fenggen GUO. Dr, Professor in the Department of Agronomy and Biotechnology, Yunnan Agricultural University. He received his PhD in Botany in 1998 from Zhejiang University, China. Publications: author/co-authors of ~60 scientific papers. Research interests: water and soil conservation and botany.

9

REMEDIATION OF SOIL CONTAMINATED WITH DECABROMINATED DIPHENYL ETHER USING WHITE ROT FUNGI

Junqin Wu, Yuechun Zhao, Lu Liu, Biao Fan, Minghua Li

Department of Applied Chemistry, South China Agricultural University, Guangzhou 510642, China

Abstract. Biodegradation of decabrominated diphenyl ether (BDE-209) in soil by white rot fungi under various experimental conditions was investigated in this study. It was found that BDE-209 in soil could be rapidly and efficiently degraded by white rot fungi, and the biodegradation fits the pseudo-first-order kinetics during a 15-day incubation period. The residues of BDE-209 in soil decreased with the increased amount of white rot fungi addition. It can be seen from the results that white rot fungi have a good degradation ability with one-step and two-step addition method. In native soil, the degradation of BDE-209 reached 52.65%, which was higher than that in sterilized soil. About 37.76–53.74% of BDE-209 degradation was observed in different soil types after 15 days. In addition, it was confirmed in this study that the presence of Cu^{2+}, Cd^{2+} could enhance the remediation of soil contaminated with BDE-209, and the residues decreased by 69.20% and 54.65% for Cu^{2+} and Cd^{2+} treatment, respectively. However, the superior ability of white rot fungi to degrade BDE-209 was not obvious at low pollution level (<0.5 mg kg^{-1}).

Keywords: white rot fungi, decabrominated diphenyl ether (BDE-209), biodegradation, soil.

Reference to this paper should be made as follows: Wu, J.; Zhao, Y.; Liu, L.; Fan, B.; Li, M. 2013. Remediation of soil contaminated with decabrominated diphenyl ether using white rot fungi, *Journal of Environmental Engineering and Landscape Management* 21(3): 171–179. http://dx.doi.org/10.3846/16486897.2012.721374

Introduction

Decabrominated diphenyl ether (BDE-209) which is a member of the class of compounds well known as polybrominated diphenyl ethers (PBDEs) is widely used as flame retardants in producing fabrics and electronic products (Liu *et al.* 2005). With fast-developing electronics industry, outdated products are frequently discarded. Unsuitable handling of e-waste would not only pose risks to the surrounding environment, but also to human health. Pearl River Delta, Taizhou area and other regions in China became e-waste recycling areas. Chen *et al.* (2011) reported the amount of PBDEs of 0.625 mg kg^{-1}, in which the content of BDE-209 was in excess of 85%. Higher concentrations were observed by Yang *et al.* (2008) in soil of the Pearl River Delta, with BDE-209 levels range from 1.31 to 6319 ng kg^{-1}. Soil contaminated with PBDEs in some Europe countries such as Norway and Sweden, were reported by Hassanin (Hassanin *et al.* 2004). Because of its low water solubility, low volatility, high lipophilicity and other properties, BDE-209 is regarded as one of the persistent organic pollutants (POPs) (Tanabe 2004). Large number of experiments proved that PBDEs including BDE-209 persist in the environment for long time (de Wit 2002), and bioaccumulate and biomagnify as they move through the food chain (Law *et al.* 2006; Wolkers *et al.* 2004). They are released to the environment and converted into several toxic compounds through their photodecomposition and biologic degradation (Schenker *et al.* 2008; Watanabe, Sakai 2003; Robrock *et al.* 2008; He *et al.* 2006). Due to the production and use of BDE-209, and potential toxicity to humans, BDE-209 has raised public safety concerns.

It is reported that current technologies used to treat contaminated soil and sediment, such as photo degradation (Crosby *et al.* 1971), thermal desorption (De Percin 1995), biodegradation by fungi, base catalyzed dechloronation (Takada *et al.* 1996), plasma arc systems, have been rapidly developed to various extent (Rubin, Burhan 2006). However, little research work on biodegradation of BDE-209 in soil has been done.

White rot fungi have the unique ability to decompose lignin, and they are also capable of degrading some compounds with similar structure. Many studies on successful fungal remediation in soil have been reported (Eggen, Majcherczyk 1998; Kubatova *et al.* 2001; Ogbo, Okhuoya 2008). Although the main mechanism involved in the fungal degradation is not yet clear, it is believed that the produce of the ligninolytic systems composed of the lignin peroxidases (LiPs), manganese dependent peroxidases (MnPs) and

Corresponding author: Yuechun Zhao
E-mail: yczhao@scau.edu.cn

laccase might be the main reason (Reddy 1995). LiPs and MnPs directly involved in most of the oxidation reaction are extracellular, which offer the potential to oxidize substrates that have low water solubility like PBDEs. A positive correlation has been found between biodegradation and ligninolytic enzyme activity in soil. Thus, it is not surprising that the ligninolytic system of white rot fungi is useful for the remediation of soil contaminated with BDE-209.

The objective of this study is to investigate the biodegradation of BDE-209 in soil by white rot fungi. Since the potential could be influenced by a range of environmental factors in the laboratory, we have been studying the effects of different conditions (the amount of white rot fungi addition, ways of white rot fungi addition, pollution levels of BDE-209, sterilized soil and native soil, soil types, the presence of heavy metals) on biodegradation kinetics.

1. Methods

1.1. Chemicals and fungi

BDE-209 used throughout this study was purchased from Jinan Enter Chemical Co., LTD. The five BDE-209 standards at different concentrations (0.2 mg kg^{-1}, 0.5 mg kg^{-1}, 1 mg kg^{-1}, 2 mg kg^{-1} and 5 mg kg^{-1}) were purchased from South China Institute of Environmental Sciences (Guangzhou, China). Hexane, acetone, dichloromethane of analytical grade were obtained from Damao Chemical Reagent Factory (China).

White rot fungi was purchased from Guangzhou Chemical Company L.T.D. of Chinese Academy of Sciences (Guangzhou, China). The fungi was incubated at 28 °C on potato dextrose agar medium containing peeled potato 200 g, dextrose 20 g, KH_2PO_4 3 g, $MgSO_4$ 1.5 g, vitamin B_1 0.01 g, and agar 20 g in 1 L distillated water for 7 days before being used in the experiments of this study. In all cases, round pieces of 10 mm-diameter from the fungal lawn were cut and tested for the effects of different conditions on the biodegradation of BDE-209.

1.2. Soil

The experimental soil was collected from different places in Guangdong Province (China). They had different organic matter contents and pH. Their physical-chemical properties are shown in Table 1. Before being used, the soil was air-dried, passed through a 2 mm sieve.

To prepare BDE-209 contaminated soil, BDE-209 was added into a volumetric flask adjusting to 250 mL with hexane. Apply the BDE-209 solution into soil. The mixture of BDE-209 solution and soil was kept at room temperature, the indoor ventilated places for a period of four weeks. Keep the water-holding capacity at 15% throughout all experiments.

1.3. Degradation of BDE-209 by fungi

The effects of different factors, including the amount of white rot fungi addition, ways of white rot fungi addition, pollution levels of BDE-209, native soil and sterilized soil, soil types, and the presence of heavy metals, on the degradation of BDE-209 in soil by fungi were investigated in the laboratory.

At the beginning, different amounts of white rot fungi (3 pieces, 6 pieces, 9 pieces in 15 g of soil) were added into 15 g of soil contaminated with BDE-209 at the same initial concentrations of 4.6256 mg kg^{-1}, while the soil without fungi was used as a control.

In the experiment with one-step addition method, 6 pieces of white rot fungi were added into 15 g of soil on the first day. The control was without white rot fungi. In the experiment with two-step addition method, 3 pieces of white rot fungi were applied to 15 g of soil on the first day and another 3 pieces were added on the 7th day. The BDE-209 concentration in samples was 4.6256 mg kg^{-1}.

White rot fungi (6 pieces in 15 g of soil) were added into different soil with three different BDE-209 pollution levels. The three levels were Level 1, Level 2, and Level 3, which represented the concentrations of BDE-209 in soil were 0.4882 mg kg^{-1}, 2.0571 mg kg^{-1} and 4.6256 mg kg^{-1}. The controls were without fungi.

Experiments on biodegradation of BDE-209 in native and sterilized soil were carried out in the laboratory. Six pieces of white rot fungi were added into 15 g of sterilized soil that was sterilized in an autoclave at 1.2 MPa for 30 min or native soil. The controls were without fungi. The BDE-209 concentration in samples was 4.6256 mg kg^{-1}.

In this study, we used lateritic red soil, laterite, red soil, paddy soil and vegetable soil at the BDE-209 concentration of 4.6256 mg kg^{-1}, 4.7123 mg kg^{-1}, 4.6348 mg kg^{-1}, 4.6982 mg kg^{-1}, 4.6537 mg kg^{-1}, respectively. For each case, one degradation experiment with fungi (6 pieces in 15 g of soil) and one control experiment without fungi were conducted.

Table 1. Physical-chemical characteristics of experimental soil

Soil	Lateritic Red Soil	Red Soil	Laterite	Paddy Soil	Vegetable Soil
Fe (mg kg^{-1})	45.00	91.21	22.77	35.00	34.00
Organic matters (g kg^{-1})	0.45	4.26	3.15	6.27	6.66
pH	4.8	5.1	5.2	5.8	5.8

To study the effects of the presence of heavy metals on soil remediation, we added 6 pieces of fungi and Cu^{2+} or Cd^{2+} at the concentration of 0.5 mg kg^{-1} into 15 g of soil separately. The BDE-209 concentration was 4.7428 mg kg^{-1} in the experiments of Cu^{2+} treatment. The BDE-209 concentration was 4.8217 mg kg^{-1} in the experiments of Cd^{2+} treatment. A set of experiments without fungi was set up as control.

All the experiments were performed in triplicate for 15 days of incubation period. The BDE-209 residues were analyzed by GC-MS on 0 d, 3 d, 6 d, 9 d, 12 d, and 15 d.

1.4. Sample pretreatment

After adding ^{13}C-PCB141 and 2 g of copper for desulfurization, the samples were extracted with hexaneacetone (1:1) solution in a Soxhlet extraction apparatus for 48 h. And then, the extract liquid was transferred to a rotary evaporator, washed three times with hexane, concentrated to 2 mL. Subsequently, the concentrate was purified with a chromatography column filled with activated silica gel. The order of wet-loaded columns from top to bottom was: 6 cm of neutral alumina, 2 cm of neutral silica gel, 5 cm of alkali silica gel, 2 cm of neutral silica gel, 8 cm of acidic silica gel, and 1 cm of anhydrous sodium sulfate. Then the elution took place with 30 mL of hexane and 70 mL of hexanedichloromethane (1:1) solution in succession. Until there was no liquid dripping, the eluent should be concentrated to 1 mL, and kept at 4 °C in darkness.

1.5. GC-MS analysis

The GC-MS analysis was conducted on Thermo Fisher TRACE-DEQ gas chromatograph-mass spectrometer with column DB-5MS (15 m × 0.25 mm.i.d, 0.1 μm), and a NCI detector at temperature of 230 °C. The column temperature was programmed to increase from 110 to 300 °C at a flow rate of 10 °C min^{-1}. The carrier gas was high purity nitrogen (99.999%). The speed of gas flow was 1 mL min^{-1}. Continuous injection (1 μL) was applied. ^{13}C-PCB141 was used as a recovery indicator to control the recovery rate during the whole operation procedure. The recovery standard was used to control the sample recovery rate. The recovery rate of the indicator range from 78% to 88.6%, and that of BDE-209 was between 88.4% and 98.7%.

2. Results and discussion

2.1. Effects of different amount of white rot fungi addition

In this study, pieces of 10 mm-diameter from the fungal lawn were cut and tested for the effects on the removal of BDE-209 in samples with different amount of white rot fungi addition. An initial BDE-209 concentration of 4.6256 mg kg^{-1} was used for all these samples and other conditions were kept constant during the whole incubation period. Figure 1 shows that the residues of BDE-209 in soil declined and reduced by 50.07%, 52.65% and 52.74% for 3 pieces, 6 pieces, and 9 pieces of fungi, respectively during the whole incubation period.

The results indicated that the removal was directly related to the amount of inoculums in this experiment. With the increasing amount of fungi addition, the residues of BDE-209 decreased. However, the residues of BDE-209 in soil declined and reduced by 52.65% and 52.74% for 6 pieces, and 9 pieces of fungi, which were very close. It can be concluded that there was an optimal amount of added fungi in the degradation process. Thus, the optimal amount should be taken into account in practice to obtain efficiency as well as cost savings.

2.2. Effects of different ways of white rot fungi addition

The effects of addition ways of fungi were shown in Figure 2. In the experiment with one-step addition method, incubation of soil with an initial addition of 6 pieces of white rot fungi led to 45.16% removal of BDE-209 after 6 d, and it increased to 52.65% after 15 d. However, the soil treated with the two-step addition approach demonstrated a different situation. Three pieces of white rot fungi were added on the first day, and the residues decreased by 23.63% from 4.6256 mg kg^{-1} to 3.533 mg kg^{-1} after 6 days. Further experiments showed that residues continued to decrease when 3 more pieces of white rot fungi were added on the 7th day. The residues were decreased by 59.01% to 2.0639 mg kg^{-1}, which was lower than that in the soil treated with one-step approach (2.2146 mg kg^{-1}) during the 15 days of incubation period.

The way of white rot fungi addition does affect the biodegradation. The research (Zhao *et al.* 2008)

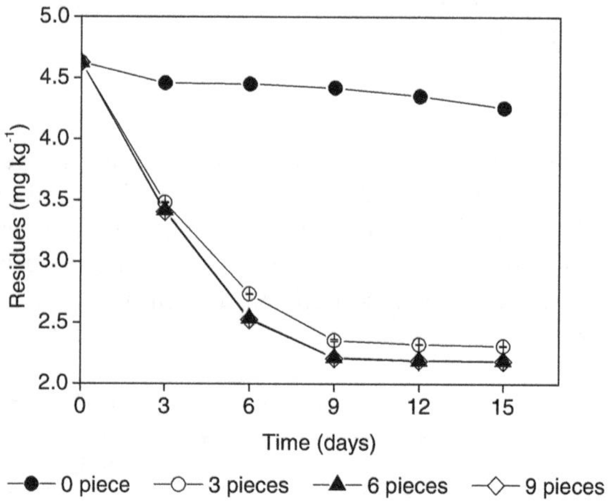

Fig. 1. Residues of BDE-209 in soil with different amount of white rot fungi addition on 0d, 3 d, 6 d, 9 d, 12 d, and 15 d. All data are means ± SE (n = 3)

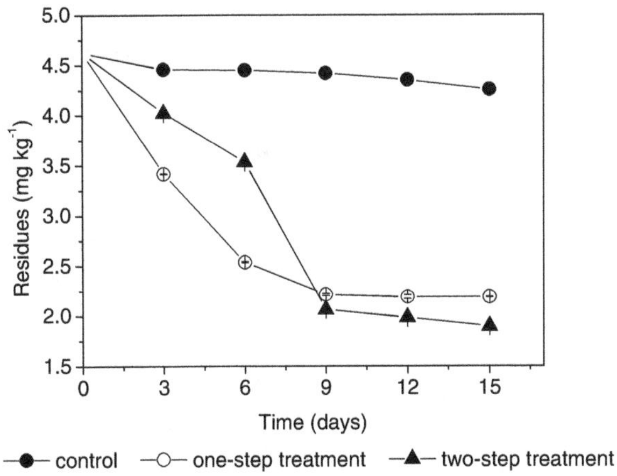

Fig. 2. Residues of BDE-209 in different ways of white rot fungi addition on 0 d, 3 d, 6 d, 9 d, 12 d, and 15 d. All data are means ± SE (n = 3)

on remediation of soil contaminated with DDTs by laccase extracted from white rot fungi showed that two-step method can bring a higher degradation rate compared with one-step method. Some other similar results were also reported (Hu *et al.* 2008). Various environmental conditions, such as soil temperature, mineral and metal ions, could affect the chemical degradation indirectly by affecting enzyme and micro-organisms' activity. The divided addition method can reduce the impact of these factors by splitting white rot fungi into 2 equal portions, and adding into soil separately. It may enhance the utilization of white rot fungi as well as biodegradation.

2.3. Degradation of BDE-209 at different pollution levels

All the data of the research based on the different concentrations of BDE-209 in soil were shown in Figure 3, which also indicated that a better biodegradation by white rot fungi occurred at higher concentration. The highest removal occurred at a BDE-209 concentration of 4.6256 mg kg^{-1} (Level 3) after 15 days, reduced by 52.65% from 4.6256 mg kg^{-1} to 2.1903 mg kg^{-1}. In the experiment of Level 2, the amount of BDE-209 decreased by 48.43% during the 15-day incubation period. In contrast, when the BDE-209 concentration was lower than 2.5 mg kg^{-1}, the degradation was worse. Figure 3 shows the removal of BDE-209 at an initial concentration of 0.4882 mg kg^{-1} (Level 1) after 15 days was reduced by only 11.04% from 0.4882 mg kg^{-1} to 0.4343 mg kg^{-1}.

It was reported that white rot fungi could effectively degrade PAHs; and the initial concentration of PAHs in 1 kg dry soil was up to 609.8 mg before any treatment (Sasek *et al.* 2003). Zhou *et al.* (2007) carried out a research on the degradation of BDE-209 in liquid culture medium at a concentration of 160 mg L^{-1}. It appears that the high concentration of pollutant is favorable for the degradation. In this study, however, the superior ability of white rot fungi to degrade BDE-209 was not obvious at low pollution level (<0.5 mg kg^{-1}).

2.4. Effects of native and sterilized soil

The experiments were carried out with native and sterilized soil. Incubation of soil led the removal of BDE-209 to decline rapidly in the first 6 days and then slowly (Fig. 4). White rot fungi induced 34.02% degradation of BDE-209 after 6 days, and then the degradation of BDE-209 increased to 44.8% with the residues of 2.5496 mg kg^{-1} in sterilized soil after 15 days. Nevertheless, an improved degradation was observed in native soil. A 6-day incubation in native soil led to 45.16% removal. After 15 days, the removal reached 52.65%, which was higher than that in sterilized soil.

Matthias (Kastner *et al.* 1998) investigated the degradation of PAHs in native and sterilized soil. The report showed that anthracene and phenanthrene degraded enormously in native soil, while only a slight

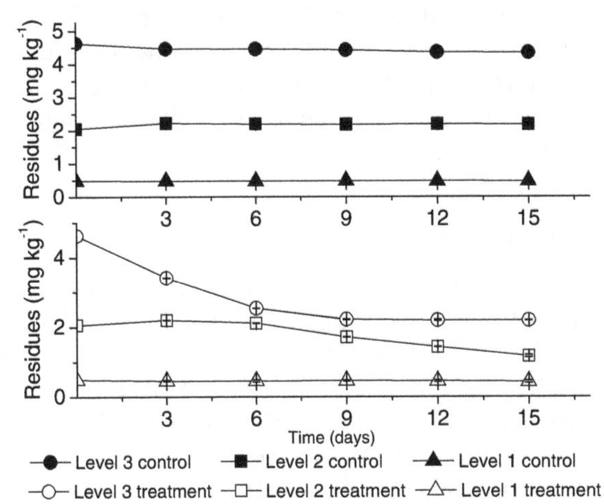

Fig. 3. Residues of BDE-209 in soil at different pollution levels on 0 d, 3 d, 6 d, 9 d, 12 d, and 15 d. All data are means ± SE (n = 3)

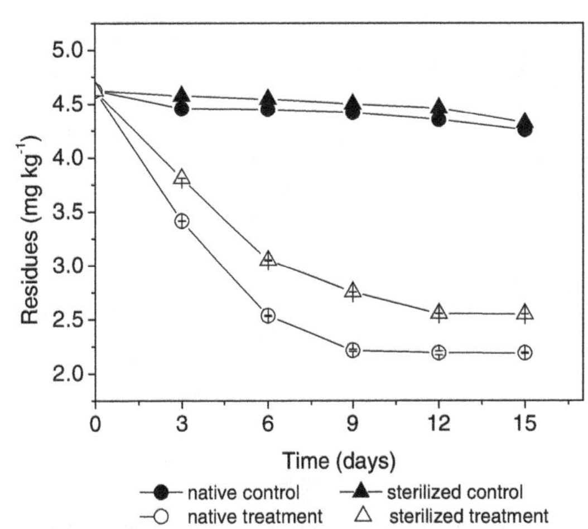

Fig. 4. Residues of BDE-209 in native and sterilized soil on 0 d, 3 d, 6 d, 9 d, 12 d, and 15 d. All data are means ± SE (n = 3)

abiotic loss was observed in sterilized soil during the whole incubation period. However, the experiments designed by Bishnoi (Bishnoi *et al.* 2008) showed that maximum degradation of PAHs in sterilized soil was higher than that in unsterilized native soil. Obviously, the conditions for biodegradation were quite different in sterilized and native soil. Although the mechanism is still unclear, we concluded that the key influencers on the biodegradation may be pH value, soil moisture, organic matter content, indigenous microorganisms and so on. In native soil, there may be competitions between introduced and indigenous microorganisms, when fungi were inoculated.

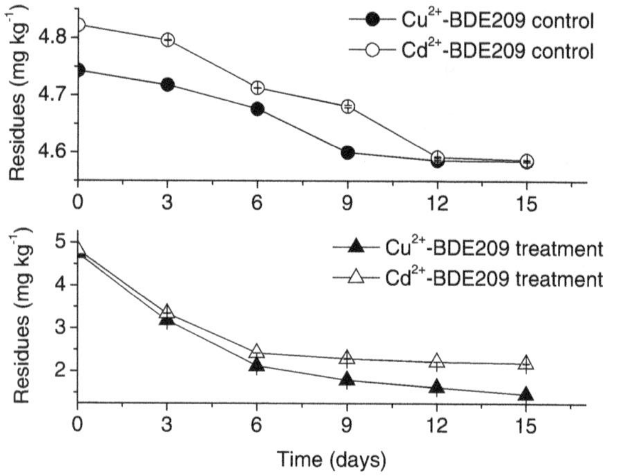

Fig. 6. Residues of BDE-209 in soil in the presence of heavy metals on 0 d, 3 d, 6 d, 9 d, 12 d, and 15 d. All data are means $\pm$SE (n $=3$)

2.5. Degradation of BDE-209 in different soils

Lateritic red soil, laterite, red soil, paddy soil and vegetable soil were used in this study. During 15 days of incubation, the residues of BDE-209 in soil decreased by 52.65%, 47.92%, 52.94%, 53.74%, 37.76% for lateritic red soil, laterite, red soil, paddy soil and vegetable soil, respectively (Fig. 5). The increasing order of BDE-209 removal was paddy soil > red soil > lateritic red soil > laterite > vegetable soil. In addition, the paddy soil was flooded for a long time, leading to a lower redox potential (Eh). The reductive soil may benefit the degradation of BDE-209 by white rot fungi.

2.6. Effects of the presence of heavy metals

A comparison of the degradation results for Cu^{2+} and Cd^{2+}-BDE209 appears in Figure 6. As shown in Figure 6, the residues of BDE-209 in soil incubated with white rot fungi decreased by 69.20% and 54.65% for Cu^{2+} and Cd^{2+} treatment, respectively.

The fungal metabolism requires trace amounts of many heavy metals. Such as copper, iron, manganese, molybdenum, zinc, and nickel are necessary for fungal growth, while chromium, cadmium, lead, mercury and silver are nonessential metals (Gadd 1994). In this study, we investigated whether the biodegradation can be enhanced as a result of the presence of copper and cadmium. In white rot fungi, the produce of the ligninolytic systems, which composed of LiPs, MnPs, and laccase, are useful for the remediation of environmental contaminants (Call, Muecke 1997). Copper, which served as a cofactor in the catalytic center of laccase, directly participates in the process of lignin degradation (Baldrian 2003). On the other hand, it has been reported that the presence of copper could improve the production of laccase (Tychanowicz *et al.* 2006; Tinoco *et al.* 2011). Thus, the presence of copper was likely to enhance the biodegradation. However, Cd^{2+} is in general the most toxic metal for white rot fungi, the presence of Cd^{2+} may inhibit fungal growth (Baldrian 2003). Atagana (2009) investigated biodegradation of PAHs by fungi in contaminated soil containing cadmium; the fungi tested were capable of removing PAHs in the presence of cadmium. One possible explanation for this is that the addition of Cd^{2+} and Cu^{2+} leads to an increase of laccase activity. The results demonstrated that the presence of Cu^{2+}, Cd^{2+} can enhance the remediation of soil contaminated with BDE-209.

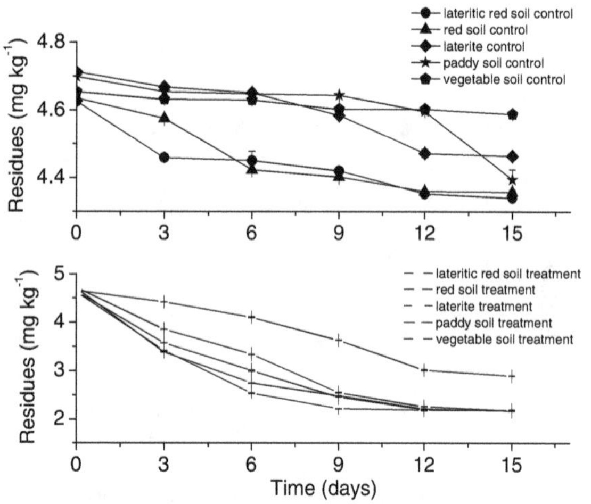

Fig. 5. Residues of BDE-209 in lateritic red soil, laterite, red soil, paddy soil and vegetable soil on 0 d, 3 d, 6 d, 9 d, 12 d, and 15 d. All data are means $\pm$SE (n $=3$)

2.7. Kinetic studies

In this study, the degradation of BDE-209 in soil fits the pseudo-first-order kinetics and follows the equation of $\ln C_t = -kt + \ln C_0$, where t is the degradation time, C_0 is the initial concentration of BDE-209, C_t is the concentration of BDE-209 on the t days, k is the degradation rate constant, and the half-life of BDE-209 is $0.693/k$ (Tables 2–7).

The residues of BDE-209 in soil treated with white rot fungi decreased rapidly during the first 6 days and then slowly during the 15-day incubation period. There was a relation between the half-life of BDE-209 and the amount of white rot fungi addition. Higher amount of white rot fungi in soil led to lower half-life. But there were no apparent declines in half-life when the addition was more than 6 pieces. In soil treated with 6 and 9 pieces of white rot fungi, the half-life was about 14.14 days, which was lower than that in soil treated with 0 and 3 pieces fungi (Table 2). Two-step method had a lower half-life than the one-step method, and their half-life values of BDE-209 were 10.21 days and 14.14 days, respectively (Table 3). The half-life values were 14.14, 15.10, 117.46 days for soil with initial BDE-209 concentration of 4.6256, 2.2573, 0.4882 mg kg^{-1}, respectively. Obviously, we cannot achieve a better biodegradation using white rot fungi at low pollution level (< 0.5 mg kg^{-1}) (Table 4). The data indicated there was a faster rate in native soil than sterilized soil, which was reflected in the half-life values, where $t_{1/2}$ values were 14.14 and 16.99 days, respectively (Table 5). This strongly supported that native soil was more favorable to remediation of BDE-209 in the laboratory. The half-life increased in the order red soil (13.46 days) < paddy soil (13.92 days) < lateritic red soil (14.14 days) < laterite (15.10 days) < vegetable soil (20.03 days) (Table 6). Compared with the half-life (13.89 days) of Cd^{2+}, a lower half-life (9.02 days) was figured out in soil treated with Cu^{2+} (Table 7).

The findings were consistent with many investigators (Tayal *et al.* 1999; Huang *et al.* 2004) who reported that the biodegradation for PCP using white rot fungi follows the pseudo-first-order kinetics, and the residues decreased sharply in the first 6 days. Studies showed

Table 2. Kinetic parameters of degradation of BDE-209 with different amount of white rot fungi addition

Degradation experiment	Pseudo-first-order kinetic equation $\ln C_t = -kt + \ln C_0$	Half-life/d	Degradation rate constant/d	Regression coefficient R
Control	$\ln C_t = -0.0047t + 1.523$	138.60	0.0047	0.910
3 pieces treatment	$\ln C_t = -0.046t + 1.399$	15.07	0.046	0.838
6 pieces treatment	$\ln C_t = -0.049t + 1.381$	14.14	0.049	0.816
9 pieces treatment	$\ln C_t = -0.049t + 1.378$	14.14	0.049	0.813

Table 3. Kinetic parameters of degradation of BDE-209 in different ways of white rot fungi addition

Degradation experiment	Pseudo-first-order kinetic equation $\ln C_t = -kt + \ln C_0$	Half-life/d	Degradation rate constant/d	Regression coefficient R
Control	$\ln C_t = -0.0047t + 1.523$	147.45	0.0047	0.910
One-step treatment	$\ln C_t = -0.049t + 1.381$	14.14	0.049	0.816
Two-step treatment	$\ln C_t = -0.0679t + 1.550$	10.21	0.0679	0.902

Table 4. Kinetic parameters of degradation of BDE-209 at different pollution levels

Degradation experiment	Pseudo-first-order kinetic equation $\ln C_t = -kt + \ln C_0$	Half-life/d	Degradation rate constant/d	Regression coefficient R
5 mg kg^{-1} control	$\ln C_t = -0.0047t + 1.523$	147.45	0.0047	0.910
5 mg kg^{-1} treatment	$\ln C_t = -0.049t + 1.381$	14.14	0.049	0.816
2.5 mg kg^{-1} control	$\ln C_t = -0.0020t + 0.808$	138.60	0.0020	0.799
2.5 mg kg^{-1} treatment	$\ln C_t = -0.0459t + 0.099$	15.10	0.0459	0.922
0.5 mg kg^{-1} control	$\ln C_t = -0.0033t + 0.720$	210.00	0.0033	0.810
0.5 mg kg^{-1} treatment	$\ln C_t = -0.0059t + 0.732$	117.46	0.0059	0.760

Table 5. Kinetic parameters of degradation of BDE-209 in native and sterilized soil

Degradation experiment	Pseudo-first-order kinetic equation $\ln C_t = -kt + \ln C_0$	Half-life/d	Degradation rate constant/d	Regression coefficient R
Native soil control	$\ln C_t = -0.0047t + 1.523$	147.45	0.0047	0.910
Native soil treatment	$\ln C_t = -0.049t + 1.381$	14.14	0.049	0.816
Sterilized soil control	$\ln C_t = -0.0029t + 1.531$	238.97	0.0029	0.997
Sterilized soil treatment	$\ln C_t = -0.0408t + 1.451$	16.99	0.0408	0.898

Table 6. Kinetic parameters of degradation of BDE-209 in various soils

Degradation experiment	Pseudo-first-order kinetic equation $\ln C_t = -kt + \ln C_0$	Half-life/d	Degradation rate constant/d	Regression coefficient R
Lateritic soil control	$\ln C_t = -0.0047t + 1.523$	147.45	0.0047	0.910
Lateritic soil treatment	$\ln C_t = -0.049t + 1.381$	14.14	0.049	0.816
Red soil control	$\ln C_t = -0.0044t + 1.527$	157.50	0.0044	0.874
Red soil treatment	$\ln C_t = -0.0515t + 1.450$	13.46	0.0515	0.935
Laterite control	$\ln C_t = -0.0039t + 1.554$	177.69	0.0039	0.940
Laterite treatment	$\ln C_t = -0.0459t + 1.487$	15.10	0.0459	0.909
Paddy soil control	$\ln C_t = -0.0035t + 1.554$	198.00	0.0035	0.690
Paddy soil treatment	$\ln C_t = -0.0498t + 1.416$	13.92	0.0498	0.894
Vegetable soil control	$\ln C_t = -0.0037t + 1.542$	187.30	0.0037	0.971
Vegetable soil treatment	$\ln C_t = -0.0346t + 1.575$	20.03	0.0346	0.963

Table 7. Kinetic parameters of degradation of BDE-209 in the presence of heavy metals

Degradation experiment	Pseudo-first-order kinetic equation $\ln C_t = -kt + \ln C_0$	Half-life/d	Degradation rate constant/d	Regression coefficient R
Cu^{2+}-BDE-209 control	$\ln C_t = -0.0026t + 1.535$	266.54	0.0026	0.919
Cu^{2+}-BDE-209 treatment	$\ln C_t = -0.077t + 1.556$	9.02	0.077	0.910
Cd^{2+}-BDE-209 control	$\ln C_t = -0.0037t + 1.451$	187.30	0.0037	0.782
Cd^{2+}-BDE-209 treatment	$\ln C_t = -0.0499t + 1.395$	13.89	0.0499	0.963

that the kinetics of biodegradation was greatly improved by the presence of manganese ions, H_2O_2 and glucose in the medium. In our experiments, the presence of Cu^{2+} and Cd^{2+}, which has positive effect on the production of laccase, enhanced the biodegradation. This strongly suggests the involvement of ligninolytic systems mechanism in the process of biodegradation.

Conclusions

1. By using white rot fungi, a rapid and efficient remediation of soil contaminated with BDE-209 could be expected. The biodegradation fits the pseudo-first-order kinetics under different experimental conditions. The residues of BDE-209 in soil treated with white rot fungi decreased rapidly during the first 6 days and then slowly during the 15-day incubation period.

2. In this study, 6 pieces of added fungi for 15 g of soil are optimal amounts in the degradation process.

3. White rot fungi have a good degradation ability with one-step and two-step addition method.

4. In native soil, the removal rate reached 52.65%, which was higher than that in sterilized soil.

5. For 15 days of incubation, 52.65%, 52.94%, 47.92%, 53.74%, 37.76% of BDE-209 was degraded by white rot fungi for lateritic red soil, laterite, red soil, paddy soil and vegetable soil, respectively.

6. The presence of Cu^{2+}, Cd^{2+} can both enhance the remediation of soil contaminated with BDE-209. The residues of BDE-209 decreased by 69.20% and 54.65% for Cu^{2+} and Cd^{2+} treatment, respectively.

Acknowledgements

This project was supported by Pollution Prevention and Control New Technology Development Projects of Guangzhou Environmental Protection Bureau of China (Suihuanyanfa 2010 No. 002).

References

Atagana, H. I. 2009. Biodegradation of PAHs by fungi in contaminated-soil containing cadmium and nickel ions, *African Journal of Biotechnology* 8: 5780–5789.

Baldrian, P. 2003. Interactions of heavy metals with white-rot fungi, *Enzyme and Microbial Technology* 32: 78–91. http://dx.doi.org/10.1016/S0141-0229(02)00245-4

Bishnoi, K.; Kumar, R.; Bishnoi, N. R. 2008. Biodegradation of polycyclic aromatic hydrocarbons by white rot fungi Phanerochaete chrysosporium in sterile and unsterile soil, *Journal of Scientific & Industrial Research* 67: 538–542.

Call, H. P.; Muecke, I. 1997. History, overview and applications of mediated lignolytic systems, especially laccase-mediator-systems (Lignozymprocess), *Journal of Biotechnology* 53: 163–202. http://dx.doi.org/10.1016/S0168-1656(97)01683-0

Chen, T.; Zhou, C.; Mou, Y. J.; Yu, B. B. 2011. PBDEs pollution of soils in a typical e-waste disposal site and its surrounding area, *Journal of Ecology and Rural Environment* 3: 20–24.

Crosby, D. G.; Wong, A. S.; Plimmer, J. R.; Woolson, E. A. 1971. Photodecomposition of chlorinated dibenzo-p-dioxins, *Science (New York, N.Y.)* 173(3998): 748–749. http://dx.doi.org/10.1126/science.173.3998.748

De Percin, P. R. 1995. Application of thermal desorption technologies to hazardous waste sites, *Journal of Hazardous Materials* 40(2): 203–209. http://dx.doi.org/10.1016/0304-3894(94)00085-U

De Wit, C. A. 2002. An overview of brominated flame retardants in the environment, *Chemosphere* 46: 583–624. http://dx.doi.org/10.1016/S0045-6535(01)00225-9

Eggen, T.; Majcherczyk, A. 1998. Removal of polycyclic aromatic hydrocarbons (PAH) in contaminated soil by white rot fungus Pleurotus ostreatus, *International Biodeterioration and Biodegradation* 41: 111–117. http://dx.doi.org/10.1016/S0964-8305(98)00002-X

Gadd, G. M. 1994. Interactions of fungi with toxic metals, *New Phytol* 124: 25–60. http://dx.doi.org/10.1111/j.1469-8137.1993.tb03796.x

Hassanin, A.; Breivik, K.; Meijer, S. N.; Steinnes, E.; Thomas, G. O.; Jones, K. C. 2004. PBDEs in European background soils: levels and factors controlling their distribution, *Environmental Science & Technology* 38: 738–745. http://dx.doi.org/10.1021/es035008y

He, J.; Robrock, K. R.; Alvarez-Cohen, L. 2006. Microbial reductive debromination of polybrominated diphenyl ethers (PBDEs), *Environmental Science & Technology* 40: 4429–4434. http://dx.doi.org/10.1021/es052508d

Hu, L. F.; Yao, J.; Lou, B.; He, R.; Shen, D. S. 2008. Fenton technique for oxidation treatment of solid-waste containing aniline, *Environmental Science* 1: 104–108.

Huang, J.; Yu, G.; Cheng, J.; Liu, L.; Zheng, Y. Q. 2004. Kinetic analysis for biodegradation process of pentachlorophenol in the presence of white rot fungus, *Journal of Agro-environmental Science* 1: 167–169.

Kastner, M.; Breuer-Jammali, M.; Mahro, B. 1998. Impact of inoculation protocols, salinity, and pH on the degradation of polycyclic aromatic hydrocarbons (PAHs) and survival of PAH-degrading bacteria introduced into soil, *Applied and Environmental Microbiology* 64: 359–362.

Kubatova, A.; Erbanova, P.; Eichlerova, I.; Homolka, L.; Nerud, F.; Sasek, V. 2001. PCB congener selective biodegradation by the white rot fungus Pleurotus ostreatus in contaminated soil, *Chemosphere* 43: 207–215. http://dx.doi.org/10.1016/S0045-6535(00)00154-5

Law, K.; Halldorson, T.; Danell, R.; Stern, G.; Gewurtz, S.; Alaee, M.; Marvin, C.; Whittle, M.; Tomy, G. 2006. Bioaccumulation and trophic transfer of some brominated flame retardants in a Lake Winnipeg (Canada) food web, *Environmental Toxicology and Chemistry* 25: 2177–2186. http://dx.doi.org/10.1897/05-500R.1

Liu, H. X.; Zhang, Q. H.; Jiang, G. B. 2005. Polybrominated diphenyl ethers and its related environmental problems, *Progress in Chemistry* 3: 554–562.

Ogbo, E. M.; Okhuoya, J. A. 2008. Biodegradation of aliphatic, aromatic, resinic and asphaltic fractions of crude oil contaminated soils by Pleurotus tuber-regium Fr. Singer – a white rot fungus, *African Journal of Biotechnology* 7: 4291–4297.

Reddy, C. A. 1995. The potential for white-rot fungi in the treatment of pollutants, *Current Opinion in Biotechnology* 6: 320–328. http://dx.doi.org/10.1016/0958-1669(95)80054-9

Robrock, K. R.; Korytar, P.; Alvarez-Cohen, L. 2008. Pathways for the anaerobic microbial debromination of polybrominated diphenyl ethers, *Environmental Science & Technology* 42: 2845–2852. http://dx.doi.org/10.1021/es0720917

Rubin, E.; Burhan, Y. 2006. Noncombustion technologies for remediation of persistent organic pollutants in stockpiles and soil, *Remediation Journal* 16: 23–42. http://dx.doi.org/10.1002/rem.20099

Sasek, V.; Cajthaml, T.; Bhatt, M. 2003. Use of fungal technology in soil remediation: a case study, *Water, Air and Soil Pollution: Focus* 3(3): 5–14. http://dx.doi.org/10.1023/A:1023990020199

Schenker, U.; Soltermann, F.; Scheringer, M.; Hungerbuehler, K. 2008. Modeling the environmental fate of polybrominated diphenyl ethers (PBDEs): the importance of photolysis for the formation of lighter PBDEs, *Environmental Science & Technology* 42: 9244–9249. http://dx.doi.org/10.1021/es801042n

Takada, S.; Nakamura, M.; Matsueda, T.; Kondo, R.; Sakai, K. 1996. Degradation of polychlorinated dibenzo-p-dioxins and polychlorinated dibenzofurans by the white rot fungus Phanerochaete sordida YK-624, *Applied and Environmental Microbiology* 62(12): 4323–4328.

Tanabe, S. 2004. PBDEs, an emerging group of persistent pollutants – editorial, *Marine Pollution Bulletin* 49: 369–370. http://dx.doi.org/10.1016/j.marpolbul.2004.07.015

Tayal, A. K.; Das, L.; Kaur, I. 1999. Biodegradation of pentachlorophenol (PCP) by white rot fungal strains screened from local sources and its estimation by high-performance liquid chromatography, *Biomedical Chromatography: BMC* 13: 220–224.

Tinoco, R.; Acevedo, A.; Galindo, E.; Serrano-Carreon, L. 2011. Increasing Pleurotus ostreatus laccase production by culture medium optimization and copper/lignin synergistic induction, *Journal of Industrial Microbiology & Biotechnology* 38: 531–540. http://dx.doi.org/10.1007/s10295-010-0797-3

Tychanowicz, G. K.; De Souza, D. F.; Souza, C. G. M.; Kadowaki, M. K.; Peralta, R. M. 2006. Copper improves the production of laccase by the white-rot fungus Pleurotus pulmonarius in solid state fermentation, *Brazilian Archives of Biology and Technology* 49: 699–704. http://dx.doi.org/10.1590/S1516-89132006000600002

Watanabe, I.; Sakai, S. 2003. Environmental release and behavior of brominated flame retardants, *Environment International* 29: 665–682. http://dx.doi.org/10.1016/S0160-4120(03)00123-5

Wolkers, H.; van Bavel, B.; Derocher, A. E.; Wiig, O.; Kovacs, K. M.; Lydersen, C.; Lindstrom, G. 2004. Congener-specific accumulation and food chain transfer of polybrominated diphenyl ethers in two Arctic food chains, *Environmental Science & Technology* 38: 1667–1674. http://dx.doi.org/10.1021/es030448a

Yang, Z. Y.; Zheng, J.; Chen, S. J.; Yuan, J. G.; Luo, X. J.; Luo, Y.; Mai, B. X. 2008. Advantage of researches on environmental pollutions from e-waste treating activities in Guangdong, China, *Asian Journal of Ecotoxicology* 6: 533–544.

Zhao, Y. C.; Fu, R.; Mo, C. H.; Yi, X. Y. 2008. Remediation of DDT contaminated soil by using laccase in different application ways, *Transactions of the Chinese Society of Agricultural Engineering* 2: 107–110.

Zhou, J.; Jiang, W.; Ding, J.; Zhang, X.; Gao, S. 2007. Effect of Tween 80 and beta-cyclodextrin on degradation of decabromodiphenyl ether (BDE-209) by white rot fungi, *Chemosphere* 70: 172–177. http://dx.doi.org/10.1016/j.chemosphere.2007.06.036

Junqin WU. Master student (since 2009), Department of Applied Chemistry, South China Agricultural University (SCAU), Guangzhou, China. Bachelor of Science (applied chemistry), Guangxi University for Nationalities, 2009.

Yuechun ZHAO. Dr, Professor, Department of Applied Chemistry, SCAU. PhD (chemical engineering), South China University of Technology, 2000. Master of Science (physical chemistry), Northeast Normal University, 1994.

Lu LIU. Master of Science (applied chemistry), SCAU, 2010.

Biao FAN. Master student (since 2010), Department of Applied Chemistry, SCAU, Guangzhou, China.

Minghua LI. Lecturer, Department of Applied Chemistry, SCAU. Master of Science (physical chemistry), Ji'nan University, Guangzhou, China, 1995.

10

CHANGES OF GROUND VEGETATION, SOIL CHEMICAL PROPERTIES AND MICROBIOTA FOLLOWING THE SURFACE FIRES IN SCOTS PINE FORESTS

Vitas Marozas[1], Kęstutis Armolaitis[2], Jūratė Aleinikovienė[3]

[1]*Department of Ecology, Aleksandras Stulginskis University, Studentų g. 11, Akademija, LT-53361 Kaunas distr., Lithuania*

[2, 3]*Department of Ecology Institute of Forestry, Lithuanian Research Centre for Agriculture and Forestry (LRCAF), Liepų g. 1, Girionys, LT-53101 Kaunas distr., Lithuania*

[3]*Department of Soil Science and Plant Nutrition, Aleksandras Stulginskis University, Studentų g. 11, Akademija, LT-53361 Kaunas distr., Lithuania*

E-mails: [1]vitas.marozas@lzuu.lt (corresponding author); [2]k.armolaitis@mi.lt; [3]j.aleinikoviene@mi.lt,

Abstract. The aim of the study was to investigate the influence of low severity surface fires on the ground vegetation, soil chemical changes and soil microbiota in Scots pine stands on sandy soils (*Arenosols*). The study was conducted in the eastern part of Lithuania (55° 35'N, 26° 07'E). The annual investigations in 1–4-year-old burned sites showed that low severity surface fires mostly affected the above-ground part of the ground vegetation. The influence of surface fire on soil chemical properties and soil microbiota was minor. Only slight increases in pH and in the content of total N in soil organic layer were detected. Concentrations of mobile K_2O and heavy metals (Cd, Cr, Cu, Mn, Zn and Fe) slightly increased in the mineral topsoil. The actinobacteria abundance increased in the soil organic layer and the mineral topsoil of the burned sites. The abundance of micromycetes decreased in the mineral topsoil after the surface fires.

Keywords: surface forest fire, *Pinus sylvestris*, ground vegetation, soil, heavy metals, soil microbiota.

Introduction

Nowadays, forest fires are recognized as an important ecological factor affecting vegetation structure and composition, energy fluxes and biogeochemical processes in boreal and hemiboreal forest landscapes (Shugart *et al.* 1992; Parviainen 1996; Angelstam 1998; Bergeron *et al.* 2002; Kuuluvainen 2002; Ryan 2002; Wallenius *et al.* 2007). Modern forest management should consider the impact of fire as the increases in the frequency and intensity of natural and anthropogenic forest fires are expected to occur in the coming decades as a consequence of global climate changes (Päätalo 1998; IPCC 2007; Flannigan *et al.* 2009).

The effect of fires on forest ecosystems depends on intensity and duration of the fire (Ice *et al.* 2004; Certini 2005). Severe crown fires can eliminate above-ground biomass, change successional rates and alter vegetation species composition, belowground physical, chemical and microbial processes. Fire changes forest soils properties including increased bulk density and altered physical structure (Boyer, Miller 1994; Arocena, Opio 2003), increased soil cation stocks (Liechty *et al.* 2005; Neff *et al.* 2005), and decreased carbon (C) and nitrogen (N) stocks in soils (Choromanska, DeLuca 2001; Carter, Foster 2004; MacKenzie *et al.* 2004; Certini 2005). In addition, during severe crown fires some heavy metals could release from the ash to the soil (Breulmanna *et al.* 2002; Pereira, Úbeda 2010). In excessive amounts they could have negative effects on forest ecosystem processes (Alloway 1995; Vasarevičius, Greičiūtė 2004; Ignatavičius *et al.* 2006; Baltrėnaitė, Butkus 2007; Pundytė *et al.* 2011).

Surface forest fires mainly affect the species composition of the ground vegetation; can promote an herbaceous flora, tree regeneration conditions and short-time increase of plant available nutrients in the soil (Parviainen 1996; Granström 2001; Gromtsev 2002; Ryoma, Laaka-Lindberg 2005; Jayen *et al.* 2006; Marozas *et al.* 2007; Parro *et al.* 2009). The effects of surface fires on soil chemical properties are generally relatively minor. In fact, the mineral soils appear to be unchanged in the face of low-intensity surface forest fires (Richter *et al.* 1982; Ferran *et al.* 2005; DeLuca, Sala 2006; Neill *et al.* 2007).

Fires have effect on soil microbial abundance and composition (Tateishi, Horikoshi 1995; Mabuhay *et al.* 2006). It is focused that forest fires induce the changes in microbial populations mainly through the shift of soil moisture content (Hanson *et al.* 2000; Pietikainen *et al.* 2000) and through the increased supply of organic C and N from the burned soil organic layer to the mineral topsoil (Klopatek *et al.* 1988; Ojima *et al.* 1994; Neary *et al.* 1999; Simard *et al.* 2001; Neff *et al.* 2005). Mineral topsoil drying followed by soil moisture retention and capacity reduction on behalf of the burned vegetation and underlying soil organic layer decreases the abundance of microbial communities (Bääth *et al.* 1995; Pietikainen *et al.* 2000). Thus, proceeding into the stationary phase, the ability of recovering and recolonizing of microbial communities is declined (Prieto-Fernandez *et al.* 1998; Wutrich *et al.* 2002).

Forest fires induce the loss of organic matter along the burning of the ground vegetation and soil organic layer. As the potential organic matter input to soil is decreased, the decrease in the abundance of microbial communities could be also estimated (Almendros *et al.* 1990). Thus, in mineral soil due to the increase of below-ground root mass the microbial abundance could increase in burned stands (Ojima *et al.* 1994). Regarding the changes in soil moisture as well as soil C and N content, the mostly reduced microbial population in the burned stands could be soil micromycetes, thus soil bacteria and actinobacteria, in contrast, could be the most abundant in the burned stands (Klopatek *et al.* 1988; Prieto-Fernandez *et al.* 1998; Joos *et al.* 2001).

Mainly the observations on surface fire impact to vegetation of pine forest ecosystems were presented in European hemiboreal forest zone (Zackrisson 1977; Marozas *et al.* 2007; Parro *et al.* 2009). Moreover, there is a lack of more complex investigations on the vegetation reestablishment, the changes in soil chemical properties and microbial population that occur after surface fires.

In Lithuania the annual number of forest fires is about 700 (from 200 to 1600 per year) (LME/SFS 2010). The total burned forest area ranges from 100 to 700 ha annually with an average burned area per one fire of 0.45 ha. Even 84% of fires emerge in Scots pine forests. The most common are surface fires (97.3%), while crown fires and underground fires amount only to 1% and 1.7%, respectively.

The aim of the study was to investigate the influence of low severity surface fires on the development of ground vegetation, soil chemical changes and soil microbiota in Scots pine (*Pinus sylvestris*) stands on sandy soils (*Arenosols*).

1. Materials and Methods

1.1. Study site

The study area was located in the eastern part of Lithuania (Zarasai district) (55° 35'N, 26° 07'E) and it falls in the transitional deciduous coniferous mixed forest hemiboreal zone of Europe (Ahti *et al.* 1968) (Fig. 1).

Fig. 1. Map of study area

The altitude above sea level is about 150–180 meters. The mean annual temperature ranges from +5.4 to +5.8 °C, with a mean January (coldest month) temperature of –6.4 °C and a mean July (warmest month) temperature of 16.9 °C. Annual mean precipitation is between 600 and 700 mm. Period with permanent snow cover continues from 100 to 110 days (Bukantis 1994). Hilly landscape, sandy soils and pure Scots pine (*Pinus sylvestris*) stands prevail in the forests of the study area.

The study was carried out in 60-year-old pure Scots pine stands with the undergrowth of Norway spruce (*Picea abies*). All studied stands were growing on nutrient-poor sandy *Arenosols* (forest type - *Vaccinio - myrtillo Pinetum*). In the ground vegetation cover prevail: *Vaccinium myrtillus*, *V. vitis-idaea*, *Calluna vulgaris*, *Festuca ovina*, *Linaria vulgaris*, *Luzula pilosa* and *Melampyrum pratense* in the dwarf shrub and herb layer, and *Dicranum polysetum*, *D. scoparium*, *Hylocomium splendens* and *Pleurozium schreberi* in the moss layer. In these stands surface fires occurred in the end of April of 2006 and 2009. Since the fires were of low severity, Scots pine trees were not damaged, while Norway spruce undergrowth and shrubs were totally killed and the ground vegetation cover was burned. The area of fires was about 60 ha in 2006 and about five ha in 2009.

1.2. Ground vegetation study and soil sampling

In total 4 permanent transects (20×1 m) with 20 sampling plots (1×1 m) were established for the ground vegetation study in the burned site (the surface fire occurred in April of 2006) and in the untouched nearby not-burned site (control) of Scots pine stand. Vegetation studies were conducted annually in June-July of 2006–2009. Each year species composition (species names according to Jankevičiene 1998) and projection cover (in per cent) of shrubs, saplings, dwarf shrubs, herbs and mosses were recorded in transects.

The soil sampling for the estimation of soil chemical properties and evaluation of the abundance of soil microbiota was carried out in July, 2009 in two burned sites in which surface fires occurred one and four years ago (in April of 2006 and 2009, respectively) and in untouched control sites. In each site, three composite samples of the organic layer (forest litter) and three samples from 0–5 cm mineral topsoil were collected at nine systematically distributed points along 16 meters transects. The organic layer was sampled using a 1000 cm^2 metallic circular frame. Then samples were dried at 105 °C and oven-dry weight (kg m^{-2}) was determined. Mineral soil was sampled with metallic auger of 2.5 cm diameter.

1.3. Analyses of soil chemical properties and soil microbiota

For chemical analyses, the organic and mineral soil samples were dried at 40 °C. Mineral soil samples were sieved through a 2×2 mm sieve. The mobile potassium (K_2O) and mobile phosphorus (P_2O_5) were determined in soil samples using the Egner-Riehm-Domingo (A-L) method (Egner *et al.* 1960); and pH was potentiometrically measured in a 1 M KCl suspension (ISO 10390). Cadmium was determined by flame and electrothermal atomic absorption spectrometric method (ISO 11047-9), whereas Cr, Pb, Ni, Cu, Mn, Zn and Fe were determined by inductively coupled plasma - atomic emission spectrometry (ICP-AES) (ISO 22036:2008). These analyses were conducted in the Agrochemical Research Laboratory of LRCAF. The total nitrogen (N) (ISO 11261) and organic carbon (org. C) (ISO 10694) concentrations were determined using the ECS 4010 analyser (COSTECH) in the Ecology Department of Institute of Forestry of LRCAF. Results were calculated on the mass of dry soil.

Soil microbiota was determined by the standard method of the sowing of organic or mineral soil suspension on the agarised nutrient medium (Thompson, Vincent 1967). Well homogenized 5 g of natural moist soil was suspended for one hour in the 50 ml of sterile water. Soil solutions for the sowing were gradually diluted by 10, 100 and 1000 times. The sowing was performed in three replicates using standard media agar (Plate-Count-Agar/standard methods) for the estimation of total soil microbiota abundance (Annand *et al.* 2003). Microbiota plates were incubated at 27 °C. After 5–7 days the morphology of the bacteria, actinobacteria and micromycetes colonies on plates was assessed by the needle strike and the optical microscopy. The count of colony forming units (CFU) of the microbiota was performed and the abundance of microbiota was counted per 1 g of the constant dry (in 90 °C) weight of the soil (according to Bradshaw 1992; Hassen *et al.* 2001).

1.4. Data statistical analysis

Vegetation data generally do not follow normal distribution (Greig-Smith 1983; Jongman *et al.* 1995). Chi-square test confirmed not normality of vegetation data; therefore we used nonparametric Wilcoxon test to test differences between pairs of data set in different years. Wilcoxon test was used, because we compared the depended data (time series) (Čekanavičius, Murauskas 2002). Soil chemical and microbiota parameters followed normal distribution (checked by chi-square test), therefore we used Student t-test. Statistical analyses were conducted using the STATISTICA 8.0.

2. Results and discussions

2.1. Changes in ground vegetation

In total 28 different species were found during four years period in the ground vegetation cover of burned and control sites of Scots pine stands. Average species number per 1 m^2 significantly ($p < 0.05$) decreased within first year after the low severity surface fire (Fig. 2).

In the subsequent years post-fire the average species number slightly increased, and in the fourth year after the surface fire it was even higher than that in the non-burned control site of pine stand. This increase in species number was mainly determined by spread out of pioneer early succesional herb species (*Calamagrostis epigejos, Equisetum hyemale, Pteridium aquilinum, Rubus idaeus, R. saxatilis, Scorzonera humilis, Solidago virgaurea, Ceratodon purpureus, Polytrichum juniperinum*).

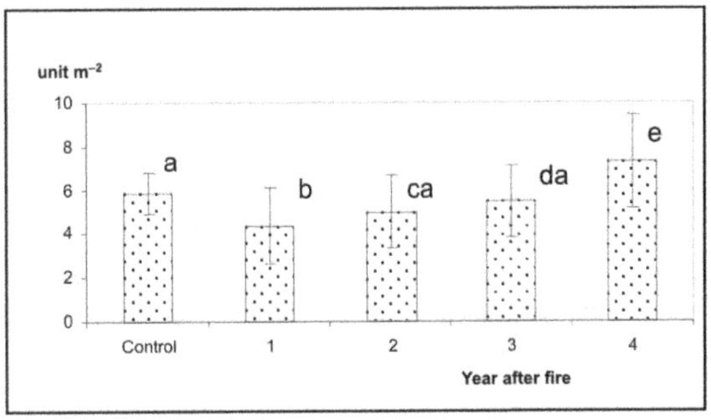

Fig. 2. Changes of average species number in the ground vegetation cover of burned and control sites in Scots pine stands. Values are given as mean ± SD. Significant differences ($p < 0.05$) are indicated by different superscript letters

The similar decline in the number of species immediately after the fire and the subsequent increase within a few further years was reported by Nuzzo (1996) and Parro *et al.* (2009).

In the cover of ground vegetation of Scots pine stand prevailed herbaceous, moss and dwarf species (Table 1).

Table 1. Changes of the average projection cover (%) of shrubs, herbaceous, dwarf shrub, herbaceous and moss species in the burned and control areas

Name of species	Control	Year after fire			
		1	2	3	4
Shrubs					
Frangula alnus	+	+	+	–	+
Sorbus aucuparia	+	–	–	–	–
Saplings					
Quercus robur	+	–	–	–	–
Picea abies	+	–	–	–	–
Herbs and dwarf shrubs					
Calamagrostis arundinacea	0.1	0.1	0.2	0.3	0.2
Calamagrostis epigejos	–	+	0.1	0.1	+
Calluna vulgaris	0.3	–	–	–	–
Convallaria majalis	0.6	+	2.6	1.9	1.8
Equisetum hyemale	–	+	+	+	+
Festuca ovina	+	+	+	0.3	0.5
Luzula pilosa	0.2	–	–	–	–
Melampyrum pratense	0.2	–	+	0.9	1.0
Peucedanum oreoselinum	–	+	0.1	+	–
Polygonatum odoratum	+	–	–	–	–
Pteridium aquilinum	–	0.9	1.5	4.2	4.5
Rubus idaeus	–	–	–	–	0.1
Rubus saxatilis	–	0.8	2.8	2.8	2.6
Scorzonera humilis	–	+	0.2	0.3	0.2
Solidago virgaurea	–	+	+	–	–
Trientalis europaea	–	0.1	1.2	+	+
Vaccinium myrtillus	17.5	2.7	23.2	50.4	47.6
Vaccinium vitis-idaea	13.4	0.4	2.0	3.9	4.4
Mosses					
Dicranum polysetum	–	–	–	–	+
Dicranum scoparium	–	–	–	–	+
Hylocomium splendens	35.7	–	–	–	–
Pleurozium schreberi	60.5	–	–	–	+
Ceratodon purpureus	–	–	–	–	0.8
Polytrichum juniperinum	–	–	–		+

+ – projection cover < 0.1%

Only two species of shrubs (*Frangula alnus* and *Sorbus aucuparia*) and two species of saplings (*Quercus robur* and *Picea abies*) occurred rarely before the fire. Besides, only *F. alnus* was found in the burned site.

In total nine herbaceous and dwarf shrub species occurred in the ground vegetation cover before the fire (Table 1). While, there were 15 species in the burned site. Dwarf shrub Vaccinium *myrtillus* and *V. vitis-idaea* were the most abundant in the control site. These dwarf shrubs were declined by surface fire, however, there were recovered, especially *V. myrtillus*, within 3–4 years.

Herbaceous species of *Calluna vulgaris, Luzula pilosa* and *Polygonatum odoratum* were found only in the control site and still did not occur in the 4-year-old fire site (Table 1). Meanwhile *Rubus saxatilis, Scorzonera humilis, Calamagrostis epigejos, Pteridium aquilinum, Trientalis europaea, Peucedanum oreoselinum, Solidago virgaurea, Equisetum hyemale, Rubus idaeus* occurred only in the burned site.

The reduction of the average projection cover of herbaceous and dwarf shrub species due to the surface fire was observed only during the first year. Projection cover recovered in the second year after the surface fire. Moreover, in the third year the projection cover of herbaceous and dwarf shrub species was even higher than that in the control site (Fig. 3). This increase was mainly because in the burned sites the coverage of *V. myrtillus* (near 50%) was almost 3-folds higher than its coverage in Scots pine stand before the fire.

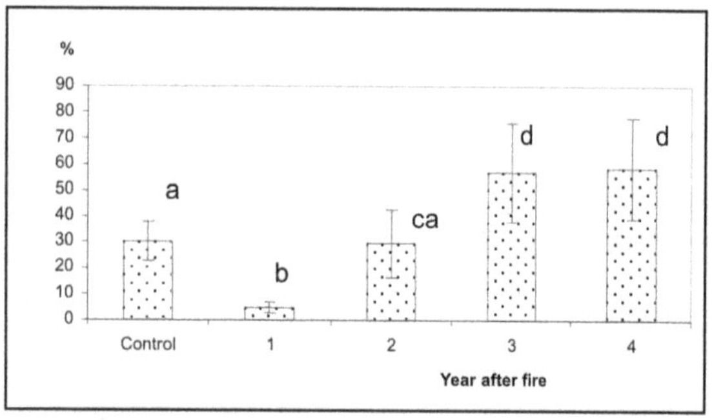

Fig. 3. Changes of average projection cover (%) of herb and dwarf shrub layer in burned and control areas. Values are given as mean ± SD. Significant differences ($p < 0.05$) are indicated by different superscript letters

Only two moss species, *Pleurozium schereberi* and *Hylocomium splendens*, occurred before the fire in observed Scots pine stand but it comprised almost 90% of ground vegetation cover (Table 1). The decline of these mosses was still in fact total in the four year after the fire. Instead of mentioned mosses, four new moss species (*Dicranum polysetum, D. scoparium, Ceratodon purpureus, Polytrichum juniperinum*) occurred in the burned sites. However, the average projection cover of the mosses was still very low (comprised only 1–2%) in 4-year-old fire sites.

Surface forest fires had considerable effect on the ground vegetation coverage. Fire destroyed above-ground part of vegetation, but the herbs and, especially, dwarf shrubs recovered quite rapidly within 3–4 years. It has been reported that the recovery of moss layer is much slower and it takes more than 10 years (Marozas *et al.* 2007; Parro *et al.* 2009). Skre (Skre *et al.* 1998) found that the biomass of *Calluna vulgaris, Polytrichum. Deschampsia flexuosa* and *Pteridium aquilinum* increased after the fire in pine forests of western Norway while regrowth of *Vaccinium myrtillus* and *V. vitis-idaea* was slower. Other investigations also suggested (Ryoma, Laaka-Lindberg 2005) that *Ceratodon, Funaria, Pohlia nutans, Polytricum* spp. appears quickly after the fire in boreal forests. Investigation of the post-fire recovery of species in Scots pine forest in the central part of the Kola Peninsula (Gorshkov, Bakkal 1996) showed that the herb and dwarf shrub layers recovered within 5–15 years after the fire while the mosses recovered within 90–140 years after the fire.

2.2. Soil chemical properties

There were no differences in the organic layer mass of *Arenosols* between the burned and control sites with average values of approximately 6.5 kgDM m^{-2} at each site (data not shown). The fact that no influence on the mass was detected in the 4-year-old burned site shows that the mineralization of organic layers was not intensified after the fire. However, fire-caused changes in chemical properties were mostly pronounced in the organic layer. Integrated parameter pH_{KCl} of organic layer was significantly higher ($p < 0.05$) in both burned sites to compare to control sites, indicating the alkaline effect of fires (Table 2).

Table 2. The chemical parameters (mean ± SD) of organic and mineral topsoil layers of *Arenosols* in control and burned sites of Scots pine stands

Soil sampling		pH_{KCl}	N, g kg^{-1}	Org. C, g kg^{-1}	C/N	P_2O_5, mg kg^{-1}	K_2O, mg kg^{-1}
Organic layer							
1-year-old burned site	control	2.8 ± 0.2	15.3 ± 0.3	431.9 ± 18.1	28.2 ± 1.2	195 ± 27	421 ± 15
	burned site	3.2 ± 0.1*	13.4 ± 0.7*	432.0 ± 23.5	32.5 ± 1.2*	283 ± 49	390 ± 36
4-year-old burned site	control	2.8 ± 0.2	12.9 ± 0.5	441.4 ± 26.8	34.3 ± 2.1	244 ± 39	434 ± 32
	burned site	3.2 ± 0.1*	14.3 ± 0.8*	438.9 ± 28.6	30.6 ± 0.9*	245 ± 13	415 ± 14
Mineral topsoil (0–5 cm in depth)							
1-year-old burned site	control	3.4 ± 0.2	2.4 ± 0.1	25.3 ± 2.7	10.6 ± 0.7	77 ± 12	48 ± 4
	burned site	3.4 ± 0.1	2.4 ± 0.1	23.2 ± 0.7	9.8 ± 0.6	88 ± 21	61 ± 5*
4-year-old burned site	control	3.7 ± 0.1	2.4 ± 0.1	25.2 ± 1.3	10.5 ± 0.7	117 ± 8	48 ± 3
	burned area	3.4 ± 0.2	2.4 ± 0.1	28.3 ± 0.7	11.9 ± 0.1	115 ± 18	64 ± 7*

* – indicates the significant ($p < 0.05$) changes in burned sites.

In the first year after surface fire total N concentrations in the soil organic layer were lower in the burned site than that in the control (Table 2). Four years after fire, total N concentrations were higher in the burned site and exceeded the pre-fire level. Since there were no significant differences in org. C concentrations in the soil organic layer, the observed changes (increase in 1-year and decrease in 4-year-old burned sites) in the C/N ratio were obviously determined by the changes in total N concentrations. We have not found significant differences in concentrations of mobile K_2O and P_2O_5 between the burned and the control sites.

Only the concentrations of mobile K_2O was slightly higher in the upper 0–5 cm mineral topsoil in both 1 and 4-year-old burned sites to compare to control sites (Table 2). Other chemical soil parameters (pH, total N, org. C, mobile P_2O_5) did not differ significantly in the mineral topsoil. The observed saturation of mineral topsoil with K_2O could be related to the distribution of the ash after surface fire. Unchanged concentrations of mobile K_2O in soil organic layers of burned sites indicated intensive leaching of K_2O from organic layers to the mineral topsoil.

Data on heavy metals show that the concentrations of Cd, Cr, Cu Mn, Zn and Fe significantly ($p < 0.05$) increased in 0–5 cm mineral topsoil after surface fire, while the concentrations of Pb and Ni did not differ between burned and control sites (Table 3). However, the increased concentrations were considerably lower to compare to critical or, even, background levels (HN 60:2004).

The effects of forest above-ground (surface and crown) fires on soil chemical properties strongly depend on fire severity. Crown fires that are severe could lead to significant chemical changes in the mineral soils (Smithwick *et al.* 2005; Hammana *et al.* 2007). The study of a short time change (one week after fire) in soil properties due to the fire was conducted in *Pinus densiflora* stands in Korea (Choonsig *et al.* 1999). It was found that high intensity forest fire increased soil pH, total N, available P, K, Ca, and Mg in the mineral topsoil (0–5 cm). However, no marked changes were observed in the deeper mineral soil layer (5–25 cm). Other studies confirmed that forest crown fires increase soil cation stocks (Liechty *et al.* 2005; Neff *et al.* 2005), but decrease carbon and nitrogen stocks in the surface layer of mineral soils (Choromanska, DeLuca 2001; Carter, Foster 2004; MacKenzie *et al.* 2004; Certini 2005).

Our study showed that the effects of surface forest fires of low severity on soil chemical properties were not significant. Surface fires did not affect the mass of organic layer, the contents of org. C and mobile P_2O_5. Only slight increase of pH and the short-term decrease of total N in the soil organic layer, as well as the increased concentrations of mobile K_2O and some heavy metals were detected in mineral topsoil. Previous studies have reported similar results. Wagle and Kitchen (1972) found no difference in extractable P among 3 years and 14 years old burned and control sites in ponderosa pine forest in the northern Arizona. Very little differences in soil properties, mainly total N decrease within first year after the fire, were found in soils within the low severity burn sites in the coniferous stands of the central and eastern Cascade Mountains of Washington State (Baird 1998; Hatten *et al.* 2005).

2.3. Soil microbiota

The surface fires mainly affect the changes of soil microbial populations in the soil organic layer and in the mineral topsoil (Klopatek *et al.* 1988). Concerning that, the abundance of soil microbiota in the organic layer and in the mineral topsoil layer (0–5 cm in depth) in Scots pine stands was determined.

In our study the mean total abundance of soil microbiota in the organic layer was 5.01 mln. and 5.91 mln. of colony forming units (CFU) per g^{-1}, respectively, in one year and 4-year-old burned sites (Table 4).

Table 3. Concentrations (mg kg^{-1}, mean ± SD) in the mineral topsoil (0–5 cm) in 4-year-old burned sites and controls

Soil sampling	Cd	Cr	Pb	Ni	Cu	Mn	Zn	Fe
Burned site	0.036 ± 0.008*	2.78 ± 0.56*	7.5 ± 0.6	1.1 ± 0.4	1.1 ± 0.1*	96.4 ± 7.1*	6.86 ± 1.25*	3488 ± 1387*
Control	0.017 ± 0.004	1.48 ± 0.38	5.5 ± 1.2	0.5 ± 0.1	0.5 ± 0.1	26.4 ± 10.1	3.94 ± 1.23	1449 ± 417

*– indicates the significant ($p < 0.05$) increase in burned sites.

Table 4. The abundance of soil microbiota in the organic layer and the mineral topsoil in Scots pine stands in control and burned sites (mean ± SD)

Soil sampling		Abundance of total microbiota, mln. CFU g^{-1}	Abundance of bacteria, mln. CFU g^{-1}	Abundance of actinobacteria, mln. CFU g^{-1}	Abundance of micromycetes, mln. CFU g^{-1}
Organic layer					
1-year-old burned site	control	4.62 ± 0.51	3.81 ± 0.32	0.28 ± 0.01	0.53 ± 0.18
	burned site	5.01 ± 0.78	4.03 ± 0.57	0.61 ± 0.18*	0.37 ± 0.03
4-year-old burned site	control	5.36 ± 0.55	4.33 ± 0.40	0.32 ± 0.03	0.71 ± 0.12
	burned site	5.91 ± 0.64	4.86 ± 0.32	0.42 ± 0.19	0.63 ± 0.13
Mineral topsoil (0–5 cm in depth)					
1-year-old burned site	control	0.71 ± 0.16	0.52 ± 0.15	0.12 ± 0.00	0.07 ± 0.00
	burned site	0.82 ± 0.15	0.60 ± 0.13	0.17 ± 0.00*	0.05 ± 0.01*
4-year-old burned site	control	0.93 ± 0.21	0.62 ± 0.18	0.20 ± 0.00	0.11 ± 0.02
	burned site	0.95 ± 0.29	0.70 ± 0.28	0.21 ± 0.00	0.04 ± 0.01*

* – indicates the significant differences ($p < 0.05$) between the control and burned sites.

In the mineral topsoil, the mean microbial counts ranged from 0.82 mln. CFU g^{-1} (1-year-old burned site) to 0.94 mln. CFU g^{-1} (4-year-old burned site). In control sites soil microbiota counts were lower than in the burned sites. Since the estimated differences were not significant ($p > 0.05$), the obtained results are indicating only the tendencies that the increase in soil microbiota counts could be fire-induced.

Crown fires, which burn the organic layers totally, decline microbial communities (Prieto-Fernandez *et al.* 1998; Wutrich *et al.* 2002). In our study low severity surface fires might have been not essentially destructive for soil microbiota since soil organic layer was not burned. On the other hand, the surface fires in studied sites seemed to produce the substances as growth promoting substrates for soil microbiota as reported by Neary *et al.* (1999). Thus, the increase of soil microbiota in burned sites might be caused by the decreased acidity of organic layers and the slight saturation of mineral topsoil mobile K_2O (see Table 2).

The distribution of soil microbiota communities is presented in Table 4 as well. Soil bacterium was dominant in the organic layer as well as in the mineral topsoil. However, mean counts of soil bacteria were not significantly ($p > 0.05$) higher in the burned sites than in the control sites. Though it was expected, no increase in soil bacteria populations was not found even in the 4-year-old burned site, where the saturation with N was increased, especially in the organic layer. Mroz *et al.* (1980) and Klopatek *et al.* (1988) reported that an increase in soil nitrogen ratios after the burning should promote an increase in total bacteria populations. It shows, as Astaraei (2008) assumed, that soil bacteria populations were not sufficiently provided with excess of the nitrogen.

Actinobacteria abundance was found to be increasing in the burned sites (Table 4). Only in 1-year-old burned site the significant 1.5- and 2.2- fold increase in actinobacteria abundance was found in the organic layer and in the mineral topsoil, respectively. Actinobacteria in the soil is sensitive to acidity and tolerate the soils with alkaline or neutral pH values (Klopatek *et al.* 1988; Joos *et al.* 2001). Since in burned sites of our study soil acidity had decreased, that likely influenced the growth of soil actinobacteria populations.

The abundance of micromycetes decreased in average by 1.4 and 2.8 times in the mineral topsoil in 1-year and 4-year-old burned sites, respectively (Table 4). It has been reported that the most of soil micromycetes are sensitive to decreased of moisture retention (Bååth *et al.* 1995; Pietikainen *et al.* 2000). In our study, mineral topsoil layers in the burned sites might have been drying and causing decreases in the soil micromycetes populations. The decrease in soil acidity also could cause the decline in micromycete abundance.

Conclusions

1. Surface fires have destroyed the above-ground part of ground vegetation in V*accinio-myrtillo Pinetum* forests. Nevertheless, ground vegetation began to recover in the subsequent years. Within 3–4 years after the fires the burned sites have had even higher number of species and ground vegetation coverage than in the unburned sites. The pioneer herb species and dwarf shrubs, mainly *Vaccinium myrtillus,* were spread out. However, the recovery of moss cover was still non-significant.

2. Surface fires did not affect the mass of soil organic layer, and the effect on soil chemical properties was minor. Only slight increases of pH and of the content of total N were detected in soil organic layer. Also, slightly increased concentrations of mobile K_2O and some heavy metals (Cd, Cr, Cu Mn, Zn and Fe) were found in the mineral topsoil.

3. Surface fires have caused slight changes of microbiota abundance in the soil organic layer and the mineral topsoil (0–5 cm in depth). Although mean abundance of soil bacteria in the burned sites were not significantly higher than in the control sites, actinobacteria abundance tended to increase in the soil organic layer and the mineral topsoil of burned sites. In opposite, the abundance of micromycetes decreased in the mineral topsoil after the surface fires.

4. Our study showed that low severity surface fires mostly affected above-ground part of ground vegetation in Scots pine stands on *Arenosols* that prevail in Lithuanian forests. The influence of surface fire on soil chemical properties and soil microbiota was minor.

Acknowledgements

The research was supported by Lithuanian Agency for Science Innovation and Technology (Award Numbers 2009 No. 31V-52; 2008 No. 31V-104) and Research Council of Lithuania (No. COST-50/2010). The research was conducted in the frame of COST action FP0701 Post-Fire Forest Management in Southern Europe and COST action ES0805 The Terrestrial Biosphere in the Earth System (TERRABITES).

References

Ahti, T.; Hämet-Ahti, L.; Jalas, J. 1968. Vegetation zones and their sections in northwestern Europe, *Annales Botanici Fennici* 5: 169–211.

Alloway, C. J. 1995. *Heavy metals in soils*. London: Springer-Verlag. 384 p. http://dx.doi.org/10.1007/978-94-011-1344-1

Almendros, G.; Gonzalez-Vila, F. J.; Martin, F. 1990. Fire-induced transformation of soil organic matter from an oak forest: an experimental approach to the effects of fire on humic substances, *Soil Science* 149: 158–168. http://dx.doi.org/10.1097/00010694-199003000-00005

Angelstam, P. K. 1998. Maintaining and restoring biodiversity by developing natural disturbance regimes in European boreal forest, *Journal of Vegetation Science* 9(4): 593–602. http://dx.doi.org/10.2307/3237275

Annand, M.; Ke-Ming, M.; Okonski, A.; Levin, S.; McCreath, D. 2003. Characterising biocomplexity and soil microbial dynamics along a smelter-damaged landscape gradient, *Science of Total Environment* 311: 247–259. http://dx.doi.org/10.1016/S0048-9697(03)00058-5

Arocena, J. M.; Opio, C. 2003. Prescribed fire-induced changes in properties of sub-boreal forest soils, *Geoderma* 113: 1–16. http://dx.doi.org/10.1016/S0016-7061(02)00312-9

Astaraei, A. R. 2008. Microbial count and succession, soil chemical properties as affected by organic debrise decomposition, *American-Eurasian Journal of Agricultural and Environmental Science* 4(2): 178–188.

Bååth, E.; Frostegård, A.; Pennanen, T.; Fritze, H. 1995. Microbial community structure and pH response in relation to soil organic matter quality in wood-ash fertilized, clear-cut or burned coniferous forest soils, *Soil Biology and Biochemistry* 27: 229–240. http://dx.doi.org/10.1016/0038-0717(94)00140-V

Baird, M. 1998. *Wildfire effects on nutrient capitals in inland coniferous forests*. Seattle, WA: College of Forest Resources, University of Washington. 46 p.

Baltrėnaitė, E.; Butkus, D. 2007. Modelling of Cu, Ni, Zn and Pb transport from soil to seedlings of coniferous and leafy trees, *Journal of Environmental Engineering and Landscape Management* 15(4): 200–207.

Bergeron, Y.; Leduc, A.; Harvey, B. D.; Gauthier, S. 2002. Natural fire regime: a guide for sustainable management of the Canadian boreal forest, *Silva Fennica* 36(1): 81–95.

Boyer, W. D.; Miller, J. H. 1994. Effect of burning and brush treatments on nutrient and soil physical properties in young longleaf pine stands, *Forest Ecolology and Management* 70: 311–318. http://dx.doi.org/10.1016/0378-1127(94)90096-5

Bradshaw, J. L. 1992. *Laboratory microbiology*. 4th ed. New York: Saunders Colege Publishing. 436 p.

Breulmanna, G.; Markertb, B.; Weckertb, V.; Herpinc, U.; Yonedad, R.; Oginoe, K. 2002. Heavy metals in emergent trees and pioneers from tropical forest with special reference to forest fires and local pollution sources in Sarawak, Malaysia, *Science of Total Environment* 285: 107–115. http://dx.doi.org/10.1016/S0048-9697(01)00899-3

Bukantis, A. 1994. *Lietuvos klimatas* [Climate of Lithuania]. Vilnius: Vilnius University Press. 188 p. (in Lithuanian).

Carter, M. C.; Foster, C. D. 2004. Prescribed burning and productivity in southern pine forests: a review, *Forest Ecology and Management* 191: 93–109. http://dx.doi.org/10.1016/j.foreco.2003.11.006

Certini, G. 2005. Effects of fire on properties of forest soils: a review, *Oecologia* 143: 1–10. http://dx.doi.org/10.1007/s00442-004-1788-8

Choonsig, K.; Won-Kyu, L.; Jae-Kyung, B.; Young-Kul, K.; Jin-Hyun, J. 1999. Short-term effects of fire on soil properties in *Pinus densiflora* stands, *Journal of Forest Research* 4: 23–25. http://dx.doi.org/10.1007/BF02760320

Choromanska, U.; DeLuca, T. H. 2001. Prescribed fire alters the impact of wildfire on soil biogeochemical properties in a ponderosa pine forest, *Soil Science Society of America Journal* 65: 232–238. http://dx.doi.org/10.2136/sssaj2001.651232x

Čekanavičius, V.; Murauskas, G. 2002. *Statistika ir jos taikymai II* [Statistics and its application II]. Vilnius: TEV. 268 p. (in Lithuanian).

DeLuca, T. H.; Sala, A. 2006. Frequent fire alters nitrogen transformations in ponderosa pine stands of the stands of the inland Northwest, *Ecology* 87: 2511–2522. http://dx.doi.org/10.1890/0012-9658(2006)87[2511:FFANTI]2.0.CO;2

Egner, H.; Riehm, H.; Domingo, W. R. 1960. Untersuchungen über die chemishe Bodenanalyse als Grundlage für Beurteilung der Nährzustandes der Böden, *Annals of the Royal Agricultural College of Sweden* 26: 199–215.

Ferran, A.; Delite, W.; Valleho, V. R. 2005. Effects of fire recurrence in *Quercus coccifera* L. Shrublands of the Valencia Region (Spain): II. Plant and soil nutrient, *Plant Ecology* 177: 71–83. http://dx.doi.org/10.1007/s11258-005-2141-y

Flannigan, M.; Stock, B.; Turetsky, M.; Wotton, M. 2009. Impacts of climate change on fire activity and fire management in the circumboreal forest, *Global Change Biology* 15: 549–560. http://dx.doi.org/10.1111/j.1365-2486.2008.01660.x

Gorshkov, V. V.; Bakkal, I. J. 1996. Species richness and structure variations of Scot pine forest communities during the period from 5 to 210 years alter fire, *Silva Fennica* 30(2–3): 329–340.

Granström, A. 2001. Fire management for biodiversity in the European boreal forest, *Scandinavian Journal of Forest Research* 3: 62–69. http://dx.doi.org/10.1080/028275801300090627

Greig-Smith, P. 1983. *Quantitative plant ecology*. Oxford: Blackwell. 359 p.

Gromtsev, A. N. 2002. Natural disturbance dynamics in the boreal forests of European Russia: a review, *Silva Fennica* 36(1): 41–55.

Hammana, S. T.; Burkea, I. C.; Strombergerc, M. E. 2007. Relationships between microbial community structure and soil environmental conditions in a recently burned system, *Soil Biology and Biochemistry* 39: 1703–1711. http://dx.doi.org/10.1016/j.soilbio.2007.01.018

Hanson, P. J.; Edwards, N. T.; Garten, C. T.; Andrews, J. A. 2000. Separating root and soil microbial contributions to soil respiration: a review of methods and observations, *Biogeochemistry* 48: 115–146. http://dx.doi.org/10.1023/A:1006244819642

Hassen, A.; Belguith, K.; Jedidi, N.; Cherif, A.; Cherif, M.; Boudabous, A. 2001. Microbial characterization during composting of municipal solid waste, *Bioresource Technology* 80: 217–225. http://dx.doi.org/10.1016/S0960-8524(01)00065-7

Hatten, J.; Zabowski, D.; Scherer, G.; Dolan, E. 2005. A comparison of soil properties after contemporary wildfire and fire suppression, *Forest Ecology and Management* 220: 227–241. http://dx.doi.org/10.1016/j.foreco.2005.08.014

Ice, G. G.; Neary, D. G.; Adams, P. W. 2004. Effects of wildfire on soils and watershed processes, *Journal of Forestry* 102: 16–20.

Ignatavičius, G.; Sakalauskienė, G.; Oškinis, V. 2006. Influence of land fires on increase on heavy metal concentrations in river waters of Lithuania, *Journal of Environmental Engineering and Landscape Management* 14(1): 46–51.

IPCC 2007. *Climate change 2007: The physical science basis*. Contribution of Working Group I to the Fourth Assessment Report of the Intergovernmental Panel on Climate Change. Cambridge: Cambridge University Press. 996 p.

Jankevičiene, R. (Ed.). 1998. *Dictionary of plant names*. Vilnius: Institute of Botany Publishers. Vilnius. 523 p.

Jayen, K.; Leduc, A.; Bergeron, Y. 2006. Effect of fire severity on regeneration success in the boreal forest of northwest Quebec, Canada, *Ecoscience* 13(2): 143–151. http://dx.doi.org/10.2980/i1195-6860-13-2-143.1

Jongman, R. H. G.; Ter Braak, C. J. F.; van Tongeren, O. F. R. 1995. *Data analysis in community and landscape ecology*. Cambridge University Press. 300 p. http://dx.doi.org/10.1017/CBO9780511525575

Joos, F. I. C.; Prentice, S.; Sitch, R.; Meyer, G.; Hoos, G. K.; Plattner, S.; Hasselmann, K. 2001. Global warming feedbacks on terrestrial carbon uptake under the Intergovernmental Panel on Climate Change (IPCC) emission scenarios, *Global Biogeochemical Cycles* 15: 891–907. http://dx.doi.org/10.1029/2000GB001375

Klopatek, C. C.; De Bano, L. F.; Klopatek, J. M. 1988. Effects of simulated fire on vesicular-arbuscular mycorrhizae in pinyon-juniper woodland soil, *Plant and Soil* 109: 245–249. http://dx.doi.org/10.1007/BF02202090

Kuuluvainen, T. 2002. Natural variability of forests as a reference for restoring and managing biological diversity in boreal Fennoscandia, *Silva Fennica* 36(1): 97–125.

Liechty, H. O.; Luckow, K. R.; Guldin, J. M. 2005. Soil chemistry and nutrient regimes following 17–21 years of shortleaf pine-bluestem restoration in the Ouachita Mountains of Arkansas, *Forest Ecology and Management* 204: 345–357. http://dx.doi.org/10.1016/j.foreco.2004.09.009

LME (Lithuanian Ministry of Environment); SFS (State Forest Service). 2010. Lithuanian Statistical Yearbook of Forestry 2009. Kaunas: State Forest Service. 152 p.

Mabuhay, J. A.; Nakagoshi, N.; Isagi, Y. 2006. Soil microbial biomass, abundance, and diversity in a Japanese red pine forest: first year after fire, *Journal of Forest Research* 11: 165–173. http://dx.doi.org/10.1007/s10310-005-0201-8

MacKenzie, M. D.; DeLuca, T. H.; Sala, A. 2004. Forest structure and organic horizon analysis along a fire chronosequence in the low elevation forests of western Montana, *Forest Ecology and Management* 203: 333–343. http://dx.doi.org/10.1016/j.foreco.2004.08.003

Marozas, V.; Racinskas, J.; Bartkevicius, E. 2007. Dynamics of ground vegetation after surface fires in hemiboreal *Pinus sylvestris* forests, *Forest Ecology and Management* 250: 47–55. http://dx.doi.org/10.1016/j.foreco.2007.03.008

Mroz, G. D.; Jurgensen, M. F.; Harvey, A. E.; Larsen, M. J. 1980. Effects of fire on nitrogen in forest floor horizons, *Soil Science Society of America Journal* 44: 395–400. http://dx.doi.org/10.2136/sssaj1980.03615995004400020038x

Neary, D. G.; Klopatek, C. C.; DeBano, L. F.; Ffolliott, P. F. 1999. Fire synthesis effects on belowground sustainability: a review and synthesis, *Forest Ecology and Management* 122: 51–71. http://dx.doi.org/10.1016/S0378-1127(99)00032-8

Neff, J. C.; Harden, J. W.; Gleixner, G. 2005. Fire effects on soil organic matter content, composition, and nutrients in boreal interior Alaska, *Canadian Journal of Forest Research* 35(9): 2178–2187. http://dx.doi.org/10.1139/x05-154

Neill, C.; Patterson, W. A.; David, W.; Craig, J. 2007. Responses of soil carbon, nitrogen and cations to the frequency and seasonality of prescribed burning in a Cape Cod oak-pine forest, *Forest Ecology and Management* 250: 234–243. http://dx.doi.org/10.1016/j.foreco.2007.05.023

Nuzzo, V. A.; McClain, W.; Strole, T. 1996. Fire impact on ground layer flora in a sand forest 1990–1994, *American Midland Naturalist Journal* 136(2): 207–221. http://dx.doi.org/10.2307/2426726

Ojima, D. S.; Schimel, D. S.; Parton, W. J.; Owensby, C. E. 1994. Long and short term effects on nitrogen cycling in tall grass prairie, *Biogeochemistry* 24: 67–84. http://dx.doi.org/10.1007/BF02390180

Päätalo, M. L. 1998. Factor influencing occurrence and impacts of fires in northern European forest, *Silva Fennica* 32(2): 185–202.

Parro, K.; Köster, K.; Jõgiste, K.; Vodde, F. 2009. Vegetation dynamics in a fire damaged forest area: the response of major ground vegetation species, *Baltic Forestry* 15(2): 206–215.

Parviainen, J. 1996. Impact of fire on Finnish forest in the past and today, *Silva Fennica* 30(2–3): 353–359.

Pereira, P.; Ubeda, X. 2010. Spatial distribution of heavy metals released from ashes after a wildfire, *Journal of Environmental Engineering and Landscape Management* 18(1): 13–22. http://dx.doi.org/10.3846/jeelm.2010.02

Pietikainen, J.; Hiukka, R.; Fritze, H. 2000. Does short-term heating of forest humus hinge its properties as a substrate for microbes?, *Soil Biology and Biochemistry* 32: 227–288. http://dx.doi.org/10.1016/S0038-0717(99)00164-9

Prieto-Fernandez, A.; Acea, M. J.; Carballas, T. 1998. Soil microbial and extractable C and N after wildfire, *Biology and Fertility of Soils* 27: 132–142. http://dx.doi.org/10.1007/s003740050411

Pundytė, N.; Baltrėnaitė, E.; Pereira, P.; Paliulis, D. 2011. Anthropogenic effects on heavy metals and macronutrients accumulation in soil and wood of *Pinus sylvestris* L, *Journal of Environmental Engineering and Landscape Management* 19(1): 34–43. http://dx.doi.org/10.3846/16486897.2011.557473

Richter, D. D.; Ralston, C. W.; Harms, W. R. 1982. Prescribed fire: effects on water quality and forest nutrient cycling, *Science* 215: 661–663. http://dx.doi.org/10.1126/science.215.4533.661

Ryan, K. C. 2002. Dynamic interaction between forest structure and fire behaviour in boreal ecosystems, *Silva Fennica* 36: 13–39.

Ryoma, R.; Laaka-Lindberg, S. 2005. Bryophyte recolonization on soil and logs, *Scandinavian Journal of Forest Research* 20(6): 5–16. http://dx.doi.org/10.1080/14004080510043361

Shugart, H. H.; Smith, T. M.; Post, W. M. 1992. The potential for application of individual-based simulation models for assessing the effects of global change, *Annual Review of Ecology Evolution and Systematics* 23: 15–38.

Simard, D. G.; Fyles, J. W.; Parer, D.; Nguyen, T. 2001. Impacts of clearcut harvesting and wildfire on soil nutrient status in the Quebec boreal forest, *Canadian Journal of Forest Research* 81: 229–237.

Skre, O.; Wielgolaski, F. E.; Moe, B. 1998. Biomass and chemical composition of common forest plants in response to fire in western Norway, *Journal of Vegetation Science* 9(4): 501–510. http://dx.doi.org/10.2307/3237265

Smithwick, E. A. H.; Turner, M. G.; Mack, M. C.; Chapin, III, F. S. 2005. Post-fire soil N cycling in northern conifer forests affected by severe, stand-replacing wildfires, *Ecosystems* 8: 163–181. http://dx.doi.org/10.1007/s10021-004-0097-8

Tateishi, T.; Horikoshi, T. 1995. Microbial biomass in the soils of burned and unburned Japanese Red Pine forests in the Setouchi District, Western Japan, *Bulletin of Japan Society of Microbial Ecology* 10: 9–20. http://dx.doi.org/10.1264/microbes1986.10.9

Thompson, J. A.; Vincent, J. M. 1967. Methods of detection and estimation of rhizobia in soil, *Plant and Soil* 26: 72–84. http://dx.doi.org/10.1007/BF01978676

Vasarevičius, S.; Greičiutė, G. 2004. Investigation of soil pollution with heavy metals in Lithuanian military grounds, *Journal of Environmental Engineering and Landscape Management* 12(4): 137–137.

Wagle, R. F.; Kitchen, J. H. 1972. Influence of fire on soil nutrients in a ponderosa pine type, *Ecology* 53: 118–125. http://dx.doi.org/10.2307/1935716

Wallenius, T. H.; Lilja, S.; Kuuluvainen, T. 2007. Fire history and tree species composition in managed *Picea abies* stands in southern Finland: implications for restoration, *Forest Ecology and Management* 250: 89–95. http://dx.doi.org/10.1016/j.foreco.2007.03.016

Wutrich, C.; Schaub, D.; Weber, M.; Marxer, P.; Condera, M. 2002. Soil respiration and soil microbial biomass after fire in a sweet chestnut forest in southern Switzerland, *Catena* 48: 201–215. http://dx.doi.org/10.1016/S0341-8162(01)00191-6

Zackrisson, O. 1977. Influence of forest fires on the north Swedish boreal forest, *Oikos* 29(1): 13–32. http://dx.doi.org/10.2307/3543289

Vitas MAROZAS. Dr, head of Ecology Department, Aleksandras Stulginskis University. Doctor of biomedical sciences (forestry). Publications: author of over 40 scientific publications, participant of over 20 international conferences. Research interests: forest vegetation ecology, biodiversity conservation.

Kęstutis ARMOLAITIS. Dr, senior research scientist of Ecology Department, Institute of Forestry of Lithuanian Research Centre for Agriculture and Forestry (LRCAF), part-time senior scientist of Perloja Experimental station of LRCAF and part-time associate professor of Šiaulių University. Doctor of biomedical sciences (forestry), Byelorussian Technology University (Minsk), 1984. More than 100 scientific publications. Research interests: forest decline, forest soil chemical condition, soil renaturalization, carbon and nitrogen turnover in forest ecosystems.

Jūratė ALEINIKOVIENĖ. Dr, junior research scientist of Ecology Department, Institute of LRCAF, part-time lecturer of Aleksandras Stulginskis University. Doctor of biomedical sciences (ecology and environmental science), Lithuanian Forest Research Institute and Vytautas Magnus University, 2009. Author of 5 scientific publications, participant of over 20 international and national conferences. Research interests: abundance and diversity of soil microbiota, carbon and nitrogen in microbiota biomass, soil biological renaturalization.

11

FACTORS AFFECTING SEASONAL AND SPATIAL PATTERNS OF WATER QUALITY IN LITHUANIAN RIVERS

Mindaugas Gudas[1], Arvydas Povilaitis[2]

[1]*Environment Protection Agency, Juozapavičiaus g. 9, LT-09311 Vilnius, Lithuania*

[2]*Aleksandras Stulginskis University, Studentų g. 11, LT-53361 Kaunas, Lithuania*

E-mails: [1]elmindow@yahoo.com (corresponding author); [2]arvis@hidro.lzuu.lt

Abstract. In this study the main processes influencing water quality of the Lithuanian rivers have been identified. Factor analysis was tested on river hydrochemical data from 108 sites for the period of 1999–2004. It enabled the identification of main factors determining water quality each season. As a result, monitoring stations were grouped into clusters each representing the influence of relevant factor.

The tested multivariate statistical procedures can be applied in practice when the reasons for water quality impairments are to be investigated. The findings reveal that Wastewater factor is prominent in small rivers downstream larger towns; Agro-geological factor – in Northern Lithuania's rivers of heavy carbonated soils and intensive agriculture lands as well as in South eastern Lithuania's rivers of more acidic soils; Hardly degradable organics factor – in Northern and Middle Lithuania's rivers of heavy-textured and fertile agricultural soils. The photosynthesis-vegetation and Aeration factors are predominant in major rivers.

Keywords: factor analysis, multivariate statistics, water quality, Lithuanian rivers.

Introduction

In the year 2000 the EU Directive 2000/60/EC (further referred to as WFD) came into force obliging all member states to carefully identify water status problems, the reasons behind them and to elaborate programs of measures to solve those problems (Directive 2000/60/EC… 2000). Implementation of measures is a very costly activity, therefore it is very important to have a correct identification of problems and their causes and thus avoid unnecessary costly actions. Ideal reasoning for water status problems could be achieved by the use of good accounting data on point source pollution, diffuse pollution inputs and detailed process-based water quality models. Unfortunately, in many cases and in Lithuania in particular the aforementioned information is far from complete and the run of detailed models is impossible or not feasible in terms of precision of the results (Vaikasas 2010). One of the solutions to help find causes with limited information could be the use of proper statistical methods, applied on water monitoring data. The methods could be employed in conjunction with modeling to create a broader picture on possible problems and the reasons behind them. However, monitoring data contains a huge array of parameters (in this work – 27), some of which are very much intercorrelated. Consequently, conventional univariate statistical methods are unfit to analyze their complexity and detect meaningful structures in data, which could shed light on processes affecting water status. Luckily, multivariate statistical methods are perfectly designed to solve such kind of problems. Among them Factor analysis (FA) is one of the most reliable tools to deal with such tasks and therefore it has been utilized by a number of authors (Alberto *et al.* 2001; Aruga *et al.* 1995; Azzellino *et al.* 2006; Boyacioglu *et al.* 2007; Boyacioglu, H., Boyacioglu, H. 2008; Brodnjak-Voncina *et al.* 2002; Charkhabi, Sakizadeh 2006; Fitzpatrick *et al.* 2007; Kannel *et al.* 2007; Koklu *et al.* 2010; Kowalkowski *et al.* 2006; Kunwar *et al.* 2005; Marques da Silva *et al.* 2001; Mendiguchia *et al.* 2004; Morales *et al.* 1999; Omo-Irabor *et al.* 2008; Ouyang *et al.* 2006; Papatheodorou *et al.* 2006; Qadir *et al.* 2008; Razmkhah *et al.* 2010; Reisenhofer *et al.* 1998; Santos-Roman *et al.* 2003; Schaefer *et al.* 2010; Simeonov *et al.* 2002; Simeonov *et al.* 2003; Singh *et al.* 2006; Sojka *et al.* 2008; Spanos *et al.* 2003; Tarrado *et al.* 2006; Vega *et al.* 1998; Vidal *et al.* 2000; Voutsa *et al.* 2001; Zhang *et al.* 2009; Zhou *et al.* 2007). The strength of the method hides in its ability to extract meaningful structures from measurement data of numerous parametres that after qualified examination can be interpreted as particular factors or processes affecting water chemistry.

In a number of occasions factor analysis is combined with cluster and even discriminant function analysis, where the latter only pinpoint to parametres that differ the most among clusters (discriminate best) (Alberto *et al.* 2001; Boyacioglu H., Boyacioglu H. 2008; Kannel *et al.* 2007; Koklu *et al.* 2010; Kowalkowski *et al.* 2006; Kunwar *et al.* 2005; Qadir *et al.* 2008; Santos-Roman *et al.* 2003; Simeonov *et al.* 2002; Simeonov *et al.* 2003; Singh *et al.* 2006; Sojka *et al.* 2008; Zhou *et al.* 2007). Finally, factor analysis is sometimes used to complement modelling, like it was done in Spain with QUAL2E model (Azzellino *et al.* 2006).

Although factor analysis was found to be a suitable tool in many applications, the situation with its use in Lithuania is a bit different. Apart from some exceptions (Povilaitis 2003), the method has almost not been used in Lithuania in the context of the analysis of physico-chemical data.

Therefore, the goal of the research was to assess the impacts different factors have on the state of Lithuanian rivers water and identify their spatial patterns by using system-oriented approach. This work aims to uncover external landuse, landcover and internal ecosystem processes and their affects on water physico-chemical parameters. Although this work is targeted at rivers, the results are important to the Curonian Lagoon and Baltic Sea as well, since river-born pollution is the main cause of water quality problems there.

1. Methodology

State monitoring physico-chemical data from the Lithuanian Environmental Protection Agency for 108 river sites have been used in this study (Fig. 1; Table 1).

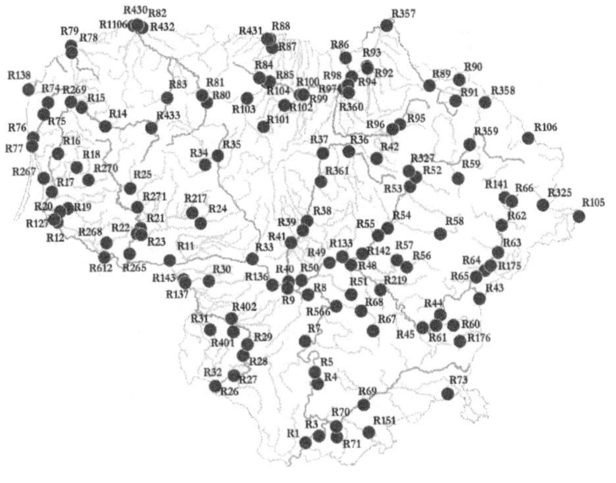

Fig. 1. Location of river water quality monitoring stations

The water samples were collected monthly, however for the analysis not always raw data was used – for significant part of the analysis raw data was aggregated as multi-annual seasonal data. The measured parameters include flow velocity (V), discharge (Q), suspended matter (SM), transparency (SK), pH, dissolved oxygen (O_2), oxygen saturation (O_2%), biochemical oxygen demand in 7 days (BOD_7), chemical oxygen demand using dichromate (COD_{Cr}) or permanganate (COD_{Mn}), total organic carbon (TOA), ammonia nitrogen (NH_4-N), nitrite nitrogen (NO_2-N), nitrate nitrogen (NO_3-N), organic nitrogen (N_{org}), phosphate phosphorus (PO_4-P), organic and adsorbed phosphorus (P_{org}), calcium (Ca^{2+}), magnesium (Mg^{2+}), sodium (Na^{+}), potassium (K^{+}), silica (Si), bicarbonate (HCO_3^{-}), sulphate (SO_4), chlorine (Cl) and total iron (Fe). The selection of parameters for this study was based on the idea to uncover as many water quality determining processes and as precisely as possible. In this sense, the guiding principle for parametre selection was the more of relevant parameters are included, the better. In addition, GIS layers of river basin boundaries and rivers (Lithuanian georeference database and the data of the Lithuanian Environmental Protection Agency) have been used.

The aforementioned data was subject to a multivariate statistical technique called Factor analysis, which have been applied by the use of Statsoft STATISTICA software as well as the PAST software. There were two types of Factor analysis used in this study. First of all, the spatial Factor analysis was employed, which was based on seasonal averages of physico-chemical data of all 108 monitoring stations combined for the whole study period of 1999–2004. Autumn-winter season was represented by the period from September to February months, spring – by March to April and summer – by May to August. Spatial FA was applied to identify main factors affecting river water quality and spatially group them according to the affecting factor and its strength. The other type of FA - local FA was employed only for a limited number of selected monitoring stations representing different groups, formed by means of spatial FA. Local FA was performed using the dataset of all data for particular selected station, covering the period of 1992–2004. Local FA results were utilized for more detailed description of processes affecting the aforementioned groups of stations.

The idea of factor analysis is to reduce intercorrelating variables into a few new representative uncorrelated integrated variables or the so-called principal components, which take a form of the 1st-order equation. The components then are orthogonally rotated in space to transform into latent factors that are explained by variables that correlate with them. The FA mathematical model is based on the assumption that the behaviour of each variable X_i is influenced by m common factors (F_1, F_2, ..., F_m). The interrelationship between factors and variables (X_1, X_2, ..., X_k) is expressed by the 1st-order equations:

$$X_i = \sum_{j=1}^{m} \lambda_{ij} F_j ; \quad (1)$$

$$X_k = \sum_{j=1}^{m} \lambda_{kj} F_j , \quad (2)$$

where λ_{ij} ($i=1$, ..., k; $j=1$, ..., m) = $cov(X_i,F_j)$ – factor loadings (the higher the loading, the more a variable is related to the factor).

2. Results and discussion

Factor analysis from Lithuanian river monitoring data discovered six factors for each season that have been affecting river water quality indices. Some of them are common for all seasons, whereas some of them are season specific unique ones. Further in this chapter each common factor is represented only by the season in which it is most pronounced, while the unique ones are represented by the season for which they have been identified. Three factors are common for all seasons. *Wastewater factor* is the most pronounced one in all periods and in autumn-winter period in particular. In autumn-winter, for instance, it is remarkable through high positive loadings

Table 1. Water quality monitoring stations from which the hydrochemical and river basin data have been used

Code	Monitoring station	Code	Monitoring station	Code	Monitoring station
R1	Nemunas upstream Druskininkai	R41	Šušvė at the mouth	R80	Venta upstream Kuršėnai
R3	Nemunas downstream Druskininkai	R42	Juosta downstream Jackagalis	R81	Venta downstream Kuršėnai
R4	Nemunas upstream Alytus	R43	Neris at Buivydžiai	R82	Venta downstream Mažeikiai
R5	Nemunas downstream Alytus	R44	Neris upstream Vilnius	R83	Virvytė downstream Patekla
R7	Nemunas upstream Prienai	R45	Neris downstream Vilnius	R84	Mūša upstream Kulpė
R8	Nemunas upstream Kaunas	R48	Neris upstream Jonava	R85	Mūša downstream Kulpė
R9	Nemunas downstream Kaunas	R49	Neris downstream Jonava	R86	Mūša downstream Saločiai
R11	Nemunas downstream Smalininkai	R50	Neris upstream Kaunas	R87	Sidabra downstream Joniškis
R12	Nemunas upstream Rusnė	R51	Lomena downstream Kaišiadorys	R88	Sidabra at the border to Latvia
R14	Minija upstream Plungė	R52	Šventoji upstream Anykščiai	R89	Nemunėlis downstream Panemunis
R15	Minija downstream Plungė	R53	Šventoji downstream Anykščiai	R90	Juodupė downstream Juodupė
R16	Minija downstream Gargždai	R54	Šventoji upstream Ukmergė	R91	Laukupė downstream Rokiškis
R17	Minija downstream Priekulė	R55	Šventoji downstream Ukmergė	R92	Tatula upstream Biržai
R18	Veiviržas at Veiviržėnai	R56	Širvinta upstream Širvintos	R93	Tatula below Biržai
R19	Šyša upstream Šilutė	R57	Širvinta downstream Širvintos	R94	Tatula at Trečionys
R20	Šyša downstream Šilutė	R58	Siesartis-Malkėstas downstream Molėtai	R95	Lėvuo upstream Kupiškis
R21	Jūra upstream Tauragė	R59	Vyžuona downstream Utena	R96	Lėvuo downstream Kupiškis
R22	Jūra downstream Tauragė	R60	Vilnia upstream N.Vilnia	R97	Lėvuo upstream Pasvalys
R23	Šešuvis at Skirgailiai	R61	Vilnia at the mouth	R98	Lėvuo at the mouth
R24	Šaltuona downstream Raseiniai	R62	Žeimena at Kaltanėnai	R99	Daugyvėnė at the mouth
R25	Lokysta downstream Šilalė	R63	Žeimena downstream Švenčionėliai	R100	Kruoja at the mouth
R26	Šešupė at the border to Poland	R64	Žeimena upstream Pabradė	R101	Obelė downstream Radviliškis
R27	Šešupė downstream Kalvarija	R65	Žeimena downstream Pabradė	R102	Obelė at the mouth
R28	Šešupė upstream Marijampolė	R66	Būka upstream Baluošas	R103	Kulpė downstream Šiauliai
R29	Šešupė downstream Marijampolė	R67	Strėva downstream Semeliškės	R104	Kulpė at the mouth
R30	Siesartis downstream Šakiai	R68	Strėva at Liūtonys	R105	Birvėta at the border to Belarus
R31	Šeimena downstream Vilkaviškis	R69	Merkys upstream Varėna	R106	Laukesa downstream Zarasai
R32	Šelmenta at the border to Poland	R70	Merkys downstream Puvočiai	R127	Skirvytė upstream Rusnė
R33	Dubysa upstream Seredžius	R71	Skroblus downstream Dubininkai	R133	Šventoji at the mouth
R34	Kražantė upstream Kelmė	R73	Šalčia downstream Šalčininkai	R136	Nemunas downstream Kaunas at Kulautuva
R35	Kražantė downstream Kelmė	R74	Akmena-Danė at Tubausiai	R137	Šešupė at the border to the Kaliningrad region
R36	Nevėžis upstream Panevėžys	R75	Akmena-Danė downstream Kretinga	R138	Šventoji at the mouth at the Baltic Sea
R37	Nevėžis downstream Panevėžys	R76	Akmena-Danė upstream Klaipėda	R141	Švogina-Žeimena upstream Vaišniūnai
R38	Nevėžis upstream Kėdainiai	R77	Akmena-Danė at the mouth	R142	Širvinta at the mouth
R39	Nevėžis downstream Kėdainiai	R78	Bartuva upstream Skuodas	R143	Siesartis at the mouth
R40	Nevėžis upstream Raudondvaris	R79	Bartuva downstream Skuodas	R150	Jiesia at Jiestrakis

of BOD_7, NH_4-N, NO_2-N, N_{org}, P_{org}, Na^+, K^+, Cl^-, PO_4-P and high negative loadings of O_2 and O_2% (Fig. 2). Similar parameters are related to this factor in other seasons as well. The factor is especially profound in small rivers downstream larger settlements where low dilution capacity translates into high pollutant concentrations (Figs 3–4). The process was also confirmed by the results of the local factor analysis. It can be noticed from the Table 2 almost the same correlating parameters and situations, when pollutant concentrations ascent as a consequence of reduced flow and dilution capacity in drier periods.

Agro-geological factor is the second most pronounced one in terms of variables explained and is common for all periods as well. It is expressed by high positive loadings on NO_3-N, SO_4^{2-}, Ca^{2+}, Mg^{2+} and high negative ones on total Fe and Si among others (Fig. 2).

Elevated concentrations of NO_3-N, SO_4^{2-}, Ca^{2+} and Mg^{2+} are found in northern Lithuanian karstic zone with prevailing limestone, dolomite or gypsum bedrock (Figs 3–4). This zone is also characterized by fertile lowlands with sod-gley soils and wide tracts of arable lands – this is one of the most intensive agricultural regions in Lithuania. Consequently, nitrates pose an acute water quality problem there. On the other hand, geological processes also determine relatively high concentrations of total Fe and Si, although this occurs mostly in rivers flowing through eastern and southeastern Lithuanian region of podzolic soils, sandy bedrock, marshes or pastures (Figs 3–4).

Elevated total Fe and Si levels are related to more acid conditions there. Low pH guarantees greater weathering and solubility of those elements from different chemical compounds and soil minerals. The main agents here are humic and fulvic acids, which tend to create colloidal organic complexes with Fe, Al and Si that do not precipitate. Although in general eastern Lithuanian rivers are less enriched with organic matter than the rivers from other regions, the former are more abundant with acid organic compounds.

Table 2. Wastewater factor loadings at the river monitoring sites where there is a well-pronounced wastewater impact

Parameter	River monitoring stations										
	R101	R103	R104	R87	R88	R91	R30	R98	R37	R85	R31
Q	**–0.70**	–0.46							**–0.53**		
SK				**–0.51**		**–0.94**					
SM				**0.75**		**0.84**					
pH							**–0.66**				
O_2						**–0.62**					
$O_2\%$						**–0.55**					
BOD_7						**0.90**		**0.87**			
COD_{Cr}					**0.55**	**0.77**					
COD_{Mn}					0.44	**0.91**					
NH_4-N	**0.61**		**0.52**	**0.57**	**0.82**	**0.68**	**0.75**	**0.67**			**0.84**
NO_2-N		**0.74**	0.41						0.46	**0.68**	
N_{org}	**0.77**			**0.72**	**0.62**						
PO_4-P				**0.59**	**0.73**		**0.81**		**0.62**	**0.73**	**0.58**
P_{org}	**0.57**			0.46	**0.59**					**0.70**	
Ca^{2+}							**-0.56**				
Mg^{2+}					0.41						**0.71**
Na^+	**0.77**	**0.77**	**0.79**	**0.61**	**0.87**		**0.86**		**0.90**	**0.89**	**0.66**
K^+		**0.70**	**0.50**	**0.62**	**0.85**		**0.83**		**0.87**	**0.79**	**0.76**
Si							**0.83**				
HCO_3^-			**0.59**		**0.60**						
SO_4^{2-}		**0.82**	**0.74**	0.46	**0.78**					**0.87**	
Cl^-	**0.78**	**0.72**	**0.80**	**0.51**	**0.75**		**0.86**		**0.79**	**0.84**	**0.88**
Fe				**0.75**		**0.57**	**0.76**				
Max.*	–	–	–	–	S–W	–	S–W	–	–	–	–

*Max. – the season of maximum wastewater impact on river water quality; S – summer; W – autumn-winter; - no seasonal variation.

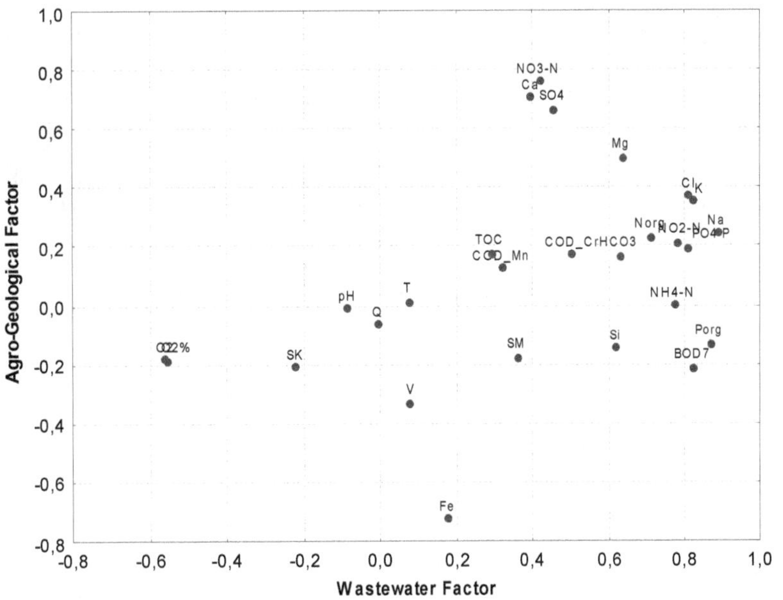

Fig. 2. The wastewater and the agro-geological factor loading plot (autumn-winter season)

The last common factor for all time periods is named as *hardly degradable organics (HDO)* one. It shows up with high loadings on TOC, COD_{Cr} and COD_{Mn} (Fig. 5).

The HDO factor is particularly strongly expressed in northern and central Lithuanian region of fertile soils where the biggest amounts of seasonal vegetation are produced (Figs 6–7). Vegetational detritus finds its way to rivers already partially decomposed on the way, leaving the remaining organic material that breaks down much slower. As expected, the process manifests itself most strongly in summer period when the bulk of vegetation is being produced.

There are also a number of factors that are characteristic only to particular seasons. The *photosynthesis-vegetation factor* is of particular relevance in the summer period. It is reflected by high positive loadings on BOD_7,

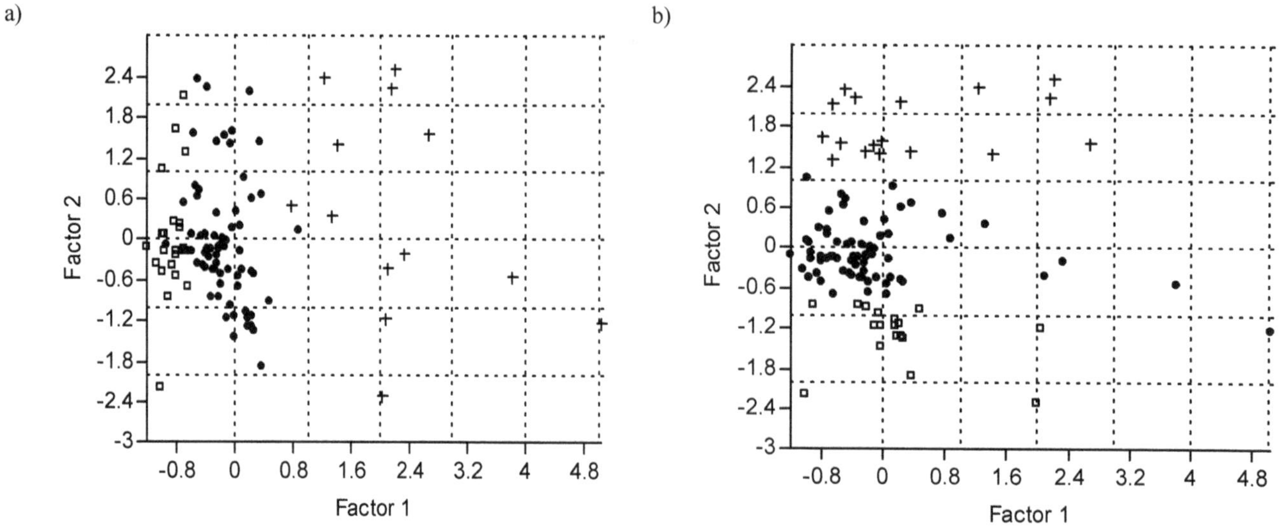

Fig. 3. Groupings of river monitoring stations along the axis of the wastewater (1st) (a) and the agro-geological (2nd) (b) factors in autumn-winter season (boxes – stations with most negative coordinates (factor scores), crosses – stations with most positive coordinates)

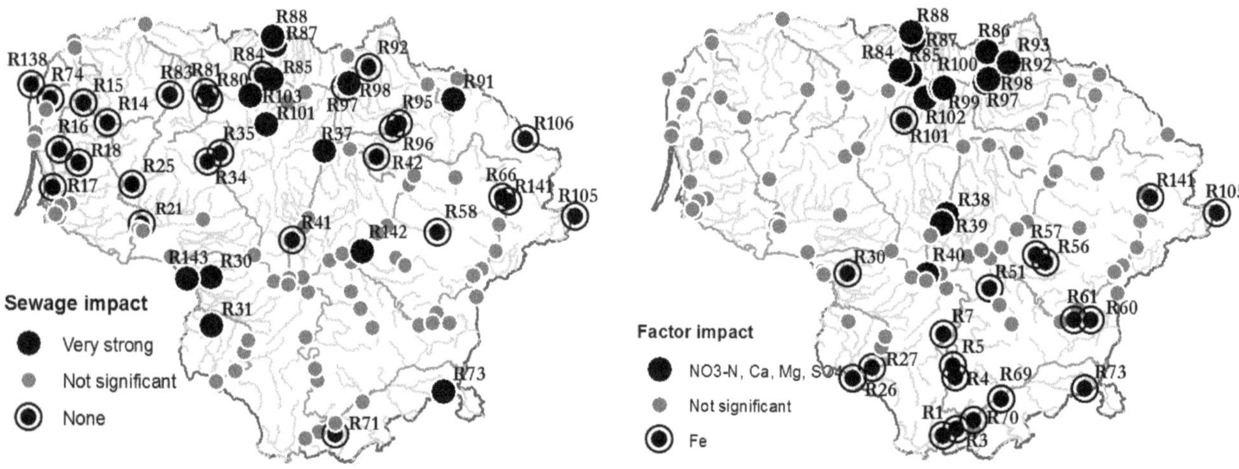

Fig. 4. River monitoring stations significantly affected by the wastewater (sewage, left) and the agro-geological factors (right) in the autumn-winter season

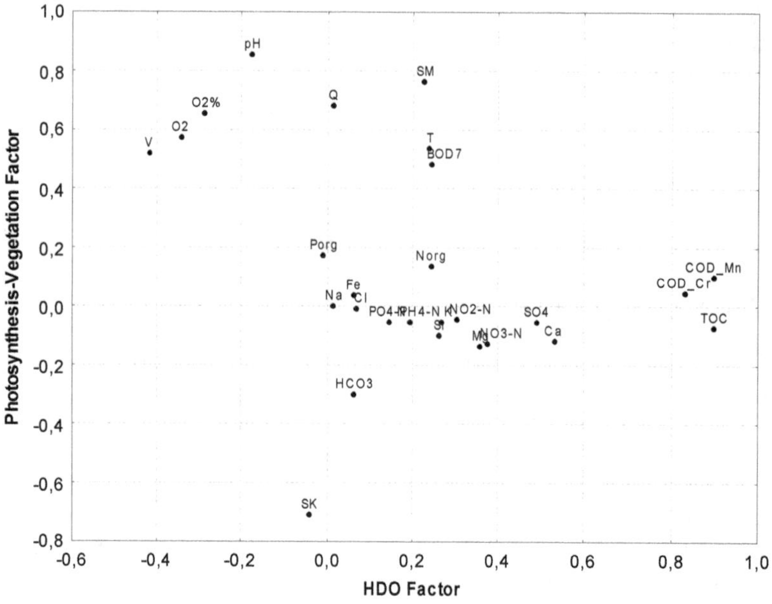

Fig. 5. The hardly degradable organics (HDO) and the photosynthesis-vegetation factor loading plot (summer season)

SM, pH, O_2, O_2%, Q, V, T and high negative loadings on SK (Fig. 5). The process of photosynthesis and production of vegetation organics is driven by intensive proliferation of phytoplankton, which manifests itself in elevated levels of SM and BOD_7 in river water. Photosynthetic reactions push up pH values as well as dissolved oxygen content. This happens despite the rule of thumb that in normal conditions when oxygen is not artificially added and the T increases, O_2 levels usually decrease because of the lower carrying capacity for this element in warmer waters. Greater Q, V and T values indicate big rivers, the waters of which warm up more in the summer. The factor is confined only to the Nemunas and Neris – Lithuanian largest rivers (Figs 6, 7). Here the conditions partly resemble lakes, where a big depth limits growth possibilities for macrophytes. Therefore, the lack of competition and warmer water creates perfect conditions for phytoplankton to thrive.

The conclusions are further confirmed by the results of the local factor analysis at separate monitoring stations in the Neris and Nemunas rivers. As seen from the Table 3, the photosynthesis-vegetation factor is also very much visible – upsurge of BOD_7, SM, pH, O_2, O_2% coincides with higher temperatures, indicating warmer season.

Aeration-organic decay factor is characteristic to autumn-winter period. It scores high on pH, O_2, O_2% as well as on V and Q (Fig. 8). The factor lifts pH, O_2 and O_2% by means of increased turbulence in more water abundant and fast flowing rivers, which include the Nemunas and some western Lithuania rivers that flow from highlands. Turbulence enhances water aeration. It also make it harder for ice to form, therefore conditions for aeration are enhanced while those for organic matter accumulation and decay are significantly worsened. The opposite situation is found in smaller and slower rivers of the Lithuanian north, where weak connection with groundwater makes it even easier for ice to form.

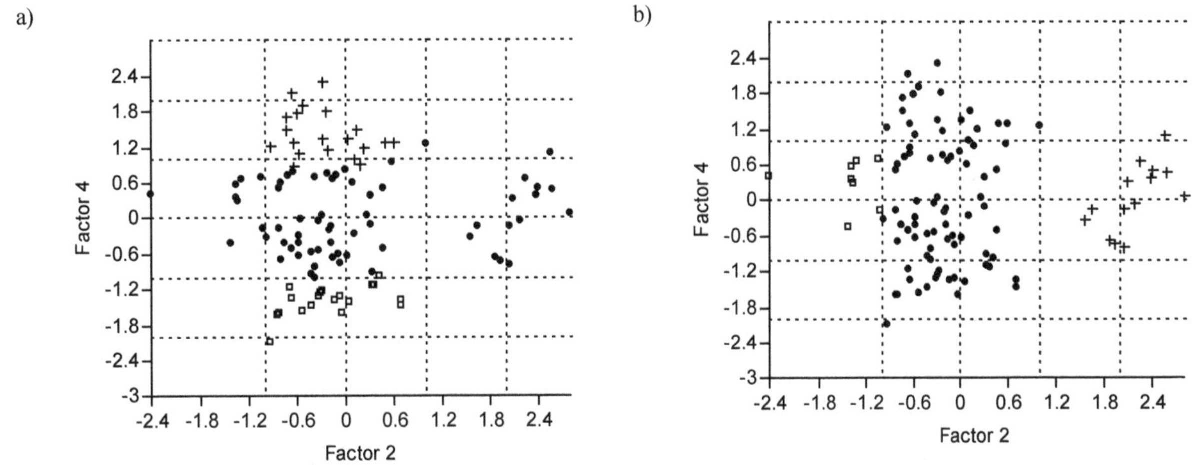

Fig. 6. Groupings of river monitoring stations along the axis of the hardly degradable organics (4th) (a) and the photosynthesis-vegetation (b) factor (2nd) in the summer season (boxes – stations with most negative coordinates (factor scores), crosses – stations with most positive coordinates)

Table 3. Photosynthesis-vegetation factor loadings at the river monitoring sites where photosynthesis processes are well pronounced in the summer period

	River monitoring stations					
Parameter	R11	R4	R48	R1	R44	R43
T	**0.81**	**–0.77**	**0.68**	**–0.79**	**–0.90**	**0.86**
SK	**–0.83**		**–0.82**	**0.82**	**0.78**	**–0.85**
SM	**0.81**	**–0.61**	**0.75**	**–0.68**	**–0.73**	**0.62**
pH	**0.80**	–0.42		**–0.75**	**–0.62**	**0.62**
O_2	**0.51**	**0.63**			–	
O_2%	**0.88**		0.44	–0.47	**–0.77**	**0.66**
BOD_7	**0.87**	**–0.85**	**0.87**	**–0.55**	**–0.83**	**0.78**
COD_{Cr}	**0.58**	**–0.52**	**0.70**		–0.45	
COD_{Mn}	**0.80**	**–0.65**	**0.65**	**–0.61**	–0.47	
NH_4-N				0.40		**–0.69**
NO_2-N	–0.43			**0.88**		**–0.55**
NO_3-N		**0.67**		**0.85**	**0.72**	**–0.85**
N_{org}				**–0.56**		
PO_4-P	**–0.77**	0.48	–0.45	**0.87**	**0.74**	**–0.64**
P_{org}	**0.75**		**0.84**			
Ca^{2+}	–0.49	**0.74**				
Si	**–0.81**				**0.75**	**–0.73**
Max.**	S–Sp	S	S	Sp–S	S–Sp	S–Sp

**Max. – Season of intensive photosynthesis process; S – summer; Sp – spring.

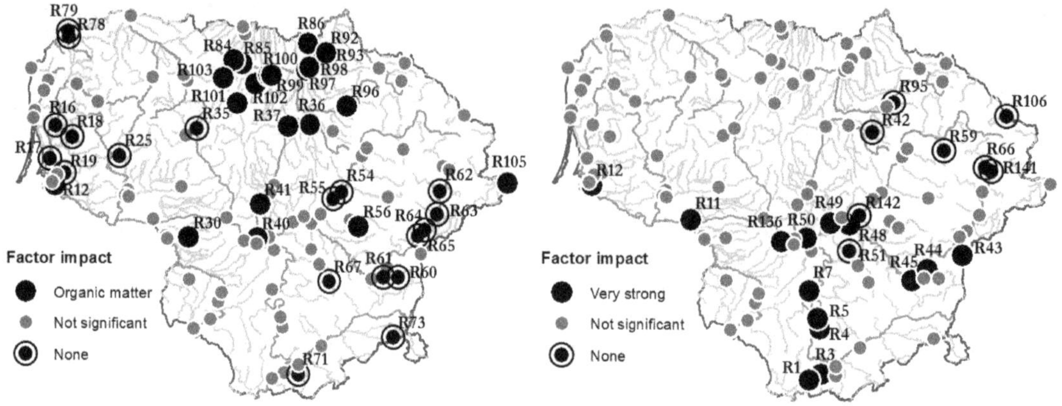

Fig. 7. River monitoring stations significantly affected by the hardly degradable organics (left) and the photosynthesis-vegetation factor (right) in the summer season

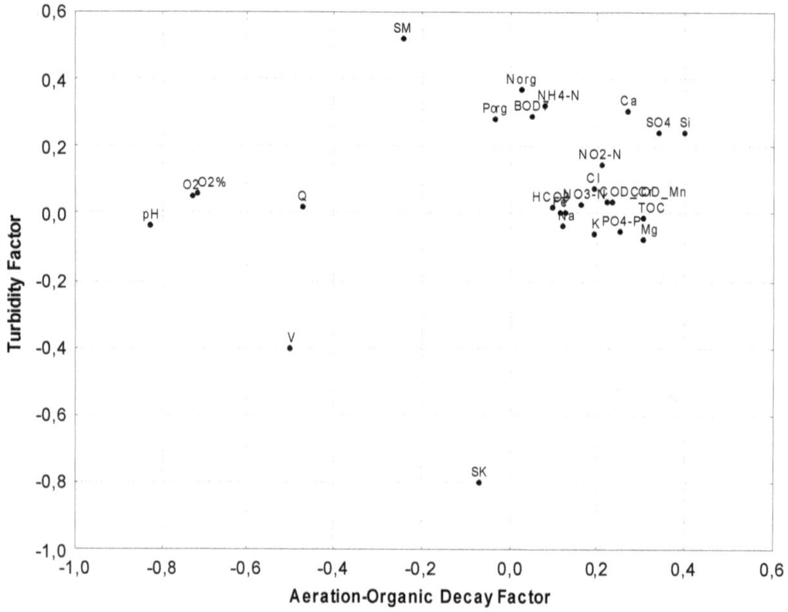

Fig. 8. The aeration-organic decay and the turbidity factor loading plot

Turbidity factor is also an autumn-winter process, which involves high positive loadings on SM and high negative loadings on SK (Fig. 8). The origin of high SM is variable. Strong factor influence in western Lithuanian rivers is related to hilly landscape and frequent floods, what translates into influx of erosion materials (Figs 9, 10). Larger factor scores in some small northern Lithuanian rivers might be due to heavy hardly permeable surfaces, leading to effective flushing of surface runoff to rivers together with SM particles. SM in the Nemunas might be related to turbulence and pick up of material from the river bed.

There are also other factors influencing water quality in different seasons, however they are of less importance. It can be said in general that in all seasons the six factor structure explained approximately 80% of common variation, which seems to be a relatively good result. Na^+, Cl^+ and K^+ are especially well explained parameters. However, there are some indicators, like Fe, Si or HCO_3^-, the explanation of which falls to the lows of 60% in some seasons. In those instances it is clear, that there are some unexplained factors behind those parameters, which are not related to all other quality indicators.

It can be seen from the results that factor analysis is capable of uncovering natural as well as human induced processes that affect water quality of rivers and can be successfully applied to Lithuanian situation.

The soundness of the FA has been subjected to a wide range of applications in water quality studies. One group of researchers (Papatheodorou *et al.* 2006; Spanos *et al.* 2003) used time series at one particular sampling location as input for factor analysis, termed as local factor analysis in this study. This makes it possible to distinguish among season dependent factors and the independent ones. The samples then are usually grouped according to affecting factors. This approach enables a detailed examination of processes in a particular location. Another group of authors (Alberto *et al.* 2001; Aruga *et al.* 1995; Boyacioglu, H., Boyacioglu, H. 2007; Brodnjak-Voncina *et al.* 2002; Charkhabi *et al.* 2006; Fitzpatrick *et al.* 2007; Kannel *et al.* 2007; Kowalkowski *et al.* 2006; Marques da Silva *et al.* 2001; Mendiguchia *et al.* 2004; Morales *et al.* 1999; Omo-Irabor *et al.* 2008; Ouyang *et al.* 2006; Qadir *et al.* 2008; Santos-Roman *et al.* 2003; Schaefer, Einax 2010; Simeonov *et al.* 2002; Simeonov *et al.* 2003; Singh

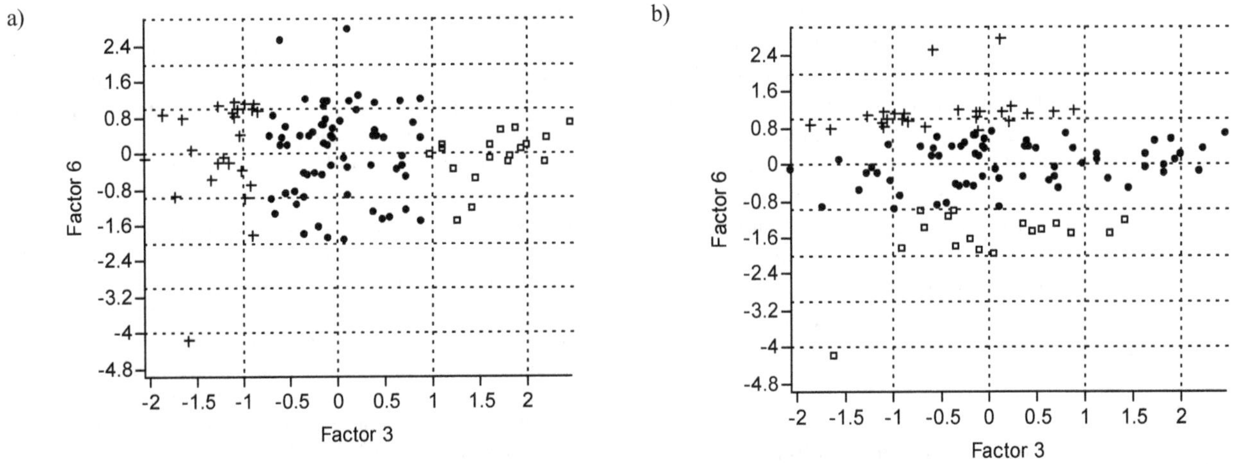

Fig. 9. Groupings of river monitoring stations along the axis of the aeration-organic (a) decay (3rd) and the turbidity (b) factor (6th) in the autumn-winter season (boxes – stations with most negative coordinates (factor scores), crosses – stations with most positive coordinates)

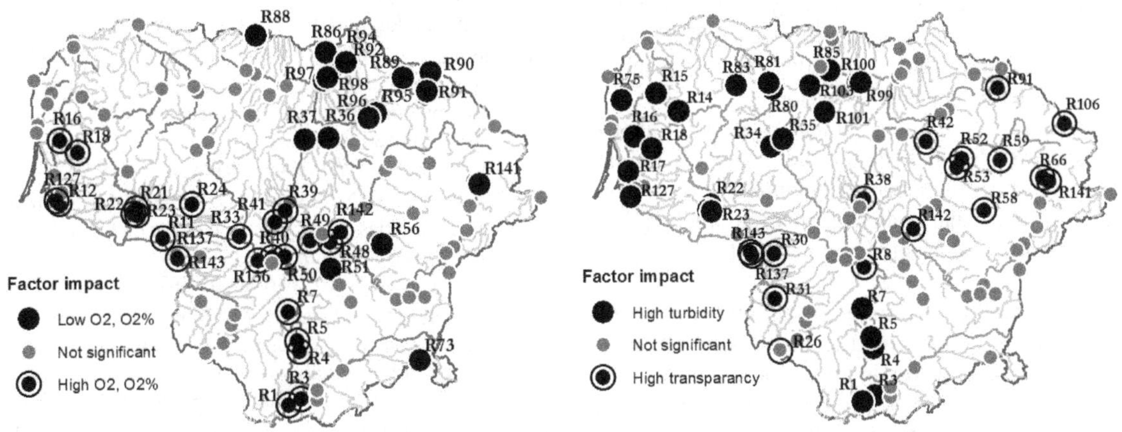

Fig. 10. River monitoring stations significantly affected by the aeration-organic decay (left) and the turbidity factor (right) in the autumn-winter season

et al. 2006; Spanos *et al.* 2003; Tarrado *et al.* 2006; Vega *et al.* 1998; Vidal *et al.* 2000; Voutsa *et al.* 2001; Zhou *et al.* 2007) pool the data from many sampling sites together to act as an input for factor analysis, termed as spatial factor analysis in this study. This type of method application is more common since it provides an opportunity to extract spatial factors acting in space and to group sampling sites and even samples according different factors. This type of method utilization prooved to be very convenient in terms of visualisation. For instance, Tarrado *et al.* (2006) visualized factor scores gradient along the river Ebro in Spain.

In addition to solitary use the method can be succesfuly applied in conjunction with other multivariate techniques. The most often combination is with cluster analysis, when factor analysis is applied to search for affecting processes in separate clusters (Fitzpatrick *et al.* 2007; Kannel *et al.* 2007; Reisenhofer *et al.* 1998; Schaefer *et al.* 2010; Simeonov *et al.* 2002; Zhang *et al.* 2009) or to confirm the designated clusters (Alberto *et al.* 2001; Boyacioglu, H., Boyacioglu, H. 2008; Mendiguchia *et al.* 2004; Omo-Irabor *et al.* 2008; Razmkhah *et al.* 2010; Simeonov *et al.* 2003; Singh *et al.* 2006; Vidal *et al.* 2000).

In practice when investigating river water status and the reasons behind it, it is recommended to apply factor analysis in combination with water quality modeling and other multivariate statistical methods. When using those methods in combination, modeling approach has an advantage in quantification of inputs and impacts of different pollution sources on river water quality – this is a feature needed for the selection of water status enhancing measures and the estimation of their potential effectiveness. However, factor analysis could potentially identify impacts and processes that the models could not reveal. This is quite a common occurrence, because modeling results are very much dependent on their structure and the input data – if there is a lack of information on particular sources of substances or it is even not expected that particular pollution sources exist in the investigated area, then this contaminant flow will not be taken into account in the modeling results. Another advantage of factor analysis is its ability to uncover situations, when water quality problems are caused by natural processes and it is not feasible to take any measures to improve water quality.

It is recommended to consider factor analysis as one of possible tools when optimizing river monitoring network and the parameters analyzed. In this case factor analysis tool would be most appropriate if the task is to select stations, representative of particular types of human impacts or natural processes as well as to select the most important processes-related parameters to measure there. However, when the task is to identify stations representative of stations groups with different levels of physico-chemical elements concentrations, the combination of cluster (CA) and discriminant analysis (DA) is recommended. CA and DA will enable to group stations according to the similarity of the concentrations of the substances as well as to identify parameters that differ the most among clusters. The identified parameters should be subject to monitoring afterwards. If there is a need to satisfy all the aforementioned needs then a combined use of FA, CA and DA methods is recommended.

Conclusions

1. The status of water ecosystems and the reasons behind it can be determined in a confident and complex way by simultaneously assessing all physico-chemical data and its interrelationships by the use of appropriate multivariate statistical methods. Such kind of the analysis empowers the understanding of not only separate water ecosystem components but also of the whole interacting system, enabling to take better water management decisions.

2. The combination of spatial and local factor analysis can be successfully applied as a tool, which can simultaneously group river monitoring sites into groups of similar values of particular water quality parameters and identify main processes that determine water quality in those groups.

3. The *Wastewater*, *Agro-geological* and *Hardly degradable organics* factors are the main ones affecting Lithuanian river water quality in all seasons.

4. *Wasterwater* factor is prominent in small rivers downstream larger towns and is reflected by low oxygen and high NH_4-N, NO_2-N, N_{org}, PO_4-P, P_{org}, COD_{Cr}, Na^+, K^+, Cl^- concentrations in water.

5. *Agro-geological* factor is pronounced in northern Lithuania's rivers of heavy carbonated soils and intensively used agricultural lands in their catchments which determine elevated Ca^{2+}, Mg^2+, HCO_3^- and SO_4^{2-} content in streamwater, as well as in south-eastern Lithuania's rivers where more acidic soils prevail in their watersheds determining higher abundance of Fe and Si.

6 *Hardly degradable organics* factor is mostly evident in northern and middle Lithuania's rivers draining heavy-textured soils in fertile agricultural areas securing increased quantities of COD_{Cr}, COD_{Mn} and TOC.

7 The *photosynthesis-vegetation* factor is acting only in the summer season and only in Lithuanian largest rivers, and is represented by abundance of suspended organics (SM, BOD_7), elevated pH and dissolved oxygen levels, resulting from intensive algal photosynthetic activity that is common in big rivers.

8. The *Aeration* and *Turbidity* factors represent the processes that take place in autumn-winter only. The former being noticeable by relatively high dissolved oxygen content in large Lithuanian rivers, while the latter being known for turbid waters of hilly region of western Lithuania.

References

Alberto, W. D.; Del Pilar, D. M.; Valeria, A. M.; Fabiana, P. S.; Cecilia, H. A.; De Los Angeles, B. M. 2001. Pattern recognition techniques for the evaluation of spatial and temporal variations in water quality. a case study: Suquia river basin (Cordoba-Argentina), *Wat. Res.* 35: 2881–2894. http://dx.doi.org/10.1016/S0043-1354(00)00592-3

Aruga, R.; Gastaldi, D.; Negro, G.; Ostacoli, G. 1995. Pollution of a river basin and its evolution with time studied by multivariate statistical analysis, *Analytica Chimica Acta* 310: 15–25. http://dx.doi.org/10.1016/0003-2670(95)00101-5

Azzellino, A.; Salvetti, R.; Vismara, R.; Bonomo, L. 2006. Combined use of the EPA-QUAL2E simulation model and factor analysis to assess the source apportionment of point and non point loads of nutrients to surface waters, *Science of the Total Environment* 371: 214–222. http://dx.doi.org/10.1016/j.scitotenv.2006.03.022

Boyacioglu, H.; Boyacioglu, H. 2007. Surface water quality assessment by environmetric methods, *Environ. Monit. Assess.* 131: 371–376. http://dx.doi.org/10.1007/s10661-006-9482-4

Boyacioglu, H.; Boyacioglu, H. 2008. investigation of temporal trends in hydrochemical quality of surface water in Western Turkey, *Bulletin of Environmental Contamination and Toxicology* 80: 469–474. http://dx.doi.org/10.1007/s00128-008-9439-0

Brodnjak-Voncina, D.; Dobcnik, D.; Novic, M.; Zupan, J. 2002. Chemometrics characterisation of the quality of river water, *Analytica Chimica Acta* 462: 87–100. http://dx.doi.org/10.1016/S0003-2670(02)00298-2

Charkhabi, A. H.; Sakizadeh, M. 2006. Assessment of spatial variation of water quality parameters in the most polluted branch of the Anzali wetland, Northern Iran, *Polish Journal of Environmental Studies* 15: 395–403.

Directive 2000/60/EC of the European Parliament and of the Council of 23 October 2000 establishing a framework for Community action in the field of water policy. *Official Journal (OJ L 327).*

Fitzpatrick, M. L.; Long, D. T.; Pijanowski, B. C. 2007. Exploring the effects of urban and agricultural land use on surface water chemistry, across a regional watershed, using multivariate statistics, *Applied Geochemistry* 22: 1825–1840. http://dx.doi.org/10.1016/j.apgeochem.2007.03.047

Kannel, R. P.; Lee, S.; Kanel, S. R.; Kahn, S. P. 2007. Chemometric application in classification and assessment of monitoring location of an urban river system, *Analytical Chimica Acta* 582: 390–399. http://dx.doi.org/10.1016/j.aca.2006.09.006

Koklu, R.; Sengorur, B.; Topal, B. 2010. water quality assessment using multivariate statistical methods-a case study: melen river system (Turkey), *Water Resources Management* 24: 959–978. http://dx.doi.org/10.1007/s11269-009-9481-7

Kowalkowski, T.; Zbytniewski, R.; Szpejna, J.; Buszewski, B. 2006. Application of chemometrics in river water classification, *Water Research* 40: 744–752. http://dx.doi.org/10.1016/j.watres.2005.11.042

Kunwar, P. S.; Malik, A.; Sinha, S. 2005. Water quality assessment and apportionment of pollution sources of Gomti river (India) using multivariate statistical techniques – a case study, *Analytica Chimica Acta* 538: 355–374. http://dx.doi.org/10.1016/j.aca.2005.02.006

Marques da Silva, A. M.; Sacomani, L. B. 2001. Using chemical and physical parameters to define the quality of Pardo river water (Botucatu – SP-Brazil), *Water Research* 35: 1609–1616. http://dx.doi.org/10.1016/S0043-1354(00)00415-2

Mendiguchia, C.; Moreno, C.; Galindo-Riano, M. D.; Garcia-Vargas, M. 2004. Using chemometric tools to assess anthropogenic effects in river water. a case study: Guadalquivir river (Spain), *Analytica Chimica Acta* 515: 143–149. http://dx.doi.org/10.1016/j.aca.2004.01.058

Morales, M. M.; Marti, P.; Llopis, A.; Campos, L.; Sagrado, S. 1999. An environmental study by factor analysis of surface seawaters in the gulf of Valencia (Western Mediterranean), *Analytica Chimica Acta* 394: 109–117. http://dx.doi.org/10.1016/S0003-2670(99)00198-1

Omo-Irabor, O. O.; Olobaniyi, S. B.; Oduyemi, K.; Akunna, J. 2008. Surface and groundwater water quality assessment using multivariate analytical methods: a case study of the Western Niger Delta, Nigeria, *Physics and Chemistry of the Earth* 33: 666–673. http://dx.doi.org/10.1016/j.pce.2008.06.019

Ouyang, Y.; Nkedi-Kizza, P.; Wu, Q. T.; Shinde, D.; Huang, C. H. 2006. Assessment of seasonal variations in surface water quality, *Water Research* 40: 3800–3810. http://dx.doi.org/10.1016/j.watres.2006.08.030

Papatheodorou, G.; Demopoulou, G.; Lambrakis, N. 2006. A long-term study of temporal hydrochemical data in a shallow lake using multivariate statistical techniques, *Ecological Modelling* 193: 759–776. http://dx.doi.org/10.1016/j.ecolmodel.2005.09.004

Povilaitis, A. 2003. Pagrindinių jonų koncentracijų karstinio regiono upėse panašumų ir skirtumų statistinis įvertinimas [Statistical assessment of similarities and differences in the concentrations of the main ions in the carstic region rivers], *Vandens ūkio inžinerija* 22(44): 5–14.

Qadir, A.; Malik, R. N.; Husain, S. Z. 2008. Spatio-temporal variations in water quality of Nullah Aik-tributary of the river Chenab, Pakistan, *Environ. Monit. Assess.* 140: 43–59. http://dx.doi.org/10.1007/s10661-007-9846-4

Razmkhah, H.; Abrishamchi, A.; Torkian, A. 2010. Evaluation of spatial and temporal variation in water quality by pattern recognition techniques: a case study on Jajrood River (Tehran, Iran), *Journal of Environmental Management* 91: 852–860. http://dx.doi.org/10.1016/j.jenvman.2009.11.001

Reisenhofer, E.; Adami, G.; Barbieri, P. 1998. Using chemical and physical parameters to define the quality of karstic freshwaters (Timavo River, North-eastern Italy): a chemometric approach, *Water Research* 32: 1193–1203. http://dx.doi.org/10.1016/S0043-1354(97)00325-4

Santos-Roman, D. M.; Warner, G. S.; Scatena, F. 2003. Multivariate analysis of water quality and physical characteristics of selected watersheds in Puerto Rico, *Journal of the American Water Resources Association* 39: 829–839. http://dx.doi.org/10.1111/j.1752-1688.2003.tb04408.x

Schaefer, K.; Einax, J. W. 2010. Analytical and chemometric characterization of the Cruces River in South Chile, *Environmental Science and Pollution Research* 17: 115–123. http://dx.doi.org/10.1007/s11356-009-0116-6

Simeonov, V.; Einax, J. W.; Stanimirova, I.; Kraft, J. 2002. Environmetric modeling and interpretation of river water monitoring data, *Analytical and Bioanalytical Chemistry* 374: 898–905. http://dx.doi.org/10.1007/s00216-002-1559-5

Simeonov, V.; Stratis, J. A.; Samara, C.; Zachariadis, G.; Voutsa, D.; Anthemidis, A.; Sofoniou, M.; Kouimtzis, T. 2003. Assessment of the surface water quality in Northern Greece, *Water Research* 37: 4119–4124. http://dx.doi.org/10.1016/S0043-1354(03)00398-1

Singh, K. P.; Malik, A.; Singh, V. K. 2006. Chemometric analysis of hydro-chemical data of an alluvial river - a case study, *Water, Air, and Soil Pollution* 170: 383–404. http://dx.doi.org/10.1007/s11270-005-9010-0

Sojka, M.; Siepak, M.; Ziola, A.; Frankowski, M.; Murat-Blazewska, S.; Siepak, J. 2008. Application of multivariate statistical techniques to evaluation of water quality in the Mala Welna River (Western Poland), *Environ. Monit. Assess.* 147: 159–170. http://dx.doi.org/10.1007/s10661-007-0107-3

Spanos, T.; Simeonov, V.; Stratis, J.; Xristina, X. 2003. Assessment of water quality for human consumption, *Microchim Acta* 141: 35–40. http://dx.doi.org/10.1007/s00604-002-0921-9

Tarrado, M.; Barcelo, D.; Tauler, R. 2006. Identification and distribution of contamination sources in the Erbo river basin by chemometrics modeling coupled to geographical information systems, *Talanta* 70: 691–704. http://dx.doi.org/10.1016/j.talanta.2006.05.041

Vaikasas, S. 2010. Mathematical modelling of sediment dynamics and their deposition in Lithuanian rivers and their deltas (case studies), *Journal of Environmental Engineering and Landscape Management* 18(3): 207–216. http://dx.doi.org/10.3846/jeelm.2010.24

Vega, M.; Pardo, R.; Barrado, E.; Deban, L. 1998. Assessment of seasonal and polluting effects on the quality of river water by exploratory data analysis, *Water Research* 32: 3581–3592. http://dx.doi.org/10.1016/S0043-1354(98)00138-9

Vidal, M.; Lopez, A.; Santoalla, M. C.; Valles, V. 2000. Factor analysis for the study of water resources contamination due to the use of livestock slurries as fertilizer, *Agricultural Water Management* 45: 1–15. http://dx.doi.org/10.1016/S0378-3774(99)00073-6

Voutsa, D.; Manoli, E.; Samara, C.; Sofoniou, M.; Stratis, I. 2001. A study of surface water quality in Macedonia, Greece: speciation of nitrogen and phosphorus, *Wat., Air. Soil. Poll.* 129: 13–32. http://dx.doi.org/10.1023/A:1010315608905

Zhang, Y.; Guo, F.; Meng, W.; Wang, X. 2009. Water quality assessment and source identification of Daliao river basin using multivariate statistical methods, *Environmental Monitoring and Assessment* 152: 105–121. http://dx.doi.org/10.1007/s10661-008-0300-z

Zhou, F.; Guo, H.; Liu, Y.; Jiang, Y. 2007. Chemometrics data analysis of marine water quality and source identification in Southern Hong Kong, *Marine Pollution Bulletin* 54: 745–756. http://dx.doi.org/10.1016/j.marpolbul.2007.01.006

Mindaugas GUDAS. Doctor of Environmental Engineering and Landscape Management (degree awarded in 2010 at the Lithuanian University of Agriculture), Director of the Environmental Status Assessment Department at the Lithuanian Environmental Protection Agency. Publications: author of 5 research papers. Main research areas: water quality, water quality improvement measures, statistical methods.

Arvydas POVILAITIS. Dr (HP), Professor of the Department of Water Management, Aleksandras Stulginskis University (ASU). Publications: author of over 60 research papers. Main research areas: hydrology, water quality, statistical methods, mathematical modelling.

FRACTIONATION OF HEAVY METALS IN SEWAGE SLUDGE AND THEIR REMOVAL USING LOW-MOLECULAR-WEIGHT ORGANIC ACIDS

Ernestas Zaleckas[1], Valdas Paulauskas[2], Eglė Sendžikienė[3]

[1,3]*Institute of Environment, Aleksandras Stulginskis University, Studentų g. 11, LT-53361, Akademija, Kauno r., Lithuania*

[2]*Department of Chemistry, Aleksandras Stulginskis University, Studentų g. 11, LT-53361, Akademija, Kauno r., Lithuania*

Abstract. The total concentration and the concentrations of individual chemical species of selected heavy metals were estimated in primary and anaerobically digested sewage sludge. The concentration of Zn (1503 mg/kg) was highest and was followed by Cu (201 mg/kg), Cr (196 mg/kg), Pb (56 mg/kg), Ni (44 mg/kg) and Cd (3.6 mg/kg). The metal was divided into 5 fractions (exchangeable (F1), adsorbed (F2), organically bound (F3), bound to carbonates (F4), and residual (F5)) via sequential extraction. The sludge treatment procedure had no significant effect on the fractionation results. In both the primary and anaerobically digested sewage sludge, the heavy metals were ranked according to their mobilities (fractions F1 and F2) in the following order: Ni > Zn > Cu > Cd > Pb > > Cr. Metal stability in the environment was evaluated by the sulphide and residual fraction F5, and the following ranking order was identified: Cr > > Pb ≈ Ni > Cd > Zn ≈ Cu. A leaching experiment with low-molecular-weight organic acids (oxalic, acetic and citric acid) revealed that the metal-removal efficiency varied depending on the number of carboxyl groups in the extracting agent, the chemical speciation of the metal (Ni, Zn or Cu) in the sludge and the concentration and pH change of the extracting solution. Acid solutions with a 0.5 M concentration, ranked according to their Zn-removal efficiency, are ranked as follows: citric acid (100%) > acetic acid (78%) > oxalic acid (71%). In all of the cases, citric acid showed the best capacity for the removal of metal from the sludge, with an extraction efficiency ranging from 30–100%, while the Ni and Cu removal efficiencies with the acetic and oxalic acid were less than 40%.

Keywords: sewage sludge, heavy metals, sequential extraction, heavy metal removal, organic acids.

Reference to this paper should be made as follows: Zaleckas, E.; Paulauskas, V.; Sendžikienė, E. 2013. Fractionation of heavy metals in sewage sludge and their removal using low-molecular-weight organic acids, *Journal of Environmental Engineering and Landscape Management* 21(3): 189–198. http://dx.doi.org/10.3846/16486897.2012.695734

Introduction

The safe disposal of municipal sewage sludge (SS) remains a major environmental problem both in Lithuania and worldwide. Lithuanian and EU environmental requirements encourage the development of modern and effective wastewater treatment technologies; therefore, the quantities of sludge are steadily increasing. The current sludge-processing methods pose a threat to the environment and contradict both long-term public interests and the principles of sustainable development. The safe disposal of sewage sludge is encouraged by the chief principles of the EU waste-management policy: avoidance of waste, volume and toxicity reduction and waste recycling.

Environmentally sound sewage sludge disposal in Lithuania is still an open question, and the amount of unutilised wastewater sewage sludge grows constantly at a rate of more than 500,000 t per year. Every year, approximately 20 thousand tonnes of sewage sludge are composted and applied to arable lands in Lithuania (Podgaiskytė, Vaitiekūnas 2009). Over 90% of the sewage sludge generated in Lithuania is stored in special deposits that are not always properly equipped. This type of sludge utilisation requires tens of hectares of land annually, which could be used for other public needs. In addition, such sludge landfilling poses a substantial threat to groundwater and surface-water quality.

Of the various methods for sewage sludge disposal, low-cost and highly effective on-land techniques have been widely applied (Bode, Imhoff 1996; Bowen *et al.* 1992; Mossakowska *et al.* 1998; Paulsrud, Nedland 1997). In most EU countries, at least half of the sludge that is produced in wastewater treatment plants is applied to agricultural land. The application of sewage

Corresponding author: Ernestas Zaleckas
E-mail: ernestas.zaleckas@gmail.com

sludge to agricultural soils has many advantages, which include providing a full array of nutrients for plant growth (e.g. phosphorus and nitrogen) and organic matter, thereby increasing the number of beneficial soil organisms, reducing the need for fertilisers and pesticides, and improving the physical and biological properties of the soil (Kvasauskas, Baltrėnas 2009; Su, Wong 2003). However, a wide variety of undesirable elements, such as heavy metals (HMs), can be found in sewage sludge and have adverse effects on the environment (Alloway 1995; Jaskelevičius, Lynikienė 2009; Idzelis *et al.* 2010). Many countries have imposed legal restrictions concerning HM content when sludge is used for on-land applications (e.g. agricultural use as an organic fertiliser or soil amendment) because high concentrations in the sludge may result in heavy metal accumulation in cultivated soils. In the European Union, the maximum acceptable total HM concentrations are fixed by Directive 86/278/EEC of the European Commission (CEC 1986); however, most EU countries enforce more stringent local legislation. In Lithuania, the HM limits in sewage sludge are fixed in LAND 20:2005 ("Requirements for Sewage Sludge Usage for Fertilisation and Remediation"). Furthermore, several studies have shown that the available fraction of heavy metals mainly influences the mobility, bioavailability or phytotoxicity of heavy metals in soils (Filgueiras *et al.* 2004; Rauret 1998). Therefore, the quantification of the various chemical fractions of heavy metals in sludge is necessary for a more detailed assessment of metal mobility, metal bioavailability and phytotoxicity.

The behaviours of heavy metals in soils and their plant uptake are strongly dependent on the nature of the metal, the physico-chemical properties of the sludge/soil and the plant species (McBride 2003). The mobilities of trace metals, their bioavailability and their related toxicity to plants strongly depend on their specific chemical forms or the way that they are bound. Consequently, these factors, rather than the total elemental content, are the parameters that must be determined to assess toxic effects and study geochemical pathways (Legret 1993; Pueyo *et al.* 2001). Sludge-derived metals are generally organically bound and less available for plant uptake than the more mobile metal salt impurities found in commercial fertilisers. In recent decades, a large number of extraction schemes have been developed for fractionation purposes and, although a number of methods have been widely used (Emmerich *et al.* 1982; Förstner, Salomons 1980; Tessier *et al.* 1979), none has been unreservedly accepted by the scientific community. In the study described in this paper, a five-step sequential extraction method was applied to Cd, Cr, Cu, Fe, Ni, Pb and Zn fractionation in accordance with the scheme proposed by Emmerich *et al.* (1982).

Heavy metals in sludge cannot be removed by common sludge treatment methods, such as aerobic or anaerobic digestion (Chipasa 2003). Some recently studied advanced sludge treatment processes aimed to reduce the potential environmental risks of sludge. According to the literature (Jankaitė, Vasarevicius 2005; Jankaitė 2009; Mench *et al.* 1994; Paulauskas *et al.* 2006), the remediation methods for soil and sewage sludge contaminated with heavy metals include solidification/stabilisation, vitrification, electrokinetic remediation, chemical extraction, phytoextraction, phytostabilisation, and chemical stabilisation. Each of these techniques has advantages and disadvantages, depending on the cleaned object, its quantity, the pollution levels and other factors. Most of these technologies are still in the research stage. There is a need for more detailed studies to understand better remediation processes in polluted media.

One of the principal technologies used for the removal of heavy metals from contaminated sewage sludge and soils is chemical extraction. The process of heavy metal acidic leaching uses the principle that metal solubilisation occurs as a result of exchanging the protons supplied by the addition of an acidic solution. The most effective chemical extraction technique is washing sewage sludge with strong acids. However, the aggressive agents used during these chemical methods may destroy the soil structure and result in secondary pollution (Cline, Reed 1995).

The use of compounds that are common in nature, such as easily biodegradable, low-molecular-weight (LMW) organic acids, is more publicly acceptable than synthetic chelate or strong mineral acid extraction technologies. Dacera and Babel (2008) investigated the potential of using raw liquids from pineapple waste fermented by *A. niger* as a source of citric acid to extract chromium, copper, lead, nickel and zinc from anaerobically digested sewage sludge. These researchers studied the effects of pH and leaching time on the extraction efficiency (EE), and it was found that the liquid seemed to remove all of the Cr after 11 days at pH 3.73 and 3.98; Ni and Zn were removed at 94% and 100% EE, respectively, at the same leaching time but only at pH 3.73. The study did not compare the extraction efficiency of citric acid to that of other LMW organic acids. Veeken and Hamelers (1999) concluded that citric and oxalic acid have increased heavy metal extraction capacities (70% and 60%, respectively, for Cu; 90% and 70%, respectively, for Zn) at a mildly acidic pH of 3–4, as compared to HNO_3 (38% and 65% for Cu and Zn, respectively, at a pH of 1.5–2). However, these researchers paid little attention to the metal chemical species.

The objectives of this study were as follows: to chemically fractionate heavy metals in primary and anaerobically digested sewage sludge; to compare the copper, zinc and nickel extraction efficiencies from sewage sludge using three different low-molecular-weight organic acids, including citric acid (CA), oxalic acid (OA) and acetic acid (AA); to evaluate the influences of factors as concentration, pH and the metal chemical species on the removal efficiency.

1. Materials and Methods

1.1. Sludge sampling and sample pre-treatment

Primary and anaerobically digested sewage sludge were collected at a wastewater treatment plant (WTP) in Kaunas, Lithuania. The sewage sludge samples were air-dried, ground to pass through a 2-mm sieve and stored in dry conditions at room temperature without access to light (in the dark glass bottles). The SS samples were dried to a constant mass at 105 °C to determine their water content. The results of all of the measurements were expressed on a dry mass (DM) basis.

1.2. Equipment

An Agilent 7500i inductively coupled plasma spectrometer with a mass detector (ICP-MS) was used to determine the total contents of 26 different elements, including HM, in the sewage sludge samples. A Perkin Elmer AAnalyst 100 atomic absorption spectrometer (AAS) with deuterium background correction was used to determine the content of Cd, Cr, Cu, Ni, Pb and Zn during the fractionation and removal experiments. A fuel-lean air-acetylene flame was used for all of the metals, except Cr, which requires the stoichiometric conditions of flame gases. When the sensitivity of flame atomic absorption spectrometry was insufficient for determining the HM in the extracts, AAS with electrothermal atomisation (Perkin Elmer HGA 800) was applied. $NH_4H_2PO_4$ (1000 mg/l) was used as a matrix modifier to determine the presence of Cd.

1.3. Reagents

All of the reagents were of analytical grade. Suprapur acids (Merck) and deionised water obtained from a Millipore MilliQ system were used to prepare the samples, extraction solutions and standard solutions. A stock standard solution of HM (1000 ± 2 mg/dm^3 in 5% HNO_3) was obtained from Merck.

The potassium nitrate, sodium hydroxide and nitric acid used in the sequential extraction procedure were obtained from Merck, and disodium ethylenediamine tetraacetate was obtained from Fluka.

All of the standards, reagent solutions and samples were stored in glass containers that were previously cleaned with 4 M HNO_3, rinsed with deionised water and dried.

Sartorius cellulose nitrate membrane filters (pore size, 0.45 μm) were used during the filtration procedure.

1.4. The total concentration of the heavy metals

For the total HM values, the sludge solids were digested with concentrated HNO_3 according to the slightly modified ISO procedure described below. A 0.2000-g quantity of the air-dried sample was accurately added to a 50-ml polytetrafluorethylene (PTFE) crucible and digested with 5-ml of concentrated nitric acid in a closed PTFE vessel on a heat plate at 170 °C. Heating was stopped after 6 h, and the crucible was maintained overnight and allowed to digest while slowly cooling to a constant (room) temperature. Next, the residue was centrifuged and diluted to a volume of 50 ml. The solution was subsequently analysed via ICP-MS using the Agilent 7500i. The obtained data are presented as mg/kg of dry matter (DM) in Table 2.

1.5. The sequential extraction procedure

Sequential extraction was performed using a five-step fractionation procedure recommended by Emmerich *et al.* (1982) in which the metals are divided into five fractions: exchangeable (F1), adsorbed (F2), bound to organic matter (F3), bound to carbonates (F4) and residual (F5) (Table 1).

The extraction procedure was performed with 1.0000 g of air-dried sewage sludge. After each successive extraction, the mixture was centrifuged at 3000 rpm for 30 min, the supernatant was decanted into polyethylene bottles, and the metal concentration in the filtered solution was determined via AAS using a flame or graphite furnace atomiser. The amount of the extracted heavy metal in each fraction was calculated as a percentage ratio of the total metal content. This analysis was performed on three parallel sewage-sludge samples.

Table 1. Sequential extraction procedure according to Emmerich *et al.* (1982)

		Extraction conditions	
Fraction	Chemical form of heavy metal	Reagent	Time and temperature
F1	Exchangeable	25 ml 0.5M KNO_3	16 h at 20 °C
F2	Adsorbed	25 ml deionized H_2O	2 h at 20 °C
F3	Bound to organic matter	25 ml 0.5M NaOH	16 h at 20 °C
F4	Bound to carbonates	25 ml 0.05M Na_2EDTA	6 h at 20 °C
F5	Residual	30 ml 4M HNO_3	16 h at 80 °C

1.6. Heavy-metal extraction using low-molecular-weight organic acids

Heavy-metal extraction with LMW organic acids (acetic, oxalic and citric) was performed using solutions of various concentrations (0.01, 0.05, 0.1, 0.5 and 1.0 M) at a liquid (ml) /solid (g) ratio of 50/1. Extraction was performed using a Heidolph® Reax overhead shaker (60 rpm) at a temperature of 20 °C for 24 h. The pH of all of the solutions was measured before and after extraction. At the end of extraction, the samples were centrifuged and filtered, and the concentration of HM in the solution was determined by applying the flame or graphite furnace AAS procedure.

The results of all of the experiments were collected as an average value derived from three established values of the measured parameter. Any values that differed from the other two values by more than 30% were excluded. The results were statistically evaluated using MS Excel and STATENG.

2. Investigation results

2.1. The total amount of heavy metals in the sewage sludge

ICP-MS analysis of 26 elements in the sludge, including Cd, Cr, Pb, Cu, Ni and Zn (Table 2), identified a wide variation in the total concentration of heavy metals in the investigated samples. In general, the Kaunas WTP sludge samples had low concentrations of Cr, Ni, Pb and Cd but relatively high concentrations of Cu and Zn compared to the limit values fixed in LAND 20:2005. However, the total concentration of heavy metals in the sludges reported by Ščančr *et al.* (2000), Wang *et al.* (2005), Alonso Álvarez *et al.* (2002) and others were much higher than the results of this investigation.

Table 2. Elemental composition of anaerobically digested sewage sludge (ICP-MS data)

Element	Unit mg/kg DM	Element	Unit mg/kg DM
Cr	**196.00±6.8**	Ba	811.87±16.2
Cd	**3.63±1.1**	Li	14.30±2.8
Ni	**44.06±3.5**	B	196.60±5.9
Pb	**56.00±5.8**	Na	1021.47±28.9
Cu	**201.73±6.1**	K	2656.99±89.5
Zn	**1503.07±22.4**	V	16.00±2.8
Mg	3320.91±98.2	Ga	25.73±3.0
Ca	41410.82±316.4	As	2.81±0.8
Fe	26789.72±186.1	Rb	12.66±1.8
Mn	326.19±11.6	Mo	4.02±1.0
Co	10.71±0.16	Ag	14.61±2.2
Sr	105.31±6.8	Pt	0.01±0.002
Al	27339.18±156.8	Tl	3.57±0.7

In bold – heavy metals investigated in fractionation experiment

Nevertheless, the agricultural application of such sludge, with its elevated concentrations of heavy metals, is restricted. The limit values for HMs in SS applied to agricultural land (Class-I SS) according to the Lithuanian normative document LAND 20:2005 are as follows: Cr < 140 (–) mg/kg, Cd < 1,5 (20–40) mg/kg, Ni < 50 (300–400) mg/kg, Pb < 140 (750–1200) mg/kg, Cu < 75 (1000–1750) mg/kg, Zn < 300 (2500–4000) mg/kg. (The limit values for HMs in sludge that are fixed by Directive 86/278/EEC of the European Commission are in brackets.) Therefore, in an effort to increase the on-land recycling of SS and extend other sludge utilisation alternatives, research should focus on efficient methods for HM removal from sludge. Furthermore, the ICP analysis (Table 2) revealed high contents of divalent metals in the sludge, such as Ca, Mg, Fe, and Ba, which may serve as competing ions for the divalent ions of heavy metals during chemical extraction procedures.

2.2. The distribution of the sewage sludge heavy metals among the chemical fractions

A sequential extraction procedure was used to characterise the retention of the metals in the sewage sludge, depending on the sludge treatment (primary or anaerobic). The concentrations of the metals in fractions F1–F5 were summed and compared to the total metal concentrations. Metal partitioning among the various chemical fractions in the primary sewage sludge and the anaerobically treated SS is reported in Fig. 1.

The fractionation experiment showed that the distribution of HMs in the fractions depended on the solid matrix composition and its properties, but it mainly depended on the nature of the metal itself. The distribution of the metal fractions varied widely for various metals.

The exchangeable fraction of almost all of the metals only represented a small portion (less than 10%) of the total amount of metal found in the sewage sludge. This fraction shows the concentrations of the metal cations, which are non-specifically adsorbed and can be easily replaced by competition with divalent metal ions. The concentration of Ni (12%) in this fraction alone is higher than that of any other metal. The proportion of other metals in this fraction was below 8% and ranged from 2% for Cr to 8% for Zn.

The Cr and Cd contents of the F2 fraction of the primary and anaerobically digested SS were below the detection limit; therefore, these metals were not taken into consideration during the statistical analysis and are not shown in Fig. 1. It is evident from the fractionation analysis data that Cr predominantly accumulated in the F5 fraction in both samples. This

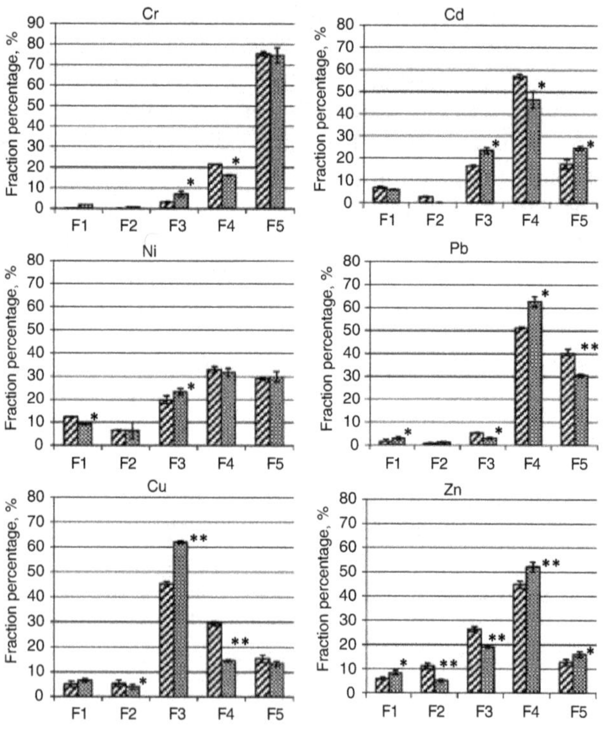

Fig. 1. The chemical fractions (F1–F5) of the heavy metals in the sewage sludge: ▨ – before anaerobical treatment; ▩ – after anaerobical treatment

Note: * – $p \leq 0.05$; ** – $p \leq 0.01$, significant differences indicated before and after treatment

metal binds exceptionally strongly to mineral SS components and does not pose an environmental risk due to its mobility.

Approximately 50% of the total Cu content is associated with the organically bound fraction. The remainder of this element is predominantly distributed between the F4 and F5 phases (22% and 14%, respectively). The preference of Cu for organic matter is supported by the high stability of Cu complexes with organic components (Pardo *et al.* 1990). The sequential extraction procedure also clearly showed that the predominant fractions of Cd and Zn (and Pb in SS after anaerobic processing) are found in fraction F4 (bound to carbonates).

The fractionation data showed that the anaerobic treatment of the sludge had no significant influence on heavy-metal mobility, except in the case of Pb (Fig. 1). The F3 fraction of Pb significantly decreased in the sludge after anaerobic treatment, and Pb was predominantly distributed in the carbonate-bound (F4) and residual (F5) fractions. In general, there were no other significant differences in the heavy-metal distribution between the fractions in the primary and anaerobically processed sludge samples.

The HM mass percentage presented in the first two fractions (F1 and F2, the exchangeable and adsorbed fractions, respectively) was less than 20% in all of the cases (Fig. 1). Because these two fractions can be characterised as highly mobile and easily bioavailable, the investigated heavy metals can be ranked in the following order according to their availability to living organisms in the environment:

$$Ni > Zn > Cu > Cd > Pb >> Cr.$$

The mobility and bioavailability of both Ni and Zn in the sludge were higher than those of the other investigated heavy metals. These results support the findings of Ščančr *et al.* (2000). Furthermore, fractions F3 and F4 (organic and carbonate–bound matter, respectively) can be described as potentially mobile, and long-term changes related to the soil acidity or intensive organic matter mineralisation can lead to increased heavy metal mobility.

Finally, fraction F5 (the sulphide and residual fraction) can be characterised as absolutely immobile. This fraction contains the portion of the HM ion mass that is bound to insoluble, stable, and poorly degradable compounds. The greater the mass percentage of the HMs that are found in this fraction, the lower the HM mobility in the environment. This observation leads to reduced HM leaching and migration potential. Therefore, in terms of their long-term stability in the environment, the investigated HMs can be ranked as follows:

$$Cr >> Pb \approx Ni > Cd > Zn \approx Cu.$$

These metal partitioning results are equally important for HM removal as stabilisation. It can be anticipated that Cr and Pb are the metals that will be the most difficult to remove from SS, while Cu, Zn and Ni will demonstrate rather high leacheability.

Given the mobility, potential bioavailability, and total concentrations of the heavy metals, it can be concluded that HM-containing sewage sludge cannot be used in agriculture because of the potentially hazardous effects of Ni, Cu and Zn on the environment. Furthermore, these elements should be easily leachable from the sludge and are expected to be characterised by high extraction efficiencies. Nevertheless, such results can be controversial because the removal process is also strongly influenced by many other factors, such as the matrix composition, the chemical properties of the extractant, the concentration and pH of the extracting solution and the temperature.

2.3. The removal of heavy metals using low-molecular-weight organic acids

For the removal of Ni, Cu and Zn from anaerobically processed sewage sludge, LMW organic acids with strong chelating properties were chosen. As shown in the literature (Yuan *et al.* 2007), the soil heavy metal removal capacity of a low-molecular-weight organic acid is affected by the number of carboxylic groups in that particular acid. Therefore, organic acids with different numbers of carboxyl groups (-COOH) were selected for this experiment:

- monocarboxylic acid – acetic acid (AA) with 1 carboxyl group;
- dicarboxylic acid – oxalic acid (OA) with 2 carboxyl groups;
- tricarboxylic acid – citric acid (CA) with 3 carboxyl groups.

The HM extraction efficiency can be strongly influenced by the acidity of the liquid medium; therefore, the pH values of the organic acid solutions were measured before and after extraction (Table 3). The data show that an increase in the organic acid concentration resulted in smaller changes in the solution pH during extraction because of the limited SS buffering capacity.

Obviously, a lower organic acid solution pH results in the increased dissolution of other competitive divalent cations, such as Ca and Fe, which can also be combined with the anions in the washing solution. Therefore, the large differences among the pH changes in the various organic acid solutions (Table 3) can explain the high variation of their extraction efficiencies. The number of carboxyl groups was also very important for the removal process; a monocarboxylic acid (acetic acid) had the lowest extraction capacity, while a tricarboxylic acid (citric acid) exhibited the strongest extraction capacity. No results were obtained using 1.0 M oxalic acid because the resulting supersaturated solution precipitated.

Additional HM removal experiments showed that the extracted metal content depends on both the nature of the organic acid and its concentration (Figs 2–4). Higher extractant concentrations resulted in more heavy metal extraction. The experimental data showed that the zinc extraction efficiency strongly depended on the organic acid concentration (Fig. 2). At the lowest concentrations of the various acid solutions (0.01 mol/L), the zinc EE was very low (0.5%, 0.9% and 1.4%, for acetic, oxalic and citric acid respectively). The pH values of all of the studied acid solutions after extraction were higher than their values before metal removal (Table 3). The data also showed that the pH values approached neutrality after extraction. This neutralisation could be one of the chief reasons for such a low heavy-metal EE. Polettini *et al.* (2007) reported that during metal extraction from sediments at a low citric acid concentration (0.02 mol/L), more than 72% of the acid combined with competing elements (Ca and Mg). It was also determined that the extraction efficiency decreased with an increasing pH.

The zinc extraction efficiency increased as the acid concentration increased. The 0.05-M LMW organic acid solutions can be ranked by their ability to remove zinc from the SS in the following order: oxalic acid (51%) > >citric acid (13%) ≈ acetic acid (12%).

Such a high oxalic acid extraction efficiency may be associated with its low initial pH of 1.6 compared to 2.3 and 3.1 for citric and acetic acid, respectively. The highest pH values were observed for acetic acid before and after extraction. Thus, AA was only able to remove a small amount of the zinc from the sludge. The solutions of 0.05 M AA and CA were able to extract all of the zinc in the highly mobile fraction (below the dotted line). However, 0.05 M oxalic acid exhibited over four times more extraction capacity than 0.05 M AA and CA. However, for all of the studied acids, this solution concentration enabled the removal of the zinc ions associated with the more or less mobile metal fractions (below the dashed line).

Citric and oxalic acid extracted nearly equal amounts of zinc at a solution concentration of 0.1 mol/L, extracting 74% and 72%, respectively, while the acetic acid extraction efficiency was only 29% (Fig. 2). Di Palma and Mecozzi (2007) found that citric acid (0.1 mol/L at a solution-to-sediment ratio of 12.5) was able to extract up to 82% of the total Zn content from the sediment. It is likely that with an increasing concentration of oxalic acid, the concentrations of competing dissolved metal ions (such as Ca^{2+} and Fe^{2+}) that can form stable complexes with this acid also increase.

The difference in the extraction efficiencies of the 0.5-mol/L acid solutions correlates well with the various pH values of the extracting media. The organic

Table 3. The pH values of solutions of LMW organic acids before and after the extraction procedure

Acid solution		pH value at concentration, mol/L				
		0.01	0.05	0.10	0.50	1.00
Acetic acid	Before extraction	3.32±0.01	3.09±0.01	2.99±0.01	2.58±0.01	2.26±0.01
	After extraction	6.89±0.02	4.71±0.01	4.58±0.02	3.82±0.02	3.35±0.01
Oxalic acid	Before extraction	1.90±0.01	1.65±0.02	1.49±0.01	1.07±0.01	n
	After extraction	7.45±0.03	7.15±0.02	1.94±0.01	1.10±0.01	n
Citric acid	Before extraction	2.80±0.01	2.30±0.01	2.26±0.01	1.98±0.01	1.78±0.01
	After extraction	6.93±0.02	3.69±0.01	3.19±0.01	2.83±0.01	2.66±0.01

Mean±standard deviation; n, no results obtained using 1.0 M oxalic acid because the resulting supersaturated solution precipitated

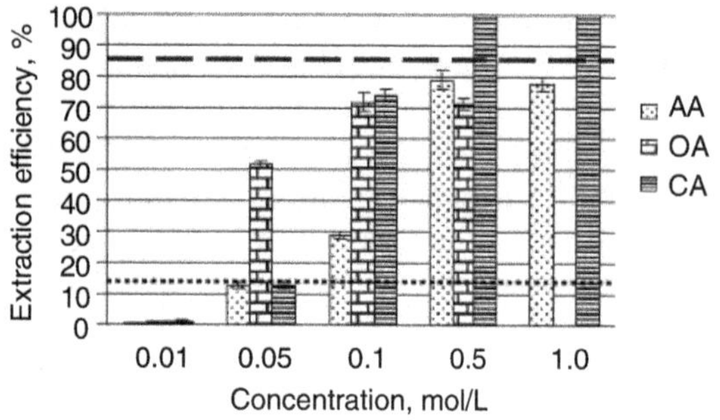

Fig. 2. The zinc-extraction efficiency (%) using various concentrations of organic acids: AA, acetic acid; OA, oxalic acid; CA, citric acid. Below the dotted line, the HM percentage in the mobile fractions (F1 + F2); between the dotted and dashed lines, the HM percentage in the potentially mobile fractions (F3 + F4); above the dashed line, the HM percentage in the immobile fraction (F5)

acids selected for this study are nonspecific acids, and thus may also react with other metal ions, rather than with Zn ions exclusively. The competition between zinc and other metal ions depends on the concentrations of the metal cations and the acid anions in the solution, the pH, and the stability constants of the corresponding metal complexes. Higher concentrations of oxalic acid produced lower pH values after the extraction procedure, which indicates that the competition of Zn with the other dissolved cations was stronger. These results (Fig. 2) show that at a 0.5-mol/L concentration, the investigated acids can be ranked according to their Zn EEs, as follows: citric acid (100%) > acetic acid (79%) > oxalic acid (71%).

Other investigations have shown that organic acids (1 mol/L), including citric acid, acetic acid, and four other investigated acids (except oxalic acid), are able to remove Zn from the soil completely. The total zinc content in the soil was 84.4 mg/kg (Khodadoust *et al.* 2005). Our results showed that 1 mol/L acetic and citric acid removes of the majority of the zinc accumulated in SS (78% and 100% respectively). The lower EE of acetic acid compared to that which has been reported in the literature may be associated with the much higher total zinc concentration in Kaunas sewage sludge and also with lower ratio of carboxyls to zinc.

Fig. 3 shows the dependence of the nickel EE (%) on the organic acid concentration; the superior extraction efficiency was determined by the higher acid concentration. The most nickel (53%) was removed by citric acid (1 mol/L), and the least nickel was removed by acetic acid. Contradictory results were found in the literature. Khodadoust *et al.* (2004) determined that organic acids (1 mol/L) effectively removed nickel, and they were ranked as follows: acetic acid > citric acid > oxalic acid. Their extraction efficiencies were 80%, 60% and 30%, respectively. It should be noted that the pH values of the studied solutions were below 3.5. Furthermore, various extraction conditions and solid matrix compositions are important factors that could lead to various extraction efficiencies. The distribution of Ni in various chemical forms (depending on the results of the sequential extraction) could also substantially influence the metal-extraction efficiency.

Fig. 4 shows the dependence of the copper extraction efficiency (%) on the organic acid concentration. Copper forms specific stable compounds with the OH groups that occur at higher pH values. The solubility constant of $Cu(OH)_2$ is 19.66 (Wuhan University 2004). The formation of this compound may result in a lower Cu extraction efficiency when the solution pH is neutral or alkaline. The fractionation results (Fig. 1) showed that Cu predominantly accumulated in the F3 fraction (bound to organic matter). These results confirm other findings that reported that copper tends to form relatively stable organic-metal complexes that can further reduce copper's leachability from sewage sludge after treatment with LMW organic acids.

The experimental data show that low-concentration organic acid solutions only removed a small amount of the copper and produced a post-extraction solution pH close to neutrality. The results of the sequential extraction confirmed the earlier findings (Fig. 1) that in the investigated sludge samples, most of the copper ions were bound to the sludge organic matter. Furthermore, the majority of the Zn was in the carbonate fraction (Fig. 1), which is easily destroyed in an acidic environment. This difference results in a lower copper extraction efficiency compared to that of zinc.

The investigated LMW organic acids may be ranked according to their Cu extraction efficiencies, as follows: citric acid > oxalic acid > acetic acid. Citric acid (0.1 mol/L) has demonstrated 2 and 10 times more copper EE than oxalic and acetic acid, respectively, with its extraction efficiency approaching 50%.

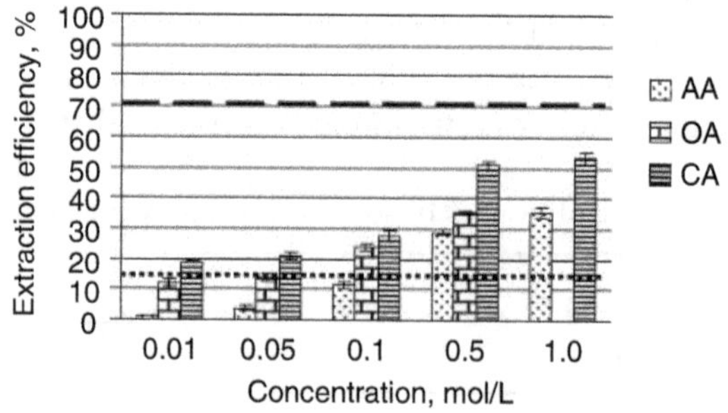

Fig. 3. The nickel extraction efficiency (%) using various concentrations of organic acids: AA, acetic acid; OA, oxalic acid; CA, citric acid. Below the dotted line; the HM percentage in the mobile fractions (F1 + F2); between the dotted and dashed lines, the HM percentage in the potentially mobile fractions (F3 + F4); above the dashed line, the HM percentage in the immobile fraction (F5)

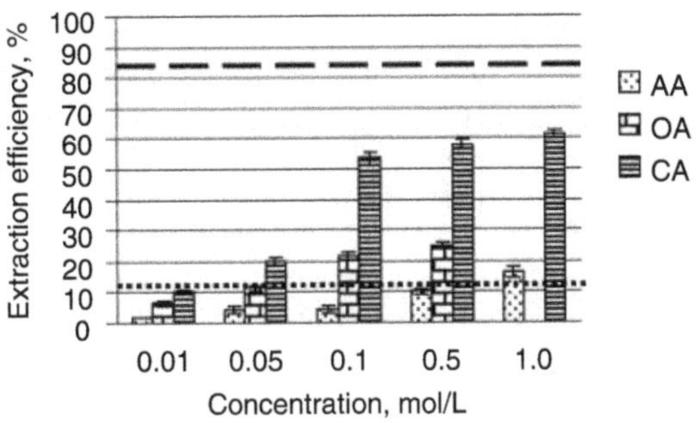

Fig. 4. The copper extraction efficiency (%) using various concentrations of organic acids: AA, acetic acid; OA, oxalic acid; CA, citric acid. Below the dotted line, the HM percentage in the mobile fractions (F1 + F2); between the dotted and dashed lines, the HM percentage in the potentially mobile fractions (F3 + F4); above the dashed line, the HM percentage in the immobile fraction (F5)

This study revealed that a tricarboxylic acid (citric acid) is a stronger extraction agent than oxalic acid (a dicarboxylic acid). In addition, oxalic acid is a strong reducing agent, and, therefore, it can be oxidised by sewage sludge organic matter, leading to fewer oxalate anions in the extraction solution.

The results of the extraction experiment show that all three investigated acids are able to remove the Cu, Zn and Ni found in the mobile metal fractions (F1 and F2). The potentially mobile metal fractions (F3 and F4) of zinc can be effectively removed, and those of nickel and copper can be removed to a lesser extent. Only a tricarboxylic acid (citric acid) was able to remove all of the zinc accumulated in the SS, including the most immobile fraction, F5. None of the investigated acids was able to remove the Ni and Cu associated with the immobile fraction (F5).

Conclusions

1. The results of sequential extraction showed that the heavy metal chemical species in sewage sludge depend less on the solid matrix composition and properties and more on the nature of the metal itself. Anaerobic digestion of the sludge did not change the mobility order of the heavy metals (Ni > Zn > Cu > Cd > Pb > > Cr). When evaluating the metal stability in the environment according to the sulphide and residual fraction (F5), the ranking order was as follows: Cr > > Pb ≈ Ni > Cd > Zn ≈ Cu.

2. The heavy-metal-extraction efficiency using low-molecular-weight organic acids depended on their solution concentration and number of carboxyl groups; a tricarboxylic acid (citric acid) had much more extraction capacity than a dicarboxylic acid (oxalic acid) or a monocarboxylic acid (acetic acid). It was determined that the optimal concentration of citric acid was 0.1 mol/L, which exhibited a metal-extraction efficiency as high as 75% for Zn.

3. The heavy-metal extraction efficiency increased with an increasing solution concentration. No results were obtained for 1.0 M oxalic acid because the supersaturated solution precipitated in this case. The 0.5-M solutions may be ranked according to the percentage of extracted zinc, as follows: citric acid (100%) > acetic acid (78%) > oxalic acid (71%). Furthermore, the extraction efficiency of copper was much less than that of zinc. The ranking order of the organic acids was also different, as follows: citric acid (58%) > oxalic acid (25%) > acetic acid (10%). In the case of nickel, the order was as follows: citric acid (51%) > oxalic acid (35%) > acetic acid (29%). In all of the cases, compared to oxalic and acetic acid, citric acid had a much greater capacity for HM extraction from sewage sludge.

Acknowledgments

This work was supported by the Lithuanian State Science and Studies Foundation.

References

Alloway, C. J. 1995. *Heavy metals in soils.* London: Springer-Verlag.

Alonso Álvarez, E.; Callejón Mochón, M.; Jiménez Sánchez, J. C.; Ternero Rodríguez, M. 2002. Heavy metal extractable forms in sludge from wastewater treatment plants, *Chemosphere* 47: 765–775. http://dx.doi.org/10.1016/s0045-6535(02)00021-8

Bode, H.; Imhoff, K. R. 1996. Current and planned disposal of sewage sludge and other waste production from the Ruhverband, *Water Science and Technology* 33(12): 219–228. http://dx.doi.org/10.1016/0273-1223(96)00476-3

Bowen, P. T.; Jacken, M. K.; Corbitt, R. A.; Conce, N. 1992. Sludge treatment, utilization and disposal, *Water Environment Research* 64: 378–386.

CEC 1986. Council of the European Communities. Council Directive of 12 June 1986 on the protection of the environment, and in particular of the soil, when sewage sludge is used in agriculture (86/ 278/EEC), *Official Journal of the European Communities* 181: 6–12.

Chipasa, K. B. 2003. Accumulation and fate of selected heavy metals in a biological wastewater treatment system, *Waste Management* 23: 135–143. http://dx.doi.org/10.1016/s0956-053x(02)00065-x

Cline, S. R.; Reed, B. E. 1995. Lead removal from soils bench scale soil washing techniques, *Journal of Environmental Engineering* 121(10): 700–705. http://dx.doi.org/10.1061/(ASCE)0733-9372(1995)121:10(700)

Dacera, D. del M.; Babel, S. 2007. Removal of heavy metals from contaminated sewage sludge using *Aspergillus niger* fermented raw liquid from pineapple wastes, *Bioresource Technology* 99(6): 1682–1689. http://dx.doi.org/10.1016/j.biortech.2007.04.002

Di Palma, L.; Mecozzi, R. 2007. Heavy metals mobilization from harbour sediments using EDTA and citric acid as chelating agents, *Journal of Hazardous Materials* 147(3): 768–775. http://dx.doi.org/10.1016/j.jhazmat.2007.01.072

Filgueiras, A. V.; Lavilla, I.; Bendicho, C. 2004. Evaluation of distribution, mobility and binding behaviour of heavy metals in surficial sediments of Louro River (Galicia, Spain) using chemometric analysis: a case study, *Science of the Total Environment* 330(1–3): 115–129. http://dx.doi.org/10.1016/j.scitotenv.2004.03.038

Förstner, U.; Salomons, W. 1980. Trace metal analyses on polluted sediments. Part I: Evaluation of environmental impact, *Environmental Technology* 1(11): 506–517. http://dx.doi.org/10.1080/09593338009384007

Idzelis, R. L.; Kesminas, V.; Svecevicius, G.; Venslovas, A. 2010. Experimental investigation of heavy metal accumulation in tissues of stone loach *Noemacheilus barbatulus* (L.) and rainbow trout *Oncorhynchus mykiss* (WALBAUM) exposed to a model mixture (Cu, Zn, Ni, Cr, Pb, Cd), *Journal of Environmental Engineering and Landscape Management* 18(2): 111–117. http://dx.doi.org/10.3846/1648-6897.2010.13

Yuan, S.; Xi, Z.; Jiang, Y.; Wan, J.; Wu, C.; Zheng, Z.; Lu, X. 2007. Desorption of copper and cadmium from soils enhanced by organic acids, *Chemosphere* 68(7): 1289–1297. http://dx.doi.org/10.1016/j.chemosphere.2007.01.046

Jankaitė, A. 2009. Soil remediation from heavy metals using mathematical modeling, *Journal of Environmental Engineering and Landscape Management* 17(2): 121–130. http://dx.doi.org/10.3846/1648-6897.2009.17.121-129

Jankaitė, A.; Vasarevičius, S. 2005. Remediation technologies for soils contaminated with heavy metals, *Journal of Environmental Engineering and Landscape Management* 13(2): 109a–113a.

Jaskelevičius, B.; Lynikienė, V. 2009. Investigation of influence of Lapės landfill leachate on ground and surface water pollution with heavy metals, *Journal of Environmental Engineering and Landscape Management* 17(3): 131–139. http://dx.doi.org/10.3846/1648-6897.2009.17.131-139

Khodadoust, A. P.; Reddy, K. R.; Maturi, K. 2004. Removal of nickel and phenanthrene from kaolin soil using different extractants, *Environmental Engineering Science* 21: 691–704. http://dx.doi.org/10.1089/ees.2004.21.691

Khodadoust, A. P.; Reddy, K. R.; Maturi, K. 2005. Effect of different extraction agents on metal and organic contaminant removal from a field soil, *Journal of Hazardous Materials* B117(1): 15–24. http://dx.doi.org/10.1016/j.jhazmat.2004.05.021

Kvasauskas, M.; Baltrėnas, P. 2009. Research on anaerobically treated organic waste suitability for soil fertilization, *Journal of Environmental Engineering and Landscape Management* 17(4): 205–211. http://dx.doi.org/10.3846/1648-6897.2009.17.205-211

LAND 20:2005. Nuotekų dumblo naudojimo tręšimui bei rekultivavimui reikalavimai [Lithuanian Environmental Normative Document. Requirements of Sewage Sludge Application in Fertilisation and Recultivation], *Valstybės žinios* 142–5135.

Legret, M. 1993. Speciation of heavy metals in sewage sludge and sludge-amended soil, *International Journal of Environmental Analytical Chemistry* 51(1–4): 161–165. http://dx.doi.org/10.1080/03067319308027621

McBride, M. B. 2003. Toxic metals in sewage sludge-amended soils: has promotion of beneficial use discounted the risk?, *Advances in Environmental Research* 8(1): 5–19. http://dx.doi.org/10.1016/s1093-0191(02)00141-7

Mench, M. J.; Didier, V. L.; Loffler, M.; Gomez, A.; Masson, P. 1994. A mimicked in-situ remediation study of metal-contaminated soils with emphasis on cadmium and lead, *Journal of Environmental Quality* 23: 58–63. http://dx.doi.org/10.2134/jeq1994.00472425002300010010x

Mossakowska, A.; Hellström, B. G.; Hultman, B. 1998. Strategies for sludge handling in the Stockholm region, *Water Science and Technology* 38(2): 111–118. http://dx.doi.org/10.1016/s0273-1223(98)00436-3

Pardo, R.; Barrado, E.; Perez, L.; Vega, M. 1990. Determination and speciation of heavy metals in sediments of the Pisuerga River, *Water Research* 24: 373–379. http://dx.doi.org/10.1016/0043-1354(90)90016-y

Paulauskas, V.; Sabiene, N.; Zaleckas, E. 2006. Sewage sludge leaching and alkaline stabilization, *Ecology* 4: 18–23.

Paulsrud, B.; Nedland, K. J. 1997. Strategy for land application of sewage sludge in Norway, *Water Science and Technology* 36(11): 283–290. http://dx.doi.org/10.1016/s0273-1223(97)00693-8

Podgaiskytė, V.; Vaitiekūnas, P. 2009. Determination of cadmium in a municipal sewage sludge based compost by spectrophotometric method, *Journal of Environmental Engineering and Landscape Management* 17(4): 219–225. http://dx.doi.org/10.3846/1648-6897.2009.17.219-225

Polettini, A.; Pomi, R.; Rolle, E. 2007. The effect of operating variables on chelant-assisted remediation of contaminated dredged sediment, *Chemosphere* 66(5): 866–877. http://dx.doi.org/10.1016/j.chemosphere.2006.06.023

Pueyo, M.; Rauret, G.; Lück, D.; Yli-Halla, M.; Muniau, H.; Quevauville, Ph.; López-Sánchez, J. F. 2001. Certification of the extractable contents of Cd, Cr, Cu, Ni, Pb and Zn in a freshwater sediment following a collaboratively tested and optimised three-step sequential extraction procedure, *Journal of Environmental Monitoring* 3(2): 243–250. http://dx.doi.org/10.1039/b010235k

Rauret, G. 1998. Extraction procedures for the determination of heavy metals in contaminated soil and sediment, *Talanta* 46(3): 449–455. http://dx.doi.org/10.1016/s0039-9140(97)00406-2

Su, D. C.; Wong, J. W. C. 2003. Chemical speciation and phytoavailability of Zn, Cu, Ni and Cd in soil amended with fly ash-stabilized sewage sludge, *Environment International* 29(7): 895–900. http://dx.doi.org/10.1016/s0160-4120(03)00052-7

Ščančr, J.; Milačič, R.; Stražar, M.; Burica, O. 2000. Total metal concentrations and partitioning of Cd, Cr, Cu, Fe, Ni and Zn in sewage sludge, *The Science of the Total Environment* 250(1–3): 9–19. http://dx.doi.org/10.1016/s0048-9697(99)00478-7

Tessier, A.; Campbell, P. G. C.; Bisson, M. 1979. Sequential extraction procedure for the speciation of particulate trace metals, *Analytical Chemistry* 51: 844–851. http://dx.doi.org/10.1021/ac50043a017

Veeken, A. H. M.; Hamelers, H. V. M. 1999. Removal of heavy metals from sewage sludge by extraction with organic acids, *Water Science Technology* 40(1): 129–136. http://dx.doi.org/10.1016/s0273-1223(99)00373-x

Wang, C.; Hu, X.; Chen, M. L.; Wu, Y. H. 2005. Total concentrations and fractions of Cd, Cr, Pb, Cu, Ni and Zn in sewage sludge from municipal and industrial wastewater treatment plants, *Journal of Hazardous Materials* 119(1–3): 245–249. http://dx.doi.org/10.1016/j.jhazmat.2004.11.023

Wuhan University 2004. *Analytical Chemistry.* Beijing: High Education Press. 322 p.

Ernestas ZALECKAS. Dr Junior Researcher at the Laboratory of Chemical and Biochemical Research for Environmental Technology, Institute of the Environment, Aleksandras Stulginskis University (ASU). Doctor of Sciences (Environmental Engineering), LUA, 2009. Publications: author of more than 10 research papers. Research interests: environmental protection, heavy metals, remediation technologies, waste utilisation.

Valdas PAULAUSKAS. Head of the Department of Chemistry, Dr Prof. at the Department of Chemistry, Aleksandras Stulginskis University (ASU). Doctor of Sciences (Organic chemistry), Kaunas University of Technology, 1987. Publications: author of more than 50 research papers. Research interests: waste, utilisation, soil remediation, heavy metal accumulation, removal and stabilisation, organic synthesis.

Eglė SENDŽIKIENĖ. Dr Senior Research Fellow at the Laboratory of Chemical and Biochemical Research for Environmental Technology, Institute of the Environment, Aleksandras Stulginskis University (ASU). Doctor of Sciences (Environmental Engineering), LUA, 2005. Publications: author of more than 20 research papers. Research interests: environment protection, waste utilisation, renewable energy.

13

REDUCTION OF FOODBORNE PATHOGENS DURING CATTLE MANURE COMPOSTING WITH ADDITION OF CALCIUM CYANAMIDE

Huasai Simujide[a], Chen Aorigele[a], Chun-Jie Wang[b], Bai Manda[a], Ma Lina[c], Mei-Yun Wu[a], Yu Li[a], Tong-Ri-Ge Bai[a]

[a]*College of Animal Science, Inner Mongolia Agricultural University, Hohhot 010018, PR China*
[b]*College of Veterinary Medicine, Inner Mongolia Agricultural University, Hohhot 010018, PR China*
[c]*College of Life Sciences, Inner Mongolia Agricultural University, Hohhot 010018, PR China*

Abstract. Inactivation of four species of foodborne pathogens (*E. coli*, *Salmonella*, *E. faecalis*, and *S. aureus*) was investigated during laboratory-scale composting of fresh cow manure with addition of calcium cyanamide ($CaCN_2$) at constant temperature conditions, and the effects of temperature and additive content of $CaCN_2$ on the efficacy of inactivation were evaluated. At different composting temperatures (20, 30, 37 and 50 °C), a significant inhibition or quick inactivation of pathogens were observed during 10 d composting with addition of 2.0% $CaCN_2$, and the effect was more obvious at mesophilic temperatures compared to thermophilic temperature. Therefore, the ideal additive content of $CaCN_2$ was determined at 30 °C through mixing 2.0%, 2.5%, and 3.0% $CaCN_2$ with manure. With increase in additive content of $CaCN_2$, the efficacies of pathogen inactivation also increased. However, the result indicates that, no less than 2.5% $CaCN_2$ should be mixed with the manure to entirely eliminate the pathogens during composting.

Keywords: foodborne pathogen; calcium cyanamide; cow manure; composting; waste management technologies.

Reference to this paper should be made as follows: Simujide, H.; Aorigele, C.; Wang, C.-J.; Manda, B.; Lina, M.; Wu, M.-Y.; Li, Y.; Bai, T.-R.-G. 2013. Reduction of foodborne pathogens during cattle manure composting with addition of calcium cyanamide, *Journal of Environmental Engineering and Landscape Management* 21(2): 77–84. http://dx.doi.org/10.3846/16486897.2012.721373

Introduction

Manure of livestock and poultry carry some of the important environmental pollutants like organic materials, sediments and microorganisms. In this case, the contamination from pathogens in animal manure is becoming more and more dominant and serious with the explosive development of animal husbandry industry and with the increase in environmental protection awareness. Several livestock species and poultry carry large populations of foodborne pathogens, such as Escherichia, Salmonella and Campylobacters in their manure (Doyle, Erickson 2006; Viazis, Diez-Gonzalez 2011). These pathogens originate from the animal's intestinal tract and typically shed in the tract asymptomatically (Doyle, Erickson 2006). Pathogens carried in manure can be easily transmitted by other vectors from manure to animals, produce, or humans (You *et al.* 2006; Martens, Böhm 2009). Therefore, the factors in this transmission pathway need be intervened to prevent the fecal pathogen's contaminations or infections. And in these interventions, sanitation of manure is considered as an effective way to control the contaminations on site.

Composting is not only the most efficient process to produce an agronomically advantageous soil organic amendment, but also one of the most environmentally friendly treatments to inactivate pathogenic organisms or reduce them to an acceptable levels (Fernández *et al.* 2007; Wichuk, McCartney 2007). Up to now, much work has been done by the researchers to investigate the pathogen destructive and disease suppressive effect of composting worldwide. Temperature plays an important role in the fate of pathogens. The efficiency of pathogen inactivation depends not only on the degree of temperature, but also on the exposure time of pathogens to heat. However, some foodborne pathogens have a great capacity to survive for long periods in manure. In a laboratory-scale bioreactor, *E. coli* O157:H7 still survived in manure composting at 21 °C after 36 d, while not detected 14 d post-composting at 50 °C

Corresponding author. Chen Aorigele
E-mail: aori60@yahoo.com.cn

(Jiang *et al.* 2003). *Salmonella* serovar Newport even surprisingly persisted for a long time, 184 d in dairy cow manure at ambient temperature (24.5±1.4 °C) in laboratory (You *et al.* 2006). And *Salmonella typhimurium* was culturable for up to 28 d and 42 d in swine manure in thermophilic composting at 55 °C and aerated liquid storage, respectively (Grewal *et al.* 2007). These results demonstrated that even under thermal conditions, some strains of pathogens could exist for a long time in manure. Such great persistence increases the chances of an outbreak occurring and further increases the environmental load. Therefore, it is necessary to produce more reliable and realistic methods feasible in both sanitation and recycling of manure.

Calcium cyanamide ($CaCN_2$) is not only a good fertilizer. It has also shown good performance in the suppression of soilborne pathogens. $CaCN_2$ effectively suppressed *Fusarium solani* f.sp. *cucurbitae* in greenhouse cucumber (Bourbos *et al.* 1997), and was also fungicidal to *Fusarium oxysporum* f. sp. *Cucumberinum* (Shi *et al.* 2009). However, very little is known about its effect on zoonotic microorganisms. Our research thus studied the impact of $CaCN_2$ on foodborne pathogens from dairy cow manure origins during composting, attempting to remedy the defect about pathogen inactivation in manure treatment.

1. Materials and methods

1.1. Raw composting materials

Fresh manure containing 87.3% moisture content, 37.3% carbon and 1.6% nitrogen was obtained from healthy dairy cows at a local dairy farm. The manure was sent to laboratory immediately after collection in portable ice boxes and processed for experiment on the same day. $CaCN_2$ was an imported product of Guangzhou WeiBo Chemical Co., Ltd., China, containing 21.3% N and 58% CaO, and granule size was 0.2–2 mm diameters.

1.2. Composting temperature

In the experimental group, $CaCN_2$ was thoroughly mixed with 1200 g fresh manure at an additive content of 2.0% (weight percentage), and then four equal portions of mixtures (each 200 g) were respectively transferred into a 400 ml glass bottle and composted at different constant temperatures for 10 d in thermostatic incubators (JC–SPJ–480, Jinan Jingcheng Experimental Instrument Co., Ltd., China). The control group was with no $CaCN_2$ mixed and ran the same procedures as the experimental group. In details, the experimental design used was a 2-factor factorial design of four temperatures (T) × two additive contents (D) (Liang *et al.* 2003). The four temperature settings were 20, 30, 37 and 50 °C, and the two additive contents were 0% and 2.0%. Every treatment was at one temperature setting with two additive contents settings.

1.3. Additive content of $CaCN_2$

In the experimental group, $CaCN_2$ was thoroughly mixed with 200 g fresh manure at the additive contents of 2.0%, 2.5%, and 3.0% in 400 ml glass bottles, respectively. And the mixtures were composted for 5 d at a constant temperature of 30 °C, which was selected as a representative composting temperature for $CaCN_2$ inhibition against pathogens. The control group was with no $CaCN_2$ mixed and ran the same procedures as the experimental group.

1.4. Sampling and pathogen enumeration

For sampling, 1.0 g of composting sample in duplicate was collected from each treatment at 0, 1, 3, 5, 7 and 10 d in the determination of a suitable composting temperature and at 0, 1, 2, 3 and 5 d in the determination of a suitable additive content. Then, the samples were analyzed for the changes of pathogen population during composting. The tested pathogens were *E. coli*, *Salmonella*, *Enterococcus faecalis* (*E. faecalis*), and *Staphylococcus aureus* (*S. aureus*).

For pathogen enumeration, serial dilutions (1:10) of composting samples with sterile physiological saline were performed and homogenized for 10min in a rotatory stirrer. 0.1 ml aliquots of each dilution were spread onto the appropriate agar plates in triplicate. All agar plates were incubated at 37 °C for 24 h, and colonies were counted. Thirty to three hundred colonies per plate were considered to be optimal countable numbers. *E. coli* was investigated from eosin-methylene blue agar (EMB) plates and confirmed by Indole Test. XLD (xylose-lysine-deoxycholate) agar was used for detection of *Salmonella*, while KF streptococci agar and Baird-Parker agar were for *E. faecalis* and *S. aureus*, respectively. When the concentration of pathogen dropped below the quantitative detection limit (100 CFU/g) of direct plating method, selective enrichment was used to assay for the presence of pathogens. The used enrichment broths were Luria-Bertani (LB) broth for *E. coli*, Rappaport-Vassiliadis (RV) enrichment broth and Selenite cystine (SC) enrichment broth for *salmonella*, Azide glucose (AD) broth for *E. faecalis*, and Soy bean casein digest broth with lecithin and polysorbate medium (SCDLP) and Manitol salt (MS) broth for *S. aureus*. After incubation for 24 h at 37 °C, the enrichment cultures were inoculated onto the surface of the selective agar plates as mentioned above and streaked for isolated colonies.

1.5. Statistical analysis

Data were analyzed using ANOVA for variances and Duncan's procedure from SAS (version 9.0) for multiple comparisons.

2. Results

2.1. Influence of composting temperature on $CaCN_2$ inhibition against foodborne pathogen

The populations of four pathogens in initial manure and the changes during composting at different temperatures are shown in Tables 1 and 2.

In the experimental group (Table 1), *E. coli* was never detected from 5, 7 and 7 d at 20, 30 and 37 °C, respectively. But, *Salmonella*, *E. faecalis*, and *S. aureus* were still detected until the end of experiment. At these three temperatures, the populations of three microorganisms respectively decreased 59.28, 62.30 and 41.58% for *salmonella*, 28.71, 44.12 and 57.80% for *E. faecalis*, and 30.72, 55.88, and 39.18% for *S. aureus* at the end of composting compared with the initial values. At 50 °C, no growth of *E. coli* and *salmonella* was observed from 1 d, while that of *E. faecalis* was from 3 d. However, *S. aureus* was thoroughly undetectable by enrichment culture just at 10 d when it did not decrease the detection limit of direct plating until 7 d.

In the control group (Table 2), the survival of all four pathogens was observed at the end of experiment at 20, 30 and 37 °C. At these three temperatures, the populations of *E. coli* respectively decreased 6.58, 9.30 and 4.01% at the end of composting compared with the initial values. For *salmonella*, the corresponding decreases were 33.48, 30.17 and 27.30%, while that were 37.17, 39.17 and 50.00% for *E. faecalis*, and 24.43, 17.81 and 16.56% for *S. aureus*. At 50 °C, *E. coli*, *Salmonella*, and *E. faecalis* were thoroughly undetectable from 3 d except *S. aureus* was still observed at the end of composting. Among them, *E. coli* and *Salmonella* were not detected from 1 d and *E. faecalis* was from 3 d.

Log removal for four pathogens during composting at different temperatures was calculated. At 50 °C, *E. coli* and *salmonella* either in the experimental group or in the control group was all not observed from 1 d. So, their log removals should be in the highest levels, and to exhibit this law and calculate average log removal, pathogen concentration (log CFU/g) in the sampling days in which pathogen was not detected or under the detection limit was taken a value of zero. In the experimental group, an average log removal for *E. coli* was respectively 4.72, 3.62, 3.95 and 5.74 at 20, 30, 37 and 50 °C, when correspondingly 3.29, 3.35, 3.18 and 6.95 for *Salmonella*, 1.52, 1.55, 1.86 and 4.39 for *E. faecalis*, and 0.90, 1.45, 1.27 and 2.94 for *S. aureus*. In the control group, the average log removal was −0.33, −0.12, 0.24 and 6.99 for *E. coli* at 20, 30, 37 and 50 °C, respectively, which was 1.56, 1.26, 1.36 and 6.96 for *Salmonella*, 1.48, 1.27, 1.79 and 4.89 for *E. faecalis*, and 0.58, 0.38, 0.17 and 0.95 for *S. aureus*.

Summing up, an average log removal for four pathogens was in the range of 0.90–4.72 at 20 °C, 1.45–3.62 at 30 °C, 1.27–3.95 at 37 °C and 2.94–6.95 at 50 °C in the experimental group, while it was correspondingly −0.33 to 1.56, −0.12 to 1.27, 0.17 to 1.79, and 0.95 to 6.99 in the control group. As to a general average log removal for all pathogens at different temperatures, it was respectively 2.61, 2.49, 2.57 and 5.00 at 20, 30, 37 and 50 °C in the experimental group, and was 0.82, 0.70, 0.89 and 4.95 in the control group.

2.2. Influence of $CaCN_2$ on foodborne pathogen during mesophilic composting

Among the mesophilic temperatures (20 °C, 30 °C, and 37 °C), 30 °C was selected to determine an ideal additive content of $CaCN_2$ in composting against the tested pathogens. The selection of the temperature was based on the results of experiment 3.1. and the consideration of initial temperature degree of composting process. As shown in Table 3, the treatments mixed with 2.5% and 3.0% $CaCN_2$ revealed good inhibitory effects on the growth of four microorganisms. And with increase inadditive content, the inhibitory efficacy was correspondingly improved. In the treatments mixed with 3.0% $CaCN_2$, the survival of *E. coli* was not observed from 1 d, when that of *Salmonella*, *E. faecalis*, and *S. aureus* was from 2 d. In the treatments mixed with 2.5% $CaCN_2$, *E. coli* was not detected from 2 d; *Salmonella*, *E. faecalis*, and *S. aureus* were all from 3 d. By contrast, a long persistence of four microorganisms appeared in the treatments mixed with 2.0% $CaCN_2$ and in the control group, that the survival of all microorganisms was still observed at the end of the experiment. The population of *E. coli* decreased below the detection limit of direct plating at 5 d in the treatments mixed with 2.0% $CaCN_2$. But in the control group, it was always detected in large quantities during the entire time of composting, decreasing only 6.36% at the end of composting in comparison to the initial value. *Salmonella*, *E. faecalis*, and *S. aureus* were all detected in direct plating in the two groups. At the end of the composting, the populations of *Salmonella* in comparison to the initial value decreased 67.15% and 22.63% in the treatments mixed with 2.0% $CaCN_2$ and control group, respectively, while the values were 42.18% and 35.68% for *E. faecalis*, and 57.40% and 19.27% for *S. aureus*.

Log removal for four pathogens during composting with addition of different contents of $CaCN_2$ at 30 °C is presented in Table 4. In the experimental group, an average log removal for *E. coli* was rescpectively 2.97, 5.34 and 6.45 in the treatments

Table 1. The survival and persistence of *E. coli*, *Salmonella*, *E. faecalis*, and *S. aureus* during cattle manure composting with addition of 2.0% $CaCN_2$ at different temperatures

Microorganism	Composting temperature (°C)	Mean[a] ±SD (log CFU/g) at sampling time (d)					
		0	1	3	5	7	10
E. coli	20	5.74±0.07A[b]a[c]	5.08±0.02Cb	<2.0[d]	ND	ND	ND
	30	5.74±0.07Ab	5.91±0.11Aa	4.68±0.02c	<2.0	ND	ND
	37	5.74±0.07Aa	5.38±0.04Bb	<2.0	3.57±0.16c	ND	ND
	50	5.74±0.07A	ND	ND	ND	ND	ND
Salmonella	20	6.95±0.18Aa	4.35±0.24Db	4.09±0.16Ab	3.90±0.16Ab	3.11±0.15Ab	2.83±0.35Ab
	30	6.95±0.18Aa	4.62±0.31Bb	3.96±0.18Bb	3.86±0.12Bb	2.96±0.20Bb	2.62±0.16Bc
	37	6.95±0.18Aa	5.53±0.17Cb	2.72±0.15Db	2.57±0.14Db	3.97±0.14Da	4.06±0.19Ca
	50	6.95±0.18A	ND	ND	ND	ND	ND
E. faecalis	20	5.19±0.03Aa	4.14±0.33Ab	3.68±0.05Cc	3.39±0.06Bd	3.44±0.04Acd	3.70±0.16Ac
	30	5.19±0.03Aa	4.08±0.11Ab	4.25±0.04Ab	3.69±0.07Ac	3.29±0.05Bd	2.90±0.14Ac
	37	5.19±0.03Aa	4.11±0.08Ab	3.97±0.08Bb	3.68±0.27Ac	2.70±0.05Cd	2.19±0.10Ac
	50	5.19±0.03Aa	4.01±0.03Ab	ND	ND	ND	ND
S. aureus	20	4.85±0.01Aa	4.70±0.03Cb	4.21±0.04Ac	3.99±0.07Ad	3.51±0.02Ae	3.36±0.11Af
	30	4.85±0.01Aa	4.18±0.07Bb	3.74±0.02Cd	3.81±0.03Bc	3.12±0.04Ce	2.14±0.00Cf
	37	4.85±0.01Aa	4.13±0.02Db	3.89±0.05Bc	3.64±0.02Cd	3.28±0.08Be	2.95±0.03Bf
	50	4.85±0.01Aa	3.89±0.03Bb	3.56±0.02Dc	2.11±0.05Dd	<2.0	ND

ND, not detected after enrichment.
[a]The mean is the average of six replicates.
[b]For each pathogen, means with different upper case letters in a column are significantly different ($P < 0.05$).
[c]Means with different lower case letters in a row are significantly different ($P < 0.05$).
[d] <2.0 log CFU/g, positive after enrichment.

Table 2. The survival and persistence of *E. coli*, *Salmonella*, *E. faecalis*, and *S. aureus* during cattle manure composting with no $CaCN_2$ mixed at different temperatures

Microorganism	Composting temperature (°C)	Mean[a] ± SD (log CFU/g) at sampling time (d)					
		0	1	3	5	7	10
E. coli	20	6.99±0.04A[b]d[c]	6.66±0.03Ce	8.32±0.02Aa	7.88±0.08Ab	7.23±0.01Bc	6.53±0.16Bf
	30	6.99±0.04Ae	7.53±0.11Ba	7.11±0.06Bd	7.40±0.13Bb	7.15±0.09Cc	6.34±0.02Cf
	37	6.99±0.04Ab	7.58±0.16Aa	6.83±0.01Cc	5.08±0.17Ce	7.56±0.11Aa	6.71±0.01Ad
	50	6.99±0.04A	ND	ND	ND	ND	ND
Salmonella	20	6.96±0.13Ad	6.95±0.23Bd	5.86±0.17Aa	4.91±0.12Ac	4.65±0.20Ab	4.63±0.26Ae
	30	6.96±0.13Aa	6.94±0.22Cb	5.93±0.26Cc	5.80±0.21Cd	4.96±0.35Be	4.86±0.31BCf
	37	6.96±0.13Ab	7.99±0.21Aa	5.04±0.21Dc	5.45±0.23Dc	4.46±0.09Bc	5.06±0.23BCc
	50	6.96±0.13Aa	ND	ND	ND	ND	ND
E. faecalis	20	6.00±0.07Aa	5.80±0.23Aa	4.89±0.15Bb	4.22±0.23Bb	3.94±0.04Bc	3.77±0.06Ad
	30	6.00±0.07Aab	6.15±0.11Aa	4.96±0.08Ab	4.66±0.06Ac	4.23±0.15Ad	3.65±0.06Be
	37	6.00±0.07Aa	6.03±0.14Aa	4.71±0.08Bb	3.74±0.15Cc	3.58±0.06Bc	3.00±0.10Bd
	50	6.00±0.07Aa	5.55±0.07Bb	ND	ND	ND	ND
S. aureus	20	4.83±0.02Aa	4.74±0.07Cc	4.77±0.06Cb	4.25±0.05Dd	3.84±0.04De	3.65±0.01Cf
	30	4.83±0.02Ab	4.80±0.05Bc	4.91±0.03Ba	4.44±0.01Bd	4.12±0.08Be	3.97±0.04Bf
	37	4.83±0.02Ac	5.24±0.01Ab	5.28±0.03Aa	4.53±0.05Ad	4.21±0.02Ae	4.03±0.06Af
	50	4.83±0.02Aa	4.72±0.02Db	4.24±0.01Dd	4.38±0.05Cc	3.86±0.03Ce	2.21±0.07Df

ND, not detected after enrichment.
[a]The mean is the average of six replicates.
[b]For each pathogen, means with different upper case letters in a column are significantly different ($P < 0.05$).
[c]Means with different lower case letters in a row are significantly different ($P < 0.05$).

mixed with 2.0%, 2.5% and 3.0% $CaCN_2$, which was 3.21, 5.11 and 6.76 for *Salmonella*, 1.38, 4.57 and 4.80 for *E. faecalis*, and 1.58, 4.05 and 4.17 for *S. aureus*. In the control group, the average log removal was −0.12, 0.78, 0.78 and 0.71 for *E. coli*, *Salmonella*, *E. faecalis* and *S. aureus*, respectively. Generally, an average log removal for four pathogens was in the range of 1.38–3.21, 4.05–5.34 and 4.17–6.76 in the experimental treatments mixed with 2.0%, 2.5% and 3.0% $CaCN_2$, while it was only −0.12 to 0.78 in the control group. A general average log removal for all pathogens was 2.29, 4.77 and 5.54 in 2.0%, 2.5% and 3.0% $CaCN_2$ treatment when 0.54 in the control group.

3. Discussion

Thermal treatment is one of the most efficient ways to inactivate pathogens during composting. However, it has been found that some foodborne pathogens could persist for long periods in manure even under thermal composting conditions (Jiang *et al.* 2003; Grewal *et al.* 2007). The result of the present study also confirmed the observations. During composting of cow manure at 50 °C for 10 d, the survival of *S. aureus* was still observed at the end of composting with the population of 2.21 log CFU/g. Nevertheless, under the same conditions, *E. coli* and *Salmonella* were not detected from 1 d; *E. faecalis* was undetectable from 3 d. So, the results of our study and other observations indicate that survival of different pathogens in manure during composting differ from each other in some extent. For this reason, the guarantee of complete elimination of different pathogens, especially the human and animal infectious pathogens, in compost previously to its application has become a tough task to be solved immediately with the increase of the cases of foodborne diseases and the emergence of new foodborne pathogens.

$CaCN_2$ has been mainly used as a fertilizer for a long time (Oh *et al.* 2006; Takeda *et al.* 2007; Soltys *et al.* 2011). However, in recent years, several studies reported its fungicidal effect on the pathogens of the soilborne diseases (Bourbos *et al.* 1997; Shi *et al.* 2009). With consideration of both advantages and to evaluate the role of $CaCN_2$ as an antimicrobial agent during composting, we studied its effect on foodborne pathogens during cow manure composting in the present study. And the results confirmed its inhibitory effect on these microorganisms. With addition of $CaCN_2$, *E. coli*, *Salmonella*, *E. faecalis*, and *S. aureus* in manure were inactivated or reduced significantly quicker than that in the treatments with no $CaCN_2$ mixed during composting. Meanwhile, the inhibitory efficacy of $CaCN_2$ against the microorganisms was influenced by composting temperature and additive content of $CaCN_2$. The inactivation at 50 °C was more rapid and effective than at 20 °C, 30 °C, and 37 °C. The result was the same both for the experimental group and control group. Therefore, the phenomenon was

Table 3. The survival and persistence of *E. coli*, *Salmonella*, *E. faecalis* and *S. aureus* during cattle manure composting with addition of different contents of $CaCN_2$ at 30 °C

Microorganism	Additive content[a] (%)	Mean[b] ±SD (log CFU/g) at sampling time (d)				
		0	1	2	3	5
E. coli	0[c]	6.45±0.06A[d]d[e]	6.90±0.07Aa	6.76±0.11Ab	6.57±0.21Ac	6.04±0.11e
	2.0	6.45±0.06Aa	5.98±0.01Bb	4.78±0.15Bc	3.16±0.03Bd	<2.0[f]
	2.5	6.45±0.06Aa	4.43±0.09Cb	ND	ND	ND
	3.0	6.45±0.06A	ND	ND	ND	ND
Salmonella	0	6.76±0.07Aa	6.66±0.02Ab	6.25±0.13Ac	5.78±0.11Ad	5.23±0.05Ae
	2.0	6.76±0.07Aa	4.57±0.09Bb	3.74±0.01Bc	3.67±0.05Bd	2.22±0.03Be
	2.5	6.76±0.07Aa	3.75±0.16Cb	2.87±0.07Cc	ND	ND
	3.0	6.76±0.07Aa	<2.0	ND	ND	ND
E. faecalis	0	5.69±0.02Ab	5.95±0.15Aa	5.46±0.02Ad	4.58±0.14Ac	3.66±0.06Ae
	2.0	5.69±0.02Aa	5.17±0.08Bb	4.81±0.12Bc	3.96±0.08Bd	3.29±0.08Be
	2.5	5.69±0.02Aa	4.48±0.04Cb	<2.0	ND	ND
	3.0	5.69±0.02Aa	3.55±0.03Db	ND	ND	ND
S. aureus	0	4.93±0.08Aa	4.36±0.01Ac	4.44±0.00Ab	4.11±0.20Ad	3.98±0.05Ae
	2.0	4.93±0.08Aa	4.02±0.15Bb	3.89±0.02Bc	3.39±0.01Bd	2.10±0.14Be
	2.5	4.93±0.08Aa	3.51±0.03Cb	<2.0	ND	ND
	3.0	4.93±0.08Aa	3.06±0.11Db	ND	ND	ND

ND, not detected after enrichment.
[a]Additive contents of $CaCN_2$ in composting materials in weight percentage.
[b]The mean is the average of six replicates.
[c]The control group that with no $CaCN_2$ mixed.
[d]For each pathogen, means with different upper case letters in a column are significantly different ($P < 0.05$).
[e]Means with different lower case letters in a row are significantly different ($P < 0.05$).
[f] <2.0 log CFU/g, positive after enrichment.

partly related to the unsupportive tendency of high temperature for the growth of microorganisms (Lung *et al.* 2001; Gong 2007). In terms of temperature influence, the inhibitory efficacies of $CaCN_2$ against four pathogens at 20 °C, 30 °C, and 37 °C were more obvious than at 50 °C. The result indicates that it is

Table 4. Log removal for *E. coli*, *Salmonella*, *E. faecalis*, and *S. aureus* during cattle manure composting with addition of different contents of $CaCN_2$ at 30 °C

Microorganism	Additive content[a] (%)	Log removal at sampling time (d)				Average log removal (Mean± SD)
		1	2	3	5	
E. coli	0[b]	−0.45	−0.31	−0.12	0.41	−0.12±0.38
	2.0	0.47	1.67	3.29	6.45	2.97±2.59
	2.5	2.02	6.45	6.45	6.45	5.34±2.22
	3.0	6.45	6.45	6.45	6.45	6.45±0.00
Salmonella	0	0.1	0.51	0.98	1.53	0.78±0.62
	2.0	2.19	3.02	3.09	4.54	3.21±0.98
	2.5	3.01	3.89	6.76	6.76	5.11±1.94
	3.0	6.76	6.76	6.76	6.76	6.76±0.00
E. faecalis	0	−0.26	0.23	1.11	2.03	0.78±1.01
	2.0	0.52	0.88	1.73	2.4	1.38±0.85
	2.5	1.21	5.69	5.69	5.69	4.57±2.24
	3.0	2.14	5.69	5.69	5.69	4.80±1.78
S. aureus	0	0.57	0.49	0.82	0.95	0.71±0.21
	2.0	0.91	1.04	1.54	2.83	1.58±0.88
	2.5	1.42	4.93	4.93	4.93	4.05±1.76
	3.0	1.87	4.93	4.93	4.93	4.17±1.53

Note: Log removal = $\log N_0 - \log N_x$, $\log N_0$ and $\log N_x$: 0 d and sampling day concentration (log CFU/g). $\log N_x$ in the sampling days in which pathogen was not detected or under the detection limit was taken a value of zero.
ND, pathogen was not detected after enrichment or below the detection limit.
[a]Additive contents of $CaCN_2$ in composting materials in weight percentage.
[b]The control group that with no $CaCN_2$ mixed.

necessary to take full advantage of fungicidal function of $CaCN_2$ during mesophilic composting or at the beginning of composting process, especially in the composting that temperatures could not be uniform throughout the entire mass of compost which is a possibility naturally occurring in all kinds of composting. The temperatures under such conditions are suitable for the growth of microorganisms or at least do not hinder it, so it becomes one of the major difficulties to inactivate pathogens during composting. Several studies have tried to overcome this difficulty (Plachá *et al.* 2008; Avery *et al.* 2009; Wong, Selvam 2009). In the present study, the complete inactivation of four foodborne pathogens during mesophilic composting in the short term was accomplished by addition of $CaCN_2$ into composting materials. However, there was a close relationship between the efficacy of pathogen inactivation and the additive content of $CaCN_2$. In 5 d composting, four tested pathogens all survived at the end of experiment when 2.0% $CaCN_2$ was added into the manure, while they were entirely undetectable from 3 d and 2 d with addition of 2.5% and 3.0% $CaCN_2$, respectively.

Conclusions

1. Survival of *E. coli*, *Salmonella*, *E. faecalis*, and *S. aureus* in the compost piles of fresh cow manure depended on composting temperatures: at thermophilic conditions, their death occurred soon; while at mesophilic conditions, they survived in large populations.

2. With addition of $CaCN_2$, a significant reduction of foodbone pathogens during composting was observed, and the pathogens were completely undetectable in the short term, even under mesophilic conditions.

3. The result of the test indicates that the addition of $CaCN_2$ into manure should be at least 2.5% for complete inactivation of foodborne pathogens during composting.

Acknowledgements

This work was supported by the Major Special Project of National Dairy Industry of China (2006BAD04A15), the Key Project of Natural Science Fund of Inner Mongolia, China (200607010403), the Talent Fund of Inner Mongolia, China, and the National Natural Science Funds, China (31060318).

References

Avery, L. M.; Williams, A. P.; Killham, K.; Jones, D. L. 2009. Heat and lime-treatment as effective control methods for *E. coli* O157:H7 in organic wastes, *Bioresource Technology* 100(10): 2692–2698. http://dx.doi.org/10.1016/j.biortech.2008.12.044

Bourbos, V. A.; Skoudridakis, M. T.; Darakis, G. A.; Koulizakis, M. 1997. Calcium cyanamide and soil solarization for the control of *Fusarium solani* f.sp. *cucurbitae* in greenhouse cucumber, *Crop Protection* 16(4): 383–386. http://dx.doi.org/10.1016/S0261-2194(96)00110-X

Doyle, M. P.; Erickson, M. C. 2006. Reducing the carriage of foodborne pathogens in livestock and poultry, *Poultry Science* 85(6): 960–973.

Fernández, J. M.; Hernández, D.; Plaza, C.; Polo, A. 2007. Organic matter in degraded agricultural soils amended with composted and thermally-dried sewage sludges, *Science of the Total Environment* 378(1–2): 75–80. http://dx.doi.org/10.1016/j.scitotenv.2007.01.008

Gong, C. M. 2007. Microbial safety control of compost material with cow dung by heat treatment, *Journal of Environmental Sciences* 19(8): 1014–1019. http://dx.doi.org/10.1016/S1001-0742(07)60164-8

Grewal, S.; Sreevatsan, S.; Michel, F. C. Jr. 2007. Persistence of *Listeria* and *Salmonella* during swine manure treatment, *Compost Science & Utilization* 15(1): 53–62.

Jiang, X.; Morgan, J.; Doyle, M. P. 2003. Fate of *Escherichia coli* O157:H7 during composting of bovine manure in a laboratory-scale bioreactor, *Journal of Food Protection* 66(1): 25–30.

Liang, C.; Das, K. C.; McClendon, R. W. 2003. The influence of temperature and moisture contents regimes on the aerobic microbial activity of a biosolids composting blend, *Bioresource Technology* 86(2): 131–137. http://dx.doi.org/10.1016/S0960-8524(02)00153-0

Lung, A. J.; Lin, C. M.; Kim, J. M.; Marshall, M. R.; Nordstedt, R.; Thompson, N. P.; Wei, C. I. 2001. Destruction of *Escherichia coli* O157:H7 and *Salmonella enteritidis* in cow manure composting, *Journal of Food Protection* 64(9): 1309–1314.

Martens, W.; Böhm, R. 2009. Overview of the ability of different treatment methods for liquid and solid manure to inactivate pathogens, *Bioresource Technology* 100(22): 5374–5378. http://dx.doi.org/10.1016/j.biortech.2009.01.014

Oh, K.; Kato, T.; Li, Z. P.; Li, F. Y. 2006. Environmental problems from tea cultivation in Japan and a control measure using calcium cyanamide, *Pedosphere* 16(6): 770–777. http://dx.doi.org/10.1016/S1002-0160(06)60113-6

Plachá, I.; Venglovský, J.; Maková, Z.; Martinéz, J. 2008. The elimination of Salmonella typhimurium in sewage sludge by aerobic mesophilic stabilization and lime hydrated stabilization, *Bioresource Technology* 99(10): 4269–4274. http://dx.doi.org/10.1016/j.biortech.2007.08.056

Shi, K.; Wang, L.; Zhou, Y. H.; Yu, Y. L.; Yu, J. Q. 2009. Effects of calcium cyanamide on soil microbial communities and *Fusarium oxysporum* f. sp. *Cucumberinum*, *Chemosphere* 75(7): 872–877. http://dx.doi.org/10.1016/j.chemosphere.2009.01.054

Soltys, D.; Rudzińska-Langwald, A.; Kurek, W.; Gniazdowska, A.; Sliwinska, E.; Bogatek, R. 2011. Cyanamide mode of action during inhibition of onion (*Allium cepa* L.) root growth involves disturbances in cell division and cytoskeleton formation, *Planta* 234(3): 609–621. http://dx.doi.org/10.1007/s00425-011-1429-5

Takeda, T.; Hatta, N.; Kikkawa, S. 2007. Ammonia nitridation synthesis and structural change of strontium cyanamide polymorphs, *Journal of the Ceramic Society of Japan* 115(1347): 729–731. http://dx.doi.org/10.2109/jcersj2.115.729

Viazis, S.; Diez-Gonzalez, F. 2011. Chapter one-enterohemorrhagic *Escherichia coli*: the twentieth century's emerging foodborne pathogen: a review, *Advances in Agronomy* 111: 1–50. http://dx.doi.org/10.1016/B978-0-12-387689-8.00006-0

Wichuk, K. M.; McCartney, D. 2007. A review of the effectiveness of current time-temperature regulations on pathogen inactivation during composting, *Journal of Environmental Engineering and Science* 6(5): 573–586. http://dx.doi.org/10.1139/S07-011

Wong, J. W. C.; Selvam, A. 2009. Reduction of indicator and pathogenic microorganisms in pig manure through fly ash and lime addition during alkaline stabilization, *Journal of Hazardous Materials* 169(1–3): 882–889. http://dx.doi.org/10.1016/j.jhazmat.2009.04.033

You, Y.; Rankin, S. C.; Aceto, H. W.; Benson, C. E.; Toth, J. D.; Dou Z. 2006. Survival of *Salmonella enterica* serovar Newport in manure and manure-amended soils, *Applied and Environmental Microbiology* 72(9): 5777–5783. http://dx.doi.org/10.1128/AEM.00791-06

Huasai SIMUJIDE. College of Animal Science, Inner Mongolia Agricultural University, Hohhot 010018, PR China. Number of publications: 16. Number of conferences participated: 2 (China). Research interests: waste management technologies, air pollution, landscape management, veterinary microbiology.

Chen AORIGELE. College of Animal Science, Inner Mongolia Agricultural University, Hohhot 010018, PR China. Number of publications: 54. Number of conferences participated: 22 (China). Awards: Inner Mongolia Autonomous Region Science and Technology Progress Award twice, Inner Mongolia Autonomous Region Youth Science and Technology Award once. Academic groups participated and post: The executive director of Straw Resources Utilization Branch of Chinese Association of Agricultural Science Societies, the executive director of Inner Mongolia Dairy Association, the vice general secretary of Inner Mongolia Association of Animal Science, the deputy director of Inner Mongolia Breeding Professional Committee, the member of Japanese Society of Animal Science, the member of Japanese Society of Grassland Science, the member of Japanese Society of Application of Animal Behavior. Research interests: waste management technologies, air pollution, landscape management, environmental management.

Chun-Jie WANG. College of Veterinary Medicine, Inner Mongolia Agricultural University, Hohhot 010018, PR China. Number of publications: 59. Number of conferences participated: 18 (China). Awards: Inner Mongolia Autonomous Region "321 talents project" the second level people, Inner Mongolia Autonomous Region Department of Education "111 talents project" the third level people. Academic groups participated and post: The council member of Chinese Association of Animal Physiology, the council member of Northern China Association of Veterinary Pathology, the member of Chinese Association of Veterinary Pathology, the premium member of Chinese Association of Animal Science and Veterinary Medicine. Research interests: waste management technologies, landscape management, environmental management, medical microbiology.

Bai MANDA. College of Animal Science, Inner Mongolia Agricultural University, Hohhot 010018, PR China. Number of publications: 5. Number of conferences where participated: 2 (China). Research interests: waste management technologies, air pollution, landscape management.

Ma LINA. College of Life Sciences, Inner Mongolia Agricultural University, Hohhot 010018, PR China. Number of publications: 5. Number of conferences participated: 2 (China). Research interests: waste management technologies, air pollution, landscape management, food and fermentation engineering.

Mei-Yun WU. College of Animal Science, Inner Mongolia Agricultural University, Hohhot 010018, PR China. Number of publications: 6. Research interests: waste management technologies, air pollution, landscape management, veterinary microbiology.

Yu LI. College of Animal Science, Inner Mongolia Agricultural University, Hohhot 010018, PR China. Number of publications: 8. Number of conferences participated: 1 (China). Research interests: waste management technologies, air pollution, landscape management, veterinary microbiology.

Tong-Ri-Ge BAI. College of Animal Science, Inner Mongolia Agricultural University, Hohhot 010018, PR China. Number of publications: 7. Number of conferences participated: 1 (China). Research interests: waste management technologies, air pollution, landscape management, veterinary microbiology.

14

ENVIRONMENT-FRIENDLY BIODIESEL PRODUCTION BY TRANSESTERIFICATION OF RAPESEED OIL: EFFECT OF REACTION PARAMETERS

Vita Kiriliauskaitė[1], Vida Bendikienė[2], Benediktas Juodka[3]

[1, 2]Department of Biochemistry and Biophysics, Faculty of Natural Sciences, Vilnius University, M. K. Čiurlionio g. 21, LT-03101 Vilnius, Lithuania

[3]Vilnius University, Vilnius, Lithuania

E-mails: [1]vita.kiriliauskaite@chf.stud.vu.lt; [2]vida.bendikiene@gf.vu.lt (corresponding author); [3]rector@vu.lt

Abstract. The huge energy demand in the industrialized world and the pollution problems caused due to the worldwide consumption of fossil fuels has made it necessary to develop alternative sources of energy. It is estimated that pure biodiesel provides over 90% reduction in unburned hydrocarbons and 75–90% reduction in aromatic hydrocarbons. This results in almost total reduction in sulphur dioxide (which causes acid rains), 40–60% reduction in soot particles, 80% and 10–15% reduction in carbon dioxide and carbon monoxide emissions respectively. Reduction of poly-aromatic hydrocarbons is extremely important as many of these hydrocarbons are cancer causing and ozone-forming compounds. Biodiesel does not interfere with the carbon cycle or cause climate change (vegetables from which oils are extracted for biodiesel production remove carbon dioxide from the atmosphere to grow).

Due to the global drive towards renewable, non-toxic and environmentally acceptable products the ability of the commercial lipolytic enzyme Lipoprime 50T to catalyze the biotechnologically important process of biodiesel (fatty acid methyl esters) production via rapeseed oil transesterification (methanolysis) was investigated. The optimal reaction conditions were determined using titrimetric and simple and accurate thin-layer chromatography and computer analysis methods that enable to follow the changes of all reaction mixture components simultaneously. The effects of molar ratio of substrates, reaction temperature and time, lipase dosage and water content in different organic medium on the composition of the reaction mixture were analyzed.

The hydrolytic activity of lipase directly depends on the reaction solvent and the water content required for high yields of the product increases with increasing the reaction medium hydrophobicity. Under the optimal process conditions identified (n-hexane, 40 °C temperature, 66 mg/ml enzyme, water content of 26%, rapeseed oil to methanol molar ratio of 1:16 and reaction time 24 h) the highest biodiesel conversion yield of around 85% was obtained.

Keywords: biodiesel, transesterification (methanolysis), rapeseed oil, lipase, organic solvent, thin layer chromatography, environmental strategy.

Introduction

The Earth faces a variety of environmental problems that are causing long-term damage to the earth's ecosystem. The only way to control current environmental issues is to create sustainable development strategies and continue to instill conservation methods. For this reason bioeconomy is seen as the main future economy that is based on environmentally friendly products and services produced by the use of biotechnology and renewable energy sources as the basic building blocks for industry. The development of the biotechnological industry and its application to agriculture, health, chemical or energy industries are the examples of bioeconomic activity.

The major part of all energy consumed worldwide comes from fossil sources. Diesel fuels have an essential function in the industrial economy of a country as they are used in heavy trucks, city transport buses, locomotives, electric generators, farm equipment, underground mine equipment, etc. (Ganesan *et al.* 2009). The huge energy demand in the industrialized world and the pollution problems caused due to the widespread use of fossil fuels has made it necessary to develop alternative sources of energy to resolve these problems. Moreover some recent estimates of the world-wide petroleum reserves predict them to last only for 50 years (Barnwal, Sharma 2005).

An increasing environmental awareness in society gave rise to intensified efforts to bring the products of the conversion of fat-rich feedstock like various oils and waste fats into the market as an additional source of bioenergy (Rass-Hansen *et al.* 2007). Such products are defined as being renewable, non-toxic and environment-friendly components of biodiesel (Bendikienė *et al.* 2011) and biolubricants (Kiriliauskaitė *et al.* 2011). Use of bioenergy feedstocks could not only provide significant environmental benefits, but could also help to reduce reliance on foreign oil. Biofuels also offer a promising road to enhance development since they use local materials, can provide local jobs, and do not require the import of expensive equipment and expertise.

In recent years various important experimental and theoretical environmental investigations and developments of technologies for the protection of environment have been conducted in Lithuania, including the studies of ability to use fatty waste for biogas production during organic waste biodegradation process (Baltrėnas, Kvasauskas 2008; Baltrėnas *et al.* 2004), investigations of biodegradation processes in food waste with different amounts of moisture (Baltrėnas *et al.* 2006).

The global biodiesel industry grew very fast in the last years and even a faster growth rate is foreseen in the near future as it shows clear advantages over traditional fuel from an economical and environmental point of view as it is a biodegradable, renewable, and non-toxic fuel (Fjerbaek *et al.* 2009).

Biodiesel can be produced by transesterification of fat-rich feedstock like various oils and waste fats. Vegetable oils present higher viscosity than diesel, normally 25 times. For this reason these mixtures of free fatty acids (FFA) and triacylglycerols (TAG) need to be chemically altered to fatty acid alkyl esters to be used as biodiesel fuel for currently used diesel engines. Catalysts investigated for transesterification are either acids, bases, both liquid and heterogenous, as well as free or immobilized enzymes (Haas *et al.* 2006; Meher *et al.* 2006). Enzymes are an interesting prospect for industrial-scale production for the development of environmentally friendly production process. The number of enzymes commercially available and the range of applications are gradually increasing. Of all known enzymes, lipases are gaining more attention due to their ability to catalyze a wide range of reactions. This makes lipases the enzymes of choice for potential applications in various fields of industry. Lipolytic enzymes work at low temperature, atmospheric pressure and are not aggressive to reactors and accessories which are closely associated with energy cost from the economic perspective (Al-Zuhair 2007).

Industrial production of biodiesel today is based on vegetable oils (Fjerbaek *et al.* 2009). The choice of alcohol has some influence on the properties of the biodiesel produced, though this does not draw as much attention as the price of the alcohol used for biodiesel production (Fjerbaek *et al.* 2009). This makes methanol and ethanol the alcohols of choice for biodiesel production.

The main problem to overcome is an inactivation of the biocatalyst due to the inhibitory effect of excess of alcohol. Short chain alcohols, especially methanol, have a low solubility in oils; therefore a new liquid phase appears in the system leading to an inactivation of the enzyme and decreased yields of alkyl ester. This problem could be removed by selection of a suitable reaction medium for lipase (Adamczak *et al.* 2009; Lu *et al.* 2008). Thus the main parameters that cause the reaction effectiveness are organic solvent and water content. Although water is not involved as a reactant or a product in transesterification reactions, water content is very important as it favors the expression of the full enzymatic activity (Salis *et al.* 2008). The use of solvents in the enzymatic production of biodiesel is mentioned as being inconvenient, but solvent recovery is a common practice in the chemical production of biodiesel and on an industrial scale, solvents can be recovered together with methanol after the enzymatic reaction (Adamczak *et al.* 2009; Royn *et al.* 2007). In the enzymatic oil transesterification (alcoholysis) other factors can influence the yield of the product and the reaction rate as well. These parameters include the reaction time and temperature, molar ratio of oil to alcohol, type of alcohol, nature and amount of enzyme, purity of reactants and mixing intensity (Fjerbaek *et al.* 2009; Adamczak *et al.* 2009).

In the present study, Lipoprime 50T-catalyzed transesterification (methanolysis) of rapeseed oil (RO) in n-hexane and in *t*-butyl alcohol was investigated in order to obtain high degrees of transesterification and high yields of environment-friendly biodiesel-fatty acid methyl esters. The effects of molar ratio of substrates, reaction temperature and time, lipase dosage and water content in different organic medium on the composition of the reaction mixture were analyzed. Methyl ester (ME) production and RO disappearance were followed by thin-layer chromatography (TLC).

1. Investigation methods

LipoPrime® 50T (hereafter Lipoprime) was kindly provided by Biopolis Ltd, the distributor of Novozymes A/S in Lithuania. Lipoprime is classified in the Chemical Abstracts Service Registry as "Lipase, triacylglycerol, CAS no. 9001- 62-1". Lipoprime is a protein-engineered lipase produced by submerged fermentation of a genetically modified *Aspergillus* microorganism (the product sheet of Novozymes). The enzyme has a broad substrate specificity. The product is produced by a non-pathogenic microorganism and is classified as non-toxic. The product preparations are biodegradable. The components of Lipoprime are listed in the relevant inventories, e.g., in EINECS and TSCA.

All chemicals used in the study were products of analytical grade.

1.1. The standard spectrophotometric assay

The hydrolytic activity of lipase on *p*-nitrophenyl butyrate solution in 2-propanol was investigated measuring the change of optical density at 400–410 nm within 3–6 min at 30 °C and pH 7.0–10.0, 100 mM universal buffer (Britton – Robinson buffer; composed of acetic, *ortho*-boric and *ortho*-phosphoric acids at a ratio of 1:1:1 providing buffering capacity over a wide range of pH) (Surinėnaitė *et al.* 2002; Bendikienė *et al.* 2004). One unit of lipase hydrolytic activity corresponds to the amount of the enzyme releasing 1 μmol of *p*-nitrophenol per minute under standard conditions.

1.2. Procedure for enzymatic transesterification of oil

Enzymatic processes are more environmentally friendly than traditional chemical methods as enzymatic reactions require milder conditions, less solvent, and give significantly cleaner products. Reactions were carried out in closed 20 ml batch reactors at a constant stirring (180 rpm), coupled to condensers in order to avoid the loss of reaction mixture components by volatilization. RO and methanol were mixed in a molar ratio of 1:4 unless specified otherwise. Enzyme was used with an amount equivalent to 44 mg/ml unless specified otherwise. Lipoprime 50T declared activity is 50 KLU/g. The activity is determined relative to an enzyme standard under the assay conditions: hydrolysis of tributyrin, 30 °C, pH 7.0 (the product sheet of Novozymes). 1 LU is the amount of enzyme, which releases 1 pmol butyric acid per minute under the given standard conditions. 1 KLU = 1000 LU. The reaction progress was followed by extracting 50 μl of reaction mixture aliquots at definite time intervals and analyzing by TLC. For analysis by the TLC method, samples were diluted with diethyl ether (v/v ratio 1:1), mixed vigorously and kept at –20 °C until chromatographic analysis (Bendikienė *et al.* 2008a).

1.3. Chromatographic analysis

TLC analysis was carried out on silica gel G-25 plates. System of solvents for the elution was chosen according to methods described by Yadav (1998) and Bendikienė (2008a). Chromatograms were developed by iodine vapor. Pure methyl oleate (MO), FFA, TAG, diacylglycerols (DAG), monoacylglycerols (MAG) and RO solutions in diethyl ether were used as standards.

1.4. Quantitative analysis

Quantitative analysis (%) of reaction products separated by TLC (average of 3 assays) was performed with a Uvitec Cambrige Fire-reader imaging system and Uvitec Fire-reader software by photodensitometry assessing spot area and color intensity. Verification of the products yield was performed also using Micro image 4.0 program which had been successfully applied in our previous experiments. The accuracy of this method was tested by determination of released FFA content by titration against sodium hydroxide using phenolphthalein as an indicator. The yields determined by TLC and titrimetry were in good agreement. The validation of the analysis had been described elsewhere (Bendikienė *et al.* 2008b; Kiriliauskaitė *et al.* 2011). Each experiment was carried out in triplicate. The results are presented as means ± S.E.M.

2. Investigation results

It was well known that enzyme activity has a relationship with the log P value of the organic solvent used as the reaction medium and that much higher enzymatic activity could be obtained in hydrophobic organic solvent that has a high log P value (Lu *et al.* 2008; Janssen *et al.* 1993). The octanol-water partition coefficient, P, is a measure of the differential solubility of a neutral substance between these immiscible liquids and thereby, an indicator of hydrophobicity of this substance. It is typically used in its logarithmic form, log P. It has been estimated that for the production of monoesters during enzymatic acylglycerol synthesis, a polar solvent with log P value below 1 is favorable, while for the production of triesters it is better to choose a non-polar solvent with a high log P value. Only the group of tertiary alcohols is an exception. Lu *et al.* (2008) studied twelve different solvents for triolein transesterification (methanolysis) catalyzed by immobilized lipase from *Candida* sp. 99–125 and investigated the effect of water content added to the reaction mixture in the range of 0–10%. They noticed that there is no correlation between the MO yield or conversion efficiency with solvents hydrophobicity (log P), dielectric constant (ε) and Hildebrant solubility parameter (δ). However, the water content needed to obtain the maximum yield of the products and the highest activity of lipase depends on solvents hydrophobicity significantly. More hydropilic solvents needed relatively less water to keep its maximum activity, while the hydrophobic solvents needed more water to maintain higher transesterification (methanolysis) yield (Lu *et al.* 2008).

To optimize reaction conditions for enzymatic RO transesterification, two solvents *t*-butanol and n-hexane with different log P value (0.79 and 3.5 respectively) were compared. n-Hexane is one of the most commonly used solvents in synthetic reactions using lipases. Bernardes *et al.* (2007) reported that the use of n-hexane helped the biodiesel production. It has been shown that there are many benefits in using *t*-butanol as a solvent in comparison to other organic solvents in lipase-catalyzed processes (Royn *et al.* 2007). *t*-Butanol dissolves both methanol and the glycerol and is not a substrate for the lipases because it does not act on tertiary alcohols. Moreover, *t*-butanol is a non-toxic solvent of relative low cost.

2.1. Effect of different reaction medium

The hydrolysis was more favorable in *t*-butanol, while transesterification reactions were more effective in n-hexane. Optimal conditions for the methanolysis in *t*-butanol were determined (40 °C, rapeseed oil to methanol molar ratio 1:6 and 1:8 and reaction time 1.5 h) in our earlier investigations (Bendikienė *et al.* 2011), and since the reaction manner was investigated to be dramatically different in studied organic solvents (Fig. 1), analogous experiments were carried out in n-hexane (Fig. 2, a) and in the mixture of *t*-butanol and n-hexane (1:1, v/v) (Fig. 2, b). At least three moles of methanol are required for the transesterification (methanolysis) reaction to accomplish a complete conversion of RO into its ME. An alcohol to oil molar ratio of six is commonly used in industrial processes to obtain higher yields of esters (Barnwal, Sharma 2005; Van Gerpen 2005). Considering the reports of inhibitory effect of excess alcohol (Adamczak *et al.* 2009), methanol to RO molar ratios of 2, 4, 6 and 8 were tested in the reactions conducted at different temperatures.

As shown in Fig. 2, changing the reaction conditions no significant differences were observed in n-hexane, indicating that Lipoprime 50T lipase retained its activity and stability within the studied range of temperatures. However, it was shown that a higher amount of methanol in the reaction mixture causes the decrease of the hydrolysis effectiveness. This could be due to the inhibition of lipase with excess methanol (Jeong, Park 2008).

Although the reaction in *t*-butanol and n-hexane mixture (1:1, v/v) was slower as compared with the processes in *t*-butanol (the traces of RO were still noticed after 48 h), the highest ME yield of about 30% was obtained under the same optimal conditions in both cases (40 °C temperature and RO to methanol molar ratio 1:6). Above 40 °C temperature, the process effectiveness decreased, possibly owing to some enzymatic deactivation.

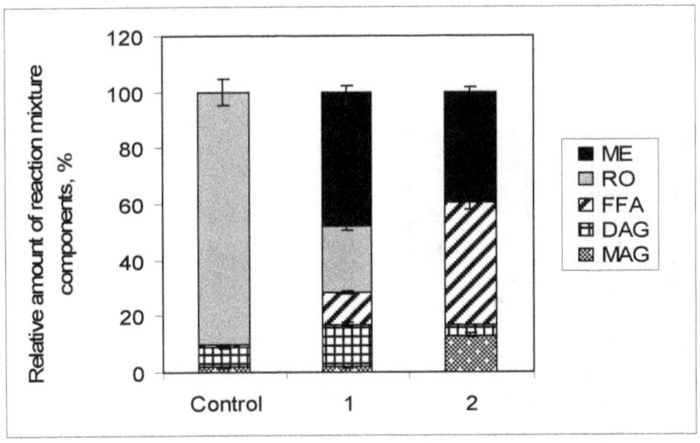

Fig. 1. Effect of organic solvent on Lipoprime 50T-catalyzed rapeseed oil transesterification. Reaction was performed at 30 °C with 44 mg/ml enzyme, RO to methanol molar ratio 1:4 and reaction time 120 h in n-hexane (1) and in *t*-butanol (2). Control – rapeseed oil. ME – methyl esters of FA; RO – rapeseed oil; FFA – free fatty acids; DAG – diacylglycerols; MAG – monoacylglycerols (MAG and DAG contents are expressed as the sum of their regioisomers)

2.2. Effect of solvent quantity

As shown in Fig. 3, the hydrolytic activity directly depends on the quantity of *t*-butanol in the reaction mixture.

The highest yields of FFA were determined in the reaction mixture when n-hexane and *t*-butanol volumetric ratio was in range of 1:2 to 1:4 (Fig. 3, 6–8 columns) and in pure *t*-butanol (Fig. 3, 9 column). As it was noticed earlier, the hydrolysis was more favorable in *t*-butanol in comparison to n-hexane and the processes in n-hexane are much slower (ME yields are higher in pure n-hexane rather than in pure *t*-butanol after 120 h from the reaction beginning, Fig. 1). After 24 h not only hydrolysis effectiveness but also the ME yields were determined to be lower in the reaction mixtures with the bigger portion of n-hexane (Fig. 3, 1–4 columns) than in the ones where *t*-butanol dominated (Fig. 3, 5–9 columns).

a)

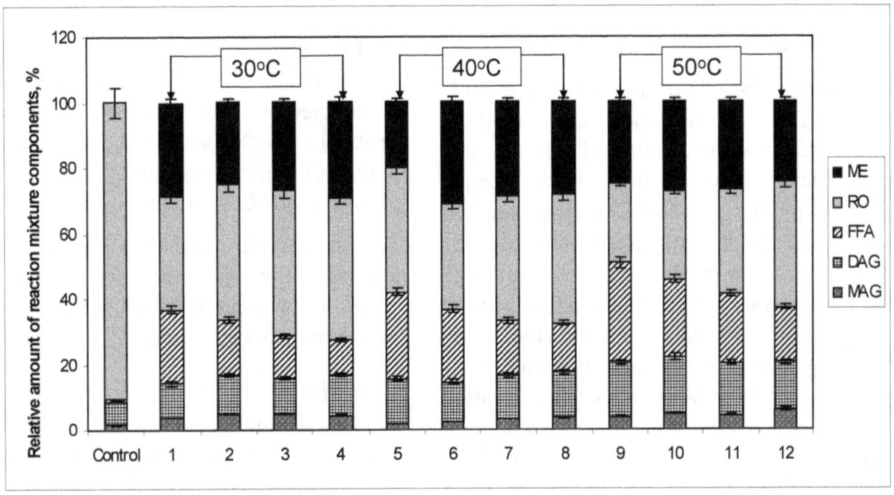

b)

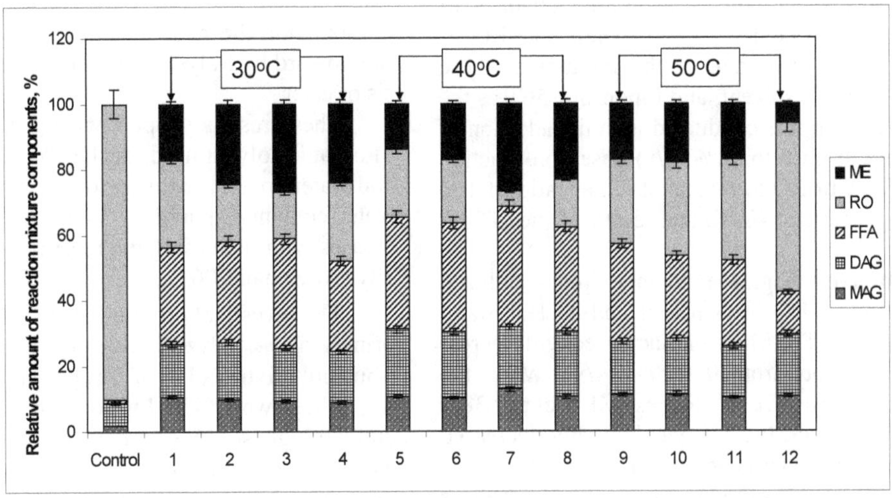

Fig. 2. Effect of molar ratio of substrates and reaction temperature on Lipoprime 50T-catalyzed rapeseed oil transesterification in n-hexane (a) and in n-hexane:*t*-butanol mixture (1:1, v/v) (b). Reaction was performed with 44 mg/ml enzyme, RO to methanol molar ratio **1:2** (1, 5, 9), **1:4** (2, 6, 10), **1:6** (3, 7, 11), **1:8** (4, 8, 12) and reaction time 48 h. Symbols are the same as in Fig. 1

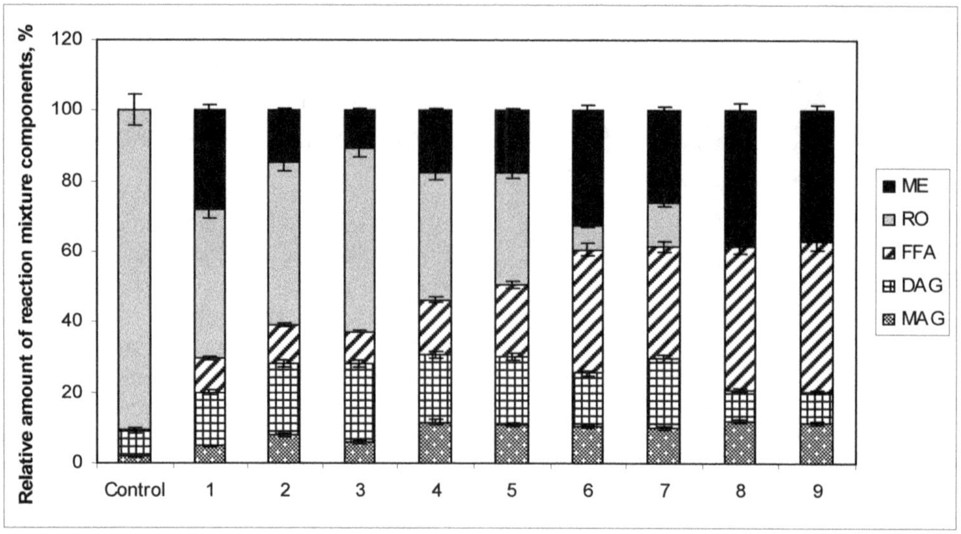

Fig. 3. Effect of solvents on Lipoprime 50T-catalyzed rapeseed oil transesterification in n-hexane:*t*-butanol mixture of different molar ratio. Reaction was performed at 40 °C with 44 mg/ml enzyme, and RO to methanol molar ratio 1:6 and reaction time 24 h. 1 – n-hexane (pure); n-hexane:*t*-butanol (v/v) ratio in the reaction mixture: 2 – 4:1; 3 – 3:1; 4 – 2:1; 5 – 1:1; 6 – 1:2; 7 – 1:3; 8 – 1:4; 9 – *t*-butanol (pure). Symbols are the same as in Fig. 1

2.3. Effect of `water content

Water content is known to be one of the main factors that affect the activity of lipase in non-aqueous medium (Adamczak *et al.* 2009). A small amount of water is necessary to keep the active conformation of the enzyme (Al-Zuhair 2007; Noureddini *et al.* 2005). However, excessive water promotes the hydrolysis of substrate and decrease the yield of the ester production. To obtain high RO conversions, the aqueous phase should be reduced to avoid ME hydrolysis and aggregation of enzyme molecules (Iso *et al.* 2001). On the other hand, the water content must be large enough to prevent enzyme inactivation by the methanol (Oliveira, Rosa 2006).

In this study, the effects of water content on the lipase-catalyzed transesterification (methanolysis) of RO was investigated via the addition of a definite quantity of water to the reaction solution of methanol and RO as substrates, *t*-butanol, n-hexane or the mixture of these solvents (1:1, v/v) as a solvent, and Lipoprime 50T as the enzyme. The reaction was conducted with the addition of water in a range from 0 to 18% (v/v) based on reaction mixture volume with RO to methanol molar ratio of 1:6, a constant temperature of 40 °C, and reaction time of 24 h (Fig. 4 a, b).

It was observed (Fig. 4 a, b) that lipase-catalyzed transesterification of RO in *t*-butanol had a clear water content dependence. The ME yield increased with increasing the water content from 0 to 9% (v/v). When the water content used was 9%, the highest ME yield of 38% was obtained. However, the ME yield sharply decreased when water content used was above 9%. This is because a higher water content shifts the reaction toward hydrolysis of substrate rather than the synthesis of ester.

Lipoprime 50T-catalyzed transesterification of RO in n-hexane had a clear water content dependence as well. However, the optimal conditions could not be determined due to too narrow range of concentrations. Moreover analogous reactions in *t*-butanol and n-hexane mixture had no clear water content dependence in this range of concentrations. Thus, these reactions were repeatedly conducted with the addition of water in a wider range from 0 to 40% (v/v) based on reaction mixture volume with RO to methanol molar ratio of 1:6, a constant temperature of 40 °C, and reaction time of 24 h.

It was observed (Fig. 5) that lipase-catalyzed transesterification (methanolysis) of RO in n-hexane had a clear water content dependence within the range of concentrations investigated.

The ME yield increased with increasing the water content from 0 to 26% (v/v). When the water content used was 26%, the highest ME yield of 68% was obtained. However, the ME yield sharply decreased when water content used was above 26%. As well as in the case of *t*-butanol, the higher content of water shifts the reaction toward hydrolysis of substrate rather than the synthesis of ester.

These results support widely discussed dependence of organic solvent used for the transesterification reaction and water content required for maximum conversion. The water content required for high yields of the product increases with increasing the reaction medium hydrophobicity (Al-Zuhair 2007).

The investigation of effect of water content on Lipoprime 50T-catalyzed RO methanolysis in hexane: *t*-butanol mixture (1:1, v/v) is presented in Fig. 6 (a-c).

The lowest (4%, Fig. 6, a, b and lane 3, Fig. 6, c) and the highest (35% and 40%, Fig. 6, a, b and lanes 10, 11, Fig. 6, c) water contents have the similar effect on the production of esters during the reaction in *t*-butanol and n-hexane mixture as a reaction medium, when the other reaction conditions are the same. When the water content was increased from 9 to 30% not only the ME yield decreased significantly (Fig. 6, a, b and lanes 4–9, Fig. 6, c), the rate of RO hydrolysis decreased as well, especially in the case of water content of 30% (v/v).

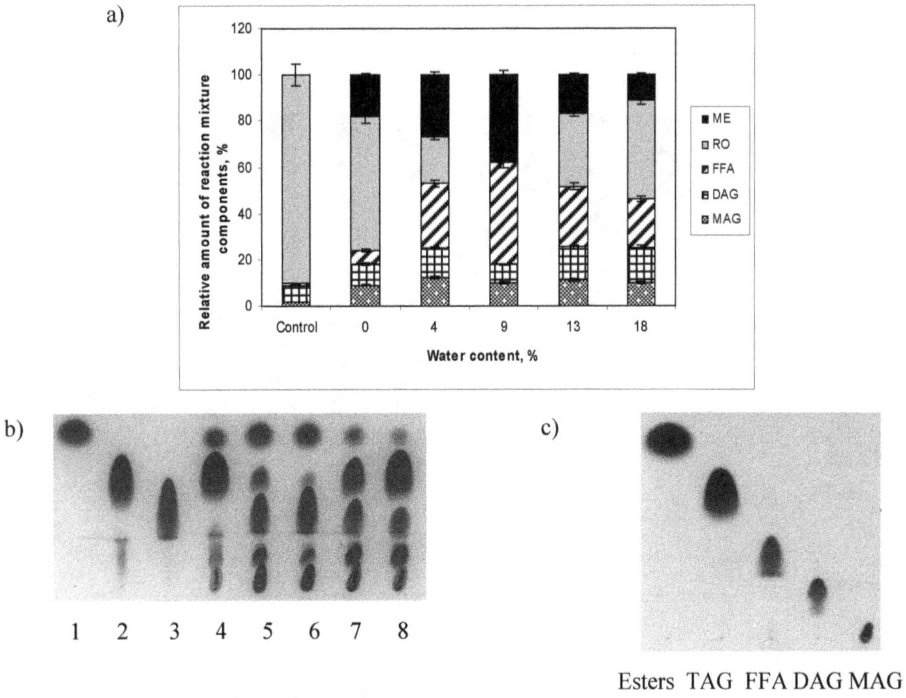

Fig. 4. Effect of water content on Lipoprime 50T- catalyzed rapeseed oil transesterification in *t*-butanol (a). TLC plate showing the products of lipase–catalyzed rapeseed oil methanolysis in *t*-butanol (b): controls: 1 – MO; 2 – RO; 3 – oleic acid (OA); 4 – components of reaction mixture without water; 5 – 4% (v/v) addition of water; 6 – 9% (v/v) addition of water; 7 – 13% (v/v) addition of water; 8 – 18% (v/v) addition of water. Reaction was performed at 40 °C with 44 mg/ml enzyme, RO to methanol molar ratio 1:6 and reaction time 24 h. The view of control samples (c). Symbols are the same as in Fig. 1

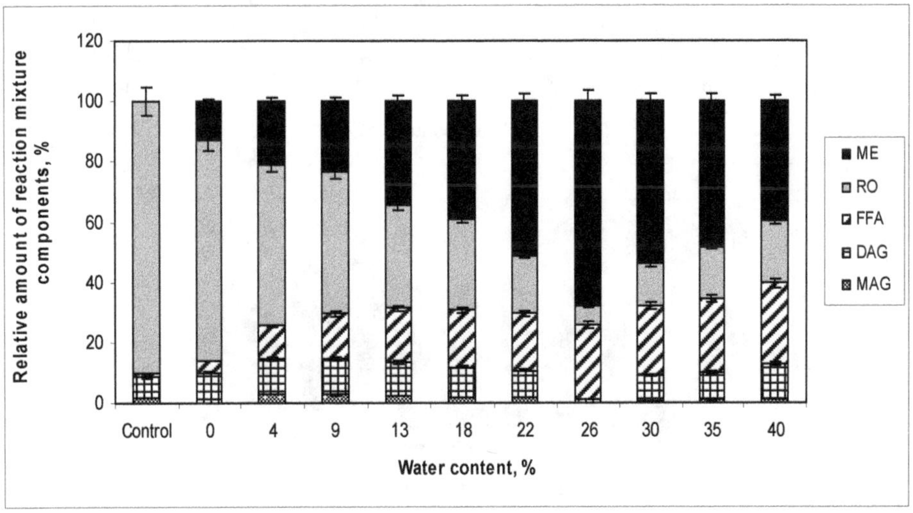

Fig. 5. Effect of water content on Lipoprime 50T-catalyzed rapeseed oil transesterification in n-hexane. Reaction was performed at 40 °C with 44 mg/ml enzyme, RO to methanol molar ratio 1:6 and reaction time 24 h. Symbols are the same as in Fig. 1

It should be noted that when the content of water was the highest (35 and 40%), only ME and FFA were observed in the reaction mixture after 168 h from the reaction beginning. Considering these results analogous reactions with water content of 35 and 40% were carried out in *t*-butanol in order to obtain a higher process rate as in earlier experiments reaction in *t*-butanol was found to be faster in comparison to n-hexane and *t*-butanol mixture. However, no significant increase of the reaction effectiveness was observed (data not presented).

2.4. Effect of substrate molar ratio

When the reactions were carried out in n-hexane with a water content of 26%, the highest ME yield of about 68% was obtained. Therefore, new transesterification (methanolysis) reactions were carried out to determine the optimal molar ratio of substrates in n-hexane in order to obtain even higher yield of the desired product (Fig. 7).

a)

b)

c)

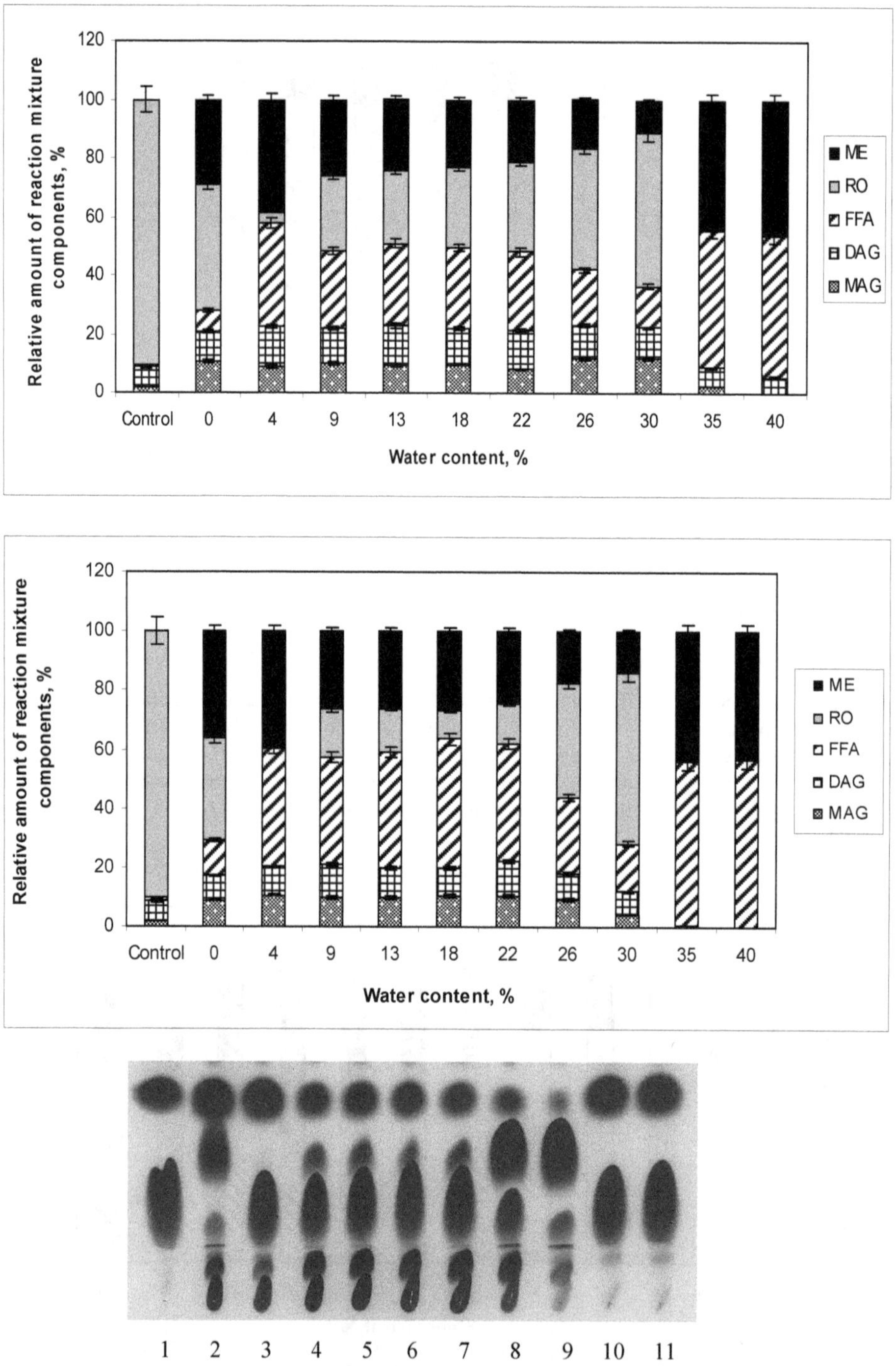

Fig. 6. Effect of water content on Lipoprime 50T-catalyzed rapeseed oil transesterification in hexane:*t*-butanol mixture (1:1, v/v). Reaction was performed at 40 °C with 44 mg/ml enzyme, RO to methanol molar ratio 1:6 and reaction time 24 h (a) and 168 h (b). TLC plate showing the products of lipase–catalyzed rapeseed oil methanolysis n n-hexane:*t*-butanol mixture (1:1, v/v), (c): control: 1 – MO and OA mixture (1:1); 2 – components of reaction mixture without water; 3 – 4% (v/v) addition of water; 4 – 9% (v/v) addition of water; 5 – 13% (v/v) addition of water; 6 – 18% (v/v) addition of water; 7 – 22% (v/v) addition of water; 8 – 26% (v/v) addition of water; 9 – 30% (v/v) addition of water; 10 – 35% (v/v) addition of water; 11 – 40% (v/v) addition of water. Symbols are the same as in Fig. 1

The molar ratio of RO to methanol ranged from 1:2 to 1:16 was studied to evaluate the effect of methanol on lipase activity in n-hexane. From the results of RO and methanol molar ratio study, it was determined that when the molar ratio of oil to methanol was in range of 1:10 – 1:16, high conversions were achieved (ME yields of around 75%) after 24 h, and no significant differences were detected with different oil to methanol molar ratios within this range, indicating that Lipoprime 50T lipase was tolerant to methanol presence within this range and maintained its activity when the water content was 26%. Considering the substantial operational stability of lipase, a RO to methanol molar ratio of 1 to 16 was utilized in the rest of the study in order to obtain higher ME yields.

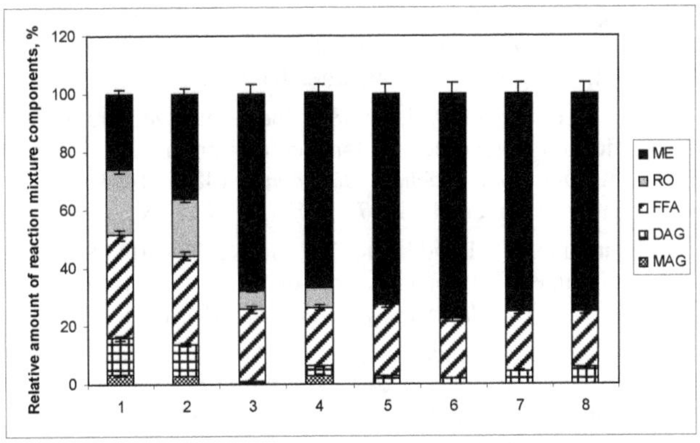

Fig. 7. Effect of rapeseed oil:methanol molar ratio on Lipoprime 50T-catalyzed rapeseed oil transesterification (methanolysis) in n-hexane. Reaction was performed at 40 °C with 44 mg/ml enzyme, the water content of 26% (v/v) and reaction time 24 h. RO : methanol molar ratio: 1 – 1:2; 2 – 1:4; 3 – 1:6; 4 – 1:8; 5 – 1:10; 6 – 1:12; 7 – 1:14; 8 – 1:16. Symbols are the same as in Fig. 1

2.5. Effect of lipase dosage

Considering earlier determined optimal conditions for the reaction the effects of Lipoprime 50T lipase quantity on transesterification (methanolysis) of rapeseed oil for biodiesel production with n-hexane used as a solvent has been evaluated. Fig. 8 depicts the effect of lipase dosage on rapeseed oil transesterification (methanolysis) at 40 °C at rapeseed oil and methanol molar ratio of 1:16 at water content of 26% (v/v).

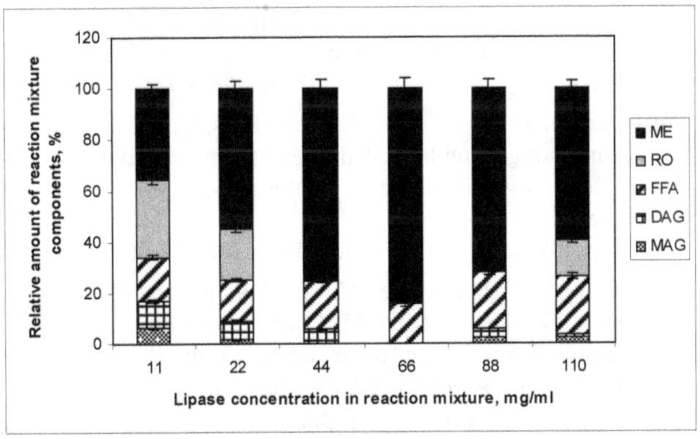

Fig. 8. Effect of enzyme amount on Lipoprime 50T-catalyzed rapeseed oil transesterification (methanolysis) in n-hexane. Reaction was performed at 40 °C with 44 mg/ml enzyme, water content of 26% and RO to methanol molar ratio 1:16 and reaction time 24 h. Symbols are the same as in Fig. 1

It was found that lower dosage of enzyme led to lower ME yield. The ME yield increased with increasing the lipase dosage from 11 to 66 mg/ml. When the lipase dosage reached 66 mg/ml, the ME yield of around 85% was obtained within 24 h. However, further increase in lipase dosage above 66 mg/ml was not capable of enhancing the ME yield, possibly owing to some enzymatic deactivation because of the formation of enzyme excess agglomerates. In enzyme-catalyzed transesterification (methanolysis), the optimal lipase and its usage concentration should be selected in terms of optimal concentration rather than high conversion conditions, as a consequence of the significant cost associated with higher enzyme amounts. Therefore, Lipoprime 50T lipase at 44 mg/ml of reaction volume could be defined as the optimal parameter for the reaction under studied conditions as the yield of ME was about 75% in this case. However, when the lipase dosage reached 66 mg/ml, the reaction mixture consisted only of ME and FFA. Taking into account the fact that a certain amount of FFA is permissible as an impurity of biodiesel, these results are really promising for the production of biodiesel in the future.

Conclusions

1. The ability of commercial lipolytic enzyme Lipoprime 50T to catalyze the biotechnologically important process of environment-friendly biodiesel production by rapeseed oil transesterification (methanolysis) was studied, the optimal process conditions were determined and the effects of water content and different organic solvent were analyzed. The reaction manner was investigated to be dramatically different in studied organic solvents. It was shown that the hydrolytic activity directly depends on the quantity of *t*-butanol in the reaction mixture and the water content required for high yields of the product increases with increasing the reaction medium hydrophobicity. When the water content is large enough the problem of the enzyme inactivation by the methanol excess can be removed in n-hexane.

2. Under the optimal conditions identified (n-hexane, 40 °C temperature, 66 mg/ml enzyme, 26% water content, rapeseed oil to methanol molar ratio 1:16 and reaction time 24 h) the highest biodiesel conversion yield of around 85% was obtained. These results indicate a great potential of producing biodiesel during Lipoprime 50 T lipase-catalyzed methanolysis under relatively mild and environment-friendly conditions.

Acknowledgements

The Research Council of Lithuania (Contract No. N-01/2009) is gratefully acknowledged for financial support and Biopolis Ltd, the distributor of Novozymes A/S in Lithuania for the kindly provided enzyme.

References

Adamczak, M.; Bornscheuer, U. T.; Bednarski, W. 2009. The application of biotechnological methods for the synthesis of biodiesel, *Eur. J. Lipid Sci. Technol.* 111: 808–813. http://dx.doi.org/10.1002/ejlt.200900078

Al-Zuhair, S. 2007. Production of biodiesel: possibilities and challenges, *Biofuels Bioprod. Bioref.* 1: 57–66. http://dx.doi.org/10.1002/bbb.2

Baltrėnas, P.; Jankaitė, A.; Raistenskis, E. 2006. Experimental investigation of biodegradation processes in food waste with different amounts of moisture, *Journal of Environmental Engineering and Landscape Management* 14(4): 173–181.

Baltrėnas, P.; Kvasauskas, M. 2008. Experimental investigation of biogas production using fatty waste, *Journal of Environmental Engineering and Landscape Management* 16(4): 178–187. http://dx.doi.org/10.3846/1648-6897.2008.16.178-187

Baltrėnas, P.; Raistenskis, E.; Zigmontienė, A. 2004. Organinių atliekų biodestrukcijos proceso metu išsiskiriančių biodujų eksperimentiniai tyrimai [Experimental investigation of biogas emissions during organic waste biodegradation processes], *Journal of Environmental Engineering and Landscape Management* 12(1 priedas): 3–9.

Barnwal, B. K.; Sharma, M. P. 2005. Prospects of biodiesel production from vegetable oils in India, *Renew. Sustain. Energy Rev.* 9: 363–378. http://dx.doi.org/10.1016/j.rser.2004.05.007

Bendikienė, V.; Juodka, B.; Surinėnaitė, B.; Dienys, G. 2008a. Lipases in conversion of oils. Experimental screening of enzymes and substrates for the practical approach of biodiesel production, *Biologija* 4: 247–252. http://dx.doi.org/10.3846/16486897.2011.579448

Bendikienė, V.; Kiriliauskaitė, V.; Juodka, B. 2011. Production of environmentally friendly biodiesel by enzymatic oil transesterification, *Journal of Environmental Engineering and Landscape Management* 19(2): 123–129.

Bendikienė, V.; Šinkūnienė, D.; Kiriliauskaitė, V.; Juodka, B. 2008b. Properties and immobilization of *Enterobacter aerogenes* 13 lipase, *Rom. Biotechnol. Lett.* 13(5): 1–8.

Bendikienė, V.; Surinėnaitė, B.; Juodka, B.; Safarikova, M. 2004. Insights into catalytic action mechanism of *Pseudomonas mendocina* 3121-1 lipase, *Enzyme Microb. Technol.* 34: 572–577. http://dx.doi.org/10.1016/j.enzmictec.2004.01.006

Bernardes, O. L.; Bevilaqua, J. V.; Leal, M. C. M. R.; Freire, D. M. G.; Langone, M. A. P. 2007. Biodiesel fuel production by the transesterification reaction of soybean oil using immobilized lipase, *Appl. Biochem. Biotechnol.* 136–140: 105–114. http://dx.doi.org/10.1007/s12010-007-9043-5

Fjerbaek, L.; Christensen, K. V.; Norddahl, B. 2009. A review of the current state of biodiesel production using enzymatic transesterification, *Biotechnol. Bioeng.* 102(5): 1298–1315. http://dx.doi.org/10.1002/bit.22256

Ganesan, D.; Rajendran, A.; Thangavelu, V. 2009. An overview on the recent advances in the transesterification of vegetable oils for biodiesel production using chemical and biocatalysts, *Rev. Environ. Sci. Biotechnol.* 8: 367–394. http://dx.doi.org/10.1007/s11157-009-9176-9

Haas, M. J.; McAloon, A. J.; Yee, W. C.; Foglia, T. A. 2006. A process model to estimate biodiesel production costs, *Bioresour. Technol.* 97: 671–678. http://dx.doi.org/10.1016/j.biortech.2005.03.039

Iso, M.; Chen, B.; Eguchi, M.; Kudo, T.; Shrestha, S. 2001. Production of biodiesel fuel from triglycerides and alcohol using immobilized lipase, *J. Mol. Catal. B: Enzym.* 16: 53–58. http://dx.doi.org/10.1016/S1381-1177(01)00045-5

Janssen, A. E. M.; Van der Padt, A.; Van Sonsbeek, H. M.; Van't Riet, K. 1993. The effect of organic solvents on the equilibrium position of enzymatic acylglycerol synthesis, *Biotechnol. Bioeng.* 41: 95–103. http://dx.doi.org/10.1002/bit.260410113

Jeong, G-T.; Park, D-H. 2008. Lipase-catalyzed transesterification of rapeseed oil for biodiesel production with *tert*-butanol, *Appl. Biochem. Biotechnol.* 148: 131–139. http://dx.doi.org/10.1007/s12010-007-8050-x

Kiriliauskaitė, V.; Bendikienė, V.; Juodka, B. 2011. Synthesis of trimethylolpropane esters of oleic acid by Lipoprime 50T, *J. Ind. Microbiol. Biotechnol.* 38(9): 1561–1566.

Lu, J.; Nie, K.; Wang, F.; Tan, T. 2008. Immobilized lipase *Candida* sp. 99–125 catalyzed transesterification (methanolysis) of glycerol trioleate: Solvent effect, *Bioresource Technol.* 99: 6070–6074. http://dx.doi.org/10.1016/j.biortech.2007.12.045

Meher, L. C.; Sagar, D. V.; Naik, S. N. 2006. Technical aspects of biodiesel production by transesterification – a review, *Renew. Sustain. Energy Rev.* 10: 248–268. http://dx.doi.org/10.1016/j.rser.2004.09.002

Noureddini, H.; Gao, X.; Philkana, R. S. 2005. Immobilized *Pseudomonas cepacia* lipase for biodiesel fuel production from soybean oil, *Bioresource Technol.* 96: 769–777. http://dx.doi.org/10.1016/j.biortech.2004.05.029

Oliveira, A. C.; Rosa, M. F. 2006. Enzymatic transesterification of sunflower oil in an aqueous-oil biphasic system, *J. Am. Oil Chem. Soc.* 83(1): 21–25. http://dx.doi.org/10.1007/s11746-006-1170-6

Rass-Hansen, J.; Falsig, H.; Jørgensen, B.; Christensen, C. H. 2007. Bioethanol: fuel or feedstock?, *J. Chem. Technol. Biotechnol.* 82: 329–333. http://dx.doi.org/10.1002/jctb.1665

Royn, D.; Daz, M.; Ellenrieder, G.; Locatelli, S. 2007. Enzymatic production of biodiesel from cotton seed oil using *t*-butanol as a solvent, *Bioresource Technol.* 98: 648–653. http://dx.doi.org/10.1016/j.biortech.2006.02.021

Salis, A.; Pinna, M.; Monduzzi, M.; Solinas, V. 2008. Comparison among immobilised lipases on macroporous polypropylene toward biodiesel synthesis, *J. Mol. Catal. B: Enzym.* 54: 19–26. http://dx.doi.org/10.1016/j.molcatb.2007.12.006

Surinėnaitė, B.; Bendikienė, V.; Juodka, B.; Bachmatova, I.; Marcinkevičienė, L. 2002. Characterization and physicochemical properties of a lipase from *Ps. mendocina* 3121–1, *Biotechnol. Appl. Biochem.* 36: 47–55.

Van Gerpen, J. 2005. Biodiesel processing and production, *Fuel Process. Technol.* 86: 1097–1107. http://dx.doi.org/10.1016/j.fuproc.2004.11.005

Yadav, P.; Saxena, R. K.; Gupta, R.; Davidson, W. S. 1998. Purification and characterization of regiospecific lipase from *Aspergillus terpeus*, *Biotechnol. Appl. Biochem.* 28: 243–249.

Vita KIRILIAUSKAITĖ. Phd student (biochemistry). Dept of Biochemistry and Biophysics, Vilnius University (VU). Publications: co-author of 4 research papers and 6 extended thesis of Lithuanian and international conferences. Research interests: biotechnology, applied enzymology, biodiesel, lipases, environmentally friendly enzymatic processes.

Vida BENDIKIENĖ. Dr Habilitation Procedure (HP), Chief Research Fellow of Dept Biochemistry and Biophysics, Vilnius University (VU).
Doctor of Science, Habilitation Procedure (biochemistry), VU, 2008. Doctor of Science (biochemistry), Byelorussian State University (Minsk, Byelorussia), 1980.
Publications: author of 1 study-guide, over 40 research papers and 12 patents. Supervisor of PhD projects and diploma projects of B. Sc. and M. Sc. levels.
Research interests: biotechnology, applied enzymology, enzymes-catalyzed biotechnologically important processes optimization methods.

Benediktas JUODKA. Rector of Vilnius University (VU). Dr Habil, 1981, Moscow State University, Russia; Head of the Department of Biochemistry and Biophysics, Vilnius University, 1971–2002; Docent (Associate Professor) of biochemistry, Department of Biochemistry and Biophysics, VU, 1971; Professor of biochemistry, Department of Biochemistry and Biophysics, VU, 1982; Prorector responsible for research of VU, 1991–2001; President of Lithuanian Academy of Sciences, 1992–2002; Adviser for Research and Education to Lithuania government 1993–1994.
Member of Lithuanian Academy of Sciences (1987); Foreign member of Latvian Academy of Sciences (1994); Foreign member (Atti della academia dei Georgofili, Italy 2003); Member of European Academy of Sciences and arts (1994); Member of European Academy of Arts, Science and Humanities (2002); Lithuanian State awards in Science (1985, 2003).
Publications: author of 2 monographs, 1 text-book, 1 patent, 3 reviews, over 150 research papers.
Research interests: covalent nucleic acid and protein complexes and it's chemical modeling; processing of pre-tRNR from eukaryotic cells; mechanism of action of RNA ligase; immobilization of enzymes upon ferromagnetic particles; enzymes-catalyzed biotechnologically important processes; molecular mechanisms of photodynamic oncotherapy.

15

BETTERMENT OF ECOLOGICAL PARAMETERS OF A DIESEL ENGINE USING BROWN'S GAS

Alfredas Rimkus, Saugirdas Pukalskas, Jonas Matijošius, Edgar Sokolovskij

Department of Automobile Transport, Vilnius Gediminas Technical University, J. Basanavičiaus g. 28, LT-03224 Vilnius, Lithuania

Abstract. Hydrogen could become an important element, allowing us to accumulate and transfer energy in a clean way. Hydrogen can be used in cars as a fuel additive which increases the combustion efficiency of the fuel-air mixture. A small amount of hydrogen gas could also be produced in a car by decomposing water by means of electrolysis, using for this purpose the energy produced by the car's electric power generator. The hydrogen and oxygen (HHO) mixture obtained, which is also known as Brown's gaseous mixture, is supplied to the engine's intake manifold.

Tests have been performed with 1.6 TD (SB) diesel engine. The automobile was tested on a chassis dynamometer running at a different speed. The engine was tested using fuel-air combustion mixture and fuel-air-HHO gas combustion mixture without additional adjustment of the fuel supply system. The test results have revealed that additional injection of HHO gas into combustion mixture resulted in up increase of fuel consumption, but the CO, the HC, the PM amount has decreased insignificantly. At few engine loads the amount of NO_x decreased, however increasing the engine load resulted in a gradual increase.

Having analysed test results we came to a conclusion that additional supply of HHO gas into combustion mixture resulted in improvement of the combustion quality of fuel-air mixture and ecological performance of the engine. This is especially relevant for the automobiles which are not equipped with a supplementary exhaust gas toxicity decreasing system.

Keywords: ecological parameters; diesel engine; Brown's gas; hydrogen.

Reference to this paper should be made as follows: Rimkus, A.; Pukalskas, S.; Matijošius, J.; Sokolovskij, E. 2013. Betterment of ecological parameters of a diesel engine using Brown's gas, *Journal of Environmental Engineering and Landscape Management* 21(2): 133–140. http://dx.doi.org/10.3846/16486897.2012.679661

Introduction

Oil reserves on planet Earth are running out. This gives rise to a necessity for finding substitutes for conventional internal combustion engine fuels. Gas and bio-gas are used and future-oriented hydrogen power generation is developed (Mažeika, Matijošius 2010). Hydrogen, the most common element on the planet, is not found as a pure substance, it is rather found in the environment as a chemical compound. The usual method used for obtaining pure hydrogen from water is electrolysis, whereas other chemical reactions are used for obtaining hydrogen from hydrocarbonates or other compounds. Providing that electric power necessary for electrolysis of H_2 is derived from renewable energy sources, the hydrogen obtained could become an important element for clean accumulation and transfer of energy (Matijošius, Sokolovskij 2009).

Hydrogen, as an energy source for automobiles, can be used in the following ways:

- In automobiles with fuel elements, used for obtaining electric power from hydrogen, in which case automobile is driven by an electric motor drive;
- By burning hydrogen as fuel in internal combustion engines.

Decreasing the harmful impact of transport vehicles on the environment is still a very topical issue. The harmful environmental impact of transport vehicles is caused by toxic emissions of internal combustion engines (Juostas, Janulevičius 2009), contamination of roadside soil (Baltrėnas *et al.* 2009), noise and other factors (Baltrėnas *et al.* 2004, 2007; Gražulevičienė, Bendokienė 2009; Paulauskas, Klimas 2011). Many research papers deal with the issues related to production and application of more environmentally friendly fuel types (Bendikienė *et al.* 2011; Baltrėnas, Kvasauskas 2008; Labeckas, Slavinskas 2010; Lebedevas *et al.* 2011; Pukalskas *et al.* 2009; Matijošius, Sokolovskij 2009).

The aim of the research is to study the possibility of additional supply of hydrogen and oxygen (HHO) gas into air-and-fuel mixture and to assess the efficiency of using this gas in a compression ignition internal combustion engine.

Research objectives:

- To explore peculiarities of production of HHO gas in an automobile by means of electrolysis;
- To study the efficiency of using HHO gas in a compression ignition internal combustion engine by employing chassis dynamometer (rolling road) tests.

Corresponding author. Edgar Sokolovskij
E-mail: edgar.sokolovskij@vgtu.lt

Hydrogen has an autoignition temperature of 858K requiring an ignition source to combust in an IC engine (Lambe, Watson 1993). Diesel fuel which has an autoignition temperature of 525K can be used as a pilot to ignite hydrogen. The literature contains a body of work in which hydrogen was used in conjunction with diesel fuel to power CI engines. This dual fuel combustion is often called diesel pilot-ignited hydrogen combustion. Diesel pilot ignited hydrogen combustion at low quantities of hydrogen is beneficial since the diesel fuel is being replaced by hydrogen, which may stretch the supply of hydrocarbon fuels.

Particulate matter (PM), oxides of nitrogen (NO_X), carbon monoxide (CO), and hydrocarbons (HC) are all regulated vehicle emissions. Carbon dioxide (CO_2) is under consideration to be limited in the future, and all of these emissions have been found to be harmful to the environment; thus, their reduction is desirable. Diesel-hydrogen dual-fuel combustion has been reported to yield decreases in all of these emissions, when compared to operation of the compression-ignition engine burning diesel fuel alone.

In one experiment, small amounts of hydro-oxygen ($2H_2$, O_2) were fumigated into the intake of the engine using an electronic injector. The resultant reduced ignition delay was believed to cause a decrease in NO_X in the test engine proportional to the amount of $2H_2$, O_2 fumigated (Gjirja *et al.* 2000). In another experiment by Tomita *et al.* (2001), hydrogen was mixed with the intake air of a direct-injection diesel engine. SOI timing of the diesel fuel was varied across a wide range, holding the overall equivalence ratio equal with and without the addition of hydrogen. Their tests showed very low NO_X emissions when SOI was advanced to or beyond 40° before top-dead-centre (BTDC), with NO_X emissions rising with later injection timing. This was reasoned to have occurred because of the thorough mixing of the hydrogen/air mixture and the diesel fuel before ignition. It was also observed that CO_2 decreased proportionally to the amount of hydrogen substituted for diesel fuel, due to less carbon available in the reactants. At low loads, lower efficiency was achieved running in a dual-fuel mode compared to running with diesel alone. The hydrogen was believed to cause an increased ignition delay due to the large mole fraction of hydrogen in the air displacing oxygen. Efficiency of the engine was affected with the addition of hydrogen, such that typically the brake thermal efficiency of the engine was slightly reduced. It was observed that combustion under dual-fuel operation is controlled by flame propagation rather than autoignition (Tomita *et al.* 2001). A study of lean premixed combustion by Jacobs *et al.* (2005) in a conventional diesel engine fuelled with diesel fuel showed that NO_X was reduced as timing was retarded. This was attributed to the reduced flame temperatures caused by retarded injection (Jacobs *et al.* 2005).

An investigation into the effects of gaseous fuels burned with diesel fuel in a direct injection engine conducted by Varde K. S. and Varde L. K. (1984) used hydrogen to supplement the diesel fuel up to 15% of the total fuel energy in a naturally-aspirated diesel engine. At light loads, the addition of hydrogen reduced soot formation up to 50% over the conventional diesel-fuel mode due to the high equivalence ratio. In their study, NO_X increased as the ratio of H:C was increased by partial fuelling with hydrogen, with greater hydrogen amounts increasing NO_X proportionally (Varde K. S., Varde L. K. 1984). Tsolakis and Megaritis (2004) used a fuel reformer to produce hydrogen-rich gas to be introduced into the combustion process by way of reformed exhaust-gas recirculation (REGR). Their findings showed that this method of achieving partially premixed charge compression ignition (PCCI) yielded the potential for reduced NO_X and PM emissions, and enhanced efficiency.

Homogeneous charge compression ignition (HCCI) has been attempted with hydrogen, and in the regime where it is feasible, an efficiency gain over spark-ignited hydrogen combustion engines was observed (Stenlaas *et al.* 2004). Hydrogen has also been experimented with in the biofuels sector; it was observed in an experiment by Senthil Kumar *et al.* (2002) that hydrogen addition to the intake air of a primarily vegetable-oil fuelled diesel engine can reduce smoke and increase thermal efficiency, addressing two inherent problems with using vegetable oils as fuel. In their study hydrogen was added as a mass fraction of the total fuel, and increasing the 'mass share' of hydrogen had the following effects: decreased CO, increased NO, increased ignition delay, decreased HC, and both increased and decreased efficiency, depending on the amount of hydrogen and type of base fuel used (Senthil Kumar *et al.* 2002).

1. Comparison of the properties of diesel fuel and hydrogen

As compared to diesel fuel, hydrogen is distinguished by nearly three times higher calorific capacity, it requires less energy to ignite, burns faster, has a higher diffusion coefficient and is better mixed with air (see Table 1).

Table 1. Physical and chemical properties of diesel fuel and hydrogen

Physical and chemical properties	*Diesel fuel*	*Hydrogen*
Calorific capacity H_u, MJ/kg	42.5	120
Density ρ, kg/m^3 at $t = 0$ °C and $p = 0.1$ MPa	~850	~0.09
Relative density as compared with air	~654	~0.0695
Composition of the main elements according to their weight, %	86C, 13 H	100 H
Boiling temperature, °C	180–360	–253
Ignition temperature, °C	250	~560
Ignition energy, mJ	–	0.02
Cetane number	45–55	5–10
Research octane number	30	130
Flame propagation velocity, m/s	~0.3	~2.7
Combustion temperature, K	2289	2449
The amount of air needed for burning of 1 kg of material, kg/kg	14.6	34.5
Possible ignition boundaries of the air-and-fuel mixture, λ	0.95–12	0.14–9.85
Possible composition of the air-and-fuel mixture according to the air volume, %	0.6–7.5	4–77
Diffusion coefficient, cm^2/s	–	0.63

When burning hydrogen CO_2 emission is equal to zero. As an inflammable mixture, hydrogen is characterized by wider ignition boundaries. Engine can run on a mixture of diesel fuel, air and hydrogen. This mixture could burn more efficiently, ensuring more complete burning of hydrocarbons contained in fuel.

2. Production of HHO gas and its application in an automobile

Accumulating and storing hydrogen is difficult. Hydrogen can be produced locally in an automobile by means of a water electrolysis device. Hydrogen electrolysis generators are known to be of two types:

- Decomposing water into separate hydrogen H_2 and oxygen O_2 gases. In an electrolysis unit oxygen is disengaged next to anode, while hydrogen is disengaged next to cathode.
- Producing HHO gas (Brown's gas) – an inflammable mixture of hydrogen and oxygen.

Water electrolysis from a chemical perspective is a reaction:

$$2H_2O + 2E = 2H_2 + O_2, \qquad (1)$$

where: E – the amount of energy theoretically necessary to produce one mole (2 g) of hydrogen.

$E = 240$ kJ. The same amount of energy is theoretically released upon burning one mole of hydrogen (Yilmaz *et al.* 2010).

Brown's gas generator is filled with electrolytic solution, composed of 97% distilled water and 3% KOH solution. Using an electric current of U = 12 V voltage and I = 25 A strength the generator produces $Q_{HHO} \approx 1.5\ l$ HHO gas in $t = 60$ sec. The volume ratio of hydrogen and oxygen gas in Brown's gas is 2:1. A gas mixture consisting of 1240 l of hydrogen and 620 l of oxygen, the total volume of which constitutes 1860 l, is produced from 1 kg of water. At 0°C temperature and ρ = 0.1 MPa pressure the density of oxygen gas ρ_{O_2} = 1.43 kg/m^3, the density of hydrogen is ρ_{H_2} = 0.09 kg/m^3, the density of Brown's gas is ρ_{HHO} = 0.54 kg/m^3. One litre of HHO gas contains 0.67 l of H_2 gas, the mass of which is $m_{H_2/l}$ = 0.06 g.

Efficiency ratio of HHO gas generator:

$$\eta_{HHO} = \frac{E_{hydrogen}}{E_{electric}}, \qquad (2)$$

where: $E_{electric}$ – the amount of electric power consumed for production of one litre of HHO gas, J:

$$E_{electric} = \frac{U \cdot I \cdot t}{Q_{HHO}}; \qquad (3)$$

$E_{electric}$ = 12000 J.

$E_{hydrogen}$ – the amount of energy released by burning of the hydrogen contained in 1 litre of HHO gas, J:

$$E_{hydrogen} = H_u \cdot m_{H_2/l}; \qquad (4)$$

$E_{hidrogen}$ = 7200 J. η_{HHO} = 0.6.

The remaining part of electric power in the gas generator turns into heat.

Upon burning of hydrogen in an engine merely about 30% of its energy is converted into mechanical work and the total net efficiency of HHO production and use constitutes $\eta_{e_{HHO}} \approx 0.18$.

The electric power used for electrolysis is obtained from the automobile's battery (Fig. 1), which is charged by an electric power generator when the engine is running. Brown's gas generator is supplied with a modulated electric current. By changing the gain coefficient the strength of the current is altered and the amount of HHO produced in adjusted. The electric circuit supplying the current to the Brown's gas generator is connected to a relay switch. When shutting the internal combustion engine down this relay disconnects power supply to the gas generator and the supply of gas is discontinued.

A small amount of gas produced by the HHO generator is directly supplied into the engine's intake manifold. When flowing along the manifold the gas is mixed with the intake air drawn into the cylinders. At the end of the compression stroke diesel fuel is sprayed into the gas mixture and the high temperature of the gas causes fuel ignition. HHO gas improves the quality and velocity of the air-and-diesel-fuel mixture combustion.

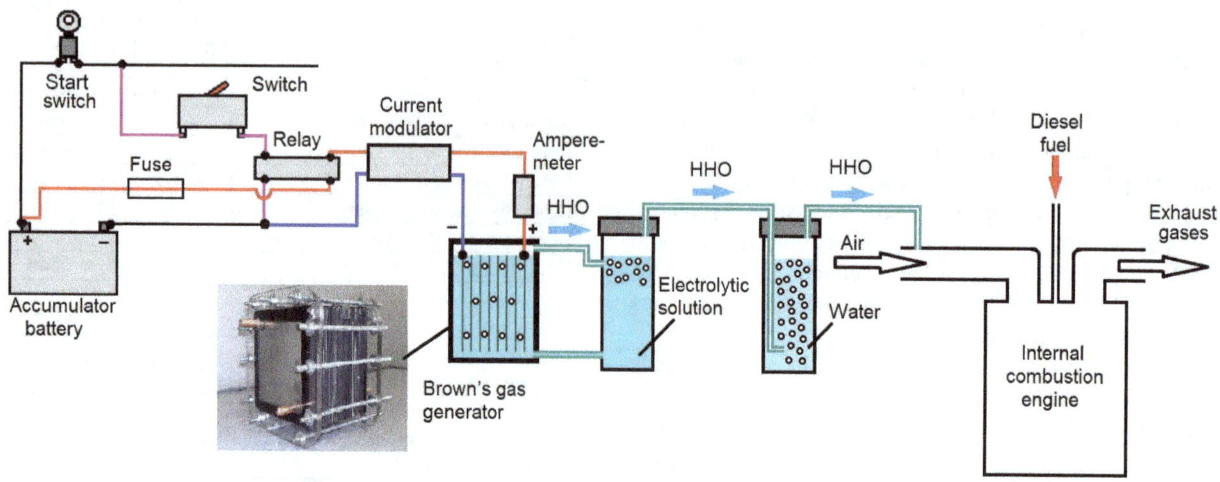

Fig. 1. Structural scheme of HHO gas generator installation in an automobile

3. Test methodology

Tests have been performed using the automobile Volkswagen Passat with a 1.6 TD (SB) diesel engine. The automobile has a HHO gas generator installed. The automobile has been tested using MAHA company's dynamometer rolling road LPS 2000. Composition of engine exhaust gases (CO, CO_2, HC, NO_x) was measured by the gas analyser TECHNOTEST 488, smokiness was measured by AVL DiCom 4000. For the purposes of measuring fuel consumption an electronic fuel consumption meter connected to a dynamometer rolling road has been installed in the automobile. Tests have been performed when the automobile was running on the fourth gear and developing various established speeds v (Table 2). When driving at the specified constant speeds the automobile is affected by particular loads of resistance to motion and air resistance. In case of testing the automobile on a dynamometer rolling road the calculated load is imposed on the traction wheels. Taking into consideration transmission losses, it has been calculated that in order to achieve the set speeds the engine has to develop the set brake horsepower P_e (Table 2).

With engine running under different operating conditions fuel consumption per hour B_d, comparative net fuel consumption b_e and engine net efficiency ratio η_e have been established according to the measured test values.

Tests have been performed three times under all operating conditions; average measured and calculated values are presented in the results.

When testing the engine under different operation conditions, the proportion of HHO gas volume, %, in the drawn in air and Brown's gas mixture is changing (Table 2):

$$P_{V_{HHO}} = \frac{Q_{HHO} \cdot 100}{Q_{air}}, \tag{5}$$

where: Q_{HHO} – the amount of HHO gas supplied into the engine, l/min. Q_{HHO} = 1.5 l/min.

Q_{air} – the amount of air drawn into the engine, l/min:

$$Q_{air} = \frac{2 \cdot V_H \cdot \eta_v \cdot n}{\tau}, \tag{6}$$

V_H – engine litre capacity, l; V_H = 1.6 l; η_v – Volumetric efficiency of cylinders; n – Engine shaft rotation frequency, min^{-1}; τ – Engine timing frequency, $\tau = 4$.

The proportion of hydrogen gas as compared to the mass of the injected diesel fuel, % (Table 2):

$$P_{mH_2} = \frac{m_{H_2} \cdot 100}{m_{diesel}}, \tag{7}$$

where: m_{H_2} – the mass of pure hydrogen in the HHO gas delivered to the cylinder, kg:

$$m_{H_2} = \frac{2 \cdot V_h \cdot \eta_v \cdot P_{V\,HHO} \cdot \rho_{H_2}}{3 \cdot 100}, \tag{8}$$

m_{diesel} – the mass of diesel fuel injected into the cylinder, kg:

$$m_{diesel} = \frac{V_h \cdot \eta_v \cdot \rho_{air}}{14.5 \cdot \lambda}. \tag{9}$$

It is stated in the announced scientific research papers that significant improvement of engine operational characteristics is noticed in the cases when the proportion of hydrogen as compared to the mass of fuel injected into the cylinder amounts to 1–5%. The tests have been performed with only 0.1–0.2% of additionally supplied hydrogen. Increase of HHO gas production using the energy generated by the test automobile's battery results in higher internal energy loss. It is rational to supply larger amounts of hydrogen from a separate tank.

Table 2. Engine operational characteristics of an automobile driving at various speeds

Automobile speed v, km/h	50	60	70	80	90
Engine shaft rotation frequency n, min^{-1}	1640	1970	2300	2620	2940
Engine brake power P_e, kW	4.7	6.5	8.5	11.4	14.9
Proportion of HHO gas volume in the sucked in mixture $P_{V\,HHO}$, %	0.14	0.12	0.10	0.09	0.08
Proportion of H_2 gas as compared to the mass of the injected diesel fuel P_{mH_2}, %	0.19	0.15	0.13	0.11	0.09
Water consumption kg per 100 km run	0.097	0.081	0.069	0.061	0.054

Addition of H_2 gas into combustible mixture requires correction (delay) of fuel injection advance angle, whereas hydrogen increases mixture burning speed.

High pressure fuel pump plunger stroke to the piston top dead centre (BTDC – before top dead centre) was decreased from 0.9 mm to 0.85 mm.

4. Test results

Supply of HHO gas decreases the amount of carbon monoxide gas in exhaust gases by 7–12.5% (Fig. 2). The amount of CO decreases because hydrogen improves the combustion process in ensures complete burning of carbon, which turns into carbon dioxide.

This is confirmed by 2.5–4.5% decrease of oxygen amount and up to 2% increase of CO_2 amount in exhaust gases (Fig. 3).

Additional supply of HHO gas caused insignificant decrease in the amount of hydrocarbons in exhaust gases (up to 1.5%) (Fig. 4). This reveals that hydrogen additive also improves combustion reaction in hydrocarbons.

With supply of HHO gas at 50 km/h automobile speed (4.7 kW engine load) the amount of NO_x is 5.5%

smaller than without additional supply of hydrogen (Fig. 5).

With increase of automobile speed this difference decreases gradually (up to 70 km/h), whereas further increase of speed and engine load with additional supply of HHO gas leads to greater emission of NO_x than in cases with no HHO supplied. Upon reaching the speed of 90 km/h and 14.9 kW engine load NO_x emissions using HHO gas are 3.5% larger than in cases with no gas supplied. This reveals that using hydrogen at small loads ensures more efficient combustion which results in decrease of the amount of free oxygen and decrease of NO_x amount in exhaust gases. Increase of engine load and rotation speed creates a richer mixture and the amount of NO_x decreases, thus additional supply of hydrogen gas increases the combustion temperature of the combustible mixture and increases the amount of nitrogen oxides.

A small amount of HHO gas drawn into engine cylinders together with air improves combustion of diesel fuel and PM amount decreases by 4–8% (Fig. 6).

Additional supply of HHO gas increases fuel consumption up to 1.7% and decreases engine net efficiency ratio up to 2.5% (Fig. 7), since efficiency ratio of hydrogen production and use in an engine constitutes only $\eta_{ehho} \approx 0.18$.

Engine economic efficiency parameters can be improved by more efficient production of HHO gas and by precise setting of the optimum fuel injection moment.

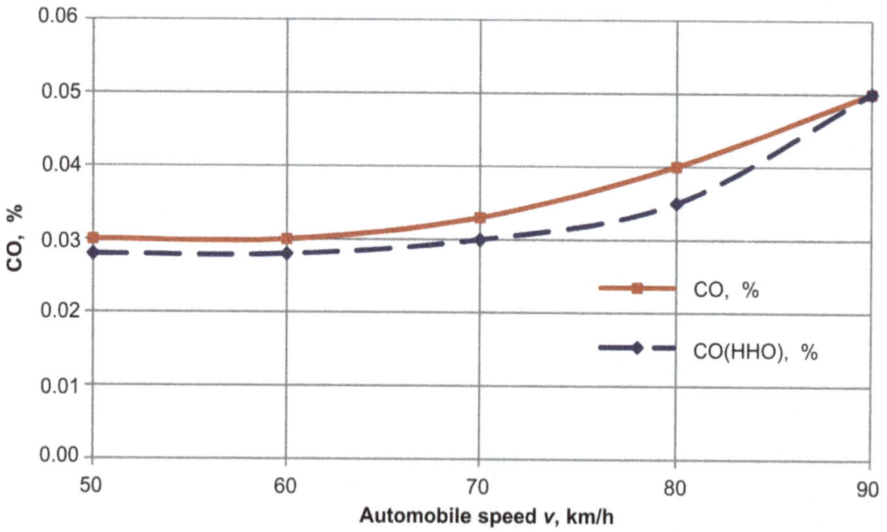

Fig. 2. Amount of CO in exhaust gases

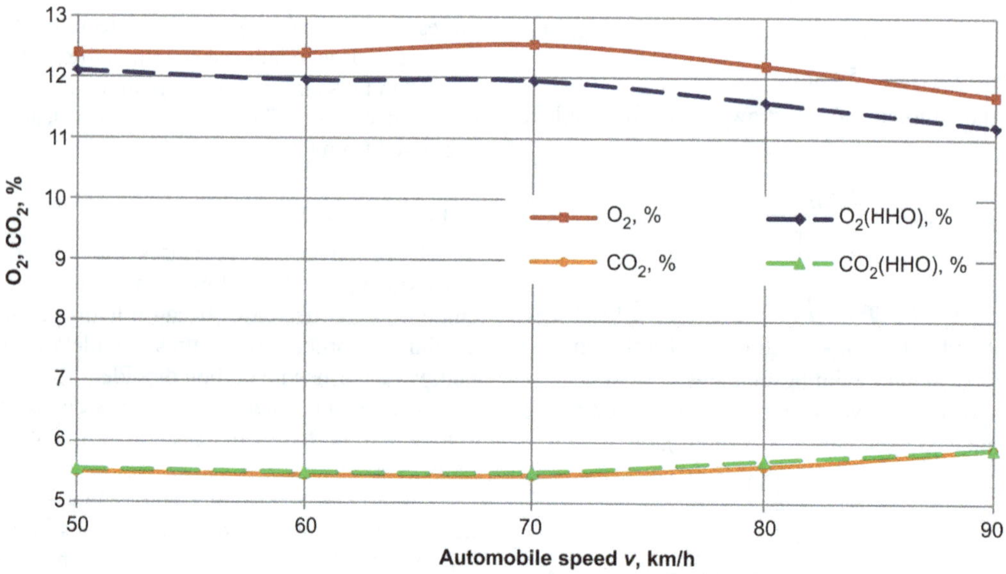

Fig. 3. Amounts of CO_2 and O_2 in exhaust gases

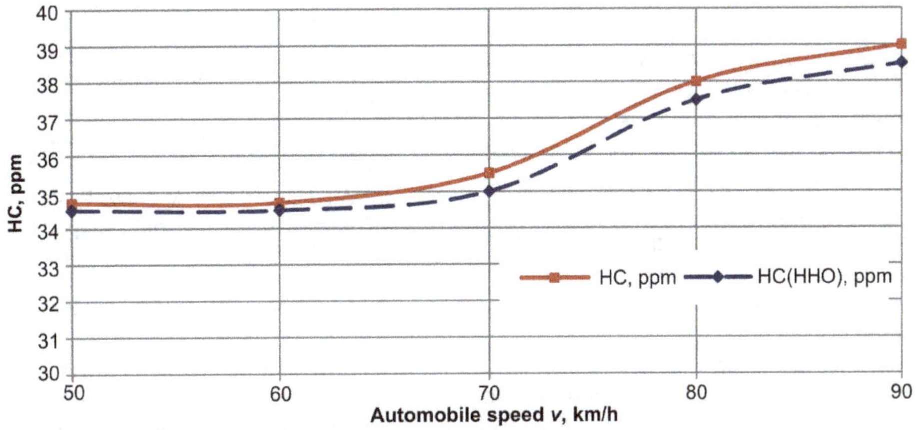

Fig. 4. Amount of HC in exhaust gases

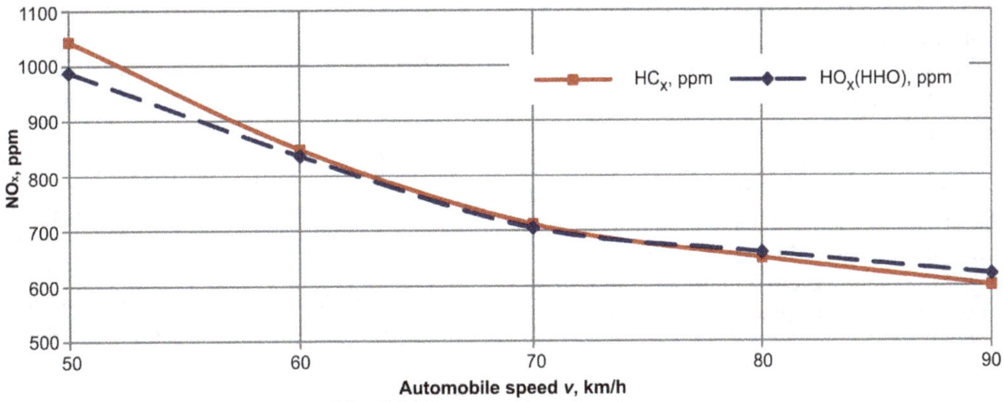

Fig. 5. NO_x amount in exhaust gases

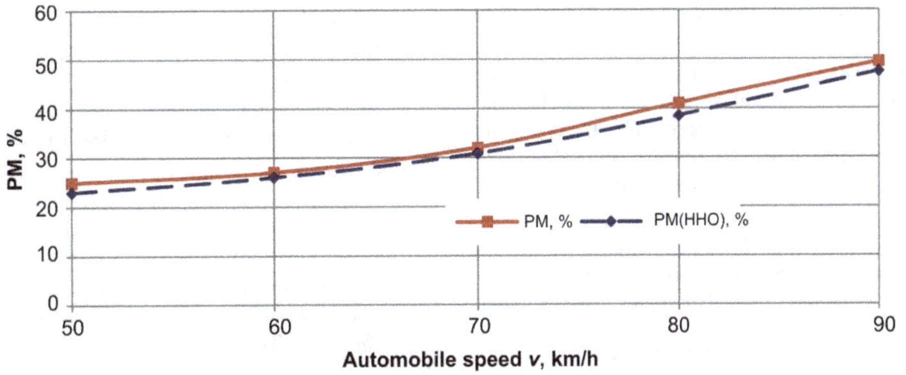

Fig. 6. Amount of particulate mater in exhaust gases

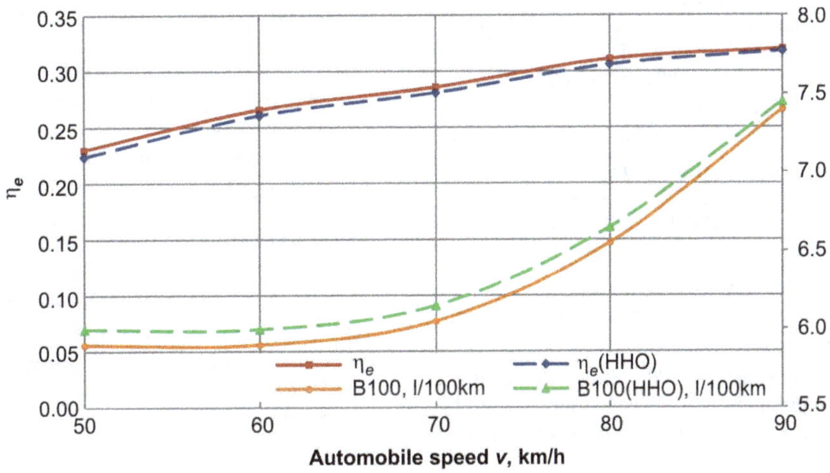

Fig. 7. Engine fuel consumption and net efficiency ratio

Conclusions

1. Dynamometer rolling road tests have revealed that a small amount of HHO gas (1.5 l/min) drawn into engine cylinders together with air improves combustion of diesel fuel under various engine speed and load conditions and decreases concentration of toxic substances in exhaust gases: O_2 gas amount decreases by 2.5–4.5%; CO gas amount decreases by 7–12.5%; CO_2 gas amount increases up to 2%; HC amounts decreases up to 1.5%; PM amount decreases by 4–8%; NO_x gas amount decreases up to 5% at small engine loads.

2. Additional HHO gas supply at medium engine loads leads to 3.5% increase of NO_x gas concentration in exhaust gases and with further increase of load additional supply of gas increases nitrogen oxides concentration.

3. Efficiency ratio of HHO gas production and use in an engine constitutes only $\eta_{eHHO} \approx 0.18$. Therefore, using this gas in a diesel engine running under medium loads results in up to 1.7% fuel consumption increase and up to 2.5% net efficiency ratio decrease.

4. Economic efficiency of diesel engines with additional HHO supply needs to be further investigated. In this article the environmental parameters had been analysed. No economic calculation had been done.

5. HHO gas can improve environmental characteristics of engines without catalyst exhaust gases neutralizers.

References

Baltrėnas, P.; Butkus, D.; Nainys, V.; Grubliauskas, R.; Gudaitytė, J. 2007. Efficiency evaluation of a noise barrier, *Journal of Environmental Engineering and Landscape Management* 15(3): 125–134.

Baltrėnas, P.; Kazlauskas, D.; Petraitis, E. 2004. Triukšmo automobilių stovėjimo aikštelėse tyrimai ir jo sklidimo skaitinis modeliavimas [Researches of noise in car parking places and digital modeling of its spread], *Journal of Environmental Engineering and Landscape Management* 12(2): 63–70.

Baltrėnas, P.; Kvasauskas, M. 2008. Experimental investigation of biogas production using fatty waste, *Journal of Environmental Engineering and Landscape Management* 16(4): 178–187. http://dx.doi.org/10.3846/1648-6897. 2008.16.178-187

Baltrėnas, P.; Vaitiekūnas, P.; Bačiulytė, Ž. 2009. Geležinkelio transporto taršos sunkiaisiais metalais dirvožemyje tyrimai ir įvertinimas, *Journal of Environmental Engineering and Landscape Management* 17(4): 244–251. http://dx.doi.org/10.3846/1648-6897.2009.17.244-251

Bendikienė, V.; Kiriliauskaitė, V.; Juodka, B. 2011. Production of environmentally friendly biodiesel by enzymatic oil transesterification, *Journal of Environmental Engineering and Landscape Management* 19(2): 123–129. http://dx.doi.org/10.3846/16486897.2011.579448

Gjirja, S.; Olsson, E.; Olsson, L.; Ekman, S. 2000. *Experimental investigation on the hydrogen peroxide fumigation into the inlet duct of a diesel engine.* SAE Technical Paper No. 2000-01-1919.

Gražulevičienė, R.; Bendokienė, I. 2009. Influence of truck traffic on acoustic pollution in Kaunas districts crossed by highways, *Journal of Environmental Engineering and Landscape Management* 17(4): 198–204. http://dx.doi.org/10.3846/1648-6897.2009.17.198-204

Jacobs, T. J.; Bohac, S. V.; Assanis, D. N.; Szymkowicz, P. G. 2005. *Lean and rich premixed compression ingnition combustion in a light-duty diesel engine.* SAE Technical Paper No. 2005-01-0166.

Juostas, A.; Janulevičius, A. 2009. Evaluating working quality of tractors by their harmful impact on the environment, *Journal of Environmental Engineering and Landscape Management* 17(2): 106–113. http://dx.doi.org/10.3846/1648-6897.2009.17.106-113

Labeckas, G.; Slavinskas, S. 2010. The effect of ethanol, petrol and rapeseed oil blends on direct injection diesel engine performance and exhaust emissions, *Transport* 25(2): 116–128. http://dx.doi.org/10.3846/transport.2010.15

Lambe, S. M.; Watson, H. C. 1993. Optimizing the design of a hydrogen engine with pilot diesel fuel ignition, *International Journal of Vehicle Design* 14(4): 370–389.

Lebedevas, S.; Lebedeva, G.; Bereišienė, K. 2011. Modifying mathematical models for calculating operational characteristics of diesel engines burning RME biofuels, *Transport* 26(1): 50–60. http://dx.doi.org/10.3846/16484142.2011.561528

Matijošius, J.; Sokolovskij, E. 2009. Research into the quality of fuels and their biocomponents, *Transport* 24(3): 212–217. http://dx.doi.org/10.3846/1648-4142.2009.24.212-217

Mažeika, M.; Matijošius, J. 2010. Three components fuel uses in diesel engine, *Research papers of Lithuanian University of Agriculture* 42(2–3): 232–239

Paulauskas, L.; Klimas, R. 2011. Modeling of the spread of motor transport noise in Šiauliai city, *Journal of Environmental Engineering and Landscape Management* 19(1): 62–70. http://dx.doi.org/10.3846/16486897.2011.557249

Pukalskas, S.; Bogdanovičius, Z.; Sendžikienė, E.; Makarevičienė, V.; Janulis, P. 2009. The mixture of biobutanol and petrol for Otto engines, *Transport* 24(4): 301–307. http://dx.doi.org/10.3846/1648-4142.2009.24.301-307

Senthil Kumar, M.; Ramesh, A.; Nagalingam, B. 2002. Use of hydrogen to enhance the performance of a vegetable oil fuelled compression ignition engine, *International Journal of Hydrogen Energy* 28: 1143–1154.

Stenlaas, O.; Egnell, R.; Johansson, B.; Mauss, F. 2004. *Hydrogen as homogeneous charge compression ignition engine fuel.* SAE Technical Paper No. 2004-01-1976.

Tomita, E.; Kawahara, N.; Piao, Z.; Fujita, S.; Hamamoto, Y. 2001. *Hydrogen combustion and exhaust emissions ignited with diesel oil in a dual-fuel engine.* SAE Technical Paper No. 2001-01-3503.

Tsolakis, A.; Megaritis, A. 2004. Partially premixed charge compression ignition engine with on-board H2 production by exhaust gas fuel reforming of diesel and biodiesel, *International Journal of Hydrogen Energy* 30: 731–735.

Yilmaz, A. C.; Uludamar, E.; Aydin, K. 2010. Effect of hydroxy (HHO) gas addition on performance and exhaust emissions in compression ignition engines, *International Journal of Hydrogen Energy* 3: 11366–11372. http://dx.doi.org/10.1016/j.ijhydene.2010.07.040

Varde, K. S.; Varde, L. K. 1984. Reduction of soot in diesel combustion with hydrogen and different H/C gaseous fuels, in *Proceedings of the 5th World Hydrogen Energy Conference,* 15–19 July, 1984, Toronto, Canada, 1631–1641.

Alfredas RIMKUS. Lector of Automobile Transport Department of Vilnius Gediminas Technical University (VGTU), 2009; lector of Automobile Transport Department, Vilnius College of Technologies and Design, 2002; PhD student of technological sciences (transport engineering), VGTU, 2009. A board member of Automobile Transport Teachers Association, 1999. Publications: author of over 10 scientific publications. Research interests: alternative fuels, internal combustion engines.

Saugirdas PUKALSKAS. Head of Automobile Transport Department of Vilnius Gediminas Technical University (VGTU), 2008; docent (Associate Professor) of Automobile Transport Department, VGTU, 2003; doctor of technological sciences (transport engineering), VGTU, 2002; Technical consultant – engineer of JSC "Silberauto", 2007–2008; Head of laboratories of Automobile Transport Department, 1996–1997; Assistant of Automobile Transport Department, VGTU, 2001–2003. Board Member of Lithuanian drivers' Union, 2000–2003; President of Lithuanian drivers' Union, 2003; council member of Lithuanian Association of Automotive Engineers, 2006. Publications: author of over 50 scientific publications. Research interests: alternative fuels, internal combustion engines, hybrid and electrical cars, traffic safety.

Jonas MATIJOŠIUS. Docent (Associate Professor) of Automobile Transport Department of Vilnius Gediminas Technical University (VGTU), 2009; lector, assistant of Automobile Transport Department, VGTU, 2008-2012; doctor of technological sciences (transport engineering), VGTU, 2012; Manager of JSC "Ecoservice", 2006–2007; Inspector of JSC "Lestma", 2000–2006;. Publications: author of over 15 scientific publications. Research interests: alternative fuels, internal combustion engines.

Edgar SOKOLOVSKIJ. Vice-Dean of Transport Engineering Faculty of Vilnius Gediminas Technical University (VGTU), 2010; docent (Associate Professor) of Department of Automobile Transport, VGTU, 2005; Docent (associate professor) of Department of Transport Technological Equipment, VGTU, 2004–2006; Doctor of Technological Science (Transport Engineering), VGTU, 2004; chief expert of Traffic Accidents Research Department, Forensic Science Centre of Lithuania, 2004–2006; assistant, expert, superior expert of Traffic Accidents Research Department Forensic Science Centre of Lithuania (earlier Institute), 1997–2004; assistant of Transport Science Institute, VGTU, 1998–1999. Member of Council of Lithuanian Automobile Engineers Association, 2011. Listed in Lithuanian Republic List of Experts of Forensic Examination (traffic accidents, transport trasology expert), 2003. Publications: author of over 45 scientific publications. Research interests: fuels and oils, transport environment protection, traffic safety.

16

SCREENING INDOOR AIR QUALITY EVALUATION IN THE LITHUANIAN THEATRE, MUSIC AND CINEMA MUSEUM

Susana López-Aparicio[1], Rima Grašienė[2]

[1]*NILU – Norwegian Institute for Air Research, Institettveien 18, 2027-Kjeller, Norway*

[2]*Lithuanian Theatre, Music and Cinema Museum, Vilniaus g. 41, 01119 Vilnius, Lithuania*

E-mails: [1]sla@nilu.no (corresponding author);[2]rimagrasiene@gmail.com

Abstract. Air pollution is a risk for human health and for the preservation of materials in particular cultural heritage objects. This paper presents the main results obtained in the project carried out by the Lithuanian Theatre, Music and Cinema Museum (LTMCM) and NILU-Norwegian Institute for Air Research. The main goal of the project was to perform a screening study and characterization of the indoor air pollution in the LTMCM. The results are used to assess the indoor air quality with regard to the preservation of cultural heritage objects therein. Measurement of NO_2, organic acids (i.e. acetic and formic acids), photo-oxidant effects of the environment, climate (temperature and relative humidity) and particulate matter deposition were performed in different indoor locations. Low concentration of outdoor generated pollutants such as NO_2 was found in storage rooms whereas higher concentrations were determined in the in the exhibition area. The results indicate that the building envelope does not protect effectively against infiltration of outdoor pollutants such as NO_2 and particles. Very high concentration of indoor generated pollutants such as acetic acid was found in one storage room with a significant source.

Keywords: Air pollution, Indoor air quality, dosimetry, NO_2, acetic acid, preservation, pollutant infiltration.

Introduction

Air pollution and its effects on cultural heritage (CH) objects have received an increased interest in the last decades. Air pollution constitutes a risk to materials and in particular to cultural heritage objects and artworks (Brimblecombe 1990; Kunera, Fitz 1995; Tétreault 2003; Hatchfield 2002). Some of the documented adverse effects include the fading of paintings, blackening of surfaces, corrosion of metals, softening of details, decomposition and yellowing of paper, among other problems (Tétreault 2003).

According to their sources, indoor air pollutants can be divided into two main groups, those generated outdoors and subsequently infiltrated indoors, and those generated indoors. Nitrogen dioxide (NO_2), ozone (O_3), sulphur dioxide (SO_2) and particulate matter (PM) are among the most common outdoor pollutants which enter the museum environment. Ozone is also indoor generated as for instance from automatic photocopying (Valuntaitė, Girgždienė 2008), however this activity is not common in museum environments. The pollutants emitted indoors are mainly organic (e.g. volatile organic compounds, VOCs) and usually emitted from building materials (Schieweck *et al.* 2005). Several studies have been published about air pollution in museums (Brimblecombe 1990) and the degradation effects on CH objects (e.g. Thomson 1986; Blades *et al.* 2000; Hatchfield 2002; Tétreault 2003). For instance, ozone and NO_2 are strong oxidants which effects on cultural heritage materials are well known. Some of the most documented effects are corrosion of metals (Eriksson *et al.* 1993; Leygraf, Graedel 2000), attack on calcareous stones and murals (Johansson *et al.* 1988), fading and colour change of pigments (Kadokura *et al.* 1988; Salmon, Cass 1993; Whitmore, Cass 1988; Lynn *et al.* 2000), embrittlement of photographic and paper based materials (Reilly *et al.* 2001; Blades *et al.* 2000) and/or soiling.

Additionally, organic compounds have adverse effects on materials, such as the formation of salt efflorescence on carbonate materials (Gibson *et al.* 1997), embrittlement, fading and degradation of pigments (Oikawa *et al.* 2006), depolymerisation of cellulose material (Dupont, Tetreault 2000) and the corrosion of metal (Tennent *et al.* 1993).

The valuable objects in the Lithuanian Theatre, Music and Cinema Museum constitute a wide representation of the modern Lithuanian history. However, degradation is observed on several objects both in exhibition and storage, which indicates that their preservation may be at risk. Degradation mechanisms can be symptomatic of an unacceptable indoor environment with regard to climate parameters, such as temperature and relative humidity, or pollutant concentration.

The main goal of the project was to perform a characterization of the indoor air pollution in the LTMCM. In order to achieve this goal, a screening study was performed based on measurements carried out both in the storage and exhibition areas. Dosimetry (EWO dosimeter;

Grøntoft *et al.* 2010), single pollutant concentrations (i.e. NO_2 and organic acids) determined by passive diffusion gas samplers and the evaluation of particle deposition on glass slides were the methods selected for the study.

1. Methods

1.1. Sampling location

The museum is located in the Radvilai Palace in the heart of Vilnius Old Town. The historic palace housed the Public Theatre of Vilnius between 1795 and 1810, and during the second part of 19th century, it was used as houserooms, shops and a pharmacy. The dimension of the museum is around 3718 m^2, of which more than 1500 m^2 are dedicated to exhibition area and approximately 500 m^2 to depository or storeroom.

The museum collection consists of a wide range of valuable Lithuanian CH objects, reaching over 350 thousand exhibits which are placed mainly in the storage rooms and, a small part of them, in exhibition. The main group of objects can be classified as paper based materials, for instance programme sheets, posters, newsletters, photographs, and manuscripts among others. Other representative groups are composed mainly by wooden objects, such as rare music instruments from 18th – 20th century, and metal objects. The collection includes in addition films, vinyl, paintings, sound recording and textiles, such as costumes from personal collections of renowned Lithuanian actors and actresses.

The museum is a naturally ventilated building located in the city centre of Vilnius. One of its two main facades faces a street with intense traffic whereas the other is facing a backyard (Fig. 1). Five different sampling locations were selected inside the building in order to cover different environmental conditions. Two of the sampling locations were in storage rooms (LM1 and LM2; Table 1 and Fig. 1) whereas the remaining three sampling locations were in exhibition areas (LM3, LM4 and LM5; Table 1 and Fig. 1). In addition, different physical characteristics such as presence or absence of window (Table 1) and whether the room is facing street or backyard were taken into account as criteria for the selection of the sampling location.

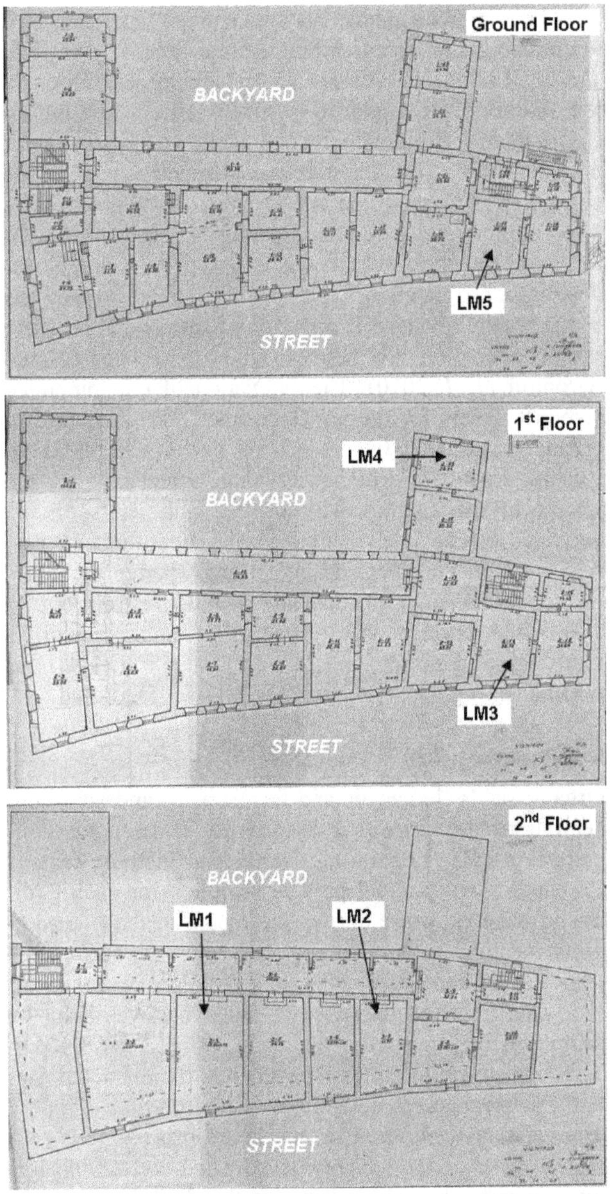

Fig. 1. Distribution of sampling locations in the Lithuanian Theatre, Music and Cinema Museum

1.2. Climate

Punctual measurements of temperature and relative humidity were carried out on a daily basis in the five sampling locations by the personnel of the museum from February to May 2010. Both temperature and relative humidity were measured with miniature thermo-hygrometers TFA with ± 1 °C and ± 5% RH accuracy.

1.3. Dosimetry

The photo-oxidant effects of the environment were measured by the Early Warning dosimeter for Organic materials (EWO) developed by NILU-Norwegian Institute for Air Research (Grøntoft *et al.* 2010).

Table 1. Description of sampling locations

Ref	Floor	Room type	Material type	Description
LM1	2nd	D	Various - mainlypaper	Scenography depository, with window to the street
LM2	2nd	D	Various - mainlywood	General depository, without window.
LM3	1st	E	MainlyWood and metal pieces	Room faces the street side
LM4	1st	E	Paper and Paintings on the wall	Room faces the backyard
LM5	Ground	E	Textile, metal, paper	Room faces the street side at street level

The EWO dosimeter is a synthetic polymer sensitive to climate parameters (i.e. temperature, relative humidity and UV Light) and NO_2 and O_3, which are usually emitted outdoors and ventilated or infiltrated into the indoor environment. The environmental effect on the dosimeter polymer film is measured by photo-spectrometry as the change in UV absorption at 340 nm from before to after three months exposure. The relation between the EWO-response and the environment is based on a non linear dose response function found from the statistical analysis of the results obtained in a measurement campaign. Dose response functions for indoor locations have been defined (Grøntoft *et al.* 2010). The evaluation of recommended response levels for the EWO dosimeter as compared to effects on organic CH objects is performed based on existing knowledge in conservation science about the effects of the single environmental parameters on the heritage objects. The results from the measurements carried out with the EWO dosimeter are reported as values of increasing environmental impact ranging from one to five, which correspond to different locations with different degree of protection, from archive (i.e. Level 1) to external storage with no control (i.e. Level 5).

1.4. Passive diffusion samplers

Organic acids (i.e. acetic and formic acid) and NO_2 concentrations were measured by passive diffusion gas samplers (Ferm 1991) from the Norwegian Institute for Air Research. Two parallel passive samplers for each compound were exposure during one month in the five sampling locations, thereafter the samplers were sent to the laboratory for analysis.

The filter of the organic acids passive diffusion sampler is impregnated in an alkali, which is dissolved in an aqueous solution after exposure and the extracted acetate and formate are determined by ion chromatography. The mean concentration during the exposure time is estimated based on the quantity of extracted acetate and formate, a constant, which contains the diffusion constant for acetic and formic acids, and a factor based on the dimensions of the passive diffusion sampler. The filter of the NO_2-passive sampler is impregnated in iodide (I-) and the formed nitrite (NO_2^-) is determined by photometry. The NO_2 average concentration for the exposure time is estimated in the same way as for organic acids. The detection limit for NO_2 after one month of exposure time is approximately 0.03 $\mu g\ m^{-3}$ and for acetic acid and formic acid it is 0.5 $\mu g\ m^{-3}$.

1.5. Particle deposition

Glass slides were exposed during three months in the sampling locations selected in the LTMCM. After exposure the glass slides were covered with GelLifter© and the deposited particles were collected and examined by microscopy. The GelLifters© were analyzed by optical microscopy using a Leitz Aristoplan microscope. Digital images where taken using a Leica DC Camera attached to the microscope and assisted by Leica IM Management system. The digital images are analyzed by the image processing software ImageJ.

2. Results and discussion

2.1. Climate

Relative humidity represents an essential environmental factor for the diagnosis and risk assessment for CH objects. Several materials (e.g. wood, paper, parchment, leather, bone, paintings, plaster or clay mineral based stones) are very sensitive to their water content, which is in equilibrium with the relative humidity in the air. Changes in the moisture content influence changes of the physical state of the materials. For instance, when moisture is removed from some materials such as wood or bone they contract favouring distortion or cracking. Other materials such as paper, parchment, leather and natural textiles become less flexible and therefore more fragile.

Temperature and relative humidity in the LTMCM are summarized in the Table 2. The temperature is found to be low in every location and does not show significant variations, whereas relative humidity shows some variations, higher in the exhibition gallery than in storage areas. Deformation of wooden instruments such as changes in volume or cracking has been observed in the LTMCM, and in particular in the instruments exhibit in the gallery. The relative humidity variations observed during four months (i.e. February 2010 to May 2010) may explain these effects. However, seasonal variations are expected to have stronger impact.

Table 2. Average temperature and relative humidity (4 months), Max and Min. Numbers in brackets are standard deviations

#	T (°C)			
	Aver		**Max**	**Min**
LM1	13.9	(1.9)	18	10
LM2	13.9	(1.7)	17	12
LM3	14.3	(1.7)	18	12
LM4	13.0	(1.5)	16	9
LM5	13.8	(1.8)	19	10
#	**RH (%)**			
	Aver		**Max**	**Min**
LM1	46.0	(4.7)	53	39
LM2	35.9	(5.3)	45	28
LM3	35.2	(12.6)	54	21
LM4	41.2	(12.7)	62	29
LM5	40.4	(9.2)	63	27

The LTMCM hold a significant collection of paper based material and climatic parameters may be concerns for the preservation of this type of materials. According to the preservation index (PI) developed by the Image Permanence Institute (IPI 2000) for paper based and photographic material, the average temperature and relative humidity obtained in the five indoor locations involve 1) slow natural aging rate (i.e. PI = 124 years); 2) there is no risk for mold germination; and 3) the conditions are recommended for the storage of photographic material. However, high values of relative humidity measured in some of the locations (i.e. 63% RH) involves lower preservation index for paper based materials (i.e. PI = 65 years). In addition, the highest relative humidity value measured in the museum involves moderate natural aging rate and conditions not recommendable for the preservation of photographic material. For paper based material and photographic materials, relative humidity above 50% is not recommendable due to fast increase of the natural aging rate.

2.2. Environmental photo-oxidant effects

Figure 2 shows the results obtained with the EWO dosimeter. Three out of five locations are classified as level 1, which means that the environment, concerning photo-oxidant effects on organic materials, is acceptable for an "Archive Store". LM1 and LM2 are storerooms, so the results are in conformity with the type of indoor location and therefore the environment is acceptable. In addition, the response of the EWO dosimeter in LM2 is cero. This room is a depository room without windows and therefore low infiltration of outdoor generated pollutants, such as NO_2 and O_3, and low levels of light. The location LM3 is part of the exhibition gallery and faces the street side of the historic building (Fig. 1). Therefore the level 1 obtained by the EWO dosimeter (Fig. 2) indicates that the gallery room LM3 has an excellent environment concerning photo-oxidant effects.

The remaining two sampling locations (i.e. LM4 and LM5) showed acceptable environment for a "Purposed Built Museum" (Level 2; Figure 2) very close to acceptable for a "Historic House" (i.e. Level 3). Both locations are used as exhibition area; one of them faces the backyard (LM4) whereas the other faces the street (LM5; Fig. 1). These locations are gallery rooms of the historic building so level 2 obtained with the EWO dosimeter indicate that the environment is very good.

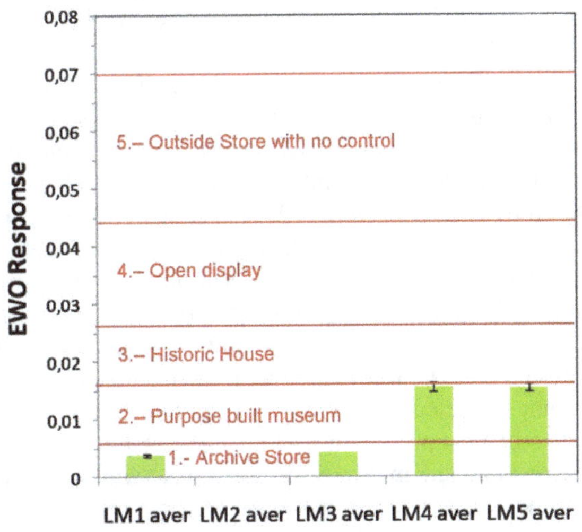

Fig. 2. EWO dosimeter responses (ΔLight absorbance) obtained in the five indoor locations in the LTMCM

2.3. Outdoor pollutant – NO_2

In the absence of indoor sources, NO_2 is an outdoor generated pollutant which infiltrates into the indoor environment. The concentration of NO_2 in the LTMCM varies from 1 $\mu g\ m^{-3}$ to approximately 30 $\mu g\ m^{-3}$ (Fig. 3). The lowest concentrations were measured in the depository or storerooms (LM1 and LM2; Fig. 3) located on the top floor of the museum. The measurements performed in the exhibition areas show values above 10 $\mu g\ m^{-3}$, and the highest level is obtained in the room located at the ground floor and facing the street side of the building (LM5; Fig. 3).

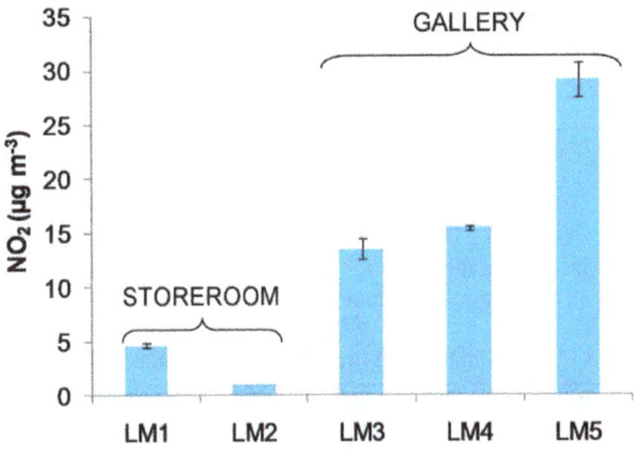

Fig. 3. NO_2 concentration obtained in the LTMCM. Values represent the average from two parallel samplers. For abbreviation and locations characteristics see Table 1

In order to evaluate the infiltration of outdoor pollutants into the museum, different locations were selected from the building. The locations LM3 and LM4 were selected from the same floor; LM3 faces the street, whereas LM4 faces the backyard. The results obtained do not show differences between both rooms (Fig. 3) and therefore the location within the building concerning the distance to traffic emissions is not a determinant parameter. The worst case scenario is observed in the room location LM5, where the concentration of NO_2 reaches 30 $\mu g\ m^{-3}$ (Fig. 3). The location LM5 is at the ground floor at approximately 2 meters from the traffic.

Outdoor NO_2 concentration in Vilnius during the NO_2 sampling period (February 2010) was available (Know your air for health 2010). Hourly NO_2 concentration data from three different monitoring stations distributed across Vilnius was collected and analysed. Average daily NO_2 concentration varies between 7 and 38 $\mu g\ m^{-3}$ in the background (i.e. Lazdynai) and industrial stations (i.e. Sabanoriu), whereas daily average NO_2 concentration is between 17 and 62 $\mu g\ m^{-3}$ in the traffic monitoring station (i.e. Zirmunai; Fig. 4).

The traffic station was selected as representing of the outdoor air quality around the LTMCM (Fig. 4). The month average concentration of NO_2 for the exposure time of interest is 39.27 $\mu g\ m^{-3}$ (standard deviation: 10.67). Taking this value into account, we obtain an indoor / outdoor ratio (I/O) between 0.02 and 0.74 (Table 3). The lowest I/O ratio is observed in the two storage rooms, LM1 and LM2, indicating very low infiltration of NO_2, whereas the highest I/O ratio is obtained in the location LM5 illustrating that the building, and specifically the room LM5, does not protect efficiently against infiltration of outdoor pollutants.

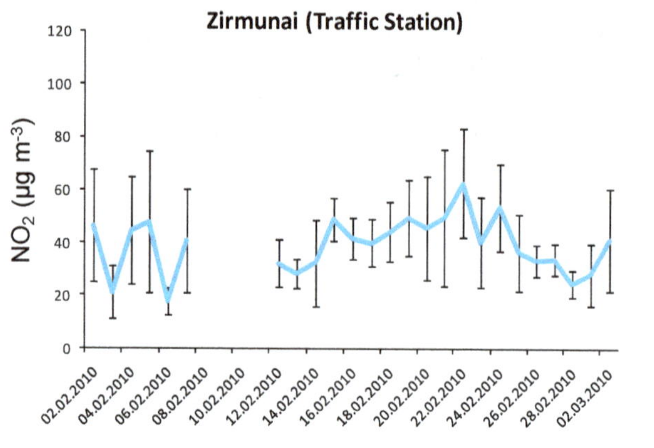

Fig. 4. Daily average outdoor NO_2 in Zirmunai monitoring station. The bars represent standard deviations

Table 3. I/O ratio obtained in the different locations

LOCATION	I/O
LM1	0.12
LM2	0.02
LM	0.34
LM4	0.39
LM5	0.74

The recommended NO_2 concentration in the museum environment for the preservation of most objects is 10 µg m^{-3} (Thomson 1986; ASHRAE 2003) and the US National Bureau of Standard (NBS 1983) suggests 4.75 µg m^{-3} of NO_2 as recommended limit value for exposure of paper based materials. The concentrations of NO_2 in the exhibition areas are found to be above 10 µg m^{-3}, indicating that the CH objects in the gallery may be under risk.

2.4. Indoor pollutant – organic acids

Organic acids were selected as indicator of indoor generated pollutants as they are mainly emitted from materials such as wood, paint and varnishes (Hatchfield 2002; Tétreault 2003). The concentration of acetic acid measured inside the LTMCM varies from 19 µg m^{-3} to approximately 280 µg m^{-3} (Fig. 5). The lowest acetic acid concentration is measured in the location which faces the street at ground level (i.e. LM5; Fig. 5), whereas the highest concentration of acetic acid is measured in one of the storeroom (i.e. LM1; Fig. 5). Formic acid follows the same pattern as acetic acid (Fig. 5). LM1 is a depository room with a large number of wooden drawers from which high emission of organic compounds may occur. The low concentration of organic acids determined in LM5 may be explained by the building materials as it has stone floor, while the others have wooden floors. Additionally, high infiltration is found to be in LM5 based on the I/O ratio of NO_2 (Table 3), which may involve high dilution of indoor generated pollutants.

Concerning the preservation of CH objects, the location LM1 is the main concern due to the high concentration of organic acids (i.e. ≈280 µg m^{-3} of acetic acid; ≈40 µg m^{-3} of formic acid). Acetic acid is known to corrode metals, in particular lead (Tennent *et al.* 1993), it degrades calcareous materials (Tennent *et al.* 1992) and reduces the degree of polymerization of cellulose in paper (Dupond, Tétreault 2000). LM1 mainly holds paper based materials which may be under risk. According to the ISO 11799 (2003) concerning the requirements for archive and library material, the recommended levels of acetic acids are below 10 µg m^{-3}. However, this value is very low as normal indoor concentrations levels commonly vary between 40 – 100 µg m^{-3} (Tétreault 2003).

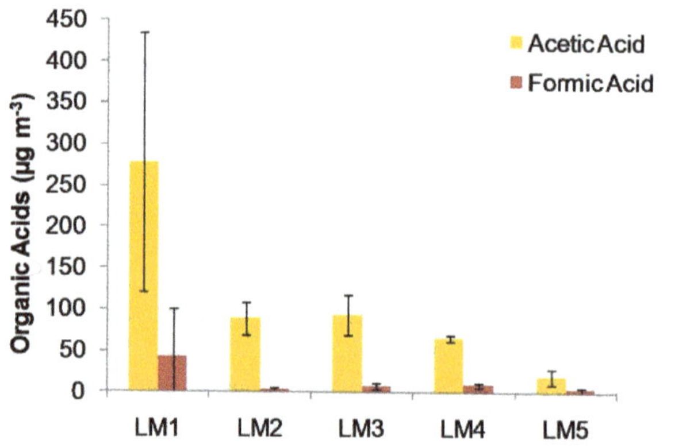

Fig. 5. Organic acids (acetic and formic acids) concentration measured in the LTMCM. Values represent the average from two parallel samplers. For abbreviations and location characteristics see Table 1

2.5. Particle deposition

Fig. 6 shows the results obtained from the study of glass slides exposed over three months in the five sampling locations. The results represent average percentage of area covered with particles obtained over a study area of approximately 8 mm^2. The highest value was obtained in the location LM5, followed by LM1, whereas the lowest value was obtained in the location LM2.

The particles deposited on the glass slides are mainly equidimensional and a low percentage of fibres were observed. In general, particles are transported into the museum by visitors or are infiltrated through the shield of the building (Nazaroff *et al.* 1993). The high value obtained in the location LM5 might correspond to infiltration from outdoor environment. The sampling station was placed on the interior windowsill at approximately 3 meters from the visitor path. The lowest obtained value (LM2; Fig. 6) is in accordance with the characteristics of the room; restricted personnel access and lack of window.

The highest percentages of area covered with particles were obtained in the three locations with windows facing the street (LM1, LM3 and LM5; Fig. 6). In addition the morphology of the particles is mainly equidimensional and low percentage of fibres brought by visitors was observed. These results indicate that particles are most probably infiltrated into the building through windows and doors. Thus the building does not protect effectively against the infiltration of particles. Particles constitute a risk for cultural heritage objects in museums as they cause soiling, are abrasive, provide sites for surface reactions and have a potential to damage artefacts due to their hygroscopic nature (Nazaroff, Cass 1991).

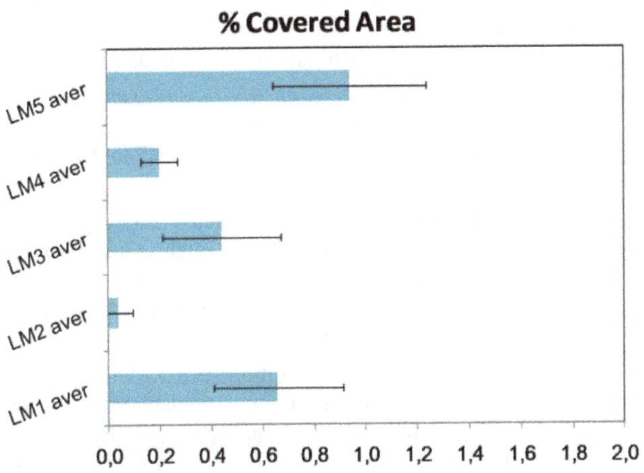

Fig. 6. Average percentage of area covered with particles obtained in the five indoor locations in the LTMCM. For abbreviations see Table 1

Conclusions

The screening study performed inside the Lithuanian Theatre Music and Cinema museum give a good indication of the indoor air quality concerning the preservation of CH objects. However, additional research is needed in order to perform a more complete evaluation concerning material exposure or even human exposure. The measurements carried out in our study have been performed only during one (single gas concentration) and three month exposure time (dosimetry). Additional information about seasonal variations of pollutant concentrations, air exchange rate of the building, particulate matter concentration and compositions, the concentration of other pollutants such as aldehydes or ozone would be of large interest.

Nevertheless, the measurements performed in our screening study allow us to establish certain conclusions and basis for further studies. The indoor air quality in the studied storage rooms of the LTMCM is "acceptable" concerning the photo-oxidant effects of the environment (i.e. synergistic effects of temperature, relative humidity, UV light, NO_2 and O_3) on organic materials. In addition, the studied exhibition rooms of the LTMCM show "very good" and "excellent" environments. However, there is reason of concern based on the results obtained from single pollutant concentration measurements. Outdoor generated pollutants such as NO_2 is a main concern in the exhibition areas of the museum as the determined concentrations are above recommended values for the preservation of most objects (i.e. 10 $\mu g\ m^{-3}$). In addition, the results obtained by the study of particle deposition may indicate high infiltration of course particles in rooms with windows facing the streets. Indoor generated pollutants such as acetic acids are a main concern for sensitive materials placed in storage rooms with a significant pollutant source.

The results obtained in the gallery of the LTMCM indicate that the building does not protect effectively against the infiltration of outdoor pollutants such as NO_2 and particles which are a risk for the preservation of cultural heritage objects in exhibition. This study contributes to the understanding of pollutant infiltration in naturally ventilated historic buildings and it constitutes essential basis for more detail characterization and evaluation of the indoor air quality in the museum environment.

Acknowledgements

This study was made possible thanks to the financial support of "Norway Grants" (NR.2004-LT0009-1NOR-02-011). Special thanks are due to all the colleagues at the Lithuanian Theatre, Music and Cinema museum, responsible for the practical work. Many thanks to Erik Andresen, Nina Dahl and colleagues at the Norwegian Institute for Air Research (NILU) for the preparation and analysis of passive diffusion gas samplers.

References

ASHRAE 2003. *Museum, libraries, and archives. Chapter 21. Applications handbook* SI edition. American Society of Heating, Refrigerating & Air-Conditioning Engineers.

Blades, N.; Oreszczyn, B.; Bordass, B.; Cassar, M. 2000. *Guidelines on pollution control in museum buildings*. London: Museum Practice.

Brimblecombe, P. 1990. The composition of museum atmosphere, *Atmospheric Environment* 24: 1–8. http://dx.doi.org/10.1016/0957-1272(90)90003-D

Dupont, A. L.; Tétreault, J. 2000. Cellulose degradation in an acetic acid environment, *Studies in Conservation* 45: 201–210. http://dx.doi.org/10.2307/1506766

Eriksson, P.; Johansson, L. G.; Strandberg, H. 1993. Initial stages of copper corrosion in humid air containing SO_2 and NO_2, *Journal of the Electrochemical Society* 140: 53–59. http://dx.doi.org/10.1149/1.2056109

Ferm, M. 1991. *A sensitive diffusional sampler*. Swedish Environmental Research Institute, Publ. IVL B-1020.

Grøntoft, T.; Odlyha, M.; Mottner, P.; Dahlin, E.; López-Aparicio, S.; Scharff, M.; Andrade, G.; Obarzanowski; M.; Ryhl-Svendsen, M.; Thickett, D.; Hackney, S.; Wadum, J.; Jakiela, S. 2010. Dosimetry for evaluation of environmental conditions for paintings in microclimate frames, *Journal of Cultural Heritage* 11: 411–419. http://dx.doi.org/10.1016/j.culher.2010.02.004

Hatchfield, P. B. 2002. *Pollutants in the museum environment*. London: Archetype Publications Ltd. 203 p.

IPI 2000, Image Permanence Institute. [Online], [cited July 2010]. Available from Internet: http://www.imagepermanenceinstitute.org/shtml_sub/dl_prescalc.asp

ISO 11799:2003. Information and documentation – document storage requirements for archive and library materials.

Johansson, L. G.; Lindqvist, O.; Mangio, R. E. 1988. Corrosion of calcareous stones in humid air containing SO_2 and NO_2, *Durability of Building Materials* 5: 439–449.

Kadokura, T.; Yoshizumi, K.; Mashiwagi, M.; Saito, M. 1988. Concentration of Nitrogen Dioxide in the Museum Environment and its effects on the Fading of Dyed Fabrics, in Mills, J. S.; Smith, P.; Yamasaki, K. (Eds.). *The conservation of Far Eastern art, preprints of the contributions to the Kyoto congress*. London, International Institute for Conservation of Historic and Artistic Works.

Kunera, V.; Fitz, S. 1995. Direct and indirect air pollution effects on materials including cultural monuments, *Water, Soil and Air Pollution* 85: 153–165. http://dx.doi.org/10.1007/BF00483697

"Know your air for health" project 2008. [online], [cited July 2010]. Available from Internet: http://www.knowyourairforhealth.eu/spip.php?article22.

Leygraf, C.; Graedel, T. E. 2000. *Atmospheric corrosion*. Electrochemical Society Series. New York: John Wiley & Sons Inc. 354 p.

Lynn, Y.; Slamon, G.; Cass, G. R. 2000. The ozone fading of traditional chinese plant dyes, *Journal of American Institute for Conservation* 39: 245–257. http://dx.doi.org/10.2307/3180094

Nazaroff, W. W.; Ligocki, M. P.; Salmon, L. G.; Cass, G. R.; Fall, T.; Jones, M. C.; Liu, H. I. H.; Ma, T. 1993. *Airborne particles in museums*. Research in conservation. The Getty Conservation Institute.

Nazaroff, W. W.; Cass, G. R. 1991. Protecting museum collections from soiling due to the deposition of airborne particles, *Atmospheric Environment* 25: 841–852. http://dx.doi.org/10.1016/0960-1686(91)90127-S

NBS 1983. National Bureau of Standards, *Air quality criteria for storage of paper-based archival records,* NBSIR 83-2795, Washington D.C: National Bureau of Standards.

Reilly, J. M.; Zinn, E.; Adelstein, P. 2001. Atmospheric pollutants aging test method development. *Final report to American Society for Testing and Materials*. Rochester: Image Permanence Institute, 52–97.

Salmon, L. G.; Cass, G. R. 1993. The fading of artists' colorants by exposure to atmospheric nitric acid, *Studies in Conservation* 38: 73–91. http://dx.doi.org/10.2307/1506460

Schieweck, A.; Lohrengel, B.; Siwinski, N.; Genning, C.; Salthammer, T. 2005. Organic and inorganic pollutants in storage rooms of the lower Saxony State Museum Hanover, Germany, *Atmospheric Environment* 39: 6098–6108. http://dx.doi.org/10.1016/j.atmosenv.2005.06.047

Tétreault, J. 2003. *Airborne pollutants in museums, galleries and archives: risk assessment, control strategies and preservation management*. Ottawa: Canadian Conservation Institute.

Tennent, N. H.; Cooksey, B. G.; Littlejohn, D.; Ottway, B. J.; Tarling, S. E.; Vickers, M. 1993. Unusual corrosion and efflorescence products on bronze and iron antiquities stored in wooden cabinets, in Tennent, N. H. (Ed.) *Conservation Science in the U.K.* Glasgow: James & James, 61–66.

Tennent, N. H.; Cooksey, B. G.; Littlejohn, D.; Ottway, B. J. 1992. Some applications of ion chromatography to the study of the deterioration of museum artefacts, Material Issues in *Art and Archaeology III. Materials research society symposium proceedings*, San Francisco, California, USA, 27 April – 1 May. 267: 869–882.

Thomson, G. 1986. *The museum environment*. London: Butterworths.

Valuntaitė, V.; Girgždienė. R. 2008. Variation of ozone and aerosol particle numerical concentrations on the working premises under different microclimatic parameters, *Journal of Environmental Engineering and Landscape Management* 16(3): 135–142. http://dx.doi.org/10.3846/1648-6897.2008.16.135-142

Whitmore, P. M.; Cass, G. R. 1988. The ozone fading of traditional Japanese colorants, *Studies in Conservation* 33: 29–40. http://dx.doi.org/10.2307/1506238

Susana LÓPEZ-APARICIO has a PhD in geology and is specialized in petrology and geochemistry. She holds a research position at the Norwegian Institute for Air Research (NILU) since January 2008. She is currently involved in several international projects dealing with air quality assessment concerning the effects on both materials and human health. She has participated in several EEA-Norway grants projects and EU projects such as PROPAINT and TeACH, relevant projects in the field of indoor air quality for the protection of cultural heritage assets.

Rima GRAŠIENĖ since 1984 is working at the Lithuanian Theatre, Music and Cinema museum and since 2005 she is the head of Counting and Preservation Department. She has participated in several projects financed by The Open Society Fund - Lithuania (OSFL) and EEA-Norway grants. She is the coordinator of the project "Environmental monitoring and air quality assessment in the Lithuanian Theatre, Music and Cinema Museum", financing from the Norwegian Financial Mechanism (EEA Norway Grants).

17

INFLUENCE OF MITOSPORIC FUNGI UPON ZINC-POLYMERIC COATINGS ON STEEL UNDER THE DIFFERENT ENVIRONMENT

Albinas Lugauskas[1], Igoris Prosyčevas[2], Algirdas Narkevičius[3], Aušra Selskienė[4], Dalia Bučinskienė[5], Alma Ručinskienė[6], Elena Binkauskienė[7]

[1, 3, 4, 5, 6, 7] *State Research Institute, Center for Physical Sciences and Technology, A.Goštauto g. 9, LT-01108 Vilnius, Lithuania*

[2] *Institute of Materials Science, Kaunas University of Technology, Savanorių pr. 271, LT-50131 Kaunas, Lithuania*

Abstract. Zinc-polymeric coating samples were exposed to different environmental conditions (for 2 years): marine – dunes on the shore of the Baltic Sea (Neringa), rural-agrarian district of Moletai and the industrial zone of Vilnius. During the whole period of the study 94 species of fungi, were isolated from the exposed samples of zinc-polymeric coating on steel. The zinc-polymeric coatings exposed to marine and rural-agrarian environments were deteriorated most markedly. A scanning electron microscope (SEM) was used to characterize the morphology of the surface. The phase composition of the zinc-polymeric coatings on the steel substrate was analysed by an X-ray diffractometer (XRD). The products formed on the zinc-polymeric coatings were detected by a Fourier transformation infrared spectrometer (FTIR). The effect of environmental conditions on the mass change of the samples was determined by the standard dissolution method.

Keywords: zinc-polymeric coating, steel, mitosporic fungi, environmental impact, changes of surface, corrosion.

Reference to this paper should be made as follows: Lugauskas, A.; Prosyčevas, I.; Narkevičius, A.; Selskienė, A.; Bučinskienė, D.; Ručinskienė, A.; Binkauskienė, E. 2013. Influence of mitosporic fungi upon zinc-polymeric coatings on steel under the different environment, *Journal of Environmental Engineering and Landscape Management* 21(3): 199–208. http://dx.doi.org/10.3846/16486897.2012.721370

Introduction

It has recently been stated, that approximately 40–50% of materials damage in the environment is connected with the activities of microorganisms, while in oil industry this index is up to 77% (Krylenkov *et al.* 2003). Such a value is determined by the environmental conditions. Processes caused by microorganisms occur in various objects (Diaz-Ballote, Ramanauskas 1999; Huang *et al.* 2006). Microorganisms deteriorate different elements inside and outside the objects, both constructive and functional (Helsen, Breme 1998; Beech 2004; Li *et al.* 2004). An active fungi growth is observed in municipal utilities, various tunnels, underground crossings, collectors, wooden, stone, metallic and multicomponental buildings and equipment, railroads and other means of transport (Hughes, Poole 1989; Roberge 1998; Marcus, Oudar 2002). The activity of microorganisms depends on their biological variety in the environment, their action specificity and ability to adapt to extreme natural environment, as well as on the composition of the unusual substrates which are formed by contamination with microorganisms and coupled with the surrounding medium (Ramanauskas *et al.* 1998, 2000, 2005; Narkevičius *et al.* 2003). Technical objects are deteriorated by bacteria, mitosporic fungi and other biota as well as their communities. The products of their activity (acids, alkalis, enzymes) and a variety of other aggressive chemical compounds soon start to contact with the components of the affected object, new chemical compounds are formed, which integrate into the destruction processes and accelerate further deeper and more diverse processes (Barton 1976; Lee *et al.* 2007; Lichušina *et al.* 2008). It should be mentioned, that small microorganisms possess vast possibilities of surface contact with the medium, which determines their destruction capability. The majority of microorganisms under natural conditions can survive for a long time in the anabiosis state and pass into an active performance only when the relative humidity of the medium reaches 60% and that of the substrate (constructions, walls, etc.) exceeds 50%. Under such conditions their performance becomes an aggressive destructive factor which can

Corresponding author: Albinas Lugauskas
E-mail: lugauskas@chi.lt

damage metallic constructions and become a threat for human health and other biota. Microbiological disturbance is promoted by the changes in temperature and medium pollution with nitrogen compounds, sulphur, carbon monoxide and dioxide, chlorides and soot, which are initially used by microorganisms for their nutrition (Domsch *et al.* 1980; Hughes, Poole 1989; Lugauskas *et al.* 2002; Marcus, Oudar 2002).

Means to limit, stabilize or completely suppress the disturbances caused by microorganisms have been searched (Ignatev, Mikhailova 1987). The surfaces of metallic constructions or equipment damaged by microorganisms get chapped and exfoliate. Their hardness and lifetime markedly diminish. Under such conditions growing mitosporic fungi are detected on metals, and the variety of them determines the corrosion behaviour (Yang *et al.* 1996). To protect metals from damage caused by microorganisms, polymeric films have been investigated (Bierwagen 2008; Binkauskienė *et al.* 2009). Recently much attention has been devoted to studies of zinc-polymeric coatings, which are widely used in automotive industry and to protect railway transport, large-block buildings, bridges and other objects made of steel to protect them from corrosion and mechanical damages (Kikuchi, Streekumari 2002; Marchebois *et al.* 2002; Tüken 2006). These coatings are composed of metallic zinc and a polymer mixture and protect from corrosion in the same manner as does the zinc coating obtained by galvanic zinc deposition. Zinc interacts with the metal surface forming a homogeneous protective film (Marchebois *et al.* 2004; Jagtap *et al.* 2007).

The aim of the present study was to isolate and identify mitosporic fungi detected on the zinc-polymeric coating deposited on steel to determine their importance in the injury processes occurring in the different ecological environments.

1. Materials and methods

Steel plates 10×15 cm and 1×3 cm in size and up to 1 mm in thickness coated with the zinc-polymeric film Dinitrol 443 L Zinkfarbe (made in EFTEC Aftermarket GmbH Pyrmonterstr 76 D-32676 Lügde) were used for studies. The density is 500 g/m^2 at a coating thickness of 100 μm. The thickness of coatings used for testing was 100 ± 15 μm. The thickness of coatings was measured by using a magnetic thickness meter MT-41NC. The thickness of every specimen was measured in 9 different points of the coating.

The zinc-polymeric coatings were obtained by paint working. The zinc-polymeric coating samples were exposed to different environmental conditions (for 2 years): marine (M) – dunes on the shore of the Baltic Sea (Neringa), rural-agrarian (R) – Kulioniu village in district of Moletai and the industrial (I) zone of Vilnius. The investigation sites were installed according to the requirements of ISO 8565:1992(E) standard. At every climatic study ground 15 steel plates were studied. The reference sample not contaminated with fungi zinc-polymeric samples were exposed under standard conditions (The Law of Lithuanian Government, HN-69: 2003). Mitosporic fungi from steel plates covered with the zinc-polymeric coating were isolated after 0.5, 1.0, 1.5 and 2.0 years of exposure. Two nutrient media were used for fungi isolation: standard malt agar (DIFCO) and dichloranglycerol agar (DG – 18, OXOID). The media were sterilized for 15 min by autoclaving at 121 °C. The zinc-polymeric coatings on steel after exposure were put straight on the agar medium for direct dissociation from the substrate (steel and silica glass plates covered with the zinc-polymeric coating). Microorganisms were transferred from the coating surface to the agar nutrient media with a sterile eyelet, a hook, a sterile pad or directly by the application method.

In order to isolate mitosporic fungi developing or present on the zinc-polymeric coating surface the samples were put in Petri dishes filled with a sterile agar medium of malt extract supplied with chloramphenicol (50 mg/l). After that the medium with the samples was sown up and cultivated in a thermostat at a temperature of 26 ± 2 °C. The extent of fungal growth was assessed by the naked eye and by a light microscope MVS-10 (Russia).

The processing of the initial data was performed using the modified methods (Booth 1971; Mirchink 1988; Bondartseva 2012). Average means of 378 assays are presented in the paper, taking 126 assay from the each environment (marine, rural-agrarian and industrial).

The changes in the surfaces of the zinc-polymeric coating were recorded using both a SEM EVO 50 EP (Carl Zeiss, SMTAG, Germany) and a light microscope MVS-10 (Russia).

The isolates were ascribed to taxonomic groups according to Ainsworth and Bisby's Dictionary of fungi (Hawksworth *et al.* 1995; Kirk *et al.* 2004). Fungi were identified according to various manuals (Carmichael *et al.* 1980; Chaverri, Samuels 2003; Domsh *et al.* 1980; Gams 1971; Klich 2002; Kozakiewicz 1989; Lugauskas *et al.* 1987, 2002; Nelson *et al.* 1983; Ramirez 1982; Samson, Frisvad 2004; Samson 1974; Oorschot 1980).

A Nicolet model 5700 FTIR in conjunction with a 10 Spec (10 Degree Specular Reflectance Accessory) was used for spectroscopy investigations of corrosion products on investigated samples. Reflectance spectra of samples were obtained in the reflectance mode. In all cases the spectral range was 4000–400 cm^{-1} (reciprocal wave length) with a 4 cm^{-1} resolution and 64 scons. Spectral manipulation included baseline correction, removal of carbon dioxide absorption bands and subtraction of water vapour interferences. The investigated samples of the zinc-polymeric coating were cleaned with dry filter paper. The atmospheric corrosion products formed on the zinc-polymeric coating surface were eliminated, collected and their FTIR spectra were recorded. In this way the products formed by a biological agent were detected.

The phase composition of the coating on the steel substrate was analysed by an XRD DRON – 3.0 Cu K_{α} ($\lambda = 0{,}154$ nm, U = 30 kV, I = 20 mA) without any additional proportions.

The mass change of coating was determined by the following method: prior to exposure coatings (plates) were washed with deionised water, dried in the air, 24 hours kept in an exicator with P_2O_5 (phosphorus pentoxide), then they were weighed. The same procedure was repeated after exposure of samples (after 0.5, 1.0, 1.5 and 2.0 years). After that the changes in mass were calculated.

To determine the coating mass loss due to corrosion, the products formed during the process were removed by using the methods and the procedure described in the ISO 8407:2009 standard: a) exposing for 3 min to 100 g/l CH_3COONH_4 (ammonium acetate) solution at a temperature of 70 °C; b) washing with deionised water; c) drying in an exicator with P_2O_5 for 24 hours; d) weighing.

2. Results and discussion

2.1. Contamination of zinc-polymeric coatings on steel by mitosporic fungi

Fungi have important biogeochemine roles in the biosphere and are intimately involved in the cycling of elements and transformations of both organic and inorganic substates. Fungi are major agents of the biodeterioration of plaster and other material (Lugauskas 1997; Gadd 2007). Little and Stahle (2001) found the chemical composition and the quantity of adsorbed organic materials were strongly depended on the nature of the substratum.

Various ecological environments were chosen for metal exposure in order to elucidate their impact on steel coating and to determine the variety of mitosporic fungi in the medium, to which the zinc-polymeric coatings were exposed. Table 1 presents average yearlong meteorological and air pollution data obtained in different environments.

From Figure 1 it is seen that zinc-polymeric coatings on steel exposed to all the conditions are contaminated with propagules of mitosporic fungi, from which fungi of various species begin to develop under favourable conditions. After the studies and identification by using the mentioned methods of fungi isolated from the zinc-polymeric coatings on steel it has been found that a part of fungi species are present on the coatings exposed to all the mentioned environmental conditions (Fig. 2). From the data presented in Figure 2 it is seen that the greatest variety of propagules detected on the zinc-polymeric coatings on steel belongs to *Cladosporium cladosporioides* fungi (M – 58; R – 73; I – 51%, respectively); *C. herbarum* (M – 32; R – 36; I – 48%); *Alternaria alternata* (M – 32; R – 43; I – 14%); *Acremonium strictum* (M – 23; R– 38; I – 27%); *Paecilomyces parvus* (M – 21; R – 63; I – 23%); *Penicillium brevicompactum* (M – 63; R – 14; I – 13%); *P. chrysogenum* (M – 21; R – 19; I – 27%); *P. verrucosum* (M – 26; R – 11; I – 22%). The frequency of detection of some fungi at different exposure sites differed markedly: *Penicillium olsonii* (M – 14; R – 2; I – 6%); *Sporotrichum aurantiacum* (M – 28; R – 7; I – 13%); *Ulocladium chartarum* (M –14; R – 8; I – 21%); *Alternaria tenuissima* (M – 3; R – 16; I – 29%); *Arthrinium phaeospermum* (M – 3; R – 22; I – 38%); *Oidiodendron echinulatum* (M – 8; R – 29; I – 16%). It has been observed that in the industrial environment where a lot of pigeons, sparrows and other birds fly most often thrush fungi are detected: *Candida albicans* (M – 1; R – 21; I – 32%); *Rhodotorula rubra* (M – 6; R – 10; I – 23%); *Exophiala jeanselmei* (M – 2; R – 6; I – 17%), *Asperlillus fumigatus* (M – 6; R – 13; I – 49%) and *Eurotium repens* (M – 3; R – 11; I – 23%). In all variants of experiments a sterile mycelium mostly of white and reddish colour is abundant on the zinc-polymeric coating (M – 89; R – 73; I – 84%).

19 species were found in the marine environment, but not in the other 2 environments. The following fungi can be considered as characteristic for marine conditions: *Blastabotrys nivea* (the frequency of detec-

Table 1. Average yearlong meteorological and air pollution data obtained in different environments

Environment	Parameters					
	Temperature, °C	Number of sunny days H/years	Amount of precipitations, mm/years	Wind speed, m/s	SO_2 concentration, µg/m^3	NO_2 concentration, µg/m^3
Marine	~ 7.0–7.5*	1850–1900*	700–750*	5.5–6.0	0.04–2.29**	0.44–4.24**
Rural-agrarian	~5.5–6.0*	1650–1700*	650–700*	3.0–3.5	~4.0–4.6***	11–12***
Industrial	~ 6.4*	1650–1700*	700*	3.5–4.0	~ 1–3***	22–30***

*Lithuanian Hydrometeorological Service
**Institute of Physics (2007)
***Office of environment protection of Lithuania

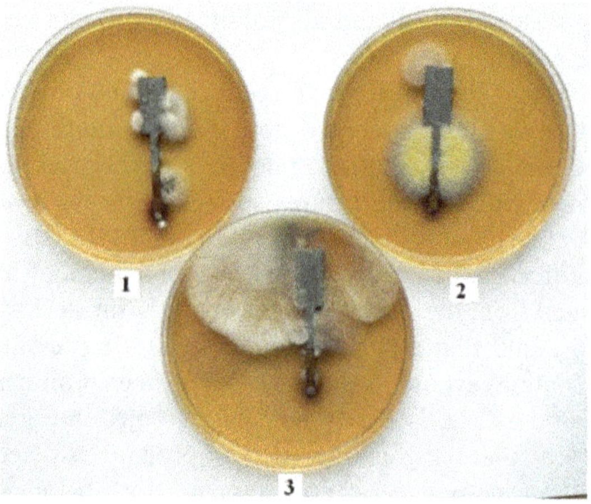

Fig. 1. Contamination with mitosporic fungi of the zinc-polymeric coatings on steel, exposed for one year to different environments: 1 – marine; 2 – rural-agrarian; 3 – industrial

tion 21%), *Fusarium solani* (19%), *Chaetomium globosum* (16%), *Penicillium atramentosum* (16%), *Acremonium murorum* (14%), *Diplococcum spicatum* (14%), *Chrysosporium pannicola* (13%), *Penicillium mirczynskii* (11%). The frequency of detection of other fungi did not exceed 10% (Table 2).

From the zinc-polymeric coatings on steel exposed to the rural-agrarian environment, alongside the fungi mentioned in Figure 2, mitosporic fungi of 20 species were isolated which were not detected under other environmental conditions. The following fungi species can be considered as the predominating ones under the conditions of agrarian environment: *Penicillium lanosum* (the frequency of detection 36%), *P.cyclopium* (34%), Sporotrichum olivaceum (29%), *Verticillium alboatrum* (24%), *Gilmaniella humicola* (21%), *Myceliophthora vellerea* (21%), *Penicillium godlewskii* (21%).

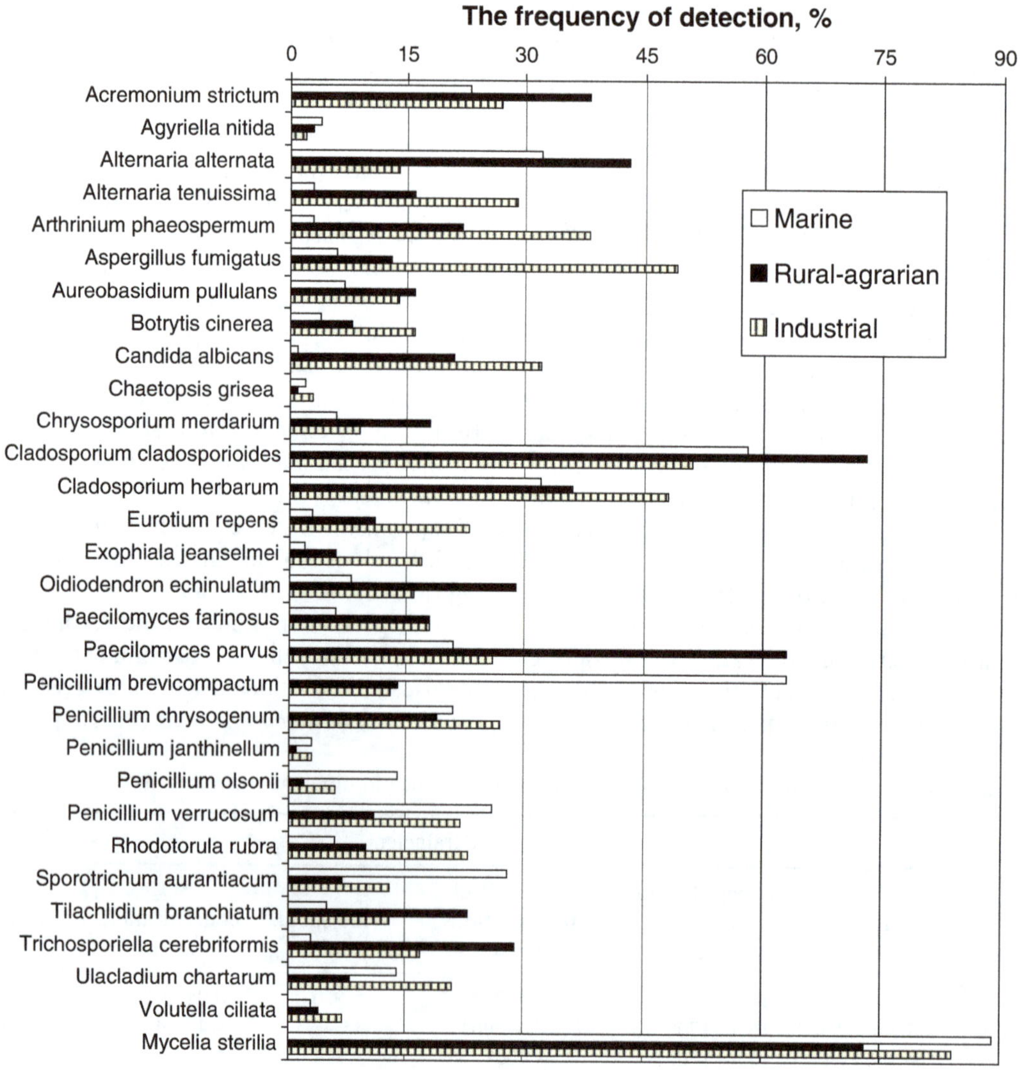

Fig. 2. Mitosporic fungi detected on the zinc-polymeric coatings on steel exposed for 2 years to different environments

Table 2. Mitosporic fungi isolated from the zinc-polymeric coatings on steel exposed to only one type of environmental conditions for 2 years

Environment	The isolated fungi and their frequency of the highest detection, %
Marine	*Acremonium domschii* – 6; *A. murorum* – 14; *Blastabotrys nivea* – 21; *Chaetomium globosum* – 16; *Chrysosporium pannicola* – 13; *Dennisographium episphaeriae* – 1; *Diplococcium spicatum* – 14; *Fusarium avenaceum* – 2; *F. moniliforme* – 8; *F. poae* – 14; *F. sambucinum* – 3; *F. solani* – 19; *F. sporotrichioides* – 1; *Heterocephalum aurantiacum* – 1; *Mortierella polycephala*– 3; *Penicillium atramentosum* – 16; *P. miczynskii* – 11; *Sphaeridium candidum* – 1; *Sphaerulomyces coralloides* – 2.
Rural-agrarian	*Alternaria dianthi* – 18; *Amblyosporium botrytis* – 7; *Chrysosporium olivaceum* – 3; *Fusarium oxysporum* – 13; *Gilmaniella humicola* – 21; *Minimidochium setosum* – 8; *Mortierella hyalina* – 14; *Myceliophthora vellerea* – 21; *Parapericonia angusii* – 18; *Phaeostalagmus cyclosporus* – 7; *Penicillium cyclopium* – 34; *P. godlewskii* – 21; *P. lanosum* – 36; *P. nalgiovense* – 15; *Sarcopodium tortuosum* – 9; *Sarocladium oryzac* – 16; *Scopulariopsis brumtii* – 11; *Sporotrichum olivaceum* – 29; *Thallospora aspera* – 10; *Verticillium alboatrum* – 24.
Industrial	*Acremonium humicola* – 10; *Acrophialophora fusispora* – 7; *Blastomyces dermatitidis* – 6; *Calcarisporium arbuscula* – 14; *Ceratocystis olivacea* – 12; *Chrysosporium pannorum* – 23; *Cochlonema megalosomum* – 17; *Curvularia lunata* – 15; *Fulvia fulva* – 34; *Geotrichum candidum* – 18; *Honsfordia ovalispora* – 12; *Michenera artocreas* – 3; *Mucor racemosus* – 21; *Oidiodendron tenuissimum* – 39; *Paecilomyces niveus* - 17; *Pagidospora amaebophila* – 3; *Phoma exigua* – 16; *Phymatotrichum funigicola* – 7; *Penicillium funiculosum* – 31; *P. nigricans* – 17; *P. sclerotiorum* – 12; *Rhizomucor pusillus* – 27; *Sclerotinia sclerotiorum* – 29; *Trichoderma polysporum* – 17; *Tubercularia vulgaris* – 8.

The biggest variety of the species of mitosporic fungi was detected on the steel specimens covered with the zinc-polymeric coating exposed to the industrial environment. 25 species of mitosporic fungi which were isolated and identified here were not found under any other environmental conditions. Here the following fungi species dominated: *Oidiodendron tenuissimum* (the frequency of detection 39%); *Fulvia fulva* (34%); *Penicillium funiculosum* (31%); *Sclerotinia sclerotiorum* (29%), *Rhizomucor pusillus* (27%), *Chrysosporium pannorum* (23%), *Mucor racemosus* (21%). The frequency of detection of other fungi species did not exceed 20%. The specific contamination of the exposure site, where organic and other non-specific contamination components are abundant can be inferred by the composition of the isolated fungi species.

It can be concluded that, the zinc-polymer coating is not a strong biocide and does not prevent the growth and formation of bioagents on its surface.

Some sorts of fungi were detected on various metals at environmental conditions by other authors (Little, Stahle 2001; Videla, Herera 2009). Therefore one can agree with Videla and Herera, that most of fungi can synthesize and emit in the environment various organic acids. The elucidation of the action of collected fungi enables to implement the means for corrosion control (Videla, Herera 2009).

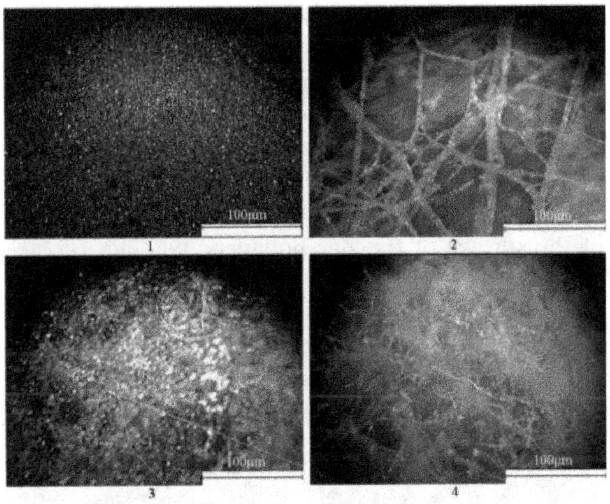

Fig. 3. Micrographs of zinc-polymeric coatings on steel under different environment conditions: 1 – reference (not exposed); 2 – marine; 3 – rural-agrarian; 4 – industrial

2.2. The changes in the surfaces of zinc-polymeric coatings on steel

The changes in the surfaces of the zinc-polymeric coatings on steel after certain exposure time under the above mentioned conditions (Fig. 3) were established by the micrographic method.

In Fig. 3–1 the image of reference sample is shown. As it was expected, the morphology of the coating surface was hardly changed and had a view of a fine-grained structure with a grain size of 1–5 μm. In Fig. 3–2 a micrograph of sample exposed to the marine environment is shown. On the surface of the sample a thick layer of mycelium-like structure is formed. The thickness of layer is ~10–15 μm, the width and length of the layer vary from ~100 to 200 μm. In Fig. 3–3 a micrograph of zinc-polymer sample exposed to the rural-agrarian environment is shown. In this case, the surface is covered with mycelium-like wires and particles of ellipse form, which chaotically cover the whole surface of the sample.

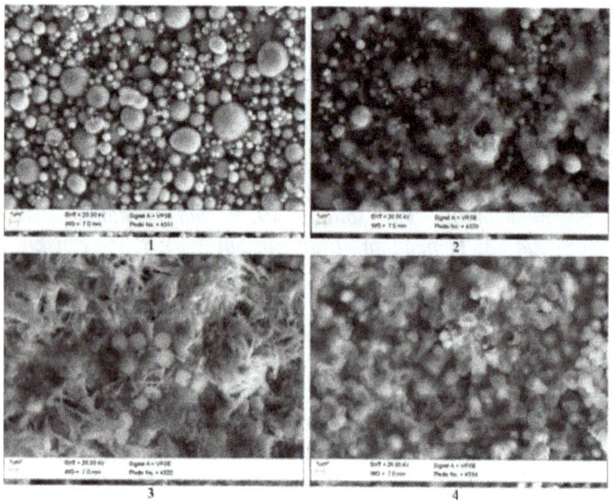

Fig. 4. The changes in the surfaces of zinc-polymeric coatings on steel after 1.5 years of exposure to different environment conditions: 1 – reference (not exposed); 2 – marine; 3 – rural-agrarian; 4 – industrial

The diameter of wires does not exceed 1–2 μm, and the ellipse-like particles diameter varies from 3 to 5 μm. In Fig. 3–4 a micrograph of the sample exposed to the industrial environment is shown. In this case the surface is covered with mycelium wires and clusters.

The SEM method made it possible to analyse surface changes, related to biocorrosion deterioration in the micrometer diapason. Figure 4 presents pictures of the surface of the zinc-polymeric coatings on steel plates exposed for 1.5 years to the marine, rural-agrarian and industrial environments.

After 1.5 years from the beginning of the experiments the surface of the zinc-polymeric coatings exposed to all the climatic conditions was damaged to a variable degree: somewhat deeper and rarer ones in the marine environment, thicker and non-uniform ones – in the other zones. Grooves were on the surface, engraved by bends of various shape and 2–4 μm in size. After 2 years of exposure on the surface of the zinc-polymeric coatings on steel developing fungi formed conidia, forming organs and started to intensively develop and spread, the deterioration of the polymeric coating became more intensive (Fig. 5). This was most clearly seen on the zinc-polymeric coatings exposed to the industrial environment, where phialospores (conidia borne on phialides) of the formed conidia of *Aspergillus fumigatus* fungi were clearly seen. On the analogous zinc-polymeric coatings specimens exposed to the marine environment the fungi of *Cladosporium cladosporioides* species developed most intensively, forming on the surface of the coating deep holes of similar size. The surface of the zinc coatings exposed to the rural-agrarian environment was covered with a rather thick fungi mycelium reminiscent of a cobweb. Clearly formed organs of propagation were not observed after 1.0 year of exposure.

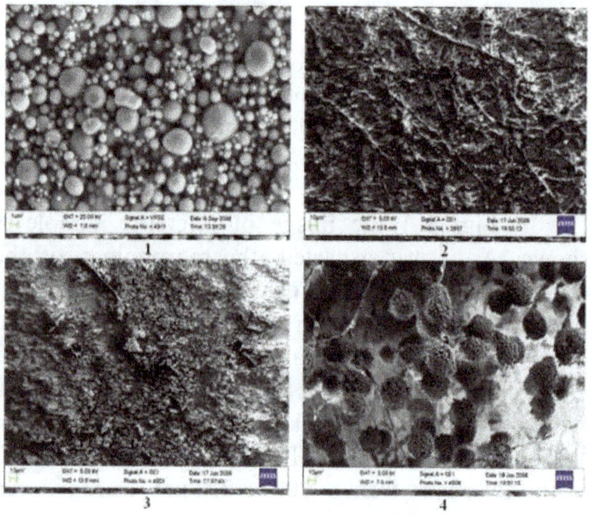

Fig. 5. A general view of mitosporic fungi development on the surfaces of zinc-polymeric coatings on steel after 2 years of exposure under different environment conditions: 1 – reference (not exposed); 2 – marine; 3 – rural-agrarian; 4 – industrial

On a part of specimens covered with the zinc coatings an almost continuous mucous bloom was seen which was formed by developing yeast-like fungi: *Exophiala jeanselmei*, *Rhodotorula rubra*, *Aureobasidium pullulans*.

2.3. Results of spectroscopy investigations of the zinc-polymeric coatings on steel

In Figure 6 FTIR reflection spectra of reference samples of zinc-polymer coatings (1) and coatings exposed to the marine, rural-agrarian and industrial environments are shown.

There is only one absorption peak at a wavelength of 580±2 cm^{-1} seen in the spectrum of reference sample, though, the spectrum of product 2 is almost identical to the spectrum of product 3, where the peaks

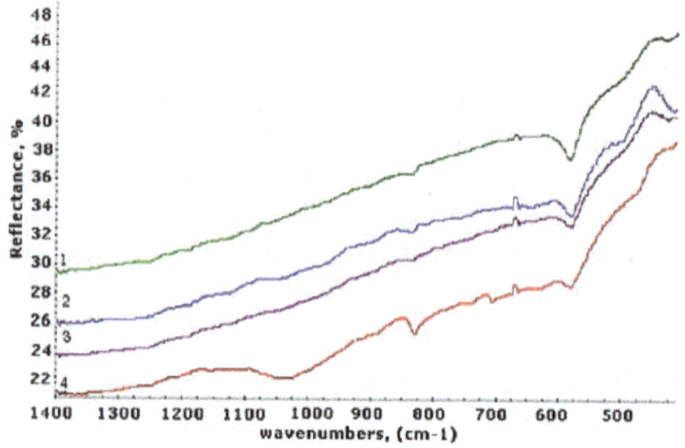

Fig. 6. Results of FTIR spectroscopy investigation of corrosion products on the zinc-polymeric coatings under different environment conditions: 1 – reference (not exposed); 2 – marine; 3 – rural-agrarian; 4 – industrial

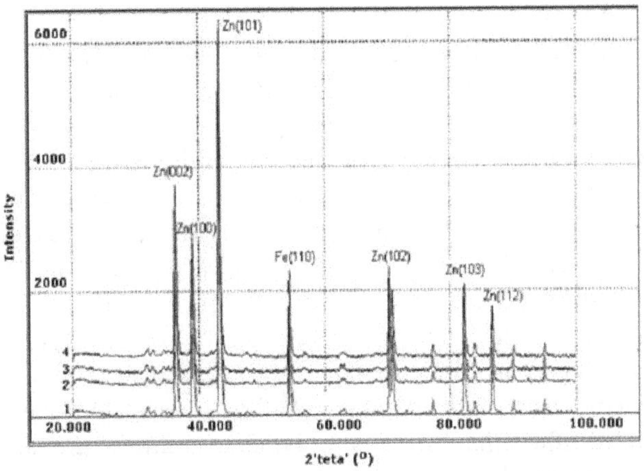

Fig. 7. Phase composition of coatings determined by XRD method under different environment conditions: 1 – reference (not exposed); 2 – marine; 3 – rural-agrarian; 4 – industrial

at wavelengths of 1150 cm^{-1}, 829,2 cm^{-1} and 662,2 cm^{-1} are almost identical in both cases, and also the shift to 578 cm^{-1} wavelength is visible in the spectra of both samples. Vibration differing from the reference spectrum may be ascribed to the vibration from bioagents. An increase in complexity of vibration spectrum of sample 4 may be explained by the complexity of bioagents and additional atmospheric pollution due to the weather conditions in the industrial environment.

Our data, which deals with the changes in zinc polymeric coatings on steel surface, does not fully agree with the data of similar investigations of other authors (Little, Stahle 2001). It can be explained by the fact, that in our tests the metal was coated with the zinc polymeric coating resistant to fungi metabolites.

In Figure 7 XRD spectra of all four samples of zinc-polymer coatings are shown. As it is visible from the spectra, the elements detected are Zn and Fe. All four zinc-polymeric coatings did not undergo phase transitions, there are almost no peaks of zinc oxide. The changes in morphology of the surface seen in Figures 4 and 5 do not necessarily lead to the changes in XRD difractogram, the effect of the bioagent may be not sufficient for the emergence of the mentioned changes, inasmuch as the XRD analysis shows the changes in the material structure and emergence of a new phase identified by the XRD method.

2.4. Mass changes of zinc-polymeric coatings

Steel plates 10 × 15 × 1 cm in size coated with the zinc-polymeric film were used for studies. The effect of mitosporic fungi and other factors on the zinc-polymeric coatings under different exposure conditions could be inferred by the changes in coating mass occurring during certain exposure time (Table 3).

Table 3. Mass changes of zinc-polymeric coatings (g/m^2) depending on the duration of exposure

	Duration of the exposure (year)			
Environment	0.5	1.0	1.5	2.0
Marine	10.1	–	13.0	3.2
Rural-agrarian	7.4	2.6	2.6	8.3
Industrial	1.2	2.0	3.2	2.1

The mass change (increase) of zinc-polymeric coatings (g/m^2) depending on the duration of exposure (Table 3) shows that at the different stations and different exposure time the tendency of unequal variation in mass is observed. This can be explained not only by the corrosion of the protective coating, but also by the removal of corrosion products due to the atmospheric impact and the activity of mitosporic fungi. As seen from the meteorological data presented in Table 1, merely greater windiness in the marine atmosphere leads to more frequent wetting and drying of the surface, as well as more frequent washing of the corrosion products from the surface of zinc-polymeric coating. During the year with moister conditions the corrosion rate does increase, corrosion products are easier cleaned away by atmospheric precipitation, whereas the decrease in coating mass and the decrease in total loss could be compensated by the activity of the mitosporic fungi present in the coating.

Mass losses of zinc-polymeric coatings (g/m^2) after removal of corrosion products depending on the exposition duration and location (Table 4), the tendency of corrosion rate does increase by increasing on the duration of exposure. The red corrosion of Fe in the marine environment was observed after 1.5 years of exposure, while in the industrial and rural-agrarian environment red corrosion is still not observed. The highest corrosion rate of Zn polymeric coating is observed in the marine environment, and the minimal one in the industrial environment.

In order to better understand the nature of deterioration processes occurring in coatings under different environment conditions and their dependence on the alternation of meteorological conditions the data obtained are shown in the diagram form (Fig. 8).

Table 4. Mass losses of zinc-polymeric coatings (g/m^2) after removal of corrosion products depending on the duration of exposure

	Duration of exposure (year)			
Environment	0.5	1.0	1.5	2.0
Marine	27.0	24.0	30.3	33.7
Rural-agrarian	24.7	15.3	25.5	33.7
Industrial	16.4	17.8	20.9	11.1

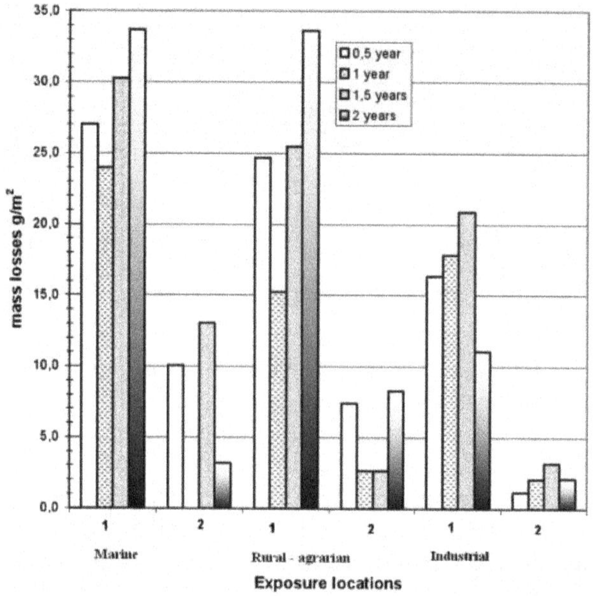

Fig. 8. Mass losses (1) after removal of corrosion products and mass change (2) after exposure of zinc-polymeric coatings versus the duration of exposure

Figure 8 presents the mass loss after 2 years at the exposure sites and the quantity of the corroded zinc in the exposed coatings, which was determined by the standard dissolution method.

Conclusions

1. Under the conditions of polluted environment mitosporic fungi are an active biological factor capable of functioning on the surface of zinc-polymeric coating and thus promote or suppress the deterioration processes occurring on it with the help of the secreted metabolites and gradually become one of the reasons of metal corrosion.

2. The zinc-polymeric coatings were contaminated with mitosporic fungi of different species. Fungi of 30 species were detected on all the specimens exposed for 2 years. Other fungi detected on the zinc-polymeric coating were not detected in the other exposure sites. The number of such fungi species was 19, 20 and 25 in the marine, rural-agrarian and industrial environments, respectively.

3. Mitosporic fungi belonging to *Cladosporium cladosporioides*, *C. herbarum*, *Alternaria alternata*, *Acremonium strictum*, *Paecilomyces parvus* are predominant and capable of developing on the zinc-polymeric coatings. Under certain environment conditions *Penicillium brevicompactum* (frequency of detection 14–63%), *Penicillium verrucosum* (22–26%), *Sporotrichum aurantiacum* (28%) can intensively develop on the zinc-polymeric coatings. In the rural-agrarian and industrial environments the following fungi were detected – *Arthrinium phaeospermum* (29 and 28%, respectively); *Candida albicans* (21 and 32%); *Oidiodendron echinulatum* (29 and 16%); *Paecilomyces parvus* (69 and 26%); *Penicillium chrysogenum* (19 and 27%); *Tilachlidium branchiatum* (29 and 13%); *Trichosporiella cerebriformis* (29 and 17%).

4. After 1.5 years of exposure the zinc-polymeric coatings exposed to the marine and rural-agrarian environments were deteriorated most markedly. The character of deterioration differed, the processes of fungi conidiogenesis were most clearly seen on the zinc-polymeric coatings exposed for 2 years to the industrial environment, where *Aspergillus fumigatus* developed most intensively, and to the marine environment, where *Cladosporium cladosporioides* developed. It testifies, that the zinc-polymeric coatings are not strong biocides and do not prevent the growth and formation of bioagents on their surface.

5. According to diffractograms all four zinc-polymeric coatings are sufficiently resistant to bioagents, although under the natural environment gradually lose their mass and the quantity of corroded zinc markedly changes, as well.

References

Barton, K. 1976. *Protection against atmospheric corrosion*. London, New York: John Wiley & Sons.

Beech, J. B. 2004. Corrosion of technical materials in the presence of the biofilms – current understanding and state-of-the-art methods of study, *International Biodeteration & Biodegradation* 53(3): 177–183. http://dx.doi.org/10.1016/S0964-8305(03)00092-1

Bierwagen, G. 2008. The physical chemistry of organic coatings revisited – viewing coatings as a materials, *Journal of Coatings Technology and Research* 5(2): 133–155. http://dx.doi.org/10.1007/s11998-007-9066-4

Binkauskienė, E.; Jasulaitienė, V.; Lugauskas, A. 2009. Effect of Aspergillus niger Tiegh. L-10 on the physical and chemical properties of a polyaniline coating in the growth substrate, *Synthetic Metals* 159(13): 1365–1368. http://dx.doi.org/10.1016/j.synthmet.2009.03.011

Bondartseva, M. A. (Ed.). 2012. *Mikroskopicheskie griby v vozdushnoj srede Sankt-Peterburga* [Microfungi in the air of Saint-Petersburg]. Sankt-Peterburg: Chimizdat (in Russian).

Booth, C. 1971. *Methods in microbiology*. London, New York: Academic Press.

Carmichael, J. M.; Kendrick, W. B.; Conners, I. L.; Sigler, L. 1980. *Genera of Hyphomycetes*. Canada: The University of Alberte Press.

Chaverri, P.; Samuels, G. J. 2003. Hypocrea/Trichoderma (Ascomycota, Hypocreales, Hypocreaceae): species with green ascospores, *Studies in Mycology* 48: 1–119.

Diaz-Ballote, L.; Ramanauskas, R. 1999. Impoving the corrosion resistance of hot dip galvanized zinc cooting by alloyng, *Corrosion Review (UK)* 17: 411–422.

Domsch, K. H.; Gams, W.; Anderson, T. H. 1980. *Compendium of soil fungi*. Volume 1. London: Academic Press.

Gadd, G. M. 2007. Geomycology: biogeochemical transformations of rocks, minerals, metals and radionuclides by fungy, bioweathering and biomediation, *Mycological Research* 111: 3–49. http://dx.doi.org/10.1016/j.mycres.2006.12.001

Gams, W. 1971. *Cephalosposporium Artige Schimmelpilze (Hyphomycetes)*. Jena: VFB G. Fischer Verlag.

Hawksworth, D. L.; Kirk, P. M.; Sutton, B. C.; Pegler, D. N. 1995. *Ainsworth & Bisby's dictionary of the fungi.* 8th ed. Wallingford: CAB International.

Helsen, J. A.; Breme, H. J. 1998. *Metals as biomaterials.* New York: John Wiley & Sons.

Lithuanian Hydrometeorological Service [online], [cited 10 April 2012]. Available from Internet: http://www.meteo.lt/klim_lt_klimatas.php

Huang, R. S.; Lin, C. J.; Isaac, H. S. 2006. Measuring streaking rates of an Al–Zn alloy using a difference imaging technique, *Corrosion Science* 48: 1867–1873. http://dx.doi.org/10.1016/j.corsci.2006.05.039

Hughes, M. N.; Poole, R. K. 1989. *Metals and microorganisms.* London, New York: Chapman and Hall.

Ignatev, R. A.; Mikhailova, A. A. 1987. *Zashchita tekhniki ot korrozii, stareniia i biopovrezhdenij* [Protection of equipment against corrosion, aging and biodamages]. Moskva: Rosselkhozizdat.

Institute of Physics 2007 [online], [cited 10 April 2012]. Available from Internet: http://gamta.lt/files/20071240915047436.pdf

Jagtap, R. N.; Nambiara, R.; Hassana, S. Z.; Malshea, V. C. 2007. Predictive power for life and residual life of the zinc rich primer coatings with electrical measurement, *Progress in Organic Coating* 58(4): 253–258. http://dx.doi.org/10.1016/j.porgcoat.2006.08.015

Kikuchi, Y.; Streekumari, K. R. 2002. Microbiallly influenced corrosion and biodeterioration of structural metals, *Journal of the Iron and Steel Institute of Japan* 88: 620–628.

Kirk, P. M.; Cannon, P. F.; David, J. C.; Stalpers, J. A. 2004. *Ainsworth & Bisby's dictionary of the fungi.* 9th ed. Wallingford: CAB International.

Klich, M. A. 2002. *Identification of common Aspergillus species.* The Netherlands: Wageningen.

Kozakiewicz, Z. 1989. Aspergillus species on stored products, *Mycological Papers* 161: 1–188.

Krylenkov, V. A.; Vlasov, D. I.; Dashko, P. E.; Startsev, S. A. 2003. Problemy sokhraneniia zhiloj i proizvodstvennoj infrastruktury gorodov ot biorazrusheniia [Problems of preservation of dwellings and industrial infrastructure of town from biodestruction], *Infstroj* 5(11): 3–13.

Lee, K.; Lee, I.; Lee, C.; Ahr, H. 2007. In situ observation in a scanning electron microscope the expliation behaviour of galvannealed Zn-Fe coating layers, *Surface and Coatings Technology* 201: 6261–6266. http://dx.doi.org/10.1016/j.surfcoat.2006.11.021

Little, B.; Stahle, R. 2001. Fungal influenced corrosion in post-tension structures, *The Electrochemical Society Interface* Winter: 44–48.

Li, Z.; Wang, F.; Liu, G. 2004. Grain size effect on the electrochemical corrosion behaviour of surface nanocrystalized low-carbon steel, *Corrosion* 60: 891–896. http://dx.doi.org/10.5006/1.3287822

Lichušina, S.; Sudavičius, A.; Juškėnas, R.; Bučinskienė, D.; Juzeliūnas, E. 2008. Deposition of cobalt–rich Zn–Co alloy coatings of a high corrosion resistance, *Transactions of Institute of Metal Finishing* 86(30): 141–147. http://dx.doi.org/10.1179/174591908X304162

Lugauskas, A.; Paškevičius, A.; Repečkienė, J. 2002. *Patogeniški ir toksiški mikroorganizmai žmogaus aplinkoje* [Pathogenic and toxic microorganisms in human environment]. Vilnius: Aldorija.

Lugauskas, A. (Ed.). 1997. Mikrobiologiniai medžiagų pažeidimai [Microbiological Biodeteriorotion of materials], *Valstiečių laikraštis* 470.

Lugauskas, A. I.; Mikulskene, A. I.; Shliauzhene, D. I. 1987. *Katalog mikromitsetov – biodestruktorov polimernykh materialov* [The catalog of micromycetes – biodestruktor of polymeric materials]. Moskva: Nauka.

Marchebois, H.; Savall, C.; Bernarda, J.; Touzaina, S. 2004. Electrochemical behaviour of zinc–rich powder coatings in artificial sea water, *Electrochimica Acta* 49(17–18): 2945–2954. http://dx.doi.org/10.1016/j.electacta.2004.01.053

Marchebois, H.; Touzain, S.; Joiret, S.; Bernard, J.; Savall, C. 2002. Zinc-rich powder coatings corrosion in sea water: influence of conductive pigments, *Progress in Organic Coatings* 45: 415–421. http://dx.doi.org/10.1016/S0300-9440(02)00145-5

Marcus, P.; Oudar, J. 2002. *Corrosion mechanism in theory and practice.* New York: Marcel Dekker Inc. http://dx.doi.org/10.1201/9780203909188

Mirchink, T. G. 1988. *Pochvennaia mikologiia* [Soil micology]. Moskva: Izdatelstvo Moskovskogo universiteta.

Narkevičius, A.; Bučinskienė, D.; Samulevičienė, M.; Ramanauskas, R. 2003. Corrosion behaviour of Zn coatings electrodeposited from alkaline and acid solutions, *Transactions of Institute of Metal Finishing* 81: 93–97.

Nelson, P. E.; Dignani, M. C.; Anaissie, E. J. 1983. Taxonomy, biology and clinical aspects of Fusarium species, *Clinical Microbiology Reviews* 7: 479–504.

Office of environment protection of Lithuania [online], [cited 10 April 2012]. Available from Internet: http://oras.gamta.lt/cms/index?rubricId=4cff26a3-ece5-46be-ad58-c8d14b94bea6

Oorschot, C. A. N. van 1980. A revision of *Chrysosporium* and allied genera, *Studies in Mycology* 20: 1–89.

Ramanauskas, R.; Juzeliūnas, E.; Narkevičius, A.; Bučinskienė, D.; Lugauskas, A.; Pečiulytė, D.; Levinskaitė, L.; Ulevičius, V.; Jasinevičienė, D. 2005. Investigation of microbiologically influenced corrosion. 1. Characterization of natural outdoor conditions in Lithuania, *Chemija* 16(1): 25–34.

Ramanauskas, R.; Mulesnikova, L.; Maldonado, L.; Dobrovolskis, P. 1998. Characterization of the corrosion behaviour of Zn and Zn alloy electrodeposits: atmospheric and accelerated test, *Corrosion Science* 40: 401–410. http://dx.doi.org/10.1016/S0010-938X(97)00144-3

Ramanauskas, R.; Quintona, P.; Bartolo–Perez, P.; Diaz–Ballote, L. 2000. Effect of corrosion product on atmospheric corrosion of electrodeposited zinc and zinc alloy coatings, *Corrosion Science* 56: 588–597. http://dx.doi.org/10.5006/1.3280562

Ramirez, C. 1982. *Manual and atlas of the penicillia.* Amsterdam: Elsevier Biomedical Press. 874 p.

Roberge, P. R. 1998. *Handbook of corrosion engineering*. New York: McGraw-Hill.

Samson, R. A.; Frisvad, J. C. 2004. Penicillium subgenus Penicillium: new taxonomic schemes, mycotoxins and other extrolites, *Studies in Mycology* 49: 1–260.

Samson, R. A. 1974. Paecilomyces and same allied Hyphomycetes, *Studies in Mycology* 6: 1–119.

The law of Lithuanian Government HN-69:2003 [online], [cited 10 April 2012]. Available from Internet: http://www3.lrs.lt/pls/inter3/dokpaieska.showdoc_l?p_id=230880

Tüken, T. 2006. Zinc modified polypyrrole coatings on mild steel, *Surface and Coatings Technology* 201(4): 2782–2790. http://dx.doi.org/10.1016/j.surfcoat.2006.05.017

Videla, H.; Herera, L. K. 2009. Understanding microbial inhibition of corrosion. A comprehensive overview, *International Biodeteriorotion and Biodegradation* 63(7): 896–900. http://dx.doi.org/10.1016/j.ibiod.2009.02.002

Yang, S. S.; Ku, C. H.; Bor, H. J.; Lin, Y. T. 1996. Effects of vapor phase corrosion inhibitor on microbial corrosion of aluminum alloys, *The Journal of Corrosion Sciences and Engineering* 29(1): 18–30.

Albinas LUGAUSKAS. Dr Habil, Professor, biology 01B. Publications: author and co-author of more than 400 research papers, 4 monographs, 2 recommendatory editions. The Guide of 19 theses for Doctoral degree, consultant of 3 theses. Research interests: the diversity and stability of the species of micromycetes, dependence of their metabolites upon functioning conditions and their role in microbial communities in soil and other substrata under changing environmental conditions and the character of technological, chemical and physical properties of raw materials.

Igoris PROSYČEVAS. Dr, senior research worker, chief of physico-chemical laboratory, Materials Science Institute of Kaunas University of Technology. Publications: co-author of about 42 research papers. Research interests: nanotechnology, materials science, spectroscopy, plasmonics.

Algirdas NARKEVIČIUS. Dr, research worker, Center for Physical Sciences and Technology. Publications: co-author of about 40 research papers. Research interests: corrosion of metals, biocorrosion, materials science, electrochemistry.

Aušra SELSKIENĖ. Dr, research worker, Center for Physical Sciences and Technology. Publications: co-author of about 23 research papers. Research interests: Chemical investigation of archaeometallurgical objects, biocorrosion, electron microscopy.

Dalia BUČINSKIENĖ. Dr, research worker, Center for Physical Sciences and Technology. Publications: co-author of about 30 research papers. Research interests: electrochemistry, corrosion of metals, biocorrosion, materials science.

Alma RUČINSKIENĖ. Dr, research worker, Center for Physical Sciences and Technology. Publications: co-author of about 30 research papers. Research interests: corrosion of metals, materials science, electrochemistry.

Elena BINKAUSKIENĖ. Dr, research worker, Center for Physical Sciences and Technology. Publications: author and co-author of about 40 research papers. Research interests: nanotechnology, conducting polymers, materials science, electrochemistry/bioelectrochemistry.

18

BIOLOGICAL AOX REMOVAL OF PULP MILL PLANT EFFLUENT BY *PSEUDOMONAS AERUGINOSA* – BENCH STUDY

Anil Kumar[a], Purnima Dhall[b], Rita Kumar[b]

[a]*National Institute of Immunology, Aruna Asaf Ali Marg, a10067 New Delhi, India*
[b]*Institute of Genomics and Integrative Biology, Mall Road, 110007 Delhi, India*

Abstract. Discharge of adsorbable organic halides (AOX) into the water bodies has resulted into many health and environmental problems such as endocrine disruption, aquatic toxicity, bioaccumulation and carcinogenicity. The already known physical, chemical and electrochemical methods are not economically viable for the control of water pollution. So this paper focuses on the biological reduction of AOX from pulp and paper mill effluent using isolated bacteria. The isolated bacteria were screened and finally *Pseudomonas aeruginosa* strain 1 was used further. The effect of various parameters such as, bacterial cell concentration, surface washing of bacterial cell and agitation were investigated and it was found that to some extend every parameter has resulted in the reduction of AOX from the effluent. It was inferred that the three time washed pellet inoculated in the ratio of 1:1 (sample: pellet) and incubated at 150 rpm at 37 °C for 24h has resulted in 78% of AOX removal.

Keywords: adsorbable organic halides, AOX, biological reduction, water pollution.

Reference to this paper should be made as follows: Kumar, A.; Dhall, P.; Kumar, R. 2013. Biological AOX removal of pulp mill plant effluent by *pseudomonas aeruginosa* – bench study, *Journal of Environmental Engineering and Landscape Management* 21(4): 296–304. http://dx.doi.org/10.3846/16486897.2012.745413

Introduction

In terms of fresh water usage pulp and paper industry ranks third in the world (Santosh 2007). Bleaching of pulp by chlorine based agents is still practiced in India and in other developing countries (Deshmukh *et al.* 2009). Waste water from bleaching unit is toxic because of the presence of chlorinated organic compounds (Savant *et al.* 2006). These organochlorine compounds are collectively termed (Saima, Izharul 2012) as adsorbable organic halides (AOX). Formation of these chlorinated organics also contributes to the color of waste water (Clifford, Noemi 2010). It has been reported that production of one ton of paper contributes 100 kg of colorimparting substances and 2–4 kg of organochlorines (Kansal *et al.* 2008). About 500 different chlorinated organic compounds have been identified in paper mill effluent (Savant *et al.* 2006); among these chlorinated hydrocarbons, benzenes, phenols, catechols, guaiacols, syringols, vanillins, chloroform, dioxins and furans etc. are of major concern (Kansal *et al.* 2008; Freire *et al.* 2003).The chemical diversity of these pollutants causes a variety of clastogenic, mutagenic, carcinogenic, endocrine effects on fishes and other aquatic communities in recipient water bodies (Karrasch *et al.* 2006). Increased awareness for the harmful effects of these pollutants has resulted in stringent regulations on black liquor discharge into the environment (CPCB 2007). The Central Pollution Control Board of India has imposed AOX discharge limit amounting to 1.5 kg AOX/ton of paper produced (MoEF 1993). This discharge limit will be further lowered to 1kg AOX/ton of paper produced in next five years for large pulp and paper industries.

A number of approaches have been explored to reduce the AOX level in paper mill effluent, which can be categorized as physico-chemical and biological. Physico-chemical techniques include adsorption, ultrafiltration, nanofiltration, reverse osmosis, neutralization of bleach effluent and supercritical water oxidation. Major limitations of above mentioned physico-chemical approach are cost and generation of secondary sludge (Patel, Suresh 2008). Whereas advanced oxidation processes (ozonation, peroxidation, photocatalysis) were developed and used as potentially powerful methods capable of transforming the pollutants into harmless substances (Biljana *et al.* 2012). Combining some oxidizing agents such as O_2/ZnO/UV, O_2/TiO_2/UV, O_3/UV, UV/H_2O_2/TiO_2, O_3/H_2O_2 has resulted in excellent color reduction and hold the

Corresponding author: Anil Kumar
E-mail: anilk@nii.ac.in

greatest promise for detoxification and mineralization of pollutants (Catalkaya, Kargi 2007; Hassan, Hawkyard 2002; Helmy *et al.* 2003; Daneshvar *et al.* 2004; Catalkaya, Kargi 2008; Ugurlu, Karaogly 2009). But none of them is industrially applicable due to high cost which is a limitation and infeasibility of the processes. However, biological treatment (Ruggaber, Talley 2006) methods for treating pulp and paper effluent are of great concern over physico-chemical methods due to their economical as well as ecofriendly impact. Biological approach involves the use of diverse kind of microorganisms like bacteria, fungi, algae and microbes of extreme habitat for reducing AOX and chromophores in pulp mill effluent. Such biological process can be performed under aerobic or anaerobic conditions or in combination of the both. The biological degradation of chlorinated organic compounds under aerobic conditions (Kim *et al.* 2002; Vasconcelos *et al.* 2006) and anaerobic conditions (Li *et al.* 2009; Maphosa *et al.* 2010; Futagami *et al.* 2008) has been reported in the literature. Researchers also employed the combination of both aerobic and anaerobic conditions like Pratibha (2007) studied the degradation of chlorophenols. The AOX removal was 84% on third day when 7 days anaerobically treated effluent was further treated by aerobic means.

Use of fungi for the treatment of pulp mill effluent was also reported by some researcher such as Basak *et al.* (2004) who reported 76% AOX reduction with 0.2 g/l feed acetate concentration and 70% AOX could be removed from chlorinated pulping waste in 7.3 h of contact with 0.2 g/l of carbon supplement by using *Penicillium camemberti*.

The major drawback associated with the fungal treatment is that fungi can effectively degrade lignin but they are unable to survive under extreme environmental conditions (high temperature, pH and presence of toxic chemicals) which usually exist in the treatment plants. In addition, fungal filaments cause structural hindrance so their utilization is not feasible for biological treatment of pulp and paper industry effluent (Amr *et al.* 2008).

Considering the above drawbacks this study explores the possibility of naturally occurring resident bacteria on the reduction of adsorbable organic halides present in the effluent emanating from pulp and paper mill located in Northern India. Various parameters were optimized in order to increase the rate of degradation and decrease the time of treatment like bacterial cell concentration, surface washing of bacterial cell and agitation. Later the isolate, exhibiting the best AOX reducing capability, was identified by 16s rRNA studies at DSMZ, Germany.

1. Material and methods

1.1. Sample collection

Sludge and Wastewater samples were collected from the inlet and outlet ETP streams of pulp and paper mill in clean amber glass containers. Samples were transported within 12 hrs. Under refrigerated conditions and stored at 2–4 °C till use.

1.2. Isolation of bacteria

Sludge sample was collected from effluent treatment plant of pulp and paper mill for the isolation of autochthonous population of bacteria. Sludge extract was prepared by dissolving 300 g of dried (50 °C) sludge in 960 ml of distilled water. Then autoclaved at 15 psi for 1h followed by centrifugation at 7000 rpm for 10 min. The supernatant (extract) was collected and used for the preparation of enrichment medium. The enrichment medium used for bacterial isolation was composed of sludge extract and nutrient broth in the ratio of 1:2. Candid B (purchased from Glenmark) was added to the medium at a concentration of 0.25 μl ml^{-1} in order to prevent fungal growth during isolation. The enrichment medium was inoculated with 5 g sludge from the said site and incubated on orbital shaker for 48 h at 30 °C, 120 rpm. The enriched sludge extract sample was serially diluted in 0.85% saline solution and spread onto plates containing enrichment medium with 2% agar. The plates were incubated at 37±2 °C for 24–96 h. Single isolated colonies were picked and streaked on fresh plates containing the NB to obtain pure cultures. The isolates obtained were then screened for their capacity to reduce AOX and color.

1.3. Screening of bacterial isolate for AOX reduction

a) Experimental setup

The AOX reduction experiment was set up to screen out effective bacterial isolate. The mother culture was prepared by inoculating one loopful of all the individual bacterial isolates separately in 25 ml of sterilized nutrient broth having 0.01% tween 80 (surfactant). The inoculated broths were incubated in an orbital shaker at 37 °C for 16–24 h so as to obtain actively growing mother cultures. The actively growing individual bacterial cultures were inoculated separately in 2000 ml of flask containing 1000 ml of sterilized nutrient broth and incubated at 37 °C, 200 rpm for 16–24 h. After the growth was achieved the cultures were centrifuged at 8000 rpm for 15 min 4 °C. Pellets were washed with phosphate buffer (pH 6.8) and stored at 4 °C till further use. The flasks containing 1 liter of wastewater sample were inoculated with the pellets and incubated in shaker at 200 rpm at 37 °C for 24 h. The ratio of wastewater: bacterial isolate was 1:0.5. After 24 h of incubation, the samples were analyzed for AOX reduction.

b) AOX (Adsorbable organic halides)

AOX was measured according to the standard method given in (APHA 98). The samples were collected (before and after treatment) and centrifuged, the supernatant was analyzed. The three main steps involved were: adsorption of organic matter onto activated carbon, organically-bound chlorine mineralization through combustion and determination of released chloride by micro-coulometric titration. AOX was estimated by using the 686 – Titroprocessor analyzer (Metrohm, Switzerland).

c) Adsorption and desorption assay

Desorption studies were carried out to the extent of adsorbed color to the bacterial biomass during the process of color removal which ultimately leads to the reduction in AOX. Different desorbing agents like 1N NaOH, 1N HCl, triple distilled water and 85% HCOOH were used for the desorption studies. Known volumes of thoroughly shaken samples were taken and centrifuged to obtain pellets. The pellets were then treated with the above chemicals and extracted twice to remove the adsorbed color, till no more color could be extracted from the biomass. The extent of color adsorbed was estimated by comparing with the actual color of control and the color removed by the bacterial biomass. Controls of pure inocula in buffer, uninoculated effluent and non-decolorizers were also run parallely and subjected to same desorption protocol, to eliminate any color being produced due to the intracellular contents of the bacteria and the chances of a false reading.

1.4. Analysis of the color adsorption data

The decolorization data was analyzed using the general equation for multilayer adsorption by Freundlich, expressed as

$$x/m = K.C^{1/n}, \quad (1)$$

where: x/m = amount of solute adsorbed per unit weight of the adsorbent;

C = is the equilibrium concentration of color remaining;

K and n = empirical constants.

The constant K is positively related to the extent of degree of adsorption, while the constant n provides a rough estimate of the intensity of adsorption. When n = 1, adsorption is considered to be linear.

A linear form of the expression will yield the constants K and n, hence:

$$\text{Log}\, x/m = \log K + 1/n \log C, \quad (2)$$

where: 'x' was calculated by subtracting the color value after decolorization, from the initial value; 'm' was calculated by estimating the dry weight of the biomass used for decolorization and C was the final color value after which there was no significant decrease.

1.5. Optimization of AOX reduction parameters

a) Bacterial cell concentration

The innoculum was prepared by inoculating one loopful of selected isolate in a 50 ml of sterilized nutrient broths having 0.01% tween 80. The inoculated broths were incubated in an orbital shaker at 37 °C for 16–24 h so as to obtain actively growing mother cultures. This actively growing culture was used to inoculate 500 ml, 1000 ml and 2000 ml of sterilized nutrient broths separately and incubated at 37 °C, 200 rpm for 16–24 h. After the growth was achieved the cultures were centrifuged on 8000 rpm at 4 °C for 15 min. The pellet of the cultures was washed with phosphate buffer (pH 6.8) and used to inoculate three 1liter of wastewater samples individually in ratios (wastewater:pellet); (1:0.5), (1:1), (1:2). These inoculated wastewater samples were incubated in shaker at 200 rpm at 37 °C for 24 h. After 24 h of incubation, the samples were analyzed for AOX reduction.

b) Surface washing of bacteria

The innoculum was prepared by inoculating one loopful of selected isolate in a 50 ml of sterilized nutrient broth having 0.01% tween 80. The inoculated broths were incubated in an orbital shaker at 37 °C for 16–24 h so as to obtain actively growing mother cultures. This actively growing culture was used to inoculate three 1000 ml sterilized nutrient broth separately and incubated at 37 °C, 200 rpm for 16–24 h. After the growth was achieved the cultures were centrifuged at 8000 rpm for 15 min at 4 °C. The pellet of first 1000 ml culture was washed only once with phosphate buffer (pH 6.8) and used to inoculate 1liter of wastewater sample. The pellet of second flask was washed twice with phosphate buffer (pH 6.8) and used to inoculate 1liter of wastewater sample. The pellet of third flask was washed three times with the same phosphate buffer (pH 6.8) and used to inoculate another 1liter of wastewater sample. These inoculated wastewater samples were incubated in shaker at 200 rpm at 37 °C for 24 h. The ratio of wastewater: bacterial isolate was 1:1. After 24 h of incubation, the samples were analyzed for AOX reduction.

c) Agitation

The actively growing culture was used to inoculate three 1000ml sterilized nutrient broth separately and incubated at 37 °C, 200 rpm for 16–24 h. After the growth was achieved the cultures were centrifuged at 8000 rpm for 15min 4 °C. Pellets were washed thrice with phosphate buffer (pH 6.8). The 3 flasks containing 1 liter of wastewater sample were inoculated with the pellets and each was incubated in shaker at different rpm i.e. (150 rpm, 200 rpm, and 250 rpm) at 37 °C for 24 h. The ratio of wastewater:pellet was 1:1. After 24 h of incubation, the sample was analyzed for AOX reduction.

1.6. Repeatability and reproducibility experiment

The innoculum was prepared by inoculating one loopful of selected bacterial isolate separately in 50 ml of sterilized nutrient broth having 0.01% tween 80. The inoculated broth was incubated in an orbital shaker at 37 °C for 16–24 h so as to obtain actively growing mother culture. The actively growing individual bacterial culture was inoculated separately in 2000 ml of flask containing 1000 ml of sterilized nutrient broth and incubated at 37 °C, 200 rpm for 16–24 h. After the growth was achieved the cultures were centrifuged at 8000 rpm for 15 min 4 °C. Pellet was washed thrice with phosphate buffer (pH 6.8) and stored at 4 °C till further use. The flask containing 1 liter of wastewater sample was inoculated with the pellet and incubated in shaker at 200 rpm at 37 °C for 24 h. The ratio of wastewater: bacterial isolate was 1:1. After 24 h of incubation, the sample was analyzed for AOX reduction.

1.7. Bacterial identification

The selected bacterium was identified by 16S rRNA studies at DSMZ, Germany. The selected bacterial isolate was identified by 16S RNA gene analysis. Genomic DNA was extracted by using kit (Real Genomics kit). 16S rRNA gene of DNA sample was amplified using universal primer 16S F = 5'- AGCAGCCGCGGTAATAC-3' and 16S R = 5'-TACGGCTACCTTGTTACG-3' synthesized by the centre for genomic application (TCGA). The PCR product purified by gel elution (Gel extraction Kit, Qiagen) was sequenced. 16S RNA gene was compared with available sequences using NCBI-BLAST tool (Johnson *et al.* 2008).

2. Results

2.1. Isolation and characteristics of bacterial population

Nine bacterial isolates were purified from all the above mentioned isolation procedure. It was hypothesized that bacteria isolated from their natural habitat have capability of surviving in harsh conditions by developing some catabolic enzymes systems, specific for particular components present in the natural habitat. The isolated colonies were diverse in their morphologies, ranging from small pin-pointed to large sized; smooth margined to wrinkled periphery; shining to dry and so on. Table 1 show the morphological characteristics of those isolated bacterial population obtained during the process of isolation.

2.2. Screening of bacterial isolate for AOX reduction

In the present study, selected (on the basis of growth rate) isolates were screened for their capability to survive and actually degrade the pollutional components present in the wastewaters to which they are exposed. The bacteria responded in varying degrees with a few bacteria not being able to survive in the given conditions. A few bacteria were able to grow in the wastewaters but were not actually able to bring down the pollutional parameters, i.e. AOX. Adsorbable organic halides (AOX) were particularly a challenging parameter. Bacterial isolates were screened individually to observe their capacity to reduce AOX of the paper mill effluent. Studies were conducted with the selected isolates obtained from pulp and paper mill. The results are depicted in Figure 1, however as can be seen from the figure, there was a marked increase in AOX values of these sample in case of some isolates some will give good response. The isolates were able to give the results after 24 h of incubation. Isolates 4 exhibited the best results giving a reduction of 2.8 mg/l (standard deviation – 0.11 with standard error – 0.05) of AOX while comparing with control showing AOX value of 10.4 mg/l after 24 h of incubation followed by isolate 7 giving reduction of 5.7 mg/l.

Table 1. Morphological characteristics of isolated bacteria

Lab Name	Configuration	Margin	Elevation	Color	Size in mm
Isolate 1	Round	Smooth	convex	bluish Cream	3 mm
Isolate 2	Irregular	Lobate	umbonate	yellowish	2–3 mm
Isolate 3	Round	Smooth	drop like	yellowish	1 mm
Isolate 4	Round	Smooth	flat	creamy white	2 mm
Isolate 5	Punchi form	Smooth	convex	yellow	Pinpoint
Isolate 6	Round	Smooth	raised	orangish	1–2 mm
Isolate 7	Round	Smooth	convex	white	1 mm
Isolate 8	Round	Smooth	drop like	yellow glossy	1 mm
Isolate 9	Spindle	Smooth	flat	thick white	2 mm

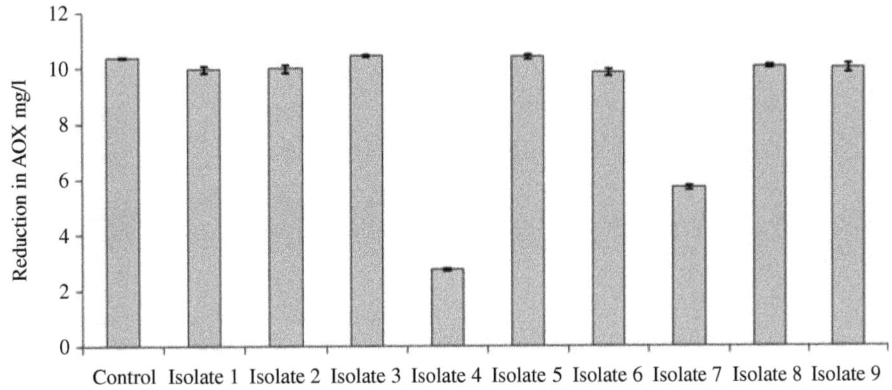

Fig. 1. Reduction in AOX level of Pulp and paper samples treated with different individual bacterial strains

2.3. Adsorption and desorption assay

Different process units of the paper manufacturing plant, contribute to the overall color of the effluent, of which Black liquor, which comes out from the pulping stage, is the major contributor. The bleaching section release organochlorine compounds, which form chromophoric compounds with lignin breakdown compounds and impart color to the effluent. This also increases the absorbable organic halide (AOX) content of the effluent. So, the bacterial isolates were screened for their capacity to reduce the color of wastewater, individually which ultimately affects the AOX level present in the effluent. It was evident from the obtained data (Fig. 2) that about 65% (standard deviation = 0.7 with standard error of = 0.31) of the color was removed by isolate 4 in comparison to other isolates which shows the reduction in the range of 30.6 to 57.2%. Isolate 4 was selected for further studies.

The colored components present in the effluent generally arise from lignin and its derivatives, which are responsible for AOX present in the effluent. They are negatively charged, in general, with mostly hydroxyl and carboxyl functional groups. In view of the results obtained during the study of various parameters, it is this charge repulsion that caused lesser decolorization in certain bacteria. Therefore, production of some exopolysaccharides or capsular elements by the best decolorizers could be actual reason for adsorption of color from the wastewater.

2.4. Optimization of parameters for AOX reduction

Effect of inoculum size: the effect of biomass on the AOX reduction ability of isolate 4 was studied. Different effluent:biomass ratios of 1:0.5, 1:1 and 1:2, were tried. Results suggested that isolate 4 produced the best results when used in the 1:1 effluent: biomass

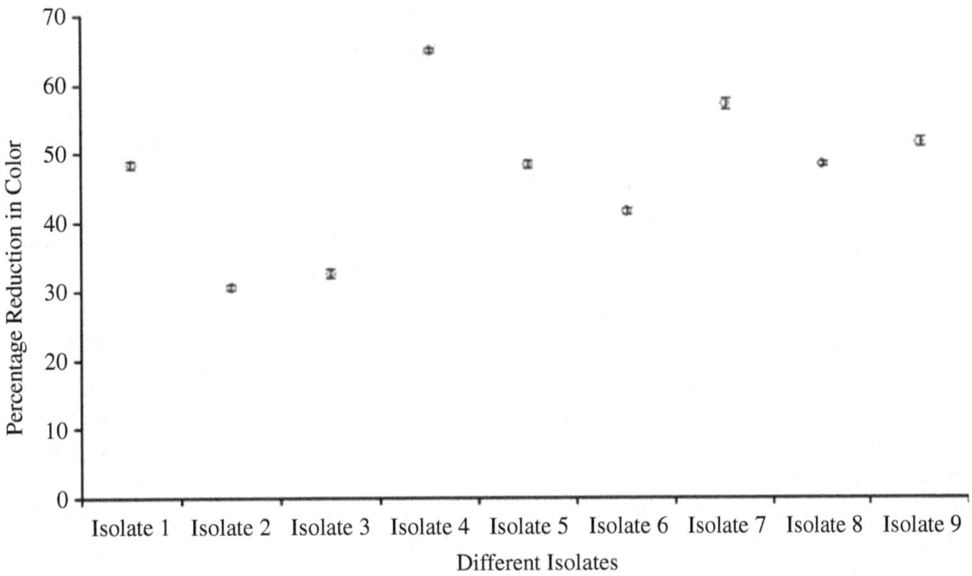

Fig. 2. Percentage reductions in color of pulp and paper mill effluent with different individual bacterial isolates

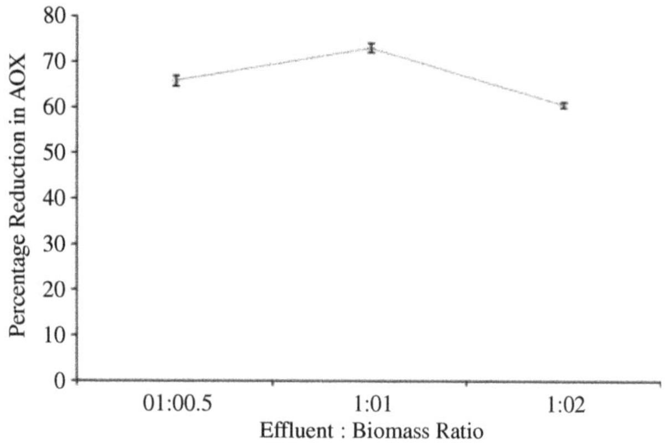

Fig. 3. Effect of rate of inoculum size on the AOX reduction of pulp and paper mill effluent

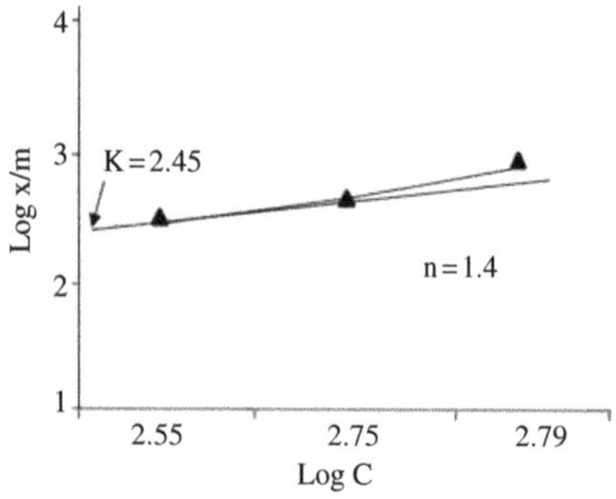

Fig. 4. Color adsorption data at different effluent: biomass ratio. 'x/m' is the amount of color adsorbed per unit biomass and C' is the amount of color remaining in the effluent after decolorization

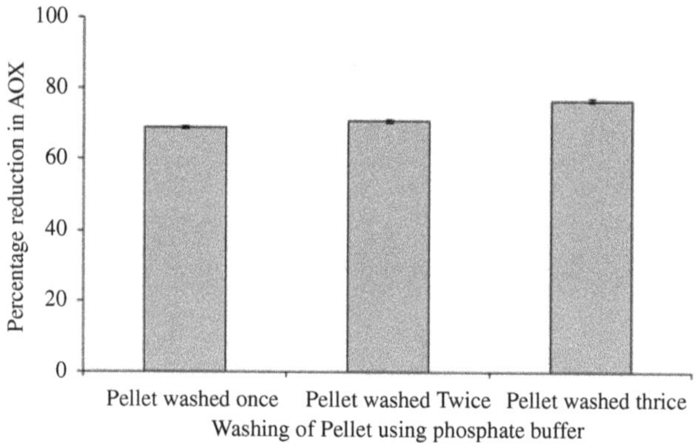

Fig. 5. Effect of surface washing of bacterial cell on the AOX reduction of pulp and paper mill effluent (n = 5)

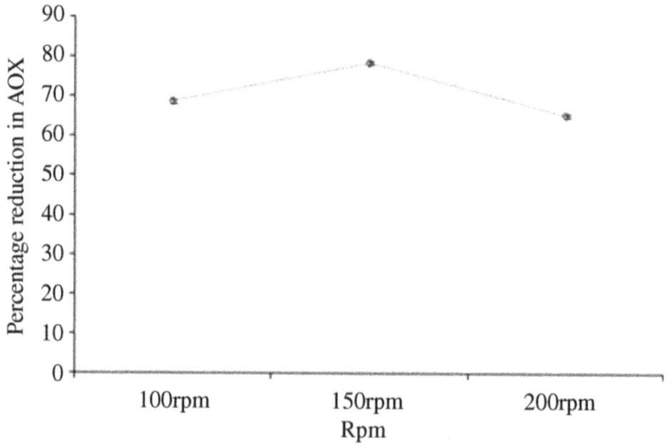

Fig. 6. Effect of rate of agitation on the AOX reduction of pulp and paper mill effluent (n = 5)

ratio. The reduction in AOX was observed up to 73% after 24 h of incubation period (Fig. 3).

2.5. Analysis of the color adsorption data

It is evident from the graph that the process of decolorization is following Freundlich's adsorption isotherm well with an n value equal to 1.4 and K value 2.54 (Fig. 4). The logarithmic version of Freundlich's equation is straight line in the form of $y = mx + b$, where, b is the y-intercept for $x = 0$. In our case, K equals x/m intercept, when the log of the concentration C (color remaining) equals one unit, the log of which is zero. The value of m, the slope of the line, is equal to 1/n. our data shows a considerable linearity till an effluent:biomass ratio 1:1, however, upon increasing the biomass proportion after this, a linear adsorption pattern is not observed. This indicates the presence of some other mechanism for color removal apart from true adsorption, which was also indicated by the adsorption desorption assay. Based on the results obtained it can be concluded that although adsorption plays an important role in the decolorization process, assimilation of some chromophoric material is also taking place.

Effect of surface washing of bacteria: effect of surface washing on AOX reduction was studied and observed that the pellet washed thrice with phosphate buffer show effective reduction in AOX, i.e. up to 76% (standard deviation of = 1.3 with standard error of = 0.5) in comparison to the pellet washed once and twice showed reduction up to 68% and 70% respectively (Fig. 5).

Effect of Agitation: the effect of rate of agitation on the AOX reduction was also investigated. A range of 100–200 rpm was selected and AOX studies conducted after a period of 24 h. It was observed that percent AOX reduction was highest, i.e. 78.4% (standard deviation = 1.1 with standard error of = 0.5) at about 150 rpm (Fig. 6).

2.6. Repeatability and reproducibility

After optimizing the various parameters the experiment was performed 5 times in order to check the reproducibility and repeatability among the results. Results showed three time washed pellet inoculated in the ratio of 1:1, incubated at 150 rpm for 24 h. remove 77–78% (standard deviation = 1.1 with standard error of = 0.5) of the AOX affectively from the pulp mill effluent.

2.7. Identification of bacterial isolates

Identification of selected bacterial strain: the isolate number 4 were identified by 16S RNA studies at DSMZ, Germany, as *Pseudomonas aeruginosa strain 1*.

3. Discussion

Color of pulp mill effluent is dark-brown and obnoxious. It is not only unaesthetic but also increases the amount of AOX in the effluent. Color of pulp mill effluent is primarily due to lignin, which is an important part of the woody raw materials utilized for paper manufacture. Physico-chemical treatments of pulp and paper wastewater were mentioned in literature but none of them is industrially applicable due to cost limitation and infeasibility of the processes. Biological treatment includes aerobic and anaerobic treatment or the combinations of both. Under aerobic conditions, reduction of AOX has been studied by Fulthorpe and Allen (1995) who compared three bacterial strains, *Pseudomonas*, *Ancylobacter* and *Methylobacter*, for their dehalogenating capabilities and reduction of AOX of bleached Kraft mill effluent from three different sources. AOX reduction was poor to negligible in the case of *Ancylobacter* and *Pseudomonas*, whereas *Methylobacterium sp.* exhibited some reduction in overall AOX levels of the effluents. Jaspers *et al.* (1994) reported reduction in AOX levels of El stage of the Kraft process by adsorption on pellets of *Phanerochaetechrysosporium,* under controlled incubation conditions. In anaerobic treatment, color (70%), lignin (25%), COD (42%), AOX (15%) and phenol (39%) were reduced in 15 days retention time. After this, such anaerobically treated effluent was treated in a bioreactor in presence of a fungal and bacterial strain separately. The reduction efficiency with fungal strain was color (95%), AOX (67%), lignin (86%), COD (88%), Phenol (63%), while with bacterial strain the reduction efficiency was: color (76%), lignin (69%), COD (75%), AOX (82%), phenol (93%). Singh and Thakur (2004) studied the sequential aerobic and anaerobic treatment of pulp and paper mill effluent and found the 15% AOX 70% color, 255 lignin and 42% COD reduction. In a study by Singh and Shekhar (2006) sequential anaerobic and aerobic treatment of pulp and paper mill effluent was done in a two-step bioreactor. Chen *et al.* (2003) designed a process by combining the co-agulation, anaerobic acidification and aeration package reactor for the treatment of bleaching effluents and observed the COD, BOD, AOX reduction efficiencies of 88.1%, 81%, 98.4% respectively. By using activated granular carbon as a medium for biofilm growth and 24 h retention time in a sequencing batch biofilm reactor, AOX removal in the range of 10–100% can be achieved with selected retention time (Mohamad *et al.* 2008). For further study fungi and algae was used for the treatment of the effluent. The effluent toxicity also reduced considerably. MyCoR process, where fungus has been used in the form of fixed film reactor that converts 70% of organic chlorides into inorganic chlorides in 48 h with 50% reduction in BOD and COD (Savant *et al.* 2006). The 67% of AOX reduction and 97% of color reduction has been achieved by using fungus *Penicilliumcamemberti* in a hemp based pulp and paper mill bleachery effluent (Taseli 2008). Mixed algal culture was used by Esra *et al.* (2002) and Tarlan *et al.* (2002) for AOX and color removal. The efficiency of algal culture for AOX and color removal was up to 70%, 80% respectively. Saravanan and Sreekrishnan (2005) studied a combination of chemical and biological process for treatment of bleached kraft pulp mill effluent. The efficiency of this combined process for AOX removal was 81%.

The present work involved the study in reduction of adsorbable organic halides of pulp and paper mill sector India. Isolation of naturally occurring bacterial stains and their utilization for the treatment of wastewaters emanating from such industrial units was the main focus of this work. Bacteria were isolated from different sources in the vicinity of a paper mill. The sources included activated sludge and old soil samples, which had been accumulating over a number of years. Over 9 pure bacterial isolates were isolated during this exercise, using selective enrichments and subculturing. After extensive screening of bacterial isolates, one of the bacteria, *Pseudomonas aeruginosa str.1* was able to reduce the approximately 65% within a period of 24 h. Various parameters were optimized like effluent: biomass ratio which results in increase in reduction of AOX up to 73% when pellet was inoculated in the ratio of 1:1 (effluent:biomass). On the other hand the surface cell washing was studied which showed that washing the pellet thoroughly will be helpful in removing the extra media attached on the surface of bacteria. Results showed pellet washed thrice will helpful in reducing the AOX further to 76%. At last different speed for shaking the flasks were studied and the flask incubated at 150 rpm shows the maximum reduction of 78.4%. Laterally the repeatability experiment was performed in which the three time washed pellet was inoculated in the ratio of 1:1 incubated at 150 rpm for 24 hrs was results in 79% of AOX removal from pulp mill effluent.

Conclusion

The results show that autochthonous bacteria isolated from the site of pulp and paper industry have the ability to remove the AOX efficiently. The AOX reduction was significantly affected by various parameters including the amount of inoculum size, rate of agitation and surface cell washing. The 78% (standard deviation = 1.1 with standard error of = 0.5) of AOX was removed when the pellet was washed thrice and inoculated in 1:1 ratio and incubated at 150 rpm for 24 h.

Acknowledgements

The authors acknowledge the financial help provided by the Department of Biotechnology, Government of India. The authors also acknowledge Mr D. P. Chandrana and Mr R. L. Lakhotia of Century pulp and paper mill, LalKuan, Uttaranchal for extending their cooperation for providing wastewater samples, whenever required and the generous hospitality offered to us upon each visit.

References

Amr, A. E.-.; Hassan, E. A.-E.; Elsayed, E.; Hafez. 2008. Molecular characterization of two native egyptian ligninolytic bacterial strains, *J. Appl. Sci. Res.* 4(10): 1291–1296.

Basak, K.; Taseli, C. F.; Gökcay, H. T. 2004. Upflow column reactor design for dechlorination of chlorinated pulping wastes by Penicillium camemberti, *J. Environ. Manage.* 72: 175–179. http://dx.doi.org/10.1016/j.jenvman.2004.04.008

Catalkaya, E. C.; Kargi, F. 2007. Color, TOC and AOX removals from pulp mill effluent by advanced oxidation processes: a comparative study, *J. Hazard. Mater. B.* 139: 244–253. http://dx.doi.org/10.1016/j.jhazmat.2006.06.023

Catalkaya, E. C.; Kargi, F. 2008. Advanced oxidation treatment of pulp mill effluent for TOC and toxicity removals, *J. Environ. Manage.* 87: 396–404. http://dx.doi.org/10.1016/j.jenvman.2007.01.016

Central pollution control board (CPCB) 2007. *Development of standards of AOX for small scale Pulp and Paper Mill*, 1–41.

Chen, Y.; Zhan, H.; Chen, Z.; Fu, S. 2003. Study on the treatment of the sulfide pulp CEH bleaching effluents with the coagulation—anaerobic acidification—aeration package reactor, *Wat. Res.* 37(9): 2106–2112. http://dx.doi.org/10.1016/S0043-1354(02)00599-7

Clifford, R. L.; Noemí, M. S. 2010. Biological and abiotic color reduction of paper mill effluents during anaerobic composting, *J. Environ. Eng.* 136(7):701–708. http://dx.doi.org/10.1061/(ASCE)EE.1943-7870.0000204

Daneshvar, N.; Salari, D.; Khataee, A. R. 2004. Photocatalytic degradation of azo dye acid red 14 in water on ZnO as an alternative catalyst to TiO2, *J. Photochem. Photobiol. A. Chem.* 162: 317–322. http://dx.doi.org/10.1016/S1010-6030(03)00378-2

Deshmukh, N. S.; Lapsiya, K. L.; Savant, D. V.; Chiplonkar, S. A.; Yeole, T. Y.; Dhakephalkar, P. K.; Ranade, D. R. 2009. Upflow anaerobic filter for the degradation of adsorbable organic halides (AOX) from bleach composite wastewater of pulp and paper industry, *Chemosphere* 75: 1179–1185. http://dx.doi.org/10.1016/j.chemosphere.2009.02.042

Dojčinović, B. P.; Roglić, G. M.; Obradović, B. M.; Kuraica, M. M.; Tosti, T. B.; Marković, M. D.; Manojlović, D. D. 2012. Decolorization of Reactive Black 5 using a dielectric barrier discharge in the presence of inorganic salts, *J. Serb. Chem. Soc.* 77(4): 535–548. http://dx.doi.org/10.2298/JSC110629179D

Esra, T.; Filiz, B. D.; Ulku, Y. 2002. Effectiveness of algae in the treatment of a wood-based pulp and paper industry wastewater, *Bioresour. Technol.* 84: 1–5. http://dx.doi.org/10.1016/S0960-8524(02)00029-9

Freire, C. S. R.; Silvestre, A. J. D.; Neto, C. P. 2003. Carbohydrate derived chlorinated compounds in ECF bleaching of hardwood pulps: formation, degradation and contribution to AOX in a bleached Kraft pulp mill, *Environ. Sci. Technol.* 37: 811–814. http://dx.doi.org/10.1021/es0200847

Fulthorpe, R. R.; Allen, D. G. 1995. A comparison of organochlorine removal from bleached Kraft pulp and paper-mill effluents by dehalogenating *Pseudomonas, Ancylobacter and Methylobacterium* strains, *Appl. Microbiol. Biotechnol.* 42(5): 782–789.

Futagami, T.; Goto, M.; Furukawa, K. 2008. Biochemical and genetic bases of dehalorespiration, *Chem. Rec.* 8: 1–12. http://dx.doi.org/10.1002/tcr.20134

Hassan, M. M.; Hawkyard, C. J. 2002. Decolourisation of aqueous dyes by sequential oxidation treatment with ozone and Fenton's reagent, *J. Chem. Tech. Biot.* 77(7): 834–841. http://dx.doi.org/10.1002/jctb.641

Helmy, S. M.; El Rafie, S.; Ghaly, M. Y. 2003. Bioremediation post-photo-oxidation and coagulation for black liquor effluent treatment, *Desalination* 158(1–3): 331–339. http://dx.doi.org/10.1016/S0011-9164(03)00472-7

Jaspers, C. J.; Jimenez, G.; Penninckx, M. J. 1994. Evidence for a role of manganese peroxidase in the decolorization of Kraft pulp bleach plant effluent by *Phanerochaete chrysosporium*: effects of initial culture conditions on enzyme production, *J. Biotechnol.* 37: 229–234. http://dx.doi.org/10.1016/0168-1656(94)90130-9

Johnson, M.; Zaretskaya, I.; Raytselis, Y.; Merezhuk, Y.; McGinnis, S.; Madden, T. L. 2008. NCBI BLAST: a better web interface, *Nucleic Acids Res.* 36: 5–9. http://dx.doi.org/10.1093/nar/gkn201

Kansal, S. K.; Singh, M.; Sud, D. 2008. Effluent quality at kraft/soda agro-based paper mills and its treatment using a heterogeneous photocatalytic system, *Desalination* 228(1–3): 183–190. http://dx.doi.org/10.1016/j.desal.2007.10.007

Karrasch, B.; Parra, O.; Cid, H.; Mehrens, M.; Pacheco, P.; Urrutia, R.; Valdovinos, C.; Zaror, C. 2006. Effects of pulp and paper mill effuents on the microplankton and microbial self-purification capabilities of the Biobio River, Chile, *Sci. Total Environ.* 359: 194–208. http://dx.doi.org/10.1016/j.scitotenv.2005.03.029

Kim, J.-H.; Oh, K.-K.; Lee, S.-T.; Kim, S.-W.; Hong, S.-I. 2002. Biodegradation of phenol and chlorophenols with defined mixed culture in shake-flasks and a packed bed reactor, *Process. Biochem.* 37(12): 1367–1373. http://dx.doi.org/10.1016/S0032-9592(02)00007-9

Li, X. M.; Li, Y. T.; Li, F. B. 2009. Interactively interfacial reaction of iron-reducing bacterium and goethite for reductive dechlorination of chlorinated organic compounds, *Chinese. Sci. Bull.* 54: 2800–2804. http://dx.doi.org/10.1007/s11434-009-0475-x

Maphosa, F.; deVos, W. M.; Smidt, H. 2010. Exploiting the ecogenomics tool box for environmental diagnostics of organohalide – respiring bacteria, *Trends. Biotechnol.* 28: 308–316. http://dx.doi.org/10.1016/j.tibtech.2010.03.005

Mohamad, A. B.; Rahman, R. A.; Kadhum, A. A. H.; Abdullah, S. R. S.; Sudin, Z. W.; Shaari, S. 2008. Removal of Adsorbable Organic Halides (AOX) from recycled pulp and paper (PandP) mill effluent using granular activated carbon–sequencing batch biofilm reactor (GAC-SBBR), *Modern. Applied. Science.* 2(5): 37–45.

Patel, U. D.; Suresh, S. 2008. Electrochemical treatment of pentachlorophenol in water and pulp bleaching effluent, *Sep. Purif. Technol.* 61: 115–122. http://dx.doi.org/10.1016/j.seppur.2007.10.004

Pratibha, S. 2007. Sequential anaerobic and aerobic treatment of pulp and paper mill effluent in pilot scale bioreactor, *J. Environ. Biol.* 28(1): 77–82.

Ruggaber, T.; Talley, J. 2006. Enhancing bioremediation with enzymatic processes: a review, *Practice Periodical of Hazardous, Toxic, and Radioactive Waste Management* 10(2): 73–85. http://dx.doi.org/10.1061/(ASCE)1090-025X(2006)10:2(73)

Saima, B.; Izharul, H. F. 2012. Pulp and paper industry – manufacturing process, wastewater generation and treatment, *Environmental protection strategies for sustainable development Startegies for Sustainability*, 397–436.

Santosh, K. S. 2007. Effect of irrigation with paper mill effluent on the nutrient status of soil, *International Journal of Soil Science* 2(1): 74–77. http://dx.doi.org/10.3923/ijss.2007.74.77

Saravanan, V.; Sreekrishnan, T. R. 2005. Bio-physico-chemical treatment for removal of colour from pulp and paper mill effluent, *J. Sci. Ind. Res.* 64: 61–64.

Savant, D. V.; Abdul-Rahman, R.; Ranade, D. R. 2006. Anaerobic degradation of adsorbable organic halides (AOX) from pulp and paper industry wastewater, *Bioresource. Technol.* 97: 1092–1104. http://dx.doi.org/10.1016/j.biortech.2004.12.013

Singh, P.; Shekhar, I. T. 2006. Colour removal of anaerobically treated pulp and paper mill effluent by microorganisms in two steps bioreactor, *Bioresource. Technol.* 97: 218–223. http://dx.doi.org/10.1016/j.biortech.2005.02.022

Singh, P.; Thakur, I. S. 2004. Removal of color and detoxification of pulp and paper mill effluent by microorganisms in two step bioreactor, *J. Sci. Ind. Res.* 63: 944–948.

Tarlan, E.; Dielk, F.; Yetis, U. 2002. Effectiveness of algae in the treatment of a wood based pulp and paper industry wastewater, *Bioresource. Technol.* 84: 1–5. http://dx.doi.org/10.1016/S0960-8524(02)00029-9

Taseli, B. K. 2008. Fungal treatment of hemp-based pulp and paper mill wastes, *Afr. J. Biotechnol.* 7(3): 286–289.

The Gazette of India: Extraordinary, Part II, Section 3, Sub-Section (i) 1993. New Delhi, India: Ministry of Environment and Forests and central Pollution Control Board.

Ugurlu, M.; Karaoglu, M. H. 2009. Removal of AOX, total nitrogen and chlorinated lignin from bleached Kraft mill effluents by UV oxidation in the presence of hydrogen peroxide utilizing TiO(2) as photocatalyst, *Environ. Sci. Pollut. Res. Int.* 16(3): 265–73. http://dx.doi.org/10.1007/s11356-008-0044-x

Vasconcelos, De L. C.; Santoyo, D.; Tepole, F.; Juarez Ramírez, C.; Ruiz-Ordaz, N.; Galíndez Mayer, C. J. J. 2006. Cometabolic degradation of chlorophenols by a strain of Burkholderia in fed-batch culture, *Enzyme. Microb. Technol.* 40: 57–60. http://dx.doi.org/10.1016/j.enzmictec.2005.10.038

Anil KUMAR is a Staff Scientist in National Institute of Immunology, New Delhi, India. He received his PhD in Botany from University of Delhi, Delhi, India, in 2008. His research interests include intellectual property rights, white biotechnology, biosensor, bioremediation and bioinformatics.

Purnima DHALL is a PhD student in Institute of Genomics and Integrative Biology, Delhi, India. She is working on development of BOD biosensor. Her research interests are BOD biosensor and bioremediation.

Rita KUMAR is a head and senior scientist in the Division of Environmental Biotechnology, Institute of Genomics and Integrative Biology. She received her PhD in medical microbiology from University of Delhi, in 1983. Her research interest includes Intellectual property rights, White biotechnology, biosensor, bioremediation and waste water treatment.

Note "Experiments mentioned in this paper were conducted at Institute of Genomics and Integrative Biology, New Delhi".

19

EMISSION OF NITROUS OXIDE AND METHANE FROM ALLUVIAL SOIL THROUGH INCUBATION

Sanak Ray, Ayusman Mohanty, Torati Sri Ramulu, Gautam Roy Chaudhury

Department of Environment and Sustainability, Institute of Minerals and Materials Technology (CSIR), Bhubaneswar 751013 Odisha, India

Abstract. Methane and nitrous oxide emission from alluvial soil under incubation using varying doses of urea and water content were studied for Tangibanta and Sorada. The N_2O emission was observed to increase with time and tapers off after attaining the peak. The theoretical emission was evaluated using empirical equations and matches well with the experimental values. The CH_4 and N_2O emission both increased with the increase in Water Filled Pore Space (WFPS) and urea concentration. The nitrification and denitrification reaction rates were determined with and without acetylene. Michaelis-Menten equation was used to evaluate the K_m and V_{max} values. The rate of emission of N_2O increased with increase of WFPS and urea concentration whereas the emission factor showed a reverse trend. The CH_4 emission rate was observed to be comparatively low with respect to N_2O emission rate around a factor of 10.

Keywords: incubation, soil contamination, emission rate, Michaelis-Menten, Principal component analysis (PCA), WFPS, Air pollution.

Reference to this paper should be made as follows: Ray, S.; Mohanty, A.; Sri Ramulu, T.; Roy Chaudhury, G. 2013. Emission of nitrous oxide and methane from alluvial soil through incubation, *Journal of Environmental Engineering and Landscape Management* 21(3): 224–232. http://dx.doi.org/10.3846/16486897.2012.745414

Introduction

The concentration of greenhouse gases (GHG) like carbon dioxide (CO_2), N_2O and CH_4 are increasing progressively. Amongst all greenhouse gases CH_4 and N_2O are the second and fourth greatest contributors towards enhancing global warming effect (IPCC 2007). The global warming potential (GWP) for CH_4 (based on 100 year time horizon) is reported to be 21 years, while that of N_2O is 310 with reference to CO_2 (Kampschreur *et al.* 2009). The gases like CH_4 and N_2O contribute around 15 and 5% to the total annual global rise respectively. Agricultural and associated sectors produce about 50 and 70% to the total anthropogenic emission of CH_4 and N_2O respectively. Besides global warming, nitrous oxide is also responsible for the destruction of stratospheric ozone (Yang *et al.* 2011). Globally, agricultural N_2O emissions have increased by nearly 17% from 1990 to 2005, and are projected to increase ~35–60% by 2030 due to increased nitrogen (N) fertilizer use (Gogoi, Baruah 2012). The biogenic emission of N_2O is due to microbial decomposition of nitrogenous compounds available in soil, municipal effluents, sediments and water bodies. The formation is either through nitrification or denitrification or nitrifier denitrification or combination of all these three processes (Horak, Siska 2006). These effects are suggestive of the fact that study of N_2O emission from soil is vital for the research on GHG management. The nitrification proceeds when NH_4^+ (Ammonium ion) or NH_3 (Ammonia) is oxidized to NO_3^- (Nitrate ion) via NO_2^- (Nitrite ion). The oxidation process passes through various intermediates like NH_2OH (Hydroxyl amine) and NO_2^-. The process of NH_3 oxidation can be arrested by using an inhibitor like C_2H_2 (Kineey *et al.* 2005). Acetylene in the concentration range of 0.1 to 10 Pa forms a very reactive unsaturated epoxide, which eventually inhibits the formation of ammonia monooxygenase through covalent bonding. Denitrification is a process where NO_3^- is reduced to N_2 (Atmospheric nitrogen) through different intermediates like NO_2^-, NO (Nitric oxide) and N_2O. The reaction is mediated by different denitrifiers (Rivett *et al.* 2008). Nitrifier denitrification is another way of nitrification where NH_3 or NH_4^+ is oxidized directly to NO_2^- followed by reduction. Methane is produced under reducing atmosphere when the redox potential reaches −180 mV. The condition obligates anaerobes to act through either CO_2 reduction or trans-methylation process (Dutta 2007; Baruah *et al.* 2010; Wang *et al.* 2001). Anaerobiosis in the soil upon submergence results in

Corresponding author: Gautam Roy Chaudhury
E-mail: gr_chaudhury@yahoo.com

the reduction of inorganic compounds and favors mineralization of organic matter through methanogenic fermentation.

Emission of CH_4 and N_2O from soil is also affected by various physico-chemical factors (Dutta 2007), i.e. soil oxidants, which act as electron accepters for organic matters, determine the emission rate (Mosier *et al.* 2004). Fields flooded over long periods or with intermittent flooding, especially with standing crops, are considered as an important source of these two important greenhouse gases due to simultaneous existence of aerobic and anaerobic environments (Synder *et al.* 2009). CH_4 emission is found to be higher in flooded organic soil, but they also emit substantial amounts of N_2O under intermittent flooding as NO_3 undergoes denitrification during temporary exposure to air (Das *et al.* 2011; Singh *et al.* 2003). Thus the quantity and quality emission of CH_4 and N_2O from different water-soil regimes are complicated but interrelated and needs in-depth analyses for planning of sustainable mitigation strategies.

GHG emission from soil through denitrification and nitrification are highly complicated. Apart from emission of N_2O, the soil also emits CH_4 by methanotrophs. Yet, ample understanding of the above processes is desired as because substantial literatures are available, but the same is somewhat scarce regarding GHG emission from agricultural field; therefore an in-depth analysis is needed to be carried out to correlate the same with various parameters. Keeping in view the importance of this subject our Institute is actively engaged in regular monitoring of GHG from various soil regimes and a part has been published (Ramulu *et al.* 2008, 2009; Sahoo *et al.* 2010). Considering the work carried out before, the objective of the present study is to measure the N_2O emission and to evaluate the rate of N_2O emission through nitrification and denitrification process with the variation of two parameters, i.e. urea and WFPS, which plays as an important aspect for N_2O emission along with microbes and organic carbon content. Some attempts have also been made to correlate the emission of CH_4 with N_2O. Michaelis – Menten type equation is also carried out to establish the relation between total N_2O emissions through nitrification and denitrification with respect to urea addition as a substrate. The interdependence of urea and WFPS for GHG emission was also studied under incubation.

1. Methods

1.1. Site description

The sampling sites were two different agricultural fields one at Tangibanta (latitude N 20° 20′ and longitude E 85° 49′) and other at Sorada (latitude N 19° 50′ and longitude E 84° 19′) situated at a distance of 20 and 250 Km from Bhubaneswar, India respectively. Both of the places have alluvial soil having area of about 700–800 m^2. The field at Tangibanta is used for pulse as well as paddy cultivation on rotation basis. The Sorada field is used exclusively for pulse cultivation. In IPCC terminology, both of the places can be classified as rain-fed, intermittently flooded soil-water regime. The farmer at Tangibanta used farmyard manure before land preparation for Kharif rice (August–December) and very little nitrogen (N) – fertilizer, usually urea (50 kg per acre) used a few weeks after transplantation or before inflorescence to boost growth. No insecticide or pesticides were used. While kharif rice is cultivated during summer monsoons, pulses are sown after a short fallow period during Rabi (January–March) season. In case of Sorada, neither chemical fertilizer nor any pesticides were used. Farmyard manure was added only before sowing.

1.2. Sampling and experimental setup

Soil was sampled during the fallow period (just before plantation) from both Tangibanta (after December) and Sorada (before August) sites. Two different sampling sites were chosen basically as the agricultural practices were completely different. All soil samples were collected in random positions at each site from 5–10 cm depth with an auger (Andert *et al.* 2011) because large portion of active root zone is present where both aerobic and anaerobic microorganisms are present. After collection, the soil samples were homogenized by the classic cone and quartering technique (Gerlach *et al.* 2002) to give one composite sample for each soil and were put inside labeled airtight polythene bags. It was air-dried, crushed and sieved with a 2 mm sieve (Fangueiro *et al.* 2008). Soil physicochemical parameters were studied fallowing the SSSA/ASA guidelines for soil study (Sparks 1996). The soil texture of both the spot was sandy clay loam (23.5–27% clay, 20.6–22.5% silt, 50.4–54.6% sand). It also contains 0.4–0.6% carbon, 0.08–0.11% total N as micronutrients. The moisture content of Sorada soil was less compared to Tangibanta. pH of the soil was slightly acidic, i.e. 5.5–6.9 for both the sites. For incubation experiments, two parameters were varied, such as external addition of N – fertilizer and variation of WFPS. 100 grams of the dried and powdered soil were taken in incubation bottles (250 ml, Borosil) with or without urea. For the set of experiment, 10, 50, 100 and 150 mg of urea per kg of soil were taken in incubation containers along with control (without urea). Urea was used as the nitrogen source due to its easy availability and wide use. The incubation bottles were closed with airtight rubber stoppers made up of polypropylene. These were kept in an incubator (REMI- CI 6S) at 37 °C during the study period. For the other variable like WFPS 10 g of weighted soil was dried at 105 °C for 48 hours to get the dry weight. Initial 100% WFPS was obtained by gravimetric method (Kinney *et al.* 2005). WFPS was varied from 30–75% gravimetrically. During the incubation experiments different WFPS were maintained by adding distilled water. The loss of evaporation was determined by taking the weight of the incubation bottles at periodic intervals (Hayakawa *et al.* 2009). Incubation experiment was also carried out using C_2H_2 (Kinney *et al.* 2005) in order to determine the nitrification and denitrification rate simultaneously. The nitrification and denitrification rates were determined as follows:

$$\text{Soil} = \text{net } N_2O \text{ produced by nitrification as well as denitrification;} \quad (1)$$

$$C_2H_2 \text{ amended soil} = \text{net } N_2O \text{ produced by denitrification only;} \quad (2)$$

$$\text{Eq. (1)} - \text{Eq. (2)} = N_2O \text{ produced by nitrification.} \quad (3)$$

1.3. Gas sampling and analysis

Air samples were drawn through disposable syringe every day at about 10 a.m. and estimated with gas chromatograph. A PerkinElmer Auto system GC with FID was used for CH_4 estimation while a Shimadzu AA30 GC with ECD detector was used to analyse N_2O. Both the GCs were dedicated and equipped with auto gas samplers, semi-micro columns and appropriate software to process the acquired data. Both NIST traceable primary and laboratory prepared secondary standards were used for quality assurance.

1.4. Soil sample analysis

Soil parameters like moisture content, total organic carbon (TOC), NH_3–N (Ammonia nitrogen), $NO_3^- - N$ (Nitrate nitrogen) and $NO_2^- - N$ (Nitrite nitrogen) were analyzed following standard methods (Andert *et al.* 2011; Reddy *et al.* 1989). All urea variation incubation experiments were carried out at 50% WFPS.

1.5. Statistical analysis

Two way analysis of variance (ANOVA) and Fisher t-test was used to determine whether the incubation parameters have any significant variance over the emission of the GHG. Using the null hypothesis technique the significance of the incubation parameters was determined. ANOVA analyses were carried out using MS-Office Excel 2007.

PCA was used to identify the correlation between GHG emission vis-à-vis the incubation parameters like days, ammonia, nitrate, TOC, moisture and urea. PCA is widely used to reduce variables and to extract a small number of latent factors in order to analyse the relationships among observed variables. To make the results more easily interpretable, the PCA with VARIMAX normalized rotation was also applied, that can maximize the factor loadings through variables for each factor. In this study, all principal factors extracted from the variables with Eigen values up to 1 were considered since the results for values ≤ 1 were not significant (Kim *et al.* 2009). When PCA with VARIMAX normalized rotation was performed, each PCA score contained information about all the variables combined into a single number, while the loadings indicated the relative contribution each variable makes to that score. PCA was evaluated using SPSS-13.

All experiments were carried out in quartet and average was used for interpreting the results. The variation was within $\pm 5\%$.

2. Results and discussion

2.1. Emission of N_2O

Figure 1 shows a typical emission curve during incubation studies under various WFPS. The graph can be divided into two parts, i.e. initial increase followed by decrease. The rate equation for N_2O emission during initial and final stages can be written as:

$$Y = Be^{-k_1 t} + Ce^{k_2 t}, \quad (4)$$

where: $Y = N_2O$ emission; T = Time in days. k_1, k_2 = Rate constant for initial and final stages respectively and B, C = Fitting parameters.

The values for B, C, k_1 and k_2 were calculated (Cardenas *et al.* 2003) using Excel-solver technique. The values are shown in Table 1.

2.2. Effect of water filled pore space

It was observed that N_2O emission efficiency increased with the increase of WFPS as shown (Fig. 1). The soil moisture content influences the oxygen diffusion, thus determining the N_2O emission efficiency. Increase

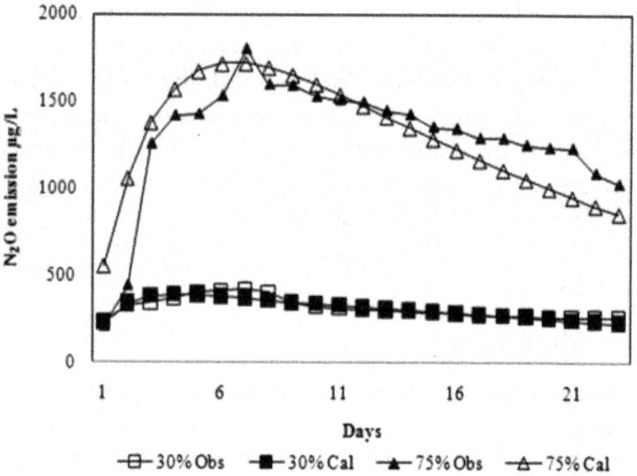

Fig. 1. Emission of N_2O with the variation of WFPS (Sorada)

Table 1. Calculation of k_1, k_2, B and C values during incubation studies using Sorada soil samples by varying WFPS

WFPS	k_1	k_2	B	C
30%	0.03	0.92	439.36	−210.28
75%	0.05	0.36	2686.61	−2134.53

of WFPS plays a vital role in interposing a barrier for O_2 diffusion thereby controlling the rate of nitrification. It was reported (Cleemput 1998) that denitrification was favored at WFPS > 60%, but it may be favored at lower concentration too if the soil has poor porosity and so retaining the moisture for a long time. Table 2 shows the emission of N_2O through nitrification and denitrification routes. This was calculated based on the difference of N_2O emission between incubation experiments carried out in presence and absence of C_2H_2. It was observed that majority of N_2O emission was due to nitrification. The N_2O emission through nitrification and denitrification routes was fitted onto a 1st order plot as shown (Fig. 2) for incubation experiments using Sorada soil. Similar results were obtained for Tangibanta soil and therefore not shown separately. The rates of emission of GHG for Sorada soil were lower than Tangibanta as in the former no chemical fertilizers were used. This suggests that chemical fertilizers play a vital role in determining the emission rates of GHG. It can be concluded from correlation of determinant values that the N_2O emission followed 1st order plot. The reaction rates for both of the soils for different WFPS are shown in Table 3 using the 1st order plot. ANOVA was applied to find out whether the WFPS played a vital role in determining the N_2O emission efficiency and for this purpose null hypothesis technique was used. Null hypothesis assumes that there is no variation of GHG emission with the change of WFPS. So if the F value (calculated) is less than or equal to F critical then it is assumed that there is no variance (Gupta 2005). On the contrary if the F calculated value is more than F critical then the null hypothesis does not hold good, i.e. there is significant variation of GHG emission with the change of WFPS. The F critical as well as F calculated along with p values is shown in Table 4. Table 4 unequivocally proves that WFPS played a significant role for GHG emission for both of the soils.

Table 2. Nitrous oxide (N_2O) emission through nitrification and denitrification routes. (Conditions: Temperature −37 °C)

Parameter	N_2O-emission in both nitrification and denitrification (μgN /Kg soil)		N_2O emission in nitrification (μgN /Kg soil)		N_2O emission through nitrification (%)	
WFPS (%)	Tangi	Sorada	Tangi	Sorada	Tangi	Sorada
30	6.6	4.7	2.5	4.7	38.4	41.9
40	14.9	10.8	7.8	10.8	52.6	56.5
60	38.6	29.4	27.1	29.4	70.2	68.1
75	58.2	52	39.7	52	68.2	69.4
Urea (mg/Kg soil)						
0	10.7	8.6	5.64	4.3	52.2	50.5
10	11.3	12	7.19	7.2	63.4	59.7
50	14.5	16	11.3	11.8	78.2	73.7
100	13.5	20.7	9.7	13.8	72.1	66.9
150	15.40	22.5	12.5	17.5	81.1	77.6

2.3. Effect of urea concentration

Urea concentration was varied from 0–15 mg/100 mg of soil in order to evaluate the N_2O emission through nitrification and denitrification routes. It was reported (Carter 2007) that urea hydrolyzed rapidly to NH_4^+ by enzymatic action vides reaction. The NH_4^+ thus produced would participate further in nitrification as well as denitrification reactions. NH_4^+ would be oxidized during nitrification to NO_3^-, which would be utilized by the denitrifiers. The extent of nitrification and denitrification processes taking place with various initial doses of urea is shown (Table 2). It was observed that the total emission of N_2O increased with the increase in urea addition. Two way ANOVA analyses were carried out and it was observed that urea played a significant role in determining the N_2O emission as shown (Table 4) using null hypothesis technique. ANOVA studies show the significant role played by urea concentration during GHG emission. Apart from two external variables one internal variable as form of soil was also considered. As mentioned earlier incubation studies were carried out using two different soils. In order to evaluate whether the soil plays an important role in emission of GHG Fisher t-test was also carried out using null hypothesis technique. The results are also shown (Table 4). The N_2O emission depends on external addition of urea up to 10 mg/Kg of soil and beyond that both the soil showed similar emission rate as at higher nitrogen concentration the initial nitrogen content in the soil was very small compared to external addition of urea. On the contrary CH_4 emission showed a significant variation between the soils suggesting thereby the emission follows a different route.

Both the variables like WFPS and urea addition followed 1st order rate kinetics. Since the emission of CH_4 and N_2O depends on WFPS as well as urea concentration therefore, the reaction can be considered as pseudo first order. So the rate equation for CH_4 or N_2O emission can be written as:

$$Rate\ of\ GHG\ emi = -dc/dt = k(WFPS)^{n1}(Urea)^{n2}, \quad (5)$$

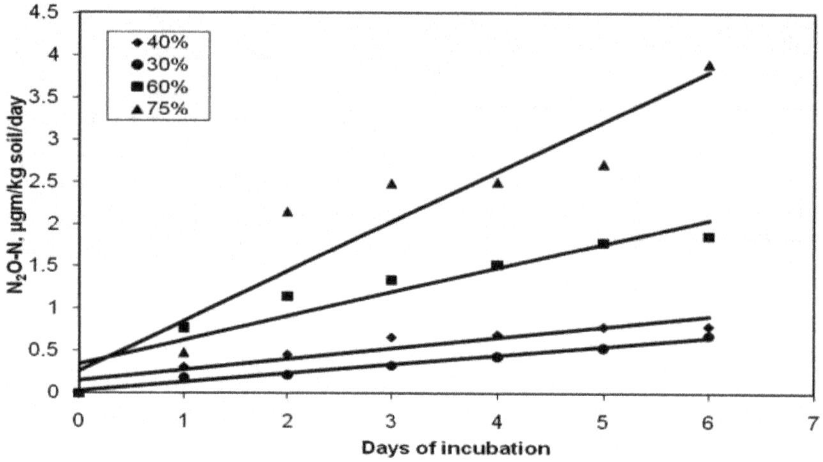

Fig. 2. Nitrification and denitrification rate of Sorada soil (Conditions: Temperature −37 °C, WFPS −30%, 40%, 60% and 75%)

where: c = concentration of GHG; n is order of reaction or dependence factor; k is specific reaction rate (Mishra *et al.* 2009). By converting the equation in logarithm form:

$$Log\,(R) = \log k + n_1 \log(WFPS) + n_2 \log(Urea). \quad (6)$$

To determine the dependence factor, experiments are arranged to fit Eq. (6). For this purpose only one parameter was varied at a time keeping the other parameter constant. The n values for N_2O and CH_4 emissions were obtained by plotting a graph between log (reaction rate) versus the log (parameter). The slope would give the n values. From the coefficient of determination values it can be concluded that the linearity in all the cases is good (shown in Table 3). Table 3 also shows the n values for two different gases using two different parameters. Using the n values the rate equations can be written using Eq. (5) which would help in determining theoretically the emission rates using set of values.

2.4. Evaluation of K_m and V_{max} values

The total N_2O emission via nitrification and denitrification routes with respect to urea addition can be fitted to a Michaelis-Menten type equation (Silvennoinen *et al.* 2008) as shown in Eq. (7).

$$1/r = (K_m/V_{max})X(1/S) + (1/V_{max}), \quad (7)$$

where: r = Reaction rate of N_2O through nitrification and denitrification route (μg/Kg soil/day); V_{max} = Maximum N_2O emission rate, μg/Kg soil / day; S = Concentration of substrate (Urea), mg/Kg of soil; K_m = Substrate concentration (mg/Kg soil) at which the reaction rate is half normal.

A straight line would be obtained if the reciprocal of N_2O emission rate is plotted against the substrate concentration. Extrapolating the linear regression to

Table 3. Emission rates of nitrous oxide (N_2O) and CH_4 with dependence factor during incubation (Conditions: Temperature −37 °C)

Parameter	Reaction Rate, μgN /Kg soil/ day (N_2O)		Dependence factor		Coefficient of determination (R^2)		Reaction Rate, μg/Kg soil/day (CH_4)		Dependence factor		Coefficient of determination (R^2)	
WFPS (%)	Tangi	Sorada	Tangi	Sorada	Tangi	Sorada	Tangi	Sorada	Tangi	Sorada	Tangi	Sorada
30	0.15	0.11					0.05	0.04				
40	0.22	0.12	1.59	1.85	0.99	0.91	0.06	0.04	0.37	0.20	0.92	0.52
60	0.42	0.28					0.07	0.04				
75	0.65	0.59					0.07	0.05				
Urea (mg/Kg soil)												
10	0.6	0.11					0.61	0.04				
50	0.76	0.21	0.15	0.43	0.90	0.91	0.54	0.04	0.07	0.05	0.98	0.31
100	0.86	0.23					0.53	0.04				
150	0.89	0.4					0.51	0.05				

Table 4. Statistical analysis using two way ANOVA and Fisher t-test to show the significance of incubation parameters

	Tangibanta				Sorada			
	WFPS (30–75%)		Urea (0–150 mg/Kg of soil)		WFPS (30–75%)		Urea (0–150 mg/Kg of soil)	
	N_2O	CH_4	N_2O	CH_4	N_2O	CH_4	N_2O	CH_4
F value	19.67	24.62	44.55	17.03	4.46	69.45	17.52	4.29
P value	2.98E − 09	4.8314	7.60E12	2.92E10	1.23E06	5.94E20	1.32E10	7.51E − 03
F critical	2.36	2.08	2.51	2.48	1.71	2.75	2.48	2.73

Fisher t-test of Tangibanta soil with urea variation

	Fisher t-test (N_2O)			Fisher t-test (CH_4)		
Urea variation(mg/Kg of soil)	t-value	P-value	t-critical	t-value	P-value	t-critical
Control	2.8	4.00E − 03	1.68	1.46	7.79E − 02	1.71
10	1.71	4.70E − 02	1.68	3.24	1.97E + 03	1.72
50	0.85	0.2	1.68	3.71	6.47E − 05	1.72
100	0.58	0.28	1.68	3.57	8.93E − 04	1.72
150	1.05	0.15	1.68	4.6	8.46E − 05	1.72

its intercept on the abscissa would give the negative reciprocal of K_m. The results are shown (Fig. 3). The K_m values for Tangibanta and Sorada soil were found to be 4.58 and 19.8 mg/Kg soil respectively while the V_{max} values were 0.83 and 0.32 µg/kg soil/day respectively. Higher V_{max} observed in case of Tangibanta soil may be due to addition of nitrogenous fertilizers during paddy cultivation whereas no fertilizer was applied to Sorada soil.

2.5. Calculation of Emission factor (EF)

The extent of N_2O formation in soil and emission is a direct function of the amount of 'N' fertilizer applied. It can be expressed as the emission factor (EF), which is the amount of N_2O emitted expressed as a fraction

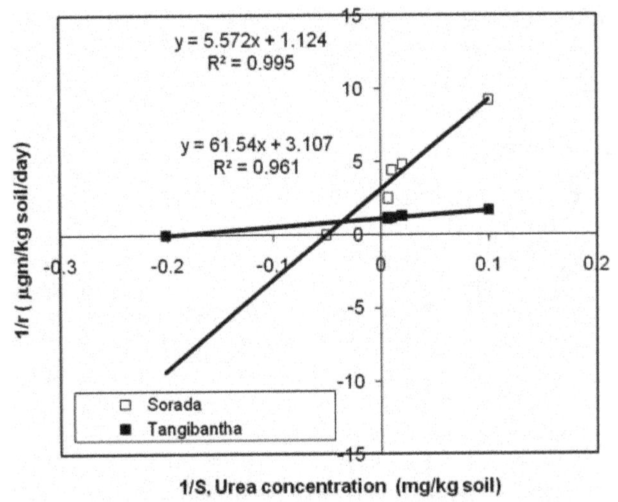

Fig. 3. Michaelis-Menten fitting curve of nitrification and denitrification routes with respect to urea addition (Conditions: Temperature −37 °C, WFPS −50%)

or percentage of the 'N' fertilizer applied to the soil. It was observed that the rate of emission increased with the increase in urea dose whereas the emission factors showed a reverse trend. The emission factor in this case varied in the range 1 to 4% of the 'N' fertilizer applied compared to 0.1–7.3% reported by other workers for similar kind of soils (Dobbie, Smith 2003). Bouwman (1996) experimentally derived a mean EF equaling $1.25 \pm 1\%$ of the 'N' fertilizer applied. This value was accepted as the (IPCC 1997) default value. The higher EF at lower 'N' fertilizer application observed in the present studies may be due to the activity of the soil microbes acting as efficient denitrifiers. Alternative explanation could be the controlled temperature and closed atmosphere prevailing in incubation experiments (Cleemput 1998). The N_2O emission increased initially but tapered off with time after two weeks of incubation. The decrease in N_2O concentration may be due to further reduction to N_2. The emission of N_2O under this condition may be either due to direct denitrification or nitrification followed by partial denitrification (Cleemput 1998). Further reduction of N_2O to N_2 suggests the dominance of denitrification towards the end, if not during the entire period.

2.6. Emission of CH_4

CH_4 emission, like N_2O, depends on the soil properties as well as microbiological activities in soil. CH_4 is generally emitted from the soil by the decomposition of mainly organic matters (Mitra *et al.* 2002). The emission of CH_4 during incubation experiments in all parameters increased progressively unlike N_2O (data not shown). The CH_4 emission rates under different incubation studies are shown (Table 3) along with dependence factor. It was further observed that CH_4

Table 5. Principal component analysis

	Factor		
	1	2	3
Days		0.94	
N_2O			0.90
CH_4		0.77	
Ammonia	0.86		
Nitrate	0.85		
Total Organic Carbon			
Moisture		−0.81	
Urea	0.88		
Eigen values	2.61	2.41	1.06
Cumulative variance	32.62	62.71	75.92

emission rate was lower compared to N_2O. The lower CH_4 emission rate may be due to lower organic carbon in both of the soils. Another possible explanation for low CH_4 may be anaerobic oxidation resulting in loss of CH_4 as shown in Eq. (8).

$$5CH_4 + 8NO_3^- + 8H^+ = 5CO_2 + 4N_2 + 14H_2O. \quad (8)$$

Table 4 shows the two way ANOVA and Fisher t-test. The results were similar to N_2O emission.

2.7. Chemical composition of the soil

Chemical composition of the soil like NH_3–N, $NO_3^- - N$ and TOC contents were analyzed at regular intervals. Out of these three parameters, NH_3–N decreased considerably with time whereas the decrease in the other two parameters was marginal. The extent of loss of NH_3–N increased with higher addition of urea from external sources. The rate of decrease varied between 0.75–5.56 μgm/day, the lowest and the highest correspond to the no addition and maximum addition of urea respectively. The $NO_2^- - N$ concentration variation was negligible during the entire period of incubation.

2.8. Principal Component Analysis (PCA)

PCA is an important tool in solving the problem of analyzing the structure of the interrelationship among a large number of variables. It defines a set of common underline dimension known as factors. In the present case we have considered the factor loading >0.5 to achieve a confidence level >95%, as in the present case the total data set is >120 (Hair *et al.* 1998). Factor-I accounted for 32.6% of cumulative variance having eigen value of 2.6. Factors-I contained three variables like ammonia, nitrate and urea. All these factors played a vital role in generation of GHG during incubation. Therefore Factor-I may be termed as 'GHG generation'. Factor-II accounts for 62.7% of cumulative variance having eigen value of 2.4. It contained variables like days, CH_4 and moisture content. Therefore, Factor-II is termed as 'CH_4 emission'. Factor-III accounts for cumulative variance of 75.9% having eigen value of 1.05. So Factor-II can be termed as 'N_2O emission'. Results are shown in Table 5.

Conclusions

1. It can be concluded that the emission of N_2O and CH_4 depends on the availability of nitrogen source, soil character and prevailing environmental conditions.
2. It was observed that N_2O emission attained a maximum and then tapered off, while CH_4 emission increased progressively with time as well as lower carbon content in both soils.
3. The emission of CH_4 was observed to be low, may be due to further oxidation by NO_3^- as well as lower carbon content in both the soil.
4. Theoretical N_2O emission was calculated using empirical equation and it explained well with the experimental values.
5. The emission of N_2O and CH_4 gases greatly depends on both nitrogen and WFPS variation used in this study.
6. Statistically it was proved that the two experimental parameters i.e. urea and WFPS played significant roles ($P \leq 0.001$) in determining the emission level.
7. The PCA classified the variables as three different influencing factors.

Acknowledgements

The authors are thankful to the Director, Institute of Minerals and Materials Technology, Bhubaneswar for his kind permission to publish this paper. The authors are also thankful to the Ministry of Environment and Forests, New Delhi for the financial support. The authors highly acknowledge the anonymous reviewers for their valuable suggestions, on the basis of which the manuscript finally came to this stage.

References

Andert, J.; Wessen, E.; Borjesson, G.; Hallin, S. 2011. Temporal changes in abundance and composition of ammonia-oxidizing bacterial and archaeal communities in a drained peat soil in relation to N_2O emissions, *Journal of Soils and Sediments* 11: 1399–1407. http://dx.doi.org/10.1007/s11368-011-0413-9

Baruah, K. K.; Gogoi, B.; Gogoi, P. 2010. Plant physiological and soil characteristics associated with methane and nitrous oxide emission from rice paddy, *Physiology and Molecular Biology of Plants* 16(1): 79–91. http://dx.doi.org/10.1007/s12298-010-0010-1

Bouwman, F. 1996. Direct emission of nitrous oxide from agricultural soils, *Nutrient Cycling in Agroecosystems* 46: 53–73.

Cardenas, L. M.; Hawkins, J. M. B.; Chadwick, D.; Scholefield, D. 2003. Biogenic gas emissions from soils measured using a new automated laboratory incubation system, *Soil Biology & Biochemistry* 35: 867–870. http://dx.doi.org/10.1016/S0038-0717(03)00092-0

Carter, M. S. 2007. Contribution of nitrification and denitrification to N_2O emission from urine patches, *Soil Biology & Biochemistry* 39: 2091–2102. http://dx.doi.org/10.1016/j.soilbio.2007.03.013

Cleemput, O. Van. 1998. Subsoils: chemo and biological dentrification, N_2O and N_2 emission, *Nutrient Cycling in Agroecosystems* 52: 187–94. http://dx.doi.org/10.1023/A:1009728125678

Das, S.; Ghosh, A.; Adhya, T. K. 2011. Nitrous oxide and methane emission from a flooded rice field as influenced by separate and combined application of herbicides bensulfuron methyl and pretilachlor, *Chemosphere* 84(1): 54–62. http://dx.doi.org/10.1016/j.chemosphere.2011.02.055

Dobbie, K. E.; Smith, K. A. 2003. Nitrous oxide emission factors for agricultural soil in Great Britain: the impact of soil water filled-pore space and other controlling variables, *Global Change Biology* 9: 204–218. http://dx.doi.org/10.1046/j.1365-2486.2003.00563.x

Dutta, A. 2007. *Mitigation of green house gas emission from tropical rice ecosystem*: PhD Thesis, Kalyani University, India.

Fangueiro, D.; Pereira, J.; Chadwick, D.; Coutinho, J.; Moreira, N.; Trindade, H. 2008. Laboratory assessment of the effect of cattle slurry on organic N degradation after soil application and N_2O and N_2 emissions, *Nutrient Cycling in Agroecosystems* 80: 107–120. http://dx.doi.org/10.1007/s10705-007-9124-4

Gerlach, R. W.; Dobb, E. D.; Raab, G. A.; Nocerino, J. M. 2002. Gy sampling theory in environmental studies. 1. Assessing soil splitting protocols, *Journal of Chemometrics* 16: 321–328. http://dx.doi.org/10.1002/cem.705

Gogoi, B.; Baruah, K. K. 2012. Nitrous oxide emissions from fields with different wheat and rice varieties, *Pedosphere* 22(1): 112–121. http://dx.doi.org/10.1016/S1002-0160(11)60197-5

Gupta, S. P. 2005. *Statistical method.* Daryaganj, New Delhi: Sultan Chand & Sons. 1020 p.

Hair, J. F.; Anderson, R. E.; Tatham, R. L.; Black, W. C. 1998. *Multivariate data analysis.* USA, New Jersey: Upper Saddle River: Prentice-Hall International, Inc. 112 p.

Hayakawa, A.; Akiyama, H.; Sudo, S.; Yagi, K. 2009. N_2O and NO emissions from an Andisol field as influenced by pelleted poultry manure, *Soil Biology & Biochemistry* 41: 521–529. http://dx.doi.org/10.1016/j.soilbio.2008.12.011

Horak, J.; Siska, B. 2006. Evaluation of N_2O emissions by DNDC model for sandy loam soils of danubian lowland, *Journal of Environmental Engineering and Landscape Management* 14(4): 165–171.

IPCC 1997. Nitrous oxide and carbon dioxide in agriculture: OECD/IPCC/IEA Phase II Development of IPCC guidelines for natural green house gas Inventory methodology. *Workshop report*, 4–6 December 1995, Geneva, OECD, IPCC, IEA.

IPCC 2007. *Climate Changes: the Scientific Basis. Contribution of Working Group 1 to the Third Assessment Report of the Intergovernmental Panel on Climatic Change.* Cambridge: Cambridge University Press.

Kampschreur, M. J.; Temmink, H.; Kleerebezem, R.; Jetten, M. S. M.; Loosdrecht, M. C. M. Van. 2009. Nitrous oxide during waste water treatment, *Water Research* 43: 4093–4103. http://dx.doi.org/10.1016/j.watres.2009.03.001

Kim, D. J.; Roy Chaudhury, G.; Ahn, J. W.; Lee, S. W. 2009. Bioleaching of complex sulfides concentrate and correlation of leaching parameters using multivariate data analysis technique, *Journal of Materials Transaction* 50: 2318–2322. http://dx.doi.org/10.2320/matertrans.M2009125

Kineey, C. A.; Mandernack, K. W.; Mosier, A. R. 2005. Laboratory investigations into the effects of pesticides mancozeb, chlorothalonil and prosulfuron on nitrous oxide production in fertilized soil, *Soil Biology and Biochemistry* 37: 837–850. http://dx.doi.org/10.1016/j.soilbio.2004.07.044

Mishra, D.; Ahn, J. G.; Kim, D. J.; Roy Chaudhury, G.; Ralph, D. E. 2009. Dissolution kinetics of spent petroleum catalyst using sulfur oxidizing acidophilic microorganisms, *Journal of Hazardous Materials* 167: 1231–1236. http://dx.doi.org/10.1016/j.jhazmat.2009.01.056

Mitra, S.; Wassmann, R.; Jain, M. C.; Pathak, H. 2002. Properties of rice soils affecting methane production potentials: 1. temporal patterns and diagnostics procedure, *Nutrient Cycling in Agroecosystems* 64: 169–182. http://dx.doi.org/10.1023/A:1021198714896

Mosier, A.; Wassmann, R.; Verchot, L.; King, J.; Palm, C. 2004. Methane and nitrogen oxide fluxes in tropical agricultural soils: sources, sinks and mechanism, *Environment, Development and Sustainability* 6: 11–49. http://dx.doi.org/10.1023/B:ENVI.0000003627.43162.ae

Ramulu, T. S.; Sahoo, S. K.; Baral, S. S.; Das, S. N.; Swamy, Y. V.; Roy Chaudhury, G. 2008. Estimation of nitrous oxide emission from pulse cultivation in rain-fed uplands, *Current Science* 95: 584–585.

Ramulu, T. S.; Sahoo, S. K.; Mohapatra, R. K; Roy Chaudhury, G.; Das, S. N. 2009. Emission of N_2O on pulse-rice crop rotation in upland, *Atmospheric Environment* 43: 2087–2090. http://dx.doi.org/10.1016/j.atmosenv.2009.01.007

Reddy, K. R.; Patrick, W. H.; Lindau, C. W. 1989. Nitrification-denitrification at the plant root sediment interface in wetlands, *Limnology and Oceanography* 34: 1004–1013. http://dx.doi.org/10.4319/lo.1989.34.6.1004

Rivett, M. O.; Buss, S. R.; Morgan, P.; Smith, J. W. N.; Bemment, C. D. 2008. Nitrate attenuation in ground water: a review of biogeochemical controlling processes, *Water Resarch* 42: 4215–4232. http://dx.doi.org/10.1016/j.watres.2008.07.020

Sahoo, S. K.; Ramulu, T. S.; Kumar, M.; Roy Chaudhury, G.; Das, S. N. 2010. Emission of N_2O from farmer's field during black gram cultivation and incubation experiments in laboratory, *Toxicological & Environmental Chemistry* 92: 1043–1052. http://dx.doi.org/10.1080/02772240903252157

Silvennoinen, H.; Liikanen, A.; Torssonen, J.; Stange, C. F.; Martikainen, P. J. 2008. Denitrification and nitrous oxide effluxes in boreal, eutrophic river sediments under increasing nitrate load: a laboratory microcosm study, *Biogeochemistry* 91: 105–116. http://dx.doi.org/10.1007/s10533-008-9262-z

Singh, S. N.; Verma, A.; Tyagi, L. 2003. Biogenic emission of GHGs from agricultural fields, *Physiology and Molecular Biology of Plants* 9: 63–73.

Sparks, D. L. 1996. *Methods of soil analysis. Part 3: chemical methods.* SSSA Book Series No. 5, Soil Science Society of America / American Society of Agronomy, Madison, Wisconsin. 1390 p.

Synder, C. S.; Bruulsema, T. W.; Jensen, T. L.; Fixen, P. E. 2009. Review of greenhouse gas emission from crop production systems and fertilizer management's effects, *Agriculture Ecosystem and Environment* 133: 247–266. http://dx.doi.org/10.1016/j.agee.2009.04.021

Wang, K.; Yu, Z.; Vermoeson, A.; Patrick, W.; Cleemput, O. Van. 2001. Nitrous oxide and methane emission from different soil suspensions: effect of soil redox status, *Biology and Fertility of Soils* 34: 25–30.

Yang, W. H.; Tej, Y. A.; Silver, W. 2011. A test of a field-based ^{15}N-nitrous oxide pool dilution technique to measure gross N_2O production in soil, *Global Change Biology* 17: 3577–3588. http://dx.doi.org/10.1111/j.1365-2486.2011.02481.x

Sanak RAY. PhD student, Department of Environment and Sustainability, CSIR – Institute of Minerals and Materials Technology. Master of Science (environmental science), Utkal University (U. U), 2010. Bachelor of Science (Botany), U. U, 2008. Scientific Publications: 1 Book Published and 4 papers communicated. Conferences attended: 2. Awards: Perusing PhD under DST-INSPIRE scheme, Government of India. Research interests: environment pollution control and management, waste water treatment, greenhouse gas emission study.

Ayusman MOHANTY. Project Assistant-II, Department of Environment and Sustainability, CSIR – Institute of Minerals and Materials Technology. Master of Science (bio-technology), Utkal University (U. U), 2010. Bachelor of Science (Bio-technology), Allahabad Agricultural Institute, 2008. Scientific Publications: 1. Research interests: biological waste water and waste treatment, constructed wetland and microbial fuel cell.

Torati SRI RAMULU. Worked as Project Assistant-II, Department of Environment and Sustainability, CSIR – Institute of Minerals and Materials Technology, 2010. Presently persuing PhD at Department of Materials Science and Engineering, Chungnam National University, South Korea. Publications: 10. Research interests: greenhouse gas emission study, bio-sensors.

Gautam ROY CHAUDHURY. Chief Scientist, Department of Environment and Sustainability, CSIR – Institute of Minerals and Materials Technology. Doctor of Chemistry, Department of Chemistry, U. U, 1980. Master in Chemistry, U. U, 1975. Publications: 110. Research Interest: recovery of metal values from lean solutions, environmental science and pollution abatement.

20

AN INTERPRETATIVE MODEL FOR THE MANAGEMENT OF CONTEMPORARY CULTURAL LANDSCAPES IN LINEAR INFRASTRUCTURE PROJECTS

Elena Berte[a], Thomas Panagopoulos[b], Bruno Zanon[a]

[a]*Department of Civil and Environmental Engineering, University of Trento, Via Mesiano 77, 38123 Trento, Italy*

[b]*Research Centre for Spatial and Organizational Dynamics (CIEO), University of Algarve, 8000 Faro, Portugal*

Abstract. Structures and potentialities of cultural landscapes have been studied mostly under a historical perspective. This article questions if contemporary societies can develop cultural landscapes. Starting from an analysis of the structure of contemporary landscapes, it investigates methods of developing cultural landscapes with contemporary knowledge and background. The theoretical assumptions are based on an ecosystem approach and consider the social and functional values in relation to the infrastructure project. Linear infrastructures are the physical basis for material and immaterial links between societies. They structure local territories, but more often reflect upper level rationalities – thus physically cross areas without providing a local service. An interpretative model to support cultural landscape dynamics within linear infrastructure projects was developed and tested in three Italian transportation infrastructures. The model was based on the analysis of the structure and the generating processes of contemporary cultural landscapes and makes use of criteria and indicators.

Keywords: contemporary cultural landscape, linear infrastructure, environmental impact assessment, landscape management, Italy.

Reference to this paper should be made as follows: Berte, E.; Panagopoulos, T.; Zanon, B. 2013. An interpretative model for the management of contemporary cultural landscapes in linear infrastructure projects, *Journal of Environmental Engineering and Landscape Management* 21(4): 248–262 http://dx.doi.org/10.3846/16486897.2012.745412

Introduction

A landscape is the outcome of relationships between the local society and the transformed environment, which is the territory (Antrop 2005). A territory is a relational concept and infrastructures are its basic constituent. In fact, they represent the backbone of the inhabited territory, providing services within and among places, and their construction and renovation have always been a factor of modernization. The landscape can therefore mirror the appropriate local knowledge of a responsible community or the superimposition of upper-level rationalities.

In the European Union, a number of government documents require to consider the surrounding landscape when building any object of infrastructure. There has been considerable research into landscape, however, this research has not been utilized by infrastructure planners and only certain aspects of the landscape are dealt within the various types of documents involved in the planning process (Antonson 2009).

The article focuses on the role of linear infrastructures in the landscape and in particular on roads and railways, stressing the profound changes interposed because of their connections with the local space. The challenge to be faced is the construction of linear infrastructures activating positive cultural landscape dynamics, thus supporting functional innovation and protecting the local identity by integrating the diverse – often conflicting – aspects and the varied scales involved.

In hard infrastructures of ancient times, linear infrastructures were a part of the physical space occupied by a local community, i.e. linear infrastructures not only attracted founders of villages, towns and cities but also served as a pivot for urban space planning. They had, therefore, a central role in the definition of a shared urban identity and in the construction of the landscape. In particular, linear infrastructures organized territories by means of configurations able to integrate natural and

Corresponding author: Thomas Panagopoulos
E-mail: tpanago@ualg.pt

human spatial dynamics, creating durable man-made ecosystems.

With the development of the industrial era, and later on with the Fordist economic model, the traditional relationships between infrastructures, local societies, and territories, became obscure. Starting in the second half of the twentieth century, the speed of the development of infrastructure networks became a central goal to support the economic development by linking together the main industrial areas, the major cities, and the harbours. The experts in charge of the elaboration of the projects were engineers and their aim was often to simplify the morphological complexity of territories and to provide fast connections. In this framework, infrastructure networks could have no relations at all with the environment they crossed. They became the initiators of new urban developments at the interchange points and provided access to previously untouched areas. In fact, the construction of networks produces a direct impact, but even more negative consequences can be brought about by the actions made possible by their presence (Antrop 2000).

In recent decades, economic restructuring and technological innovation, as well as political and economic integration among territories, have enhanced the contradictions between infrastructure lines and local territories. Multifaceted and intertwined dynamics are involved: new levels of interrelations, from large (the global, the supra-local) to focused scales; territorial integration or exclusion; competition for space; morphological changes; and social and ecosystem fragmentation.

Large-scale infrastructure lines overlap and replace preexisting ones. Usually, in long inhabited territories, infrastructure systems were based on a slow coevolution of natural and social components such as fields, woodlands, water streams, historic centres, and traditional rural areas. Linear infrastructures, most often, have negative impacts because of various factors. They produce contamination (Baltrenas *et al.* 2004), and often a disturbance of the sense of place and belonging of individuals to a peculiar and characteristic environment (Alkan *et al.* 2009). With respect to the interactions with the natural features, problems arise because of the erosion of the ecosystemic functions produced by the restructuring of the ecosystems. Moreover, apart the desertification of the directly involved soil, a barrier effect is produced, determining fragmentation, which is a phenomenon of isolation and impoverishment of many species (Bogaert *et al.* 2005). Such effects highlight the inevitable conflicts between some social and natural processes and the complexity of their relations (Farina *et al.* 2005a).

Infrastructure planning must therefore take into account these issues, appropriately managing the resources of the physical–structural, ecosystemic, and sociocultural kind, whose roles and values vary according to the particular area crossed (urban, rural, natural). It is this complexity that makes it difficult but necessary to manage the linkages between linear infrastructure and landscape dynamics. Many different local conditions are crossed by an infrastructure project and the key issue is how to transform a new proposal into a project that maintains and supports the complexity of cultural landscapes.

Many official documents have formulations concerning the landscape with which authorities must or should cope. One such document is the European Landscape Convention (ELC 2000). ELC was signed on 20 October 2000 and came into force on 1 March 2004. In February 2008, 35 countries have ratified, accepted, approved or just signed the convention. Therefore, many countries have only recently begun to apply the ELC in their social structure.

In environmental and cultural literature, there lacks a structured approach to cultural landscapes. Such landscapes are usually considered when "historic" features emerge and few reflections have been expressed about the presence of contemporary cultural landscapes and on their complex meanings. It is an important deficit, because the maintenance of the existing cultural landscapes and the creation of new ones are strictly related to a sustainable development perspective, as they are a concretization of its foundations. Cultural landscapes represent the strongest and most durable interaction between the natural and the social systems (Scazzosi 2004).

The following main questions are addressed:

Is the concept of contemporary cultural landscape relevant or are only historic cultural landscapes to be considered?

Does the planning process have a role in developing cultural landscape dynamics?

Is it possible to support and / or activate cultural landscape dynamics when planning linear infrastructure projects?

The objective of the present study was to analyze the structure of cultural landscapes and suggest a model for the development of new landscapes considering which drivers have changed and which, if adequately managed, allow new durable relations between natural and social systems to take place. To achieve this objective, a key conceptual and operational instrument was developed and then tested in the case of three linear transportation infrastructures in Italy. This interpretative model can be used to evaluate the dynamics and to support the development of contemporary cultural landscapes.

1. Structure and processes of contemporary cultural landscapes

Cultural landscapes have been defined differently according to the diverse approaches adopted and the purposes pursued. A univocal vision does not exist because of the multidimensional and multidisciplinary characteristics at the basis of the landscape structure (Naveh 2001; Tress *et al.* 2009). In addition to the UNESCO definition connected to the World Heritage Convention of 1972, geographers amongst others and more recently landscape ecologists have provided a number of definitions that form the basis of this theory (Naveh 1998; Schmitz *et al.* 2003; Hazen 2008; Tempesta 2010).

This research begins from the widespread assumption that landscapes are complex systems based on the interactions between the natural and the social systems. Within this group are cultural landscapes, which are based on the optimization of the relationships among resources, information, and use. Resources comprise natural and cultural goods and values available in a specific context; information is based on knowledge and allows identifying the natural and cultural resources and supporting know-how; use refers to the actions performed according to the resources with advantages both for the social and the natural systems (Farina *et al.* 2005a). The capability of integrating the above-mentioned three systems through up-to-date solutions produces contemporary cultural landscapes.

The structure and driving processes of landscapes have been rapidly changing in the last two centuries and many experts stress the speed of the change occurring in the recent decades (Antrop 2000; Naveh 2001). The drivers of historic landscapes consisted of economic, cultural, and ecosystem processes, which acted at the same hierarchical level. Currently, economic processes are the determinants in landscape evolution (Farina *et al.* 2003).

Complexity in contemporary cultural landscapes is destined to become even harder to grasp and manage, as the increasing number of dynamics coexisting in contemporary cultural landscapes implies even more complex structures (Fig. 1). This has a number of consequences in the relationships between

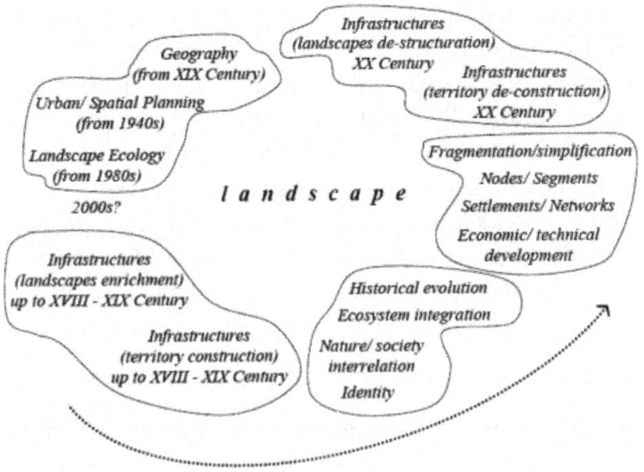

Fig. 1. Relations between landscape complexity-simplification and human drivers

the social and the natural components. As regards the social component, the awareness of local cultural specificities and uniqueness is expected to become stronger because of the comparison with external contexts. These specificities can be at the same time endangered and supported by the increased knowledge and accessibility of external resources and experiences, in many cases promoted by the normative framework (e.g. the European Landscape Convention guidelines). The mutual interactions between cultures and know-how from different backgrounds and levels may enrich the scenario of future cultural landscapes with new dynamics if safeguarding of local identities is pursued (which means site-responsive knowledge). As regards the natural component, contemporary cultural landscapes can maintain and increase their important role as biological and genetic refuges. Nevertheless, they will be characterized by an augmentation of fragility and vulnerability because of a major sensitivity to anthropic disturbance, with a consequent lowering of resilience (Farina *et al.* 2003).

According to this research, there are four major drivers to be considered: identity process, integration process, multiscale process, and innovation process.

Cultural landscapes are based on identity and integration: they are indeed the outcome of local identities and cultures as evidence of a territorial history as well as expressions of the interaction between man and nature (ELC 2000). According to Panagopoulos (2009), cultural landscapes should lead people to form emotional attachments to the land and thereby develop a greater appreciation for the sustainability goals. Multiscale dynamics using stronger and growing relationships between local, regional, national, and supra-national socioeconomic levels, together with rapid sociocultural innovations, are two determinant drivers that particularly affect contemporary landscapes (Opdam *et al.* 2006).

As far as identity is concerned, many authors (Terkenli 2001; Scazzosi 2004), together with the European Landscape Convention (ELC 2000), have analyzed and stressed the relations occurring between landscape and identity and all agree on the importance of such a relation for the development of local landscapes. Identity is related to the historical–cultural character of a place and to the sense of belonging of the local society, which means not only collective memory but also reliance on that place for constructing a future. Basic sources for local identities consist in the recognition of differences among places, in natural and cultural specificities, in the awareness of past evidence in a physical and symbolic sense, forming the collective memory and experience of a population (Scazzosi 2004). Local identity can therefore generate a reaction to modernization, homologation, and globalization. The European Landscape Convention gives primary attention to identity, assuming that landscape underpins its development (ELC 2000, art. 5). UNESCO defines identity as the cultural tradition of a context, whereas for Tishler (1982) it is the local and regional knowledge, which defines the long-standing characteristics of a place. The sociologist Castells (1992) considers identity a resource for transformation, a social process at the basis of the construction of meanings and cultural values.

Integration is the second basic aspect of a cultural landscape and it is strictly related to multifunctionality (Wiggering *et al.* 2006). Such a process is enhanced and analyzed in particular by the Landscape Ecology, which aims at understanding the interactions between natural and human dynamics occurring within landscapes. Farina (2000) defines cultural landscapes as the highest expression of integration between human activities and environmental dynamics. According to Antrop (2005), landscape is at the basis of enduring sets of linkages based on the relationships between the physical environment and human society. In such a context, people are the developers through their engagement with the world around them. The integration process is based, amongst other aspects, on a social comprehension of natural dynamics, the conservation of diversity and heterogeneity, and the use of renewable natural and cultural resources. Integration regards not only the relationships between the natural and the social system, but is an essential function within each of them (e.g. "decision making" in the social system). Analyses of cultural heritage connectivity can inform about functions of the landscape and its social and economic conditions (Antonson *et al.* 2010).

The multiscale processes underpin both the natural and the social systems, which are based on multilevel relations. According to Antrop (2000), landscapes evolve continuously by "internal" and "external" factors. Internal factors are those that may be controlled at the local level (e.g. by the direct action of the inhabitants). External factors are mostly indirect and influence the local landscape conditions through upper strategies and policies. Decisions are made on different hierarchical levels of policy-making and manifest themselves in terms of actions at different levels. The multiscale processes work on connections, junctions, and linkages between the dynamics at different levels. For instance, the local ecological networks are connected to the regional and the national and continental ones; local identities should be recognized at the regional level, at the national one, and so forth. To control such dynamics, it is helpful to use recent guidelines and tools provided at European and national levels that aim at supporting the local context under a multiscale perspective. Examples of these tools are the European Landscape Convention (ELC 2000), the Environmental Impact Assessment (EIA) and the Strategic Environmental Assessment (SEA). Under this perspective, some scholars underline the importance of a multiscale planning development, which is considered the more adequate approach to face and manage complex systems such as landscapes (Mortberg *et al.* 2007).

Currently, multilevel cultural exchanges are stronger and local contexts are enriched with new stimuli. As a consequence, new complex multilevel relationships requiring an appropriate management emerge: the risk, however, is to weaken local socioeconomic dynamics, identity values, and nature–society interactions, in the name of upper powerful relations.

Together with the multiscale process, innovation is one of the main driving forces for the development of contemporary landscapes, because they are time-dependent. The durability of a cultural landscape is connected to the ability to renovate, through appropriate actions, the interactions between society and nature. Starting from the landscape structures inherited from the past, it is essential to activate innovation processes defining new functional systems, where previous values (both natural and social) are the basis for new coherent solutions. Innovation means shifting from one phase to another and according to the geographer L. Gambi (1973), the society re-creates its living space through modalities based on evolutionary steps. Nowadays innovation happens fast. This may challenge some of the characteristics of historic cultural landscapes, based on long-time stability and nature adaptation, with few and slow changes (Farina *et al.* 2005a). This process, if not properly managed, may trigger other dynamics that are detached by identity–historical matrices, and more in general can weaken the interactions between natural and social systems.

Appropriate local knowledge was at the basis of the construction of historical cultural landscapes. Thus, the ability of a local society to provide responses of long duration to its living needs using the environmental recourses meant constructing territories functionally operating but also rich in identity and symbolic meanings. As the basis of contemporary cultural landscapes, there must be projects that are simultaneously technically sound, socially recognized, and economically viable. The role of a project is to support, develop, and strengthen the processes that are in a context. To trigger constructive dynamics and to avoid a weakening of the system, identity process, integration process, multiscale process, and innovation processes need to be planned and managed together.

2. Interpretative model for cultural landscape dynamics

The development of contemporary cultural landscapes is affected by many interrelated factors so that variations can be triggered not only by planning, but also by community values, sense of place, environmental

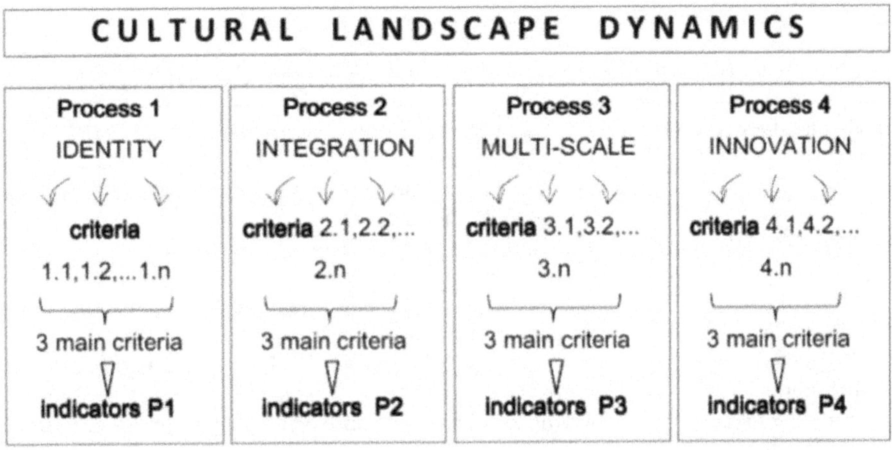

Fig. 2. Conceptual scheme of the Cultural Landscape dynamics Interpretative Model

attitudes, administrative skills, and political and economic situations.

The dynamics of a cultural landscape could be either carried on, started, restarted or weakened, interrupted, or cancelled through a planning process. To empirically support some of such dynamics, the Cultural Landscape Dynamics Interpretative Model (CLDIM) has been developed. The methodology of this conceptual and operative model has been developed bearing in mind that cultural landscapes rely on the rules of complex systems, and their evolution and change are influenced by processes of different dimensions in space and time (Farina *et al.* 2005b). The model is a result of combined criteria that are the outcomes of the elaborations of aspects, principles, and tools taken from spatial planning best practices, landscape ecology, and from the analysis of two historic cultural landscapes (the Roman Centuriazione and the historical railways). Spatial planning best practices have contributed to elaborate criteria based on cultural heritage and identity preservation, large/local actions management and integration, local development, flexibility, and participation. Landscape ecology perspective has contributed to elaborate criteria based on social–environmental systems analysis, physiognomic structural elements analysis, multifunctionality, growth limits estimation, and multiscale approach.

The four contemporary landscape processes analyzed (identity, integration, multiscalarity, innovation) underpin the four parts of the model. Each part is divided into a number of criteria and a number of indicators (Fig. 2).

Criteria (1.1 . . . 4.n) aim to analyze the state of the art of a process in a context and to understand a process in its complexity. The aim of the table of indicators is to activate a number of planning actions, i.e. to support a cultural landscape process in relation to a linear infrastructure plan. The indicators that characterize the second part of the model have been elaborated according to 12 main criteria of the interpretative grid. Each criterion is related to one or more indicators. Some derive from the elaboration of existing indicators, others have been structured ex novo. They are divided into pressure indicators and response indicators. They aim at addressing a number of actions to contrast pressure factors to support cultural landscape dynamics in planning processes, and in general to support sustainable conditions (OCS 2005).

The score system related to each indicator has been unified with the aim both at making their compilation easier and to allow a compared reading at the end of their application. The score system is a qualitative scale and is based on four levels: $+++=3$; $++=2$; $+=1$; $-=0$. This score method is useful as long as the results are interpreted with a degree of caution and not out of context. The scale has a limited number of classes, which often means groups of alternatives may hold the same position. The meaning of each score level is specified for each indicator.

The following four tables of the model were used to describe the cultural landscapes using the four processes respectively, and the criteria, pressure, and response indicators.

Cultural landscapes are related to identity. They are based on the evolution of the know-how of local societies. Loss of social interest determines landscape destructuration. Table 1 shows the three criteria of the identity process and the respective indicators related to the criteria.

Criterion 1 is about the acknowledgment of the local-social heritage and methods to sustain local processes for the valorisation of cultural heritage and has three indicators: "activism of local associations in the project elaboration"; "activism of local institutions in the project elaboration"; "use of local projects already planned in the areas involved".

Criterion 2 is about managing land quality changes and has four indicators: "expropriated ha/ km"; "repair strategies for loss of value in sensitive

Table 1. Criteria and indicators used at the identity process of the cultural landscapes

PROCESS	CRITERIA / DESCRIPTION	MEASURES / INDICATORS
I D E N T I T Y	**1. Sustaining local processes for the valorization of cultural heritage** Acknowledgment of the local-social heritage: A. INVOLVEMENT OF ASSOCIATIONS / GROUPS (related to cultural/ environmental heritage) B. INVOLVEMENT OF LOCAL INSTITUTIONS C. ENHANCE / USE EXISTING LOCAL PROJECTS already developed by local institutions	A. INVOLVEMENT OF ASSOCIATIONS / GROUPS 1.1. activism of associations / spontaneous groups in the project elaboration a. in problems identification b. in choosing options c. in project approval + + +many in the 3 phases + +many / some in 2 / 3 phases +one (few) in one phase – options that did not change the project / obstructionism because not involved; – no activism from local associations B. INVOLVEMENT OF LOCAL INSTITUTIONS 1.2. activism of local institutions in the project elaboration a. in problems identification b. in choosing options c. in project approval + + +many in the 3 phases + +many / some in 2 / 3 phases +many / some in one phase – obstructionism because not involved – no activism from local institutions C. ENHANCE / USE EXISTING LOCAL PROJECTS 1.3. use of local plans / projects already planned in the areas involved + + +acknowledged and re-elaborated where possible + +re-elaborated in some cases +re-elaborated in few cases – not considered
	2. Managing land quality changes A. CHANGES OF SURFACES comparison between agricultural land, sustainable land, urban areas reduction B. REPAIR STRATEGIES activation of repair measures (mitigations) concerning the relations with sensitive areas, loss of quality value of the fields, loss of quality life	A. CHANGES OF SURFACES 2.1. ha/ km expropriated (agricultural, urban, natural) B. REPAIR STRATEGIES 2.2. Repair strategies for loss of value in sensitive areas + + +systemic project (integrated project) + +shared projects between planners / local institutions +some interactions between planners / local institutions – no actions 2.3. Repair strategies in agricultural areas + + +systemic project + +shared projects between planners/ local institutions +some interaction between planners/ local institutions; – no actions 2.4. Repair strategies to prevent loss of quality life + + +systemic project + +shared projects between planners/ local institutions +some interaction between planners/ local institutions; – no actions
	3. Supporting the structure of local milieu The criterion aims at enhancing whether or not the infrastructure matches with the local milieu.	3.1. objectives and actions declared to support the structure of local milieu + + +main aim or integrated aim since the preliminary steps + +enhanced after the presentation to local institutions +considered only after the presentation to local institutions – no actions

areas"; "repair strategies in agricultural areas"; "repair strategies to prevent loss of life quality".

Criterion 3 is about the structure of local milieu and has one indicator: "objectives and actions declared to support the structure of local milieu".

Cultural landscapes are characterized by multi-functionality determined by a variety of uses: the outcome is an integrated landscape where natural and social processes are compatible. Contemporary cultural landscapes originate from integrated goals: physic, ecosystem, and social elements are planned together. They are based on local know-how integrated with experts' knowledge (outside tools). Table 2 shows the three main criteria of the integration process, and the respective indicators related to the criteria.

Table 2. Criteria and indicators used at the integration process of the cultural landscapes

PROCESS	CRITERIA / DESCRIPTION	MEASURES / INDICATORS
INTEGRATION	**4. Interdisciplinarity** Presence/absence of experts from different disciplines. Early or late presence? (macro-areas: agronomy, architecture, ecology, economy, engineering, geography, geology, history, sociology)	4.1. participation since the early steps of the project elaboration. + + +maximum interdisciplinarity + +medium/ high interdisciplinarity +few interdisciplinarity – absence of interdisciplinarity 4.2 calls for specific studies (late/partial participation). + + +the majority of disciplines + +half disciplines +few disciplines – one discipline
	5. Integrated land configurations Capability to carry out an integrated territorial project through integrated solutions which consider urban and environmental solutions in the territories involved	5.1. Opportunity for urban regeneration + + +widespread approach of the local institutions together with the infrastructure planning group (integrated initiative) + +approach in some areas of the local institutions together with the infrastructure planning group (integrated initiative) +sporadic cases of local initiative – no cases 5.2. Opportunity for environmental regeneration + + +widespread approach of the local institutions together with the infrastructure planning group (integrated initiative) + +delimited approach of the local institutions together with the infrastructure planning group (integrated initiative) +sporadic cases (local initiative) – no cases 5.3. Coherence / integration with the existing transportation-network + + +total coherence/integration with the network + +coherence for the majority of the infrastructure +partial coherence – no coherence
	6. Flexibility Elaboration/evaluation of alternatives projects during the planning phase	6.1. elaboration and evaluation of alternative projects + + +alternative projects undertaken by the planning group + +project re-organization after requests +acknowledgment of limited requests – no alternatives undertaken

Criterion 4 is about the interdisciplinarity involved in the planning phase and has two indicators: "participation since the early steps in the project elaboration"; "calls for specific studies (late-partial participation)".

Criterion 5 is about the capability of carrying out an integrated territorial project through integrated solutions that consider urban and environmental aspects. It has three indicators: "opportunity for urban regeneration"; "opportunity for environmental regeneration"; "coherence / integration with the existing transportation network".

Criterion 6 is about the flexibility during the planning phase and has one indicator: "elaboration and evaluation of alternative projects".

Cultural landscapes are based on multiscale relations structured in interdependencies and exchanges. Natural system: local ecosystem dynamics are related to upper dynamics through linkages (e.g. flows and migrations). Social system: in historic cultural landscapes local societies were the main actors. Upper interests that weigh on local systems lead to development of tools (European Landscape Convention – Environmental Impact Assessment – Strategic

Table 3. Criteria and indicators used at the multiscale process of the cultural landscapes

PROCESS	CRITERIA / DESCRIPTION	MEASURES / INDICATORS
MULTI-SCALE	**7. Multi-level compensations** Relations between planning infrastructure and planning local compensation.	7.1. Active compensations + + +systemic project (integrated project) + +shared projects between planners / local institutions (master plan) +local independent actions (money provided to local administrations) – no actions
	8. Managing land transformation Relations between planning infrastructure and planning local transformations.	8.1. Active territorial transformations + + +systemic project (integrated project) + +shared projects btw planners / local institutions (master plan) +local independent actions – no actions
	9. Supporting system complexity **A. Infrastructure as a component of a complex system (large scale)** Infrastructure coherency in relation to the ecological system, settlement system and agricultural areas **B. Infrastructure as a system (intermediate scale)** Infrastructure as a system: coherence of typological choices and materials **C. Single work as a system (focused scale)** Single work as a system: ecological and technical values of the single work (tunnel, viaduct, bare road, trench road) and relation with the settlement system, and ecological system.	A. INFRASTRUCTURE AS A COMPONENT OF A COMPLEX SYSTEM 9.1. Compatibility with the ecological network + + +re-structure a new ecological network after the changes provoked + +provide effective solutions for the majority of the incongruities provoked +provide only sporadic solutions – not considered 9.2. Compatibility with settlements (inhabited areas) + + +analyze and plan new effective scenarios + +analyze and plan effective scenarios in some cases +provide only sporadic solutions (mitigations) – not considered 9.3. Compatibility with land-use values (e.g. agricultural areas) + + +analyze and plan new effective scenarios + +analyze and plan effective scenarios in some cases +provide only sporadic solutions (mitigations) – not considered B. INFRASTRUCTURE AS A SYSTEM 9.4. Typological and materials coherence + + +planned and recognizable as a unique infrastructure + +recognizable the coherence for the majority of the infrastructure +recognizable partial coherence – no coherence of solutions C. SINGLE WORK AS A SYSTEM 9.5. Value 1: technical value (quality of the choices) + + +high value + +good +fair – low / not considered 9.6. Value 2: ecological value + + +high value + +good +fair – low / not considered 9.7. Value 3: relation between the single work and the context + + +planned considering each different context + +planned considering different works for a number of different contexts +planned only in sporadic contexts – no relation

Environmental Assessment) to manage such dynamics at local level. Table 3 shows the three main criteria of the multiscale process and the respective indicators related to the criteria.

Criterion 7 is about the relations between the planning of the infrastructure and the planning of the local compensations and has one indicator: "active compensations".

Criterion 8 is about the relations between infrastructure planning and the local transformations and has one indicator: "active territorial transformations".

Criterion 9 is about the complexity of the supporting system. It is divided into the subcriteria 9a – infrastructure as a component of a complex system – in relation to the ecological system, settlement system, and agricultural areas (large scale); 9b – infrastructure coherence of typological choices and materials (intermediate scale); 9c – ecological and technical values of the single work (tunnel, viaduct, bare road, trench road), and relation with the settlement system and ecological system (focused scale). Criterion 9 has 7 indicators: "compatibility with the ecological network"; "compatibility with settlements"; "compatibility with land-use values"; "typological and materials coherence"; "technical value"; "ecological value"; and "relation between the single work and the context"

Cultural landscapes are evolving systems that reflect natural and cultural processes, which guarantee renewable dynamics. The use of innovative tools compatible with local specificities may denote added value to local know-how. Contemporary cultural landscapes answer to problems of contemporary society–nature interactions.

Innovation was considered to overcome the basic mission of the infrastructure (transport).Creativities may rise at local level in relation to the project development (e.g., use of infrastructure spaces for social cohesion – alternatives to transport, creation of laboratories, exhibitions, cultural meetings). Table 4 shows the three main criteria of the innovation process and the respective indicators related to the criteria.

Criterion 10 is about the use of the infrastructure project as a guideline to other projects and has one indicator: "Acknowledgment of innovative models". Criterion 11 is about the actions and processes aimed

Table 4. Criteria and indicators used at the innovation process of the cultural landscapes

PROCESS	CRITERIA / DESCRIPTION	MEASURES / INDICATORS
INNOVATION	**10. Innovative models acknowledged** Use of infrastructure project models as guideline	10.1. Acknowledgment of innovative models + + +determinant for the project development + +partial benefits +low benefits – no acknowledgment
	11. Extraordinary institutional relations Actions and processes aimed at improving governance (social–institutional learning)	11.1. Supporting experiences of interrelation between institutions + + +yes with all / majority of the institutions involved. Long-term duration (at least during planning execution) + + +innovative early experiences of interrelation + +yes with all / the majority of the institutions involved. Medium / short-term duration (during the planning phase) +with some institutions – sporadic experiences of interrelations 11.2. Ad hoc significant studies elaborated by other institutions and / or informal bodies in itinere / ex post + + +determinant for the project development + +partial benefits +low benefits – no acknowledgment
	12. Creativity (inclusion of unusual, complementary, innovative uses / actions) Capability to overcome the basic mission of the infrastructure (transport).	12.1. Project creativity + + +ex: creative solutions (e.g. permeable motorway in urban areas – Boulevard JFK, Luxembourg, arch. T. Latz) + +ex: peculiar projects for complementary aspects i.e. stop areas, specific single works with a symbolic value +ex: call for competition (ideas) for complementary aspects – ex: only transportation 12.2. Local creativities started-up with the project + + +new / existing associations work out related creative projects. Widespread the outcomes, awaken the society + +new or existing associations (formal-informal) elaborate creative projects related to the infrastructure project +limited experiences – no reactions at local / informal level

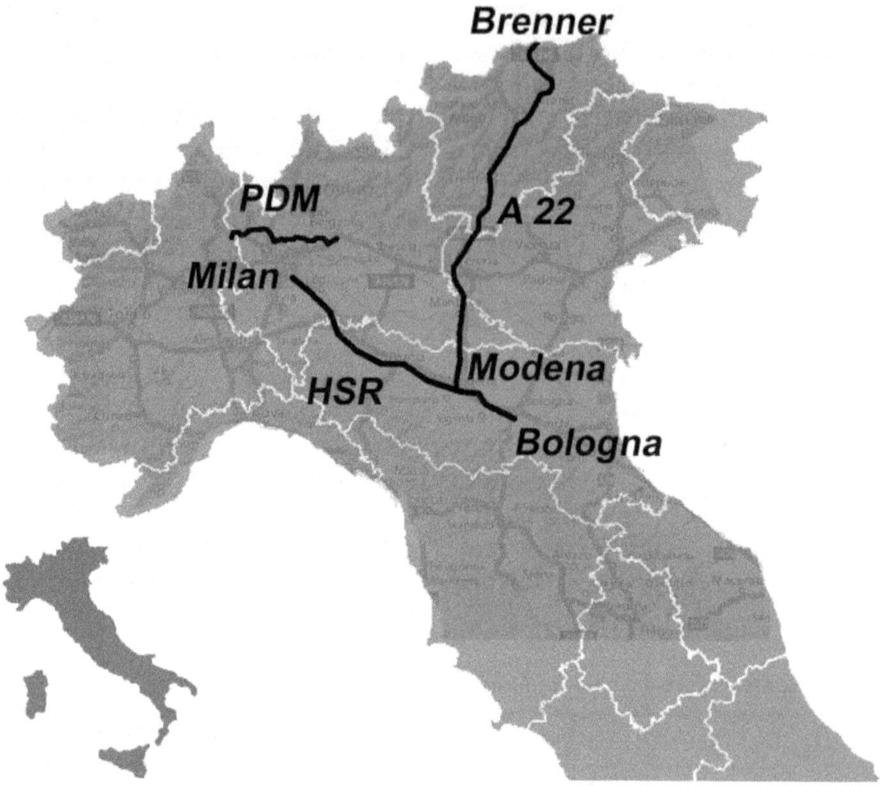

Fig. 3. Schematic map with the three case studies (A22: Brenner Motorway; HSR: high speed railway Bologna-Milan; PDM: Pedemontana Lombarda Motorway)

at social–institutional learning and has two indicators: "supporting experiences of interrelation between institutions"; "ad hoc significant studies carried out by other institutions and / or informal bodies". Criterion 12 is about the rise of creativity at local level in relation to the project development and has two indicators: "project creativity"; "local creativities started up with the project"

3. The model application to three case studies

To test the validity and improve its structure, the interpretative model has been applied to three contemporary linear infrastructure projects planned within the last 50 years (Fig. 3). This has permitted to formulate some considerations about the state of the art of contemporary cultural landscapes development in relation to infrastructure projects.

The three case studies have been chosen following four criteria:

a) selection of the projects within a national context (Italy) to understand the evolution of the approaches of linear infrastructure planning and the related legislative–cultural debate;
b) identification of the period of realization according to a peculiar phase concerning linear infrastructures development;
c) consideration of the evolution of the approaches of planners and landscape architects in relation to landscape and infrastructure planning;
d) identification of a project with an innovative character in each period.

Concerning the periods of realization, three different phases have been identified:

- the 1960s, characterized by the commencement of the Italian motorway network;
- the 1980s, characterized by a relaunch of the railway network with the high-speed railway project as an alternative to road and flight networks;
- the 2000s, representative of the current situation concerning linear infrastructure development within high-density extra-urban contexts.

The case identified in the 1960s is the Brenner Motorway, located in north-east Italy (regions of Trentino Alto Adige, Veneto, Emila Romagna). The project demonstrates the attention put in those years to the innovative technical solutions that were at the basis of the rapid development of the road system. In this case, the approach to landscape was mainly aesthetic and based on visual choices (use of arboreal and floral species). In general, landscape projects related to linear infrastructures were rare.

The second case is the High-Speed Railway Bologna–Milan located in northern Italy (Emilia Romagna, Lombardy regions). In the 1980s in Italy, the focus on the "environmental" aspect of landscape began, which culminated with the acknowledgment of the Environmental Impact Assessment (law 349/1986). In this case, a deep symbolic approach to landscape was carried out in parallel by assigning some exemplary projects to renowned architects.

The current case identified is the Pedemontana Lombarda Motorway located in north Italy. The motorway, planned within the Lombardy region, is a picture of the current Italian innovative approach to landscape development in relation to a large infrastructure plan. Particular attention has been placed to work out the compensations and mitigation aspects through a structured and organic project. A widespread involvement of the local communities has characterized the elaboration of a number of focused projects.

Information has been collected by applying the grid of criteria of the interpretative model to the three case studies. The indicators table was filed and the outcomes were outlined and compared.

4. Comparisons and trends of the processes in the three case studies

Considering that the three projects are innovative for the period in which they are planned, this comparison aims at presenting a general trend of the modalities of development that cultural landscapes are undertaking. Graphs for each criterion of the interpretative model have been elaborated to structure the comparison. They are the graphic representation of the outcomes of the indicators table, and aim both at underlying and comparing the results of the three case studies. In each figure, the x axis correspond to the three case studies: the Brenner highway (A22), the High Speed Railway Bologna-Milan (HSR) and the Pedemontana Lombarda Motorway (PDM). The y axis represents the values $(- = 0; + = 1; ++ = 2; +++ = 3)$ and corresponds to the score system used for each case in the indicators table. The rectangles represent the outcomes of each indicator. In some cases, a graphic representation has not been performed, because of the different measurement units of the indicators (i.e. binary measure, yes / no).

The results have been contextualized in each period. For example, the indicator "technical value" has obtained the maximum score both in the Brenner motorway case (1960s) and in the Pedemontana Lombarda motorway case, although this last one has been planned more than 40 years later. A comparison of the indicator outcomes follows, together with an elaboration of the results divided into the four processes.

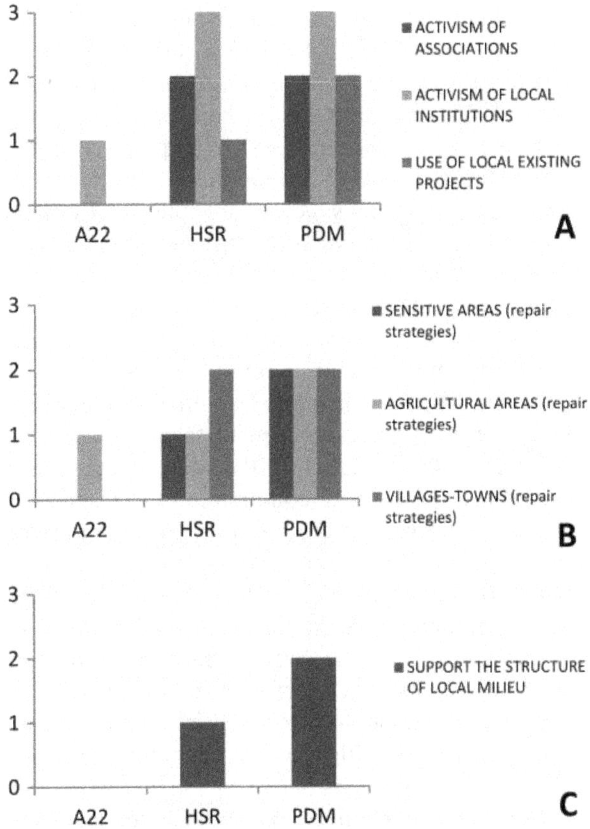

Fig. 4. Graphic representation of the identity process at the Brenner highway (A22), the High Speed Railway (HSR) and the Pedemontana Motorway (PDM)

4.1. Outcomes from the Identity process

The identity process was divided into three criteria. The graphic representation of the results obtained for the three case studies for the identity process is presented in Fig. 4 with a separate graph for each of the criteria and for each one of the related indicators.

From the comparison of the three cases concerning the identity process, what emerges is that in general this grew in importance from the 1960s and always characterizes more the planning process towards a landscape based on identity values. An almost constant development of the planning actions in the direction of the objectives represented by the criteria can be observed. This is a result of the improvement of the planning instruments as well as the acknowledgment of the importance of local communities for heritage conservation and renovation.

The adoption of the European Landscape Convention from the year 2000 and its acknowledgment by a growing number of European nations confirms such awareness. Nevertheless, as shown by the graphs, the maximum level is still to be reached in the majority of the cases. An improvement of the process can be obtained through a more active involvement of the local stakeholders, in particular the cultural and environmental associations, which still do not have any influence in the structure of the decisional bodies. A consequence of this is the lack of attention toward the residual open spaces (e.g. woods), in some cases sacrificed to safeguard other land structures (e.g. agricultural areas characterized by explicit economic values). The inclusion of local existing projects was an aspect to strengthen as well as to consider from the beginning of the planning procedure; however, the case of Pedemontana Lombarda has been undertaken only for the compensation project.

Another aspect enhanced is the relation between the expropriated hectares and the infrastructure. Further investigation is needed about the reorganization of the open areas consequent to the transformation. To provide an example of the large impact of such an aspect on the local communities, thousands of farms have been involved in those transformations consequent to the construction of the High Speed Railway Bologna-Milan. The data provided by Pedemontana Lombarda SpA, FSI SpA, and a literature overview as regards the Brenner motorway, demonstrated an increasing level of expropriations from the 1960s. Those augments of expropriations are mostly consequent to the mitigation and compensation projects.

Land quality changes have been better managed in the third case (Fig. 4b). Nevertheless, a further improvement could be carried out if the project of the infrastructure is not detached from the environmental project, and a change of perspectives is undertaken.

The strongest support to the local milieu is observable in the case of the Pedemontana project (Fig. 4c). The aim of the Pedemontana motorway was to sustain the economic structure of the productive area located in the north of the Lombardy region characterized by small-medium enterprises. In addition, the case of the high speed railway is emblematic in this sense, as its primary aim is to link faster the national and international centres; nevertheless, the activism of the local institutions enable obtaining the interconnection between the new line and the local railway network.

4.2. Outcomes from the integration process

The integration process is divided into three criteria. The graphic representation of the results obtained from the three case studies for the integration process is presented in Fig. 5 with a separate graph for each of the three criteria and for each one the related indicators.

This comparison testifies that the integration process grows following a dynamic that is less homogeneous if compared with the tendency of the previous process. As shown by the graphs, except for some indicators, the desideratum has still to be reached in all studied cases. In general, a determinant step forward is

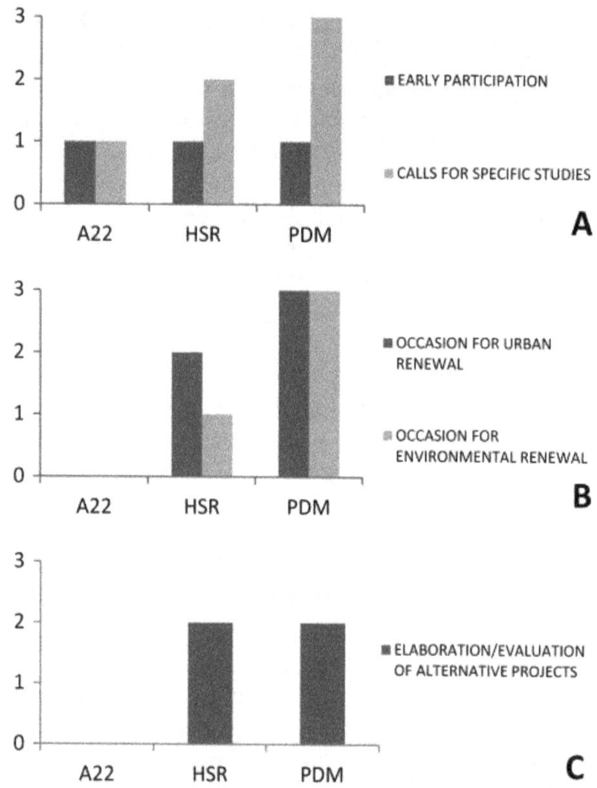

Fig. 5. Graphic representation of the integration process at the Brenner highway (A22), the High Speed Railway (HSR) and the Pedemontana Motorway (PDM)

noticeable in the case of the Pedemontana Lombarda motorway for the criteria and indicators of the integration process (Fig. 5a). The participation of multidisciplinary professionals since the very beginning of the planning phase is an aspect that has to be reconsidered and improved. Good quality levels of urban and environmental regeneration related to the infrastructure project have characterized the Pedemontana Lombarda compensation project (Fig. 5b). In this case, a high level of integration between planners and local institutions has been observed. Concerning the coherence with the existing transportation network, the three projects were developed according to the real transportation needs both at upper and at local levels.

The projects demonstrate growing levels of flexibility in relation to the elaboration and evaluation of the alternatives (Fig. 5c). Nevertheless, other steps forward need to be undertaken that will lead to diminished detachment that occurs between the early infrastructure layout and the project changes required by the regional, provincial, and local institutions.

4.3. Outcomes from the multiscale process

The multiscale process has a variable trend according to the different criteria. In general, a tendency toward the desiderata is noticeable. In some cases, there is a late development of a dynamic, in others, there is a lack of attention, and a criterion loss importance. In all

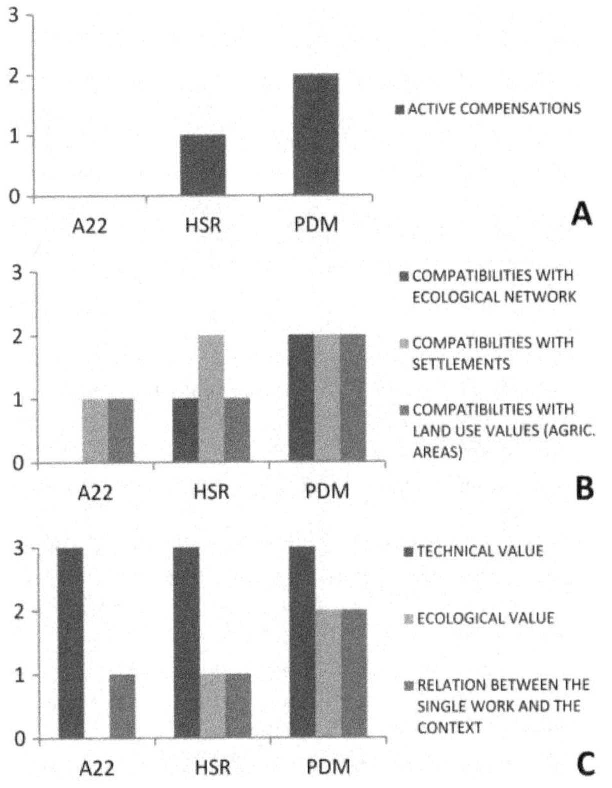

Fig. 6. Graphic representation of the multiscale process at the Brenner highway (A22), the High Speed Railway (HSR) and the Pedemontana Motorway (PDM)

the circumstances, the better condition is verifiable in the Pedemontana project. Two criteria that require some reflections emerge: "multilevel compensation" and "managing land transformation effects". The first is a prerogative mainly of the Pedemontana case (Fig. 6a). The analysis underlines a general high value of the compensation project that elaborates ex-post-remedial solutions to the infrastructure project designed in advance. To improve its outcomes, new planning perspectives should be explored toward the development of a systemic integrated project. This should be based on the inclusion and the management of the cultural and environmental dynamics.

The second criterion is related to the prevention of land exploitation in the areas involved by the infrastructure. This aspect requires deeper attention as satisfying levels of landscape management have not been achieved in the three cases analyzed. The development of an effective multiscale legal framework that engages the region, the provinces, and the municipalities could be a performing solution to manage the areas potentially included in the exploitation.

The considerations carried out for the criterion "supporting system complexity", which underlines a multiscale analysis of the projects, has reached satisfying levels mainly in relation to the intermediate scale and the focused scale. The large-scale level that was about the relations with the external structures (Fig. 6b),

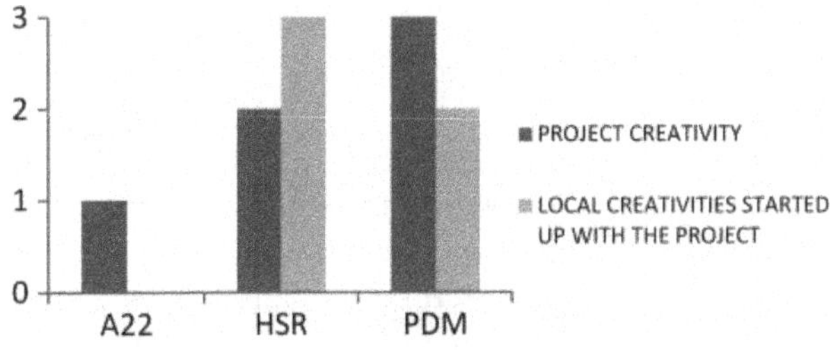

Fig. 7. Graphic representation of the innovation process at the Brenner highway (A22), the High Speed Railway (HSR) and the Pedemontana Motorway (PDM)

still have to be improved and this can be obtained through the development of a systemic project. The relation of a single work as a system: ecological and technical values of the single works such as tunnel, viaduct, trench road, and relation with, and ecological settlement system are improved at the Pedemontana case (Fig. 6c).

4.4. Outcomes from the innovation process

Innovation was one of the prerogatives that drove the choice of all three projects. Therefore, such a process was developed for the three cases. However, some differences that have characterized the development of the process from the 1960s are evident. Innovative models have been used to develop parts of each project by acknowledging outstanding projects developed in the United States, England, France, and Germany. High levels have been reached also at the three projects in relation to the development of experiences of interrelation between institutions indicating governance improvement. To this extent, the Milan Polytechnic carried out the compensation project for the Pedemontana Lombarda together with the local municipalities. Good levels of creativity have been achieved to design the single works. Regarding the novelty used to carry out the project through original solutions as well as enhancing the local creative reactions (laboratories, exhibitions), the second case demonstrated positive results (Fig. 7). The project of the high speed railway had experienced the development of local laboratories, workshops, and exhibitions. The aim was to begin reflections concerning the changing drivers of local landscapes by opening the areas of involvement in the project wide to the communities. This practice was in line with European Landscape Convention principles and guidelines. It was an interesting and innovative aspect to repropose and innovate in other contexts through the involvement of the local cultural associations.

The presented working tool has not yet been tested within the planning processes of any road or railway project. Its application to future linear infrastructure projects from the very beginning of the planning phase is required to evaluate and improve the model effectiveness. In 2010, the model was applied also to assess landscape dynamics related to the highway A22 in south Portugal (Berte, Panagopoulos 2011) in which it was proven that the model can provide a replicable method to reduce the pressure of linear infrastructures on local dynamics in similar projects and enhance cultural dynamics. This can be achieved through the management and support of cultural landscape processes in the regional planning procedures.

Conclusions

The present study provided an analysis of the structure and potentialities of contemporary cultural landscapes. The model has been worked out bearing in mind that the interacting processes between natural and human factors underpin cultural landscape development.

From the present study it can be concluded that:

1. The methodology has provided a reading key that considers four basic landscape processes: identity, integration, multiscale, and innovation.
2. The model was effective and helpful to compare the three case studies as well as to draw a number of conclusions about the presence of cultural landscape dynamics in relation to linear infrastructure projects.
3. It permitted enhancing the outcomes of the planning processes aiming to show when these have led to the development of cultural landscape dynamics.
4. It permitted also to evaluate which processes have reached a good level and which can still be ameliorated and how.
5. The application of this model will result in improved EIA documents taking a more modern and comprehensive approach towards the landscape. In addition, it can be useful at the environmental monitoring phase during and after the construction. Such work with landscape would allow approximating infrastructure sector practices to those mandated in the current EU policies that have recently entered into force in all EU nations as a result of the European Landscape Convention.

References

Alkan, H.; Korkmaz, M.; Tolunay, A. 2009. Assessment of primary factors causing positive or negative local perceptions on protected areas, *Journal of Environmental Engineering and Landscape Management* 17: 20–27. http://dx.doi.org/10.3846/1648-6897.2009.17.20-27

Antonson, H. 2009. Bridging the gap between research and planning practice concerning landscape in Swedish infrastructural planning, *Land Use Policy* 26: 169–177. http://dx.doi.org/10.1016/j.landusepol.2008.02.009

Antonson, H.; Gustafsson, M.; Angelstam, P. 2010. Cultural heritage connectivity. A tool for EIA in transportation infrastructure planning, *Transportation Research Part D: Transport and Environment* 15: 463–472. http://dx.doi.org/10.1016/j.trd.2010.05.003

Antrop, M. 2000. Background concepts for integrated landscape analysis, *Agriculture, Ecosystem and Environment* 77: 17–28. http://dx.doi.org/10.1016/S0167-8809(99)00089-4

Antrop, M. 2005. Why landscapes of the past are important for the future, *Landscape and Urban Planning* 70: 21–34. http://dx.doi.org/10.1016/j.landurbplan.2003.10.002

Baltrenas, P.; Vaitiekūnas, P.; Mincevič, I. 2004. Investigation on the impact of transport exhaust emissions on the air, *Journal of Environmental Engineering and Landscape Management* 12: 3–11.

Berte, E.; Panagopoulos, T. 2011. Applying an interpretive model to assess cultural landscape dynamics in a highway of Portugal, *WSEAS Transactions on Environment and Development* 6: 769–782.

Bogaert, J.; Farina, A.; Ceulemans, R. 2005. Entropy increase of fragmented habitats: a sign of human impact?, *Ecological Indicators* 5: 207–212. http://dx.doi.org/10.1016/j.ecolind.2005.02.002

Castells, M. 1992. The world has changed: can planning change?, *Landscape and Urban Planning* 22: 73–78. http://dx.doi.org/10.1016/0169-2046(92)90009-O

ELC 2000. *European Landscape Convention.* European Treaty Series, 176. Council of Europe, Florence.

Farina, A. 2000. *The Cultural Landscape as a Model for the Integration of Ecology and Economics.* Bioscience, 50: 313–320. http://dx.doi.org/10.1641/0006-3568(2000)050[0313:TCLAAM]2.3.CO;2

Farina, A.; Johnson, A. R.; Turner, S. J.; Belgrano, A. 2003. 'Full' world versus 'empty' world paradigm at the time of globalisation, *Ecological Economics* 45: 11–18. http://dx.doi.org/10.1016/S0921-8009(02)00255-0

Farina, A.; Bogaert, J.; Schipani, I. 2005a. Cognitive landscape and information: new perspectives to investigate the ecological complexity, *Biosystems* 79: 235–240. http://dx.doi.org/10.1016/j.biosystems.2004.09.018

Farina, A.; Santolini, R.; Pagliaro, G.; Scozzafava, S.; Schipani, I. 2005b. Eco-semiotics: a new field of competence for ecology to overcome the frontier between environmental complexity and human culture in the Mediterranean, *Israel Journal of Plant Sciences* 53: 167–175. http://dx.doi.org/10.1560/5UTK-YW53-KDQT-YU48

Gambi, L. 1973. *Una geografia per la storia.* 2nd ed. Einaudi, Torino (in Italian).

Hazen, H. 2008. "Of outstanding universal value": the challenge of scale in applying the World Heritage Convention at national parks in the US, *Geoforum* 39: 252–264. http://dx.doi.org/10.1016/j.geoforum.2007.05.007

Mortberg, U. M.; Balfors, B.; Knol, W. C. 2007. Landscape ecological assessment: a tool for integrating biodiversity issues in strategic environmental assessment and planning, *Journal of Environmental Management* 82: 457–470. http://dx.doi.org/10.1016/j.jenvman.2006.01.005

Naveh, Z. 1998. Ecological and cultural landscape restoration and the cultural evolution towards a post-industrial symbiosis between human society and nature, *Restoration Ecology* 6: 135–143. http://dx.doi.org/10.1111/j.1526-100X.1998.00624.x

Naveh, Z. 2001. Ten major premises for a holistic conception of multifunctional landscapes, *Landscape and Urban Planning* 57: 269–284. http://dx.doi.org/10.1016/S0169-2046(01)00209-2

Panagopoulos, T. 2009. Linking forestry, sustainability and aesthetics, *Ecological Economics* 68: 2485–2489. http://dx.doi.org/10.1016/j.ecolecon.2009.05.006

Opdam, P.; Steingröver, E.; Rooij, S. V.; 2006. Ecological networks: a spatial concept for multi-actor planning of sustainable landscapes, *Landscape and Urban Planning* 75: 322–332. http://dx.doi.org/10.1016/j.landurbplan.2005.02.015

OCS – Osservatorio Città Sostenibili 2005. *Monitoraggio e valutazione ambientale delle infrastrutture di trasporto.* Dal corridoio 5 alla rete regionale. Milano: Franco Angeli (in Italian).

Scazzosi, L. 2004. Reading and assessing the landscape as cultural and historical heritage, *Landscape Research* 29: 335–355. http://dx.doi.org/10.1080/0142639042000288993

Schmitz, M. F.; De Aranzabal, I.; Aguilera, P.; Rescia, A.; Pineda, F. D. 2003. Relationship between landscape typology and socioeconomic structure: scenarios of change in Spanish cultural landscapes, *Ecological Modelling* 168: 343–356. http://dx.doi.org/10.1016/S0304-3800(03)00145-5

Tempesta, T. 2010. The perception of agrarian historical landscapes: a study of the Veneto plain in Italy, *Landscape and Urban Planning* 97: 213–328. http://dx.doi.org/10.1016/j.landurbplan.2010.06.010

Tishler, W. H. 1982. Historical landscapes: an international preservation perspective, *Landscape Planning* 9: 91–103. http://dx.doi.org/10.1016/0304-3924(82)90001-6

Tress, B.; Tress, G.; Fry, G. 2009. Integrative research on environmental and landscape change: PhD students' motivations and challenges, *Journal of Environmental Management* 90: 2921–2929. http://dx.doi.org/10.1016/j.jenvman.2008.03.015

Terkenli, T. S. 2001, Towards a theory of the landscape: the Aegean landscape as a cultural image, *Landscape and Urban Planning* 57: 197–208. http://dx.doi.org/10.1016/S0169-2046(01)00204-3

Wiggering, H.; Dalchow, C.; Glemnitz, M.; Helming, K.; Müller, K.; Schultz, A.; Stachow, U.; Zander, P. 2006. Indicators for multifunctional land use—Linking socio-economic requirements with landscape potentials, *Ecological Indicators* 6: 238–249. http://dx.doi.org/10.1016/j.ecolind.2005.08.014

Elena BERTÈ. Graduated in Architecture from University of Ferrara in 2002 (Italy). She gained her Environmental Engineering PhD in 2009 from University of Trento (Italy). Publications: author / co-author of two scientific papers. Research interests: landscape and environmental planning, cultural landscapes, sustainable infrastructure.

Thomas PANAGOPOULOS. Dr, Professor at the Department of Earth, Sea and Environment, University of Algarve. He gained his Forestry and Natural Environment PhD in 1995 from Aristotle University (Greece). Director of the Master in Landscape Architecture and Executive Board member of UNISCAPE (the European Network of Universities for the implementation of the European Landscape Convention). Publications: author / co-author of ~50 scientific papers. Research interests: landscape reclamation, landscape architecture, spatial organizational dynamics, environmental impact assessment, environmental management.

Bruno ZANON. Graduated in Architecture, Professor at the Department of Civil, Environmental and Mechanical Engineering, University of Trento, Italy. Member of national and international planning associations, and Italian representative in COST Actions. Publications: author / co-author of books and articles published in Italian and international journals. Research interests: spatial organisation dynamics, urban and regional planning, urban and territorial sustainability, environmental sensitivity.

21

IMPACT OF COPPER AND ZINC MIXTURE ON BACTERIAL FLORA OF DIGESTIVE TRACT OF RAINBOW TROUT (*ONCORHYNCHUS MYKISS*)

Janina Šyvokienė, Milda Zita Vosylienė

Nature Research Centre, Akademijos g. 2, LT-08412 Vilnius, Lithuania

Abstract. The effects of 96-hour exposure and within 12 days after exposure to 0.25, 0.125, 0.06 portions of the 96-hour LC50 of Cu+Zn ions mixture on microbiological parameters of rainbow trout digestive tract were studied evaluating quantitative and functional characteristics of bacteria. The extinction of proteolytic bacteria was found in fish affected by 0.25 portion of the 96-hour LC50 of metal mixture. The recovery of proteolytic bacteria was determined in fish exposed to 0.125 and 0.06 portions of the 96-hour LC50 of Cu+Zn ions mixture after 8 days of keeping fish in clean water. The abundance of the investigated functional groups of bacteria was gradually recovering to the control level after 12 days of keeping fish in clean water.

Keywords: water pollution, environment monitoring, environmental sustainability.

Reference to this paper should be made as follows: Šyvokienė, J.; Vosylienė, M. Z. 2013. Impact of copper and zinc mixture on bacterial flora of digestive tract of rainbow trout (*Oncorhynchus mykiss*), *Journal of Environmental Engineering and Landscape Management* 21(4): 288–295. http://dx.doi.org/10.3846/16486897.2012.721372

Introduction

Heavy metals and other man-made pollutants penetrating into the aquatic ecosystems from agriculture, urban and industrial sources, negatively affect the composition of bacterial communities of the digestive system of aquatic organisms, functional activity of intestinal bacteria and the whole physiological status of an organism. Moreover, due to pollution of aquatic ecosystems, separate species of autochthonous bacterial flora or even functional groups of bacteria in the digestive tract of aquatic species can be eliminated (Sugita *et al.* 2005; Bakke-McKellep *et al.* 2007; Navarrete *et al.* 2008).

Most of toxic contaminants enter fish via a food chain and from surrounding water. Whereas heavy metals are characterized by stability, they can persist in environment. Fish is at the top of the aquatic food chain and during its life can accumulate the large amount of metals (Couture, Pyle 2011). Eventually, through the food chains heavy metals can be transmitted to the humans (Saei-Dehkordi, Fallah 2011).

Low concentrations of copper (Cu) and zinc (Zn) ions exist in water bodies naturally. Fish use these metals as microelements (Watanabe *et al.* 1997) by consuming them together with food or directly from water (Handy 1996; Wood 2011). However, both Cu and Zn ions are determined to be toxic to aquatic fish when their levels increase above the background concentrations (Clearwater *et al.* 2002).

Pollution from industry and other anthropogenic activities has heightened the level of copper (Annual report of Water Quality of Rivers of Lithuania 2004). Copper sulphate ($CuSO_4$) is one of the most widely used algicides for phytoplankton control in fish ponds, reservoirs and lakes, due to leaching and runoff water from treated areas, it contaminates aquatic habitats (Gioda *et al.* 2007).

The toxic effects of waterborne and diet borne copper and zinc on fish are rather widely described in literature (Clearwater *et al.* 2002; Handy 2003; Bagdonas, Vosylienė 2006; Kazlauskienė, Vosylienė 2008). The acute, lethal and sublethal effects of Cu and Zn ions on physiological parameters of fish were assessed (De Boeck *et al.* 2004; Kazlauskienė, Vosylienė 2008; Eyckmans *et al.* 2011).

Genotoxic effects of different waterborne Cu and Zn ions mixture concentrations in the blood erythrocytes of *O. mykiss* were observed (Andreikėnaitė 2010).

Induction of toxicity of diet borne Cu and Zn ions have been described in rainbow trout (*Oncorhynchus mykiss*), channel catfish (*Ictalurus punctatus*), common carp (*Cyprinus carpio*) (Lanno *et al.* 1985; Julshamn *et al.* 1988; Handy 1993). Reduced survival, growth and reproduction depending on diet borne metals in fish were determinated (Gundogdu *et al.* 2009). Interaction

Corresponding author: Janina Šyvokienė
E-mail: janina370@yahoo.com

between diet borne metals and gastrointestinal tract of aquatic animals in ecosystems was noted (Liao *et al.* 2011).

Most of the metals ions in water bodies are found to be in the mixtures (or complexes ATSDR, Agency for Toxic Substance and Disease Registry, Draft 2004); therefore, it is important to evaluate effects of separate heavy metals and their mixtures (Kamunde, MacPhail 2011). The acute toxicity of Cu and Zn ions mixture and responses of different functional systems or changes in physiological parameters induced by these metals on *O. mykiss* at different life stages were studied by Kazlauskienė, Vosylienė (2008). According to these authors the effects of heavy metals ions singly could differ from the effects of Cu+Zn mixtures to fish.

Food and feeding conditions significantly influence the composition of gastrointestinal microbiota of fish (Onarheim, Raa 1990; Ringø 1993; Ringø, Olsen 1999; Uchii *et al.* 2006; Korsnes *et al.* 2006; Martin-Antonio *et al.* 2007; Delzenne 2008). Many authors highlight on changes in fish bacterial flora depending on species (Bairagi *et al.* 2002), fish development study (Ringø, Birkbeck 1999; Fjellheim *et al.* 2007), feeding habitat and season (Šyvokienė 1989; Ringø *et al.* 2006, 2010; Šyvokienė *et al.* 2011).

The formation of microorganism populations in the digestive tract of fish was found to take place directly through the surrounding environment (Nayak 2010). Microorganisms compete for the feeding substrates, and with an occurrence of new substrate in the environment quantitative and qualitative changes take place in the endo-system of macroorganisms. Therefore, intensification of fermentation activity of bacteria of one or even several functional groups and at the same time inhibition or even elimination of other groups can take place. It causes changes in metabolites excluded by microorganisms, which affect the physiological status, productivity and immune system of an animal. The narrower the spectrum of qualitative composition and abundance of microorganisms the less are the possibilities to adapt to the changing feeding substrates (Šyvokienė 1989, 1990, 1991). However, the enzymes of digestive tract of fish perform the most important role in these processes (Kuzmina 2005).

In the bacterial communities of the digestive tract of fish dominating facultative bacteria are: *Vibrio, Aeromonas, Enterobacteriaceae* (Ringø, Birkbeck 1999; Al-Harbi, Uddin 2004).

Our previous studies of the impact of heavy metals ions on microorganism communities of fish digestive tract demonstrated the reliance of the changes on the composition of heavy metals or their mixtures, their concentration or exposure time (Šyvokienė, Mickėnienė 1999a, b; Mickėnienė, Šyvokienė 2001; Mickėnienė *et al.* 2007; Šyvokienė *et al.* 2003, 2006).

The purpose of this work was to investigate the effects of 96-hour exposure and within 12 days after the end of exposure to different concentrations of Cu and Zn ions mixture on the bacterial flora of the digestive tract of adult rainbow trout (*Oncorhynchus mykiss.*

1. Materials and methods

Microbiological investigations were carried out at the Laboratory of Genotoxicology of the Institute of Ecology of the Nature Research Centre. Rainbow trout adults were obtained from the Žeimena Fish Hatchery (Švenčionys district, Lithuania) and kept in the holding tanks of about 3000 l capacity supplied with flow-through artesian aerated water. For experiment, fish were transferred from the holding tanks to the aquaria of 40 l capacity and kept for 1 week in the new medium until acclimation, i.e. till they started to swim freely and feed well. The length of fish under study ranged from 15.5 ± 1.1 to 18.0 ± 1.2 cm, and the weight from 32.7 ± 1.0 to 43.1 ± 2.3 g, respectively. Artesian water of high quality was used for dilution. The average hardness of dilution water was approximately 284 mg/l ($CaCO_3$), alkalinity 244 mg/l (HCO_3^-), the mean pH was 8.0, temperature was 12 ± 0.5 °C, and the oxygen concentration ranged from 8 to 10 mg/l. Water in the aquarium was changed daily and fish were fed until satiety with the commercial DANA FEED fish food. Test fish were exposed to 0.25, 0.125 and 0.06 portion of the 96-hour LC50 of metal ions mixture. After the exposure part of fish was taken for microbiological analyses. Whereas other fish of each group were transferred to the new aquaria with constantly aerated metal-free water in which they were kept for 4, 8 and 12 days till analyses. Control fish were affected to the same manipulations (transferred to another aquaria) but were kept all the time in clean water.

The concentrations of Cu^{+2} and Zn^{+2} ions were chosen based on previous studies, which indicated that the 96-hour LC50 of copper was 0.65 mg/l and the 96-hour LC50 of zinc was 3.79 mg/l (Svecevičius, Vosylienė 1996; Svecevičius 1999; Kazlauskienė, Vosylienė 2008). The final concentrations were recalculated according to the amount of copper and zinc ions. Expressions of the metals concentrations studied are shown in Table 1 (Bagdonas, Vosylienė 2006).

Table 1. Concentrations of Cu and Zn in mixture studied

Metals	Concentrations Portion of LC50	mg/l
Copper	0.25	0.16
Zinc	0.25	0.948
Mixture of Cu and Zn	0.25+0.25	0.16+0.948
Copper	0.125	0.08
Zinc	0.125	0.474
Mixture of Cu and Zn	0.125+0.125	0.08+0.474
Copper	0.0625	0.625
Zinc	0.0625	0.238
Mixture of Cu and Zn	0.0625+0.0625	0.04+0.238

By (Bagdonas, Vosylienė 2006)

Copper and zinc sulphate ($CuSO_4{\cdot}5H_2O$), $ZnSO_4{\cdot}7H_2O$, reagent grade 99%) (<REAKHIM> Company, Russia) were used as the toxicants, and stock solutions were prepared by dissolving a necessary amount of salts in distilled water (Svecevičius 2012).

For each microbiological analysis 8–10 fish specimens were used. Populations of aerobic and facultative anaerobic heterotrophic bacteria occurring in the digestive system of fish were counted using the dilution plate technique (Hansen, Olafsen 1999). The surface of rainbow trout specimens was sterilised with 95% ethanol, and then fish were dissected to remove their digestive tract. The content of the digestive tract was removed to sterile Petri dishes. Each set of experiment involved ten specimens of fish. All digestive tract samples were weighed and placed into a test tube, and then nine volumes of diluents were added. The tenfold dilution was further done serially. The volume of 0.1 ml of least dilutions (expected to give 30 to 300 colony-forming units, CFUs), was plated in triplicate on solid media. Incubation was carried out aerobically at 20 °C for seven days, except for Mac Conkey agar plates, which were incubated at 37 °C for 7 days. The number of CFU in the digestive tract of *O. mykiss* was established on four media: tryptone soya agar (OXOID) was chosen for isolation of total heterotrophic bacteria (THB), milk agar for proteolytic bacteria (PB) as a separate group of total heterotrophic bacteria, PB were identified according to the zones of protein (casein) hydrolysis on milk agar, MacConkey agar (OXOID) was used for total coliform bacteria (TCB). Bacterial colonies appearing on each plate were counted, and a CFU per 1 g (wet weight) of the digestive tract content was obtained (Kuznetsov, Dubinina 1989).

All data were analysed using standard statistical methods. Mean values, standard deviations and 95% confidence intervals were assessed.

The average values and mean square deviations were calculated (Sakalauskas 2003).

2. Results and discussion

Normal intestinal microflora of aquatic organisms is a natural biosorbent of toxic substances penetrating from the environment and produced in the host organism. Therefore, it responds to pollutants most sensitively (Šyvokienė, Mickėnienė 1999a). The dominating heterotrophic bacteria were observed among bacterial communities of the digestive tract of control fish (Figs 1–4) in the experiment of 96-hour exposure of rainbow trout to different parts of LC50 of Cu+Zn ions mixture.

The abundance of heterotrophic bacteria constituted 5.2×10^4 (4.72 log CFU g) CFU g of cells per gram of the content of the fish digestive tract. After 96-hour exposure to 0.06 portion of the LC50 of Cu+Zn ions mixture (concentrations of metal ions in this portion of mixture were ~16 times lower than 96-hour LC50 of Cu and 96-hour LC50 of Zn, respectively) the abundance of total heterotrophic bacteria decreased 6 times; after the same duration exposure to 0.125 portion of the LC50 of Cu+Zn ions mixture (concentrations of metal ions in this portion of mixture ~8 times were lower than 96-hour LC50 of Cu and 96-hour LC50 of Zn, respectively) −26 times; and after exposure to 0.25 portion of the LC50 of Cu+Zn ions mixture (concentrations of metal ions in this portion of mixture were lower ~4 times than 96-hour LC50 of Cu and 96-hour LC50 of Zn, respectively) −18 times as compared with this Parameter of control fish.

The exposure to the least concentration −0.06 portion of LC50 Cu+Zn mixture induced two-fold elevation of the abundance of *O. mykiss* proteolytic bacteria, whereas the density of PB in digestive tract of fish lowered more than 11.5 times after exposure to medium −0.125 portion of the LC50 Cu+Zn mixture.

Exposure of fish to the highest (0.25 portion of the 96-hour of the LC50 Cu+Zn mixture) concentration induced the extinction of proteolytic bacteria from bacterial communities of the digestive tract of rainbow trouts. In contrast, the abundance of coliform bacteria increased 140 times, as compared to the controls (Fig. 1).

Our previous studies of rainbow trout exposure to a range of concentrations from 0.16 to 0.67 parts of the 96-hour LC50 of equitoxic copper-zinc mixture showed that the mixture at concentration 0.16 part of the 96-hour LC50 induced significant changes in microbiological parameters. The abundance of heterotrophic bacteria decreased more than 2 times as compared to the controls ($p < 0.05$) and that of total coliform bacteria lowered, but the difference was not significant as compared to the controls. The densities of cultivable total coliform bacteria actually increased following the addition of the toxicant, possibly as a result of the release of organic carbon from poisoned, copper-zinc sensitive organism (Šyvokienė *et al.* 2006).

During experiment the recovery of the balance of bacterial flora of the digestive tract of rainbow trout after 4, 8 and 12 days of keeping fish in clean water was observed. The recovery of proteolytic bacteria in the digestive tract of fish groups affected to 0.125 and 0.25 portions of 96-hour of the LC50 Cu+Zn after 4 days keeping fish in clean water was not determined (Fig. 2). However, the recovery of the abundance of bacteria of all functional groups in bacterial communities of the digestive tract of *O. mykiss* was detected after 8 days of keeping fish in clean water. While, the restoration of proteolytic bacteria in bacterial communities of the digestive tract was observed almost entire only after 12 days of keeping rainbow trouts in clean water (Fig. 3).

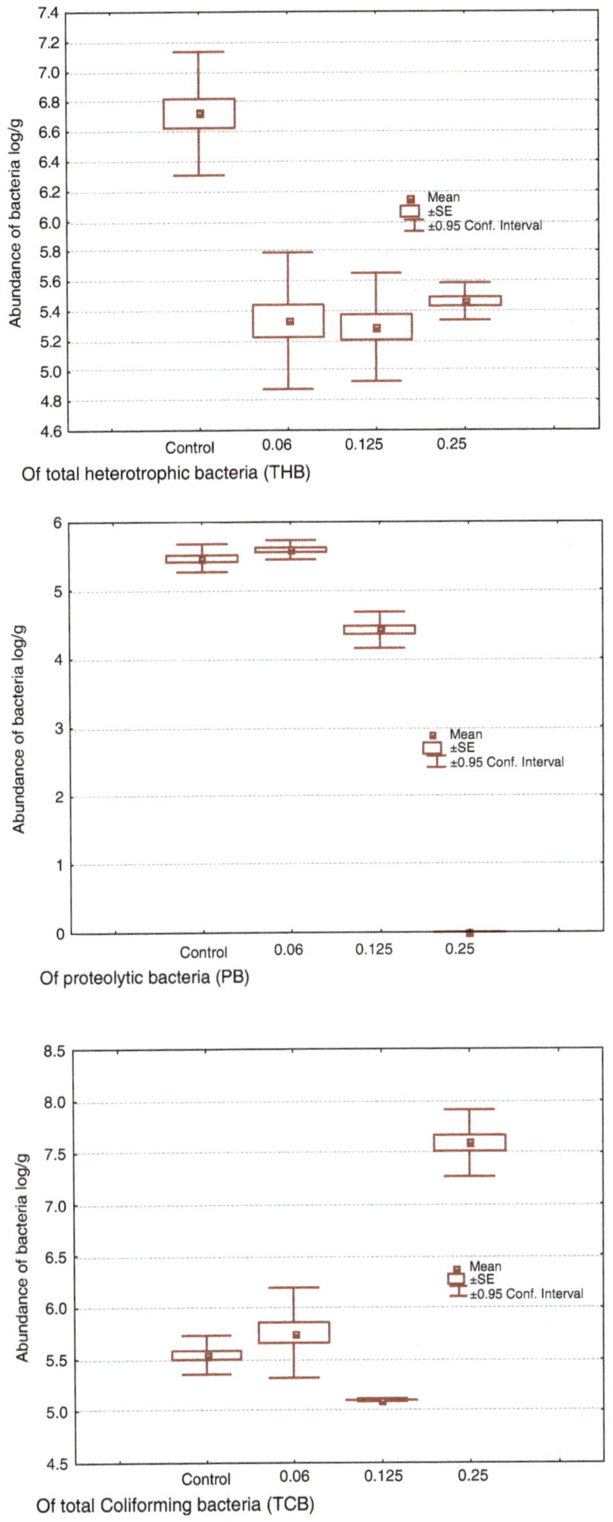

Fig. 1. The impact of the 96-hour exposure to the LC50 Cu+Zn ions mixture on the abundance of bacteria in digestive tract of rainbow trout (Oncorhynchus mykiss)
Concentration in portion of equitoxic 96-hour LC50 mixture

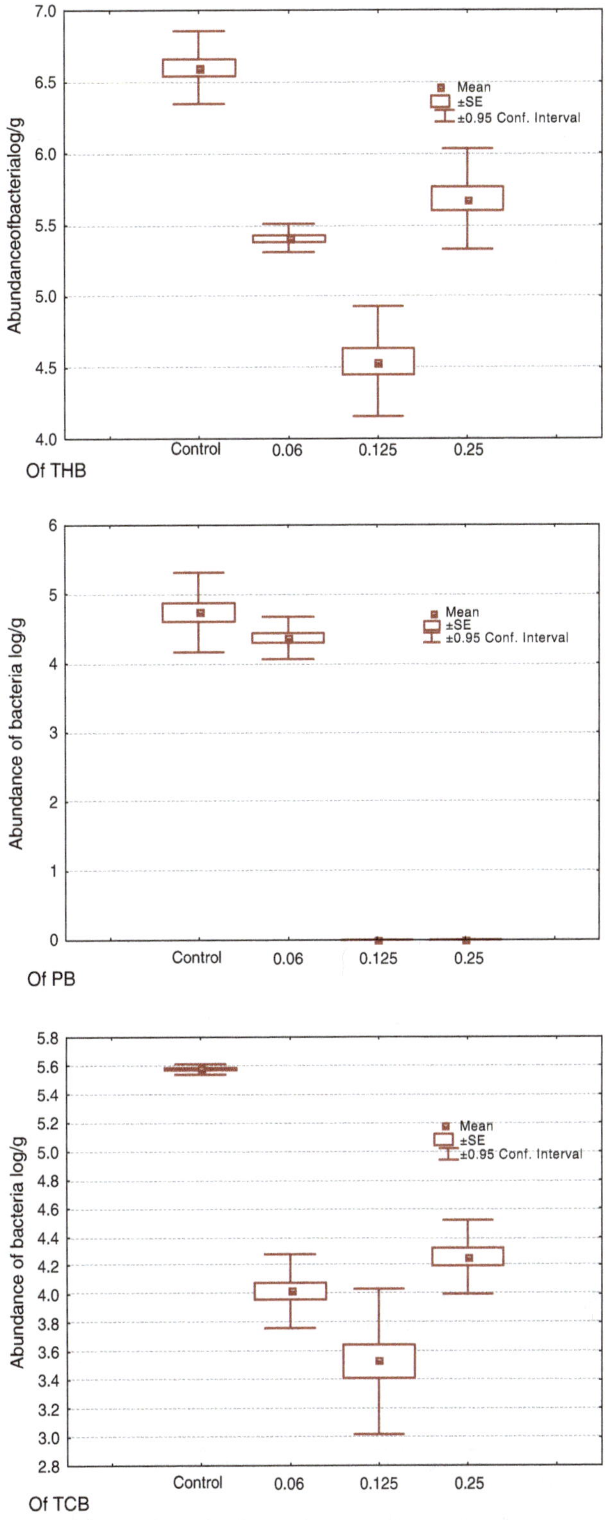

Fig. 2. The impact of the 96-hour exposure to the LC50 Cu+Zn ions mixture on abundance of bacteria in digestive tract of trouts after 4 days keeping fish in clean water
Concentration in parts of equitoxic 96-hour LC50 mixture

The constantly changing environment is the main factor controlling the growth and development of microorganisms. Experimental studies on heavy metals toxicity demonstrate the negative effects of these contaminants on the structures of bacterial communities of the digestive tract and bacterial digestion of aquatic organisms as well as existence of imbalance of bacterial communities. Contamination of environment

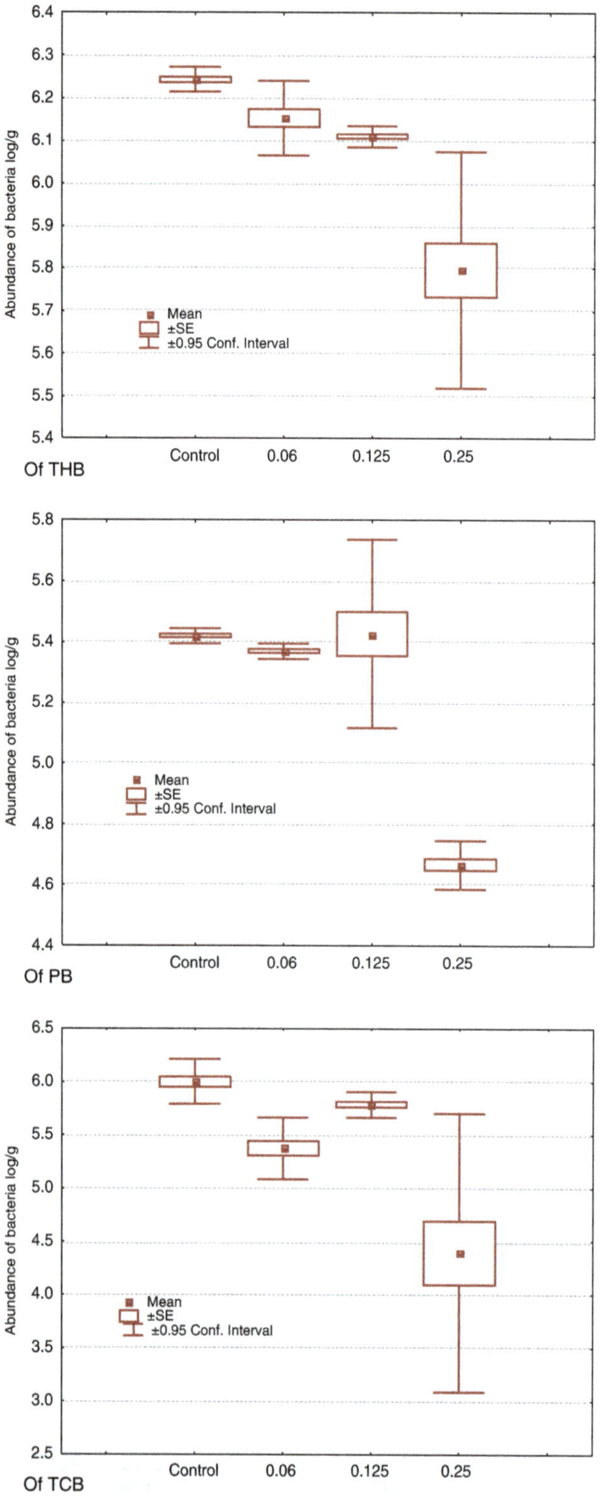

Fig. 3. The impact of the 96-hour exposure to the LC50 of Cu+Zn ions mixture on abundance of bacteria in digestive tract of trouts after 8 days keeping fish in clean water Concentration in parts of equitoxic 96-hour LC50 mixture

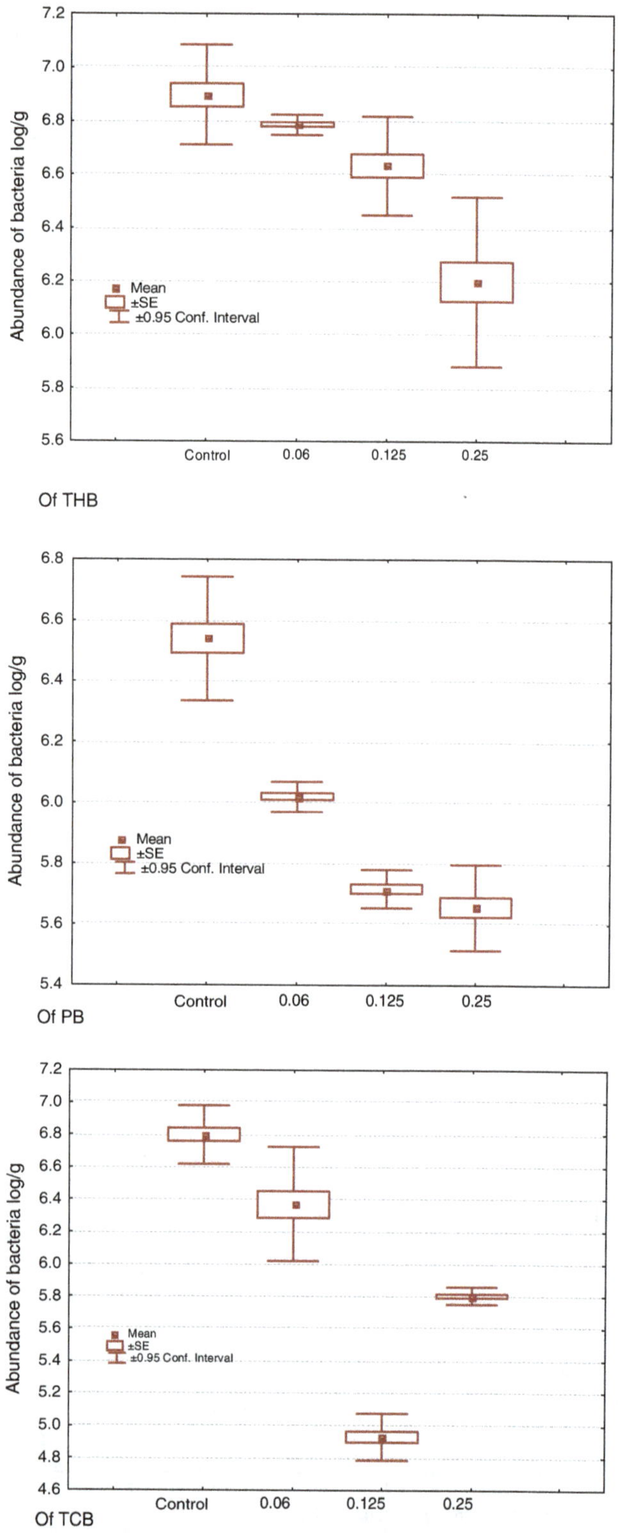

Fig. 4. The impact of the 96-hour exposure to the LC50 of Cu+Zn ions mixture on abundance of bacteria in digestive tract of trouts after 12 days keeping fish in clean water Concentration in parts of equitoxic 96-hour LC50 mixture

was found to reveal a tendency of extinction of some proteolytic bacteria beneficial for the digestion of an organism (Šyvokienė 1989, 1990, 1991; Nayak 2010). Summarising obtained data, we can consider that proteolytic bacteria of the digestive tract of rainbow trout were the most sensitive to (Cu+Zn ions) mixture exposure. Proteolytic bacteria determine the regularities of the formation and functioning of autochthonous bacterial flora in the early ontogenesis of fish (Šyvokienė *et al.* 1995, 1999b).

After exposure to higher (0.125 portion of the LC50) concentration of mixture of studied metals the disturbances in the bacterial flora of the digestive tract of rainbow trout were observed. The decrease in the abundance of vitally important functional groups of bacteria was detected (Figs 2–4), a tendency of extinction of proteolytic bacteria was observed (Figs 1, 2). The recovery of PB was observed after 8 days of keeping fish in clean water (Figs 2, 3). According to Witeska and Kosciuk (2003) waterborne heavy metals, initially bound to gills and subsequently deposited in other tissues, might affect the fish, even if toxic agents were removed from the water.

Conclusions

1. The results of this study indicate that exposure of rainbow trout to 0.06 portion of the 96-hour LC50 of Cu+Zn ions mixture induced a six-fold decrease in the abundance of total heterotrophic bacteria in digestive tract of rainbow trout. Exposure to Cu+Zn mixture at concentration 0.125 portion of the LC50 decreased the abundance of heterotrophic bacteria 26 times and to Cu+Zn mixture at concentration 0.25 portion of the 96-hour LC50 – 18 times.

2. The imbalance of trout's bacterial flora and the extinction of proteolytic bacteria from bacterial communities of the digestive tract of exposed fish were determined after treatment to the highest concentration (0.25 portion of the LC50 of Cu+Zn mixture). The abundance of total coliform bacteria group after exposure with the same metal mixture concentration increased 140 times.

3. The recovery of investigated proteolytic bacteria in bacterial cenoses of the digestive tract of *O. mykiss* was detected only after 8 days of transfering fish to clean water and the restoration of proteolytic bacteria in bacterial communities of the digestive tract was observed almost entire only after 12 days of transfering fish to clean water.

Acknowledgements

We express our sincere thanks to Dr Liongina Mickėnienė for assistance in carrying out the experiment and Dr Laura Andreikėnaitė for assistance in preparing the article for submission. Authors would like to thank the reviewers for their helpful comments to improve this article.

References

Al-Harbi, A. H.; Uddin, M. N. 2004. Seasonal variation in the intestinal bacterial flora of hybrid tiliapia (*Oreochromis niloticus x Oreochromis aureus*) cultured in earthen ponds in Saudi Arabia, *Aquaculture* 229: 37–44. http://dx.doi.org/10.1016/S0044-8486(03)00388-0

Andreikėnaitė, L. 2010. *Genotoxic and cytotoxic effects on contaminants discharged from the oil platforms in fish and mussels*: Phd Thesis. Lithuania, Vilnius.

ATSDR (Agency for Toxic Substance and Disease Registry), Draft 2004. *Evaluation of the Toxicology of Chemical Mixtures Commonly Found at Hazardous Waste Sites, US Public Health Service, US Department of Human Services.* Atlanta, GA.

Bagdonas, E.; Vosylienė, M. Z. 2006. A study of toxicity and genotoxicity of copper, zinc and their mixture to rainbow trout (*Oncorhynchus mykiss*), *Biologija* 1: 8–13.

Bairagi, A.; Ghoch, K.; Sen, S. K.; Ray, A. K. 2002. Enzyme producing bacterial flora isolated from fish digestive tracts, *Aquaculture International* 10: 109–121. http://dx.doi.org/10.1023/A:1021355406412

Bakke-McKellep, A. M.; Koppang, E. O.; Gunnes, G.; Sanden, M.; Hemre, G. I.; Landsverk, T.; Krogdahl, A. 2007. Histological, digestive, metabolic, hormonal and some immune factor responses in Atlantic salmon. *Salmon salar* L. fed genetically modified soybeans, *Journal of Fish Diseases* 30: 65–79. http://dx.doi.org/10.1111/j.1365-2761.2007.00782.x

Clearwater, S. J.; Farag, A. M.; Meyer, J. S. 2002. Bioavailability and toxicity of diet borne copper and zinc to fish, *Comparative Biochemical Physiology C* 132: 269–313.

Couture, P.; Pyle, G. 2011. Field studies on metal accumulation and effects in fish, *Fish Physiology* 31: 417–473. http://dx.doi.org/10.1016/S1546-5098(11)31009-6

De Boeck, G.; Meeus, W.; De Coen, W.; Blust, R. 2004. Tissue-specific Cu bioaccumulation patterns and differences in sensitivity to waterborne Cu in three freshwater fish: rainbow trout (*Oncorhynchus mykiss*), common carp (*Cyprinus carpio*), and gibel carp (*Carassius auratus gibelio*), *Aquatic Toxicology* 70: 179–188. http://dx.doi.org/10.1016/j.aquatox.2004.07.001

Delzenne, N. M.; Cani, P. D. 2008. Gut microflora is a key player in host energy homeostasis, *Medical Science* 24: 505–510.

Eyckmans, M.; Celis, N.; Horemans, N.; Blust, R.; De Boeck, G. 2011. Exposure to waterborne copper reveals differences in oxidative stress response in three freshwater fish species, *Aquatic Toxicology* 103: 112–120. http://dx.doi.org/10.1016/j.aquatox.2011.02.010

Fjellheim, A. J.; Playfoot, K. J.; Skjermo, J.; Vadstein, O. 2007. Vibrionaceae dominates the microflora antagonistic towards *Listonella anguillarum* in the intestine of cultured Atlantic cod (*Gadus morhua* L.) larvae, *Aquaculture* 269: 98–106. http://dx.doi.org/10.1016/j.aquaculture.2007.04.021

Gioda, C. R.; Lissner, L. A.; Pretto, A.; da Rocha, J. B. T.; Schetinger, M. R. C.; Neto, J. R.; Morhc, V. M.; Loro, V. L. 2007. Exposure to sublethal concentrations of Zn (II) and Cu (II) changes biochemical parameters in *Leporinus obtusidens*, *Chemosphere* 69: 170–175. http://dx.doi.org/10.1016/j.chemosphere.2007.04.008

Gundogdu, A.; Harmantepe, F. B.; Dogan, G.; Karsli, Z.; Yagan Asci, M. 2009. Effects of dietborne copper on accumulation in the tissues and organs, growth and feed utilization of rainbow trout (*Oncorhynchus mykiss*,

Walbaum, 1792) juvenile, *Journal of Animal and Veterinary and Advances* 8: 2495–2502.

Handy, R. D. 2003. Chronic effects of copper exposure versus endocrine toxicity: two sides of the some toxicological process?, *Comparative Biochemistry and Physiology* 135A: 25–38.

Handy, R. D. 1993. The effect of acute exposure to dietary Cd and Cu on organ toxicant concentrations in rainbow trout, *Oncorhynchus mykiss, Aquatic Toxicology* 27: 1–14. http://dx.doi.org/10.1016/0166-445X(93)90043-Z

Handy, R. D. 1996. Dietary exposure to toxic metals in fish, in Taylor, E.W. (Ed.). *Toxicology of Aquatic Pollution: Physiological, Cellular and Molecular Approaches*, Cambridge, England: Cambridge University Press, 29–60. http://dx.doi.org/10.1017/CBO9780511735516.004

Hansen, G. H.; Olafsen, J. A. 1999. Bacterial interactions in early life stages of marine cold water fish, *Microbiologycal Ecology* 38: 1–26. http://dx.doi.org/10.1007/s002489900158

Julshamn, K.; Andersen, K. J.; Ringdal, O.; Brenna, J. 1988. Effect of dietary copper on the hepatic concentration and subcellular distribution of copper and zinc in the rainbow trout (*Salmo gairdneri*), *Aquaculture* 73: 143–155. http://dx.doi.org/10.1016/0044-8486(88)90049-X

Kamunde, C.; MacPhail, R. 2011. Effect of humic acid during concurrent chronic waterborne exposure of rainbow trout (*Oncorhynchus mykiss*) to copper, cadmium and zinc, *Safety* 74: 259–269. http://dx.doi.org/10.1016/j.ecoenv.2010.10.007

Kazlauskienė, N.; Vosylienė, M. Z. 2008. Characteristic features of the effect of Cu and Zn mixtures on rainbow trout *oncorhynchus mykiss* in ontogenesis, *Polish Journal of Environmental Studies* 17: 291–293.

Korsnes, K.; Nicolaisen, O.; Skar, C. K.; Nerland, A. N.; Bergh, O. 2006. Bacteria in the gut of juvenile cod *Gadus morhua* fed live feed enriched with four different commercial diets, *ICES Journal of Marine Science* 63: 296–301. http://dx.doi.org/10.1016/j.icesjms.2005.10.012

Kuzmina, V. V. 2005. *Physiological and biochemical principles of exotrophy processes in fish*. Moscow: Nauka (in Russian).

Kuznetsov, S. I.; Dubinina, G. A. 1989. *Methods for investigation of aquatic microorganisms.* Moscow: Nauka (in Russian).

Lanno, R. P.; Slinger, S. J.; Hilton, J. W. 1985. Maximum tolerable and toxicity levels of dietary copper in rainbow trout (*Salmo gairdneri* Richardson), *Aquaculture* 49: 257–268. http://dx.doi.org/10.1016/0044-8486(85)90084-5

Liao, Ch. M.; Ju, Y. R.; Chen, W. Y.; Chen, B. Ch. 2011. Assessing the impact of waterborne and dietborne cadmium toxicity on susceptibility risk for rainbow trout, *Science of the Total Environment* 409: 503–513. http://dx.doi.org/10.1016/j.scitotenv.2010.10.044

Martin-Antonio, B.; Manchado, M.; Infante, C.; Zerolo, R.; Labella, A.; Alonso, C.; Borrego, J. J. 2007. Intestinal microbiota variation in Senegalese sole (*Solea senegalensis*) under different feeding regimes, *Aquaculture Research* 38: 1213–1222. http://dx.doi.org/10.1111/j.1365-2109.2007.01790.x

Mickėnienė, L.; Šyvokienė, J. 2001. Changes of the diversity of the bacteriocenoses in the digestive tract of fish under the impact of heavy metals, *Ekologija* 4: 11–15.

Mickėnienė, L.; Šyvokienė, J.; Stasiūnaitė, P. 2007. Effect of copper ions on growth and bacterial abundance in the intestinal tract of rainbow trout (*Oncorhynchus mykiss*) larvae, *Acta Zoologica Lituanica* 17: 16–22.

Ministry of Environment of the Lithuanian Republic 2004. *Annual report of Water Quality of Rivers of Lithuania*. Vilnius (in Lithuanian).

Navarrete, P.; Mardones, P.; Opazo, R.; Espejo, R.; Romero, J. 2008. Oxytetracycline treatment reduces bacterial diversity of intestinal microbiota of Atlantic salmon, *Journal of Aquatic Animal Health* 20: 177–183. http://dx.doi.org/10.1577/H07-043.1

Nayak, S. K. 2010. Role of gastrointestinal microbiota in fish, *Aquaculture Research* 41: 1553–1573. http://dx.doi.org/10.1111/j.1365-2109.2010.02546.x

Onarheim, A. M., Raa, J. 1990. Characterization and possible biological significance of autochthonous flora in the intestinal mucosa of seawater fish, in Lesel, R. (Ed.). *Microbiology in Poicilotherms*, Amsterdam, Netherlands: Elsevier, 197–201.

Ringø, E.; Birkbeck, T. H. 1999. Intestinal microflora of fish larvae and fry, *Aquaculture Research* 30: 73–93. http://dx.doi.org/10.1046/j.1365-2109.1999.00302.x

Ringø, E.; Olsen, R. E. 1999. The effect of diet on aerobic bacterial flora associated with intestinal of arctic char (*Salvelinus aloinus*), *Journal of Applied Microbiology* 86: 22–28. http://dx.doi.org/10.1046/j.1365-2672.1999.00631.x

Ringø, E. 1993. The effect of chronic oxide Cr203 on aerobic bacterial population associated with the intestinal epithelial mucosa of arctic char *Salvilinus alpinus* (L.), *Canadian Journal of Microbiology* 39: 1169–1173. http://dx.doi.org/10.1139/m93-177

Ringø, E.; Olsen, R. E.; Gifstadt, T. O.; Dalmo, R. A.; Amlund, H.; Hemre, G. I.; Bakke, A. M. 2010. Prebiotics in aquaculture: a review, *Aquaculture Nutrition* 16: 117–136. http://dx.doi.org/10.1111/j.1365-2095.2009.00731.x

Ringø, E.; Sperstad, S.; Myklebust, R.; Refstie, S.; Krogdahl, A. 2006. Characterisation of the microbiota associated with intestine of Atlantic cod (*Gadus morhua L.*): the effect of fish meal, standard soybean meal and a bioprocesses soybean meal, *Aquaculture* 261: 829–841. http://dx.doi.org/10.1016/j.aquaculture.2006.06.030

Saei-Dehkordi, S. S.; Fallah, A. A. 2011. Determination of copper, lead, cadmium and zinc content in commercially valuable fish species from the Persian Gulf using derivative potentiometric stripping analysis, *Microchemical Journal* 98: 156–162. http://dx.doi.org/10.1016/j.microc.2011.01.001

Sakalauskas, V. 2003. *Duomenu analizė su STATISTIKA*. Vilnius: Margi raštai, 83–148.

Sugita, H.; Kurosaki, M.; Okamura, T.; Yamamoto, S.; Tsuchiya, C. 2005. The culturability of intestinal bacteria of Japanese coastal fish, *Fisheries Science* 71: 956–958. http://dx.doi.org/10.1111/j.1444-2906.2005.01050.x

Svecevičius, G. 1999. Acute toxicity of zinc to common freshwater fishes of Lithuania, *Acta Zoologica Lituanica* 9(2): 114–118.

Svecevičius, G. 2012. Avoidance of Copper and Zinc by Rainbow Trout *Oncorhynchus mykiss* Pre-exposed to Copper, *Bulletin of Environmental Contamination and Toxicology* 88: 1–5. http://dx.doi.org/10.1007/s00128-011-0426-5

Svecevičius, G.; Vosylienė, M. Z. 1996. Acute toxicity copper to common freshwater fishes of Lithuania, *Ekologija* 2: 17–21.

Šyvokienė, J. 1990. Symbiotic digestion in hydrobionts and insects, in Lesel, R. (Ed.). *Microbiology in poecilotherms*, New York, Oxford, Amsterdam: Elsevier, 323–326.

Šyvokienė, J. 1991. Interrelation between microorganisms and its digestive tract microorganisms, *Ekologija* 4: 37–44.

Šyvokienė, J.; Mickėnienė, L. 1999a. The effect of heavy metals on microorganisms of the digestive tract of hydrobionts, in Lovejoy, D. A. (Ed.). *Heavy metals in the environment: an integrated approach*. Vilnius, 248–255.

Šyvokienė, J.; Mickėnienė, L. 1999b. Microflora of the digestive tract of rainbow trout (*Oncorhynchys mykiss*) exposed to the mixture of heavy metals, in *Proceedings of SECOTOX 99 Fifth European Conference on Ecotoxicology and Environmental Safety*, 15–17 March, Munich, Germany, 11 p.

Šyvokienė, J. 1989. *Symbiotic digestion in hydrobionts and insects*. Vilnius: Mokslas (in Russian).

Šyvokienė, J.; Mickėnienė, L.; Svecevičius, G. 2006. Acute toxicity of copper-zinc mixture to rainbow trout (*Oncorhynchus mykiss*) and its impact on the bacterial flora of the digestive tract, *Acta Zoologica Lituanica* 23: 235–240.

Šyvokienė, J.; Mickėnienė, L.; Petrauskienė, L.; Stasiūnaitė, P. 1995. Effects of heavy metals on microflora in the digestive tract of the rainbow trout (*Salmo gairdneri* Rich.), *Ekologija* 1: 75–79.

Šyvokienė, J.; Stankus, S.; Andreikėnaitė, L. 2011. Bacterioflora of digestive tract of fishes *in vitro*, *Veterinarija ir zootechnika* 56: 93–101.

Šyvokienė, J.; Stasiūnaitė, P.; Mickėnienė, L. 2003. The impact of municipal wastewater and heavy metal mixture on larvae of rainbow trout (*Oncorhynchus mykiss*), *Acta Zoologica Lituanica* 1: 372–378.

Uchii, K.; Matsui, K.; Yonekura, R.; Tani, K.; Kenzaka, T.; Nasu, M.; Kawabata, Z. 2006. Genetic and physiological characterization of the intestinal bacterial microbiota of bluegill (*Lepomis macrochirus*) with three different feeding habits, *Microbiological Ecology* 51: 277–284. http://dx.doi.org/10.1007/s00248-006-9018-z

Watanabe, T.; Kiron, V.; Satoh, S. 1997. Trace minerals in fish nutrition, *Aquaculture* 151: 185–207. http://dx.doi.org/10.1016/S0044-8486(96)01503-7

Witeska, M.; Kosciuk, B. 2003. The changes in common carp blood after short-term zinc exposure, *Environment Science and Pollution Research* 10: 284–286. http://dx.doi.org/10.1065/espr2003.07.161

Wood, C. M.; Farrell, A. P.; Brauner, C. J. 2011. Homeostasis and toxicology of essential metals, *Fish Physiology* 31: 1–497. http://dx.doi.org/10.1016/S1546-5098(11)31001-1

Janina ŠYVOKIENĖ. Dr Habil, Institute of Ecology of Nature Research Centre. During the latter decade she has published 53 scientific articles in publications included in the list of Information Sciences Institute (ISI) and in other reviewed publications abroad. 10 scientific articles have been published in reviewed editions with citing index in the database "ISI Web of Science".She has given 30 discourses in local and international conferences and symposiums. The main research trends in the Laboratory of Genotoxicology – ecotoxic effects of environmental pollution and ecological changes in freshwater and marine ecosystems. Assessment of genotoxic, cytotoxic and mutagenic effects of chemical compounds in aquatic organisms. Genotoxicity in marine and freshwater ecosystems. Peculiarities of ecotoxic effects and biodegradation of contaminants in microorganism communities of digestive tract of aquatic organism. Species and genera identification in PAH-degradation bacteria from digestive tract of aquatic organism using 16S rRNA sequencing methods. Research interests include water animals (fish, molluscs and crayfish) ecotoxicology, ecology of microorganisms, trophology, physiology of digestion, environmental researches, hydrocarbon-degrading bacteria identification using RNA sequencing methods.

Milda Zita VOSYLIENĖ. Dr, former Senior Researcher of Laboratory of Ecology and Physiology of Hydrobionts, Institute of Ecology of Nature Research Centre.
During the latter decade she has published 34 scientific articles in publications included in the list of Information Sciences Institute (ISI) and other reviewed publications abroad. 12 scientific articles have been published in reviewed editions with citing index in the database "ISI Web of Science". She has participated and given 20 discourses in local and international conferences and symposiums. In 2007 she and co-authors have been awarded the Lithuanian Science Award in the field of applied science activity (experimental development) for the cycle of works "Application of Biological Tests in Assessment of Water Environment Toxicity". The main research trends – functioning of aquatic animals under conditions of anthropogenic impact, regularities and mechanisms of adaptation in ontogenesis, experimental researches. Research of natural water and sediments ecotoxicity, effect of different chemical materials (heavy metals and their mixtures, amonium, oil products) on biomarkers of fish. Assessment of wastewater effect on morphological and physiological markers of fish. Scientific research interests include fish ecotoxicology, ecological physiology, fish physiology, ichthyohaematology, ecology, environmental research.

22

RESEARCH ON CORRELATION BETWEEN NITROGEN COMPOUNDS, IRON AND MANGANESE CONCENTRATIONS IN DRINKING WATER SUPPLY SYSTEMS

Marina Valentukevičienė[1], Agnieszka Karczmarczyk[2], Anželika Jurkienė[3], Auksė Grigaitytė[4]

[1,3,4]*Department of Water Management, Vilnius Gediminas Technical University, Saulėtekio al. 11, LT-10223 Vilnius, Lithuania*

[2]*Faculty of Civil and Environmental Engineering, Warsaw University of Life Science, ul. Nowoursynowska 159, 02-776 Warsaw, Poland*

Abstract. A wide range of material sources potentially contribute to contaminant loads in potable water. The ability of water supply systems to act as emission control barriers to tap water micro pollutants, thereby providing environmental benefits in addition to potable water savings, have not been fully explored. This paper investigates the sources, presence and potential fate of a selection of nitrogen micro pollutants in water supply systems. All of the investigated compounds are listed under the requirements to the quality of water intended for human consumption. Significant water quality changes are identified. A wide range of potential treatment trains are available for water treatment and reuse but treatment efficiency data for nitrogen substances is very limited. Nitrogen substances removal through water treatment is following to be predominantly due to ammonium ions, nitrate and nitrite limited concentrations requirement, with only minor contributions to the water supply network. The majority of conventional water treatment plants periodically supply water with nitrogen compounds residual to the potable water distribution system. Hence, it is important to ensure that other nitrogen sources control options (e.g. pipelines materials, and groundwater sources controls) for potable water supply continue to be pursued, in order that nitrogen compounds emissions from these sources are effectively reduced and/or phased out as required under the demands of the requirements to the quality of water intended for human consumption. The aim of this research was to estimate the correlation between nitrogen compounds, iron and manganese concentrations in drinking water supply pipelines in Vilnius (Lithuania) and Warsaw (Poland).

Keywords: nitrogen substances, ammonium, nitrate, nitrite, drinking water, water pollution.

Reference to this paper should be made as follows: Valentukevičienė, M.; Karczmarczyk, A.; Jurkienė, A.; Grigaitytė, A. 2013. Research on correlation between nitrogen compounds, iron and manganese concentrations in drinking water supply systems, *Journal of Environmental Engineering and Landscape Management* 21(3): 216–223. http://dx.doi.org/10.3846/16486897.2012.674036

Introduction

Drinking water is prepared only from ground water sources in all Lithuanian settlements (Jaskelevicius, Lynikiene 2009). All around the EU countries open water sources (rivers, lakes, desalinated sea water) are used (Ranieri, Swietlik 2010). Drinking water safety and quality standards of Lithuania and Poland are based on European Directive 98/83/EC on the quality of water intended for human consumption and World Health Organization Guidelines on Drinking-water Quality (2004). The main problems of water quality decrease occur during water supply and stagnation in distributing networks (Agatemor, Okolo 2007; Nawrocki *et al.* 2010), when water quality indicators and toxic parameters were increased till unacceptable values (Agatemor, Okolo 2008; Venkatesan, Swaminathan 2009). Nitrogen and its substances are associated with significant water quality decrease to water users. The most toxic nitrogen substances (Gladwin *et al.* 2004), nitrite and nitrate, generally are found in shallow wells water. It is the result of nitrogen-based fertilizers usage in agriculture. During the nitrification process nitrifying bacteria oxidize ammonium ions and ammonia into nitrite and finally into nitrate. Nitrification process is very fast, so generally detectable compound is nitrate (final stage of nitrification process). Allowed concentrations are: for nitrite – 0.50 mg/L, nitrate – 50 mg/L, most harmful of these is the nitrite.

Corresponding author: Marina Valentukevičienė
E-mail: marina.valentukeviciene@vgtu.lt

Ammonium is described as indicating variable and its regulated concentration is 0.50 mg/L.

It is known that the microorganisms that occurred in raw water sources also have favourable conditions to survive in pipelines. Ammonium and nitrate that naturally exist in the groundwater create favourable conditions for anaerobic ammonium oxidizing bacteria and nitrite reducing bacteria (Berry, Raskin 2006). Therefore, the total growth of microorganisms can increase as well (Markku *et al.* 2004).

Microbiological nitrification process can cause interference of chemical and microbiological changes of water quality at water supply network (Lee *et al.* 2005; Lipponen *et al.* 2002). Certain material of pipelines (Neville 2001), dissolved oxygen concentration, admixtures of water and other factors (Lehtola *et al.* 2005; McNeill, Edwards 2002) create favourable conditions for vital activities of microorganisms (Bonadonna *et al.* 2009).

This investigation presents comparison research on using two different water sources for the evaluation of nitrogen compounds diversity (Mazeikiene *et al.* 2010). Ground water from Vilnius (Lithuania) was chosen as it is natural and has no negative influence on water quality. Open sources water in Warsaw (Poland) was strongly affected by using physical and chemical water treatment methods. The research on comparison of water quality changes was carried out using water samples from both water distributing systems. The novelty of this research is related to all findings about the results showed that distribution networks in boreal region are potential environments for nitrifying bacteria, and in some conditions nitrite accumulation may exceed the acceptable concentration (Lipponen *et al.* 2002). Some additional water quality characteristics (ammoniumions, nitrates, iron and manganese concentrations) were absent from previous research.

The aim of this research was to estimate the correlation between nitrogen compounds, iron and manganese concentrations in drinking water supply systems of Vilnius (Lithuania) and Warsaw (Poland), in pipelines networks from different water sources (groundwater and open waters sources).

1. Materials and methods

In order to evaluate changes of nitrogen compounds in water supply network, Vilnius and Warsaw city's water supply systems were selected.

Vilnius is supplied with groundwater from water reservoir (at sampling point No. 1) then water is pumped to water supply system and is delivered to water users. During the research samples were taken from three sectors of Vilnius water supply network (Fig. 1).

Changes of nitrogen compounds were observed in pipelines which supply water from water reservoir located in Savanoriu Avenue. Water from the water reservoir is pumped into three different directions until it gets to sampling points. Three sectors of tested water supply network have the same distance between water reservoir and sampling points (Table 1). Explored part of Vilnius water network system composes cast iron and steel pipelines. Diameter of pipelines varies from 150 to 500 millimetres. Some previous research showed that cast iron and steel pipelines materials have very similar impact on water quality (Amosenkiene *et al.* 2009).

The water to Warsaw (Poland) users is supplied from 3 water treatment plants: Central, Praski and Polnocny (Fig. 2). They cover 50%, 20% and 30% of water users needs respectively. Almost all water is supplied from surface water sources taken by infiltration or bank uptakes from Vistula (Central and Praski) or from Zegrzynskie Lake (Polnocny).

At Central Water Treatment Plant water is collected in four water infiltration uptakes that are located along the Vistula River. Water is taken from under 6–13 m of the river bottom. It is collected as the water after infiltration processes. Then water is supplied to the water treatment plant and is prepared for drinking purposes (Table 1).

The water is collected by the same infiltration technology by three uptakes at Praski Water Treatment Plant, and then treated at the water treatment plant.

Surface water is taken from Zegrzynskie Lake constructed by a dam on the Bug-Narew River at the Polnocny Water Supply Plant. Collected water is treated using flotation process. This treatment method is based on introduction of microscopic air bubbles, which adhere to pollutants and float them to the water surface creating an easily removable blanket.

Water samples from part of Vilnius water network were taken monthly from four points, namely: at the users' taps (sampling points No. 2, 3, 4) and the water tank outlet (sampling point No. 1). Water samples from Warsaw water supply network were taken monthly from 9 points distributed between 3 water supply districts (Central, Praski and Polnocny). The analyses of samples were carried out at Drinking Water Laboratory of JSC "Vilniaus Vandenys" and the Laboratory of Ecotechnology-Water Center (Warsaw).

Water quality analyses were made and certain technological parameters were determined for the control and evaluation of water quality changes process using International Standard methods: total iron concentration, mg/l; manganese concentration, mg/l; ammonium ions concentration, mg/l; nitrate and nitrite concentrations, mg/l.

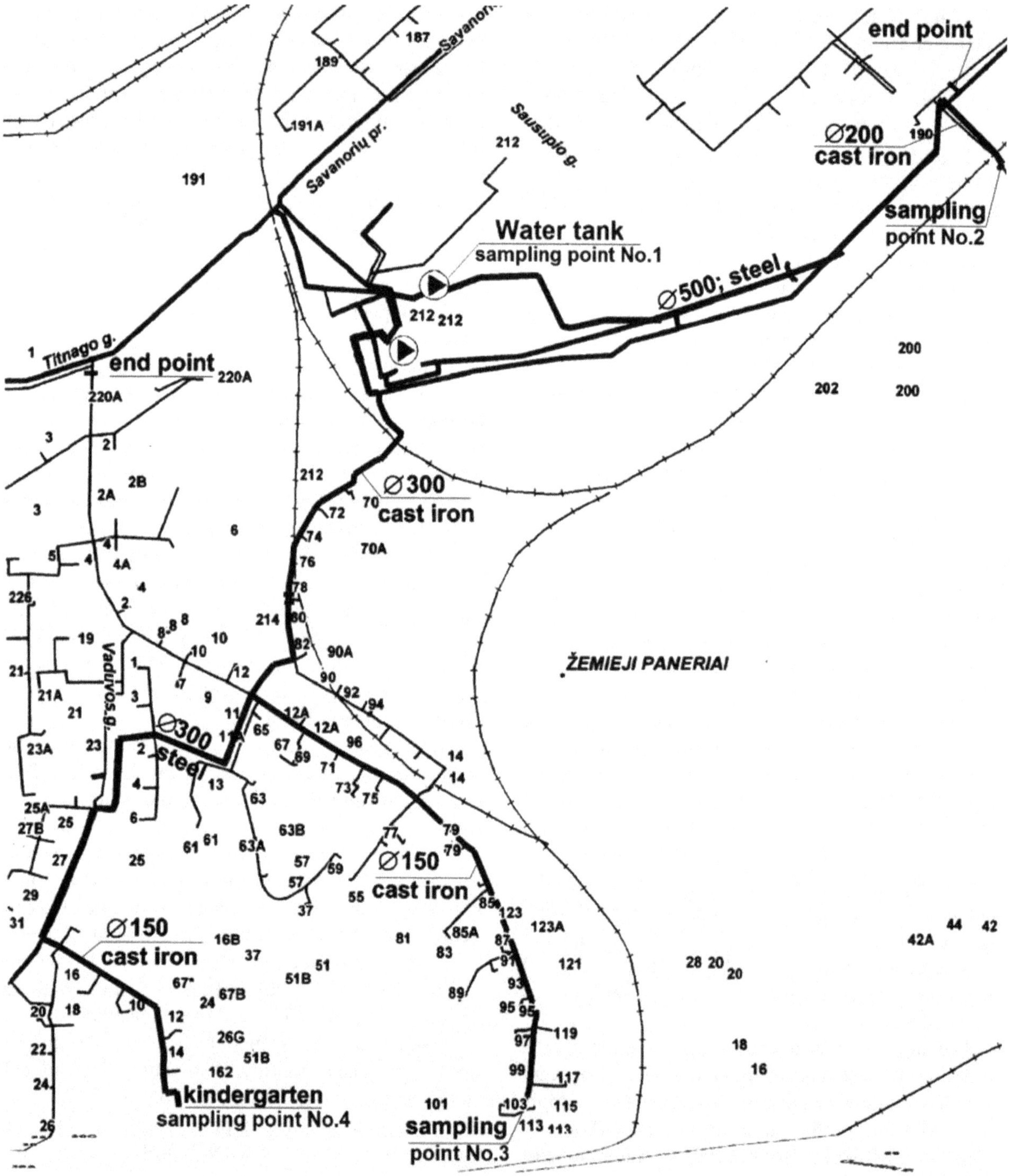

Fig. 1. Sampling scheme of Vilnius water supply network (Lithuania)

The results of the experimental investigation were statistically estimated from the registered analyses. The average concentration at the typical point was calculated on the basis of three measured values of one sample.

Collected water samples were then analyzed using International Standards (ISO) and/or European Norms (EN) approved methods for determination of water quality indicators and toxic parameters.

The quantitative results are presented as the arithmetic mean of six independent measurements ($x \pm SD$, $n = 6$). Significant differences ($p < 0.05$) were removed from the results estimations and the measurements were analyzed once again.

The data from the experimental research was statistically calculated from the registered analyses. The concentration of substances (ammonium ions, nitrate, nitrite, total iron, manganese) was measured

Table 1. Data of analysed water supply networks

	Vilnius (Lithuania)			Warsaw (Poland)		
Sampling point	No. 2	No. 3	No. 4	A	B	C
Location	Sausupio str.	Umedziu str.	Vaduvos str.	Szolc Rogozinskiego str., Nowoursynowska str., Kabacki Dukt str., Krasinskiego str.	Palestynska str.	Madalinskiego str., Kramarska str., Gierczak str., Sonaty str.
Sector of water network	Water tank-S.P. No. 2	Water tank-S.P. No. 3	Water tank-S.P. No. 4	Central Water Supply System A	Polnocny Water Supply System B	Praski Water Supply System C
Water treatment technologies	Groundwater is taken from wells field, no treatment applied, pumped to water tank and distributed to water users.			Surface Vistula water infiltration, aeration, coagulation, sedimentation and stabilization, filtration through sand and activated carbon filters, chlorination-disinfection.	Surface Vistula water infiltration, aeration, filtration through fast sand filters and chlorination-disinfection	Surface Zegrzynski reservoir water treatment based on flotation.

S.P. – Sampling point

11 times in raw water, in taps of users and in the water storage reservoirs. The average concentration at typical points was estimated following the formula, where concentration of substances at typical points, probability at the occurrence of concentration, number of days, number of different values of the concentration were evaluated.

The average concentrations of substances, mentioned above, at the characteristic point, were also calculated. The standard statistical estimation error of the arithmetic mean was approximately 11%.

2. Results and discussion

The values of the quality of water samples that were taken at the characteristic points of the pipeline of Vilnius and Warsaw water supply networks are presented in Table 2.

Nitrogen was defined to be measured in different forms, because of denitrification occurrence. Total iron concentration was chosen to be measured because it makes influence on denitrification process as nitrate reducer (Gerke *et al.* 2008). Manganese concentrations in water samples were measured as they are micronutrients needed for effective metabolism of denitrifying bacteria.

Concentrations of measured chemical compounds were analyzed in order to find dependencies between nitrogen substances and other chemical elements such as total iron and manganese (Cerrato *et al.* 2006). Also, there was reliable dependency found between total iron and ammonium ions concentrations.

The results of the present research showed the dependency between nitrogen compounds and total iron concentrations at the different sampling points of Vilnius water supply network (Lithuania). There was no dependency defined between nitrogen compounds concentrations and total iron or manganese in Warsaw water network (Poland).

Some graphics with the results from Vilnius water network are presented in Figures 3, 4, 5 and 6.

Dependency between ammonium ions and total iron concentrations was established as third order polynomial regression type ($R^2 = 0.77$):

$$C_{Fe_{total}} = -24.08 \cdot C^3_{NH_4} + 33.303 \cdot C^2_{NH_4} - 6.987 C_{NH_4} + 0.913. \quad (1)$$

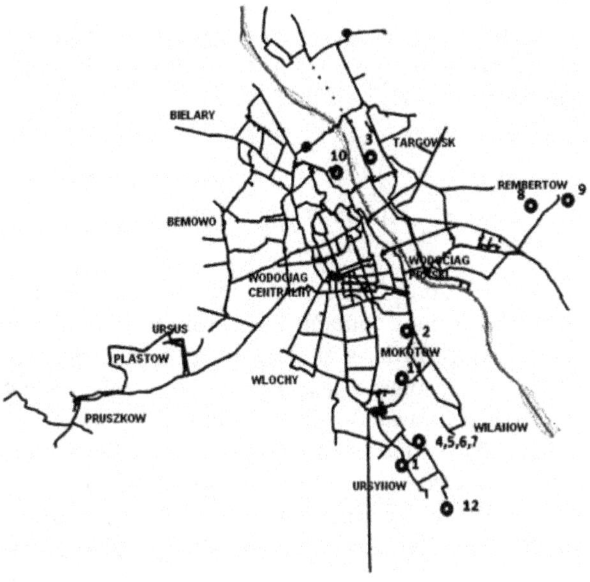

Fig. 2. Sampling scheme of Warsaw water supply network (Poland)

Table 2. The results of water quality at the characteristic point in the pipeline of water supply systems

Vilnius (Lithuania)						Warsaw (Poland)					
	Concentration of chemical compounds						Concentration of chemical compounds				
Sampling Point	NH_4, mg/l	NO_3, mg/l	NO_2, mg/l	Fe, mg/l	Mn, mg/l	Sampling point	NH_4, mg/l	NO_3, mg/l	NO_2, mg/l	Fe, mg/l	Mn, mg/l
No.1	0.18	0.30	0.03	0.69	0.17	A	0.07	10.60	0.002	<0.02	0.036
No.2	0.17	0.10	0.02	0.24	0.18	B	<0.01	11.12	<0.005	0.02	0.017
No.3	0.20	1.50	0.03	0.67	0.30	C	0.02	11.00	<0.005	0.01	0.030
No.1	0.22	1.50	0.09	0.69	0.07	A	0.11	11.20	0.004	<0.02	0.039
No.2	0.23	0.80	0.06	0.38	0.14	B	0.01	9.45	<0.005	<0.01	0.020
No.4	0.12	0.30	0.07	0.50	0.15	C	<0.01	12.00	<0.005	<0.01	0.030
No.3	0.11	0.30	0.08	0.52	0.24	A	0.06	13.70	0.002	<0.02	0.025
No.1	0.18	0.40	0.04	0.94	0.37	B	0.01	11.81	0.007	<0.01	0.025
No.2	0.10	0.50	0.03	0.34	0.19	C	<0.01	13.80	<0.005	0.07	0.030
No.4	0.04	0.70	0.03	0.75	0.49	A	<0.01	9.45	<0.002	<0.02	0.024
No.3	0.06	0.50	0.03	0.52	0.29	B	<0.01	9.68	<0.005	<0.01	0.025
No.1	0.22	0.60	0.03	0.71	0.21	C	0.01	9.30	<0.005	<0.01	0.030
No.2	0.18	0.80	0.02	0.42	0.16	A	<0.01	7.44	0.002	<0.02	0.020
No.4	0.14	0.50	0.03	0.49	0.26	B	0.01	6.79	<0.005	<0.01	0.010
No.4	0.10	1.00	0.04	0.82	0.29	C	0.03	8.70	0.005	0.02	0.030
No.3	0.11	0.50	0.03	0.68	0.29	A	0.01	6.27	<0.002	<0.02	0.029
No.1	0.31	0.30	0.03	1.23	0.39	B	<0.01	3.73	<0.005	0.02	0.014
No.2	0.08	0.20	0.03	0.41	0.23	C	0.02	6.40	<0.005	0.02	0.030
No.4	0.11	0.40	0.03	0.56	0.36	A	0.01	6.32	0.002	<0.02	0.026
No.3	0.18	0.70	0.04	2.75	0.60	B	<0.01	2.60	<0.005	<0.01	0.011
No.1	0.25	1.00	0.03	0.97	0.43	C	0.03	6.90	<0.005	<0.01	0.030
No.2	0.09	1.10	0.03	0.50	0.25	A	<0.01	5.94	0.002	<0.02	0.035
No.4	0.21	0.70	0.04	2.73	0.65	B	<0.01	2.71	<0.005	0.03	0.008
No.4	0.37*	1.10*	0.47*	4.29*	13.56*	C	0.02	6.10	<0.005	0.01	0.030
No.3	0.10	1.30	0.03	0.73	0.31	A	0.01	6.02	<0.002	<0.02	0.044
No.1	0.30	0.80	0.03	1.14	0.44	B	<0.01	7.04	<0.005	0.02	0.007
No.2	0.11	0.80	0.03	0.38	0.26	C	0.02	6.80	<0.005	0.02	0.020
No.4	0.14	1.10	0.04	1.80	0.53	A	<0.01	6.72	0.002	<0.02	0.025
No.4	0.08	1.00	0.03	2.01	0.45	B	<0.06	5.20	<0.041	0.03	0.007
No.3	0.10	1.30	0.03	0.72	0.31	C	<0.01	7.60	<0.005	<0.01	0.030

*- samples after water network brake down

Dependency between nitrate and total iron concentrations was established as third order polynomial regression type ($R^2 = 0.76$):

$$C_{Fe_{total}} = -0.135 \cdot C^3_{NO_3} - 0.292 \cdot C^2_{NO_3} + 1.045 \cdot C_{NO_3} + 0.208. \quad (2)$$

There is some evidence that ground water containing Fe^{2+} normally contains little or no nitrates (Rivett *et al.* 2008). Srinivasan *et al.* (2008) demonstrated that Fe^{2+} acts to promote denitrification process, in which Fe^{2+} reduces nitrate to nitrite, and then it is regenerated by the oxidation of organic carbon. Nitrite can then be reduced to gaseous nitrogen compounds in absence of organic compounds by the further oxidation of iron (Rivett *et al.* 2008). Mn^{2+} is also potential electron donor for autotrophic denitrification processes.

Dependency between manganese and nitrite concentrations was established as third order polynomial regression type ($R^2 = 0.86$):

$$C_{NO_2} = 0.888 \cdot C^3_{Mn} - 1.06 \cdot C^2_{Mn} + 0.412 \cdot C_{Mn} - 0.021. \quad (3)$$

Denitrifying bacteria obtain energy for metabolism and growth from the oxidation of reduced iron and manganese (Sarin *et al.* 2004). Their metabolic requirements for nitrogen can be met by available NH_4^+ or organic nitrogen in the environment, or from the direct assimilation of nitrate (Rivett *et al.* 2008). Manganese is used as micronutrient for effective metabolism.

The causality of obtained dependencies is in the case Fe^{3+} is precipitated as ferric oxide or oxyhydroxide (Rivett *et al.* 2008). This precipitation reaction releases H^+ ions into solution, balancing some of their

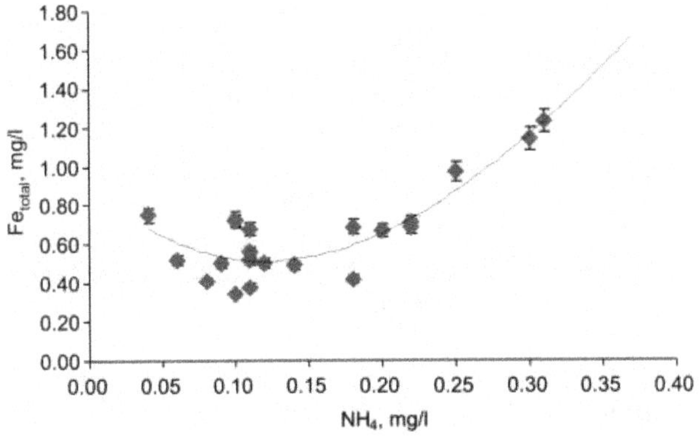

Fig. 3. Dependency between ammonium ions and total iron concentrations in Vilnius network sampling points

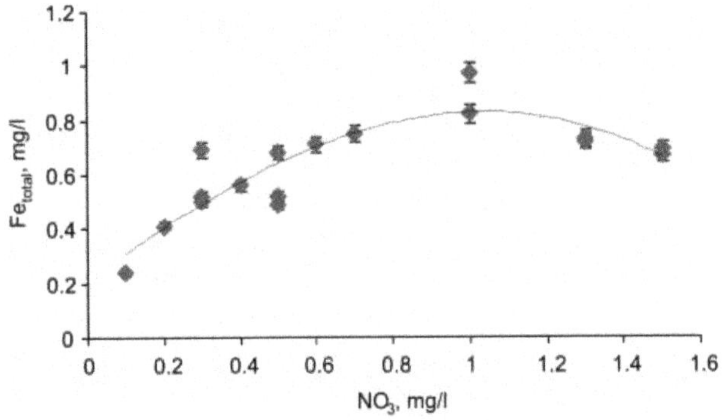

Fig. 4. Dependency between nitrate and total iron concentrations in Vilnius network sampling points

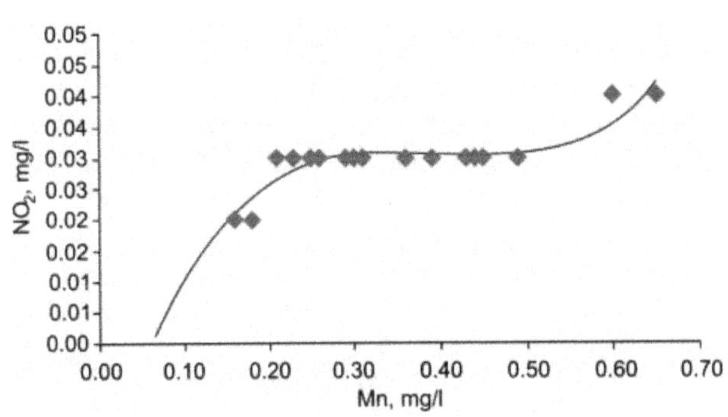

Fig. 5. Dependency between manganese and nitrite concentrations in Vilnius network sampling points

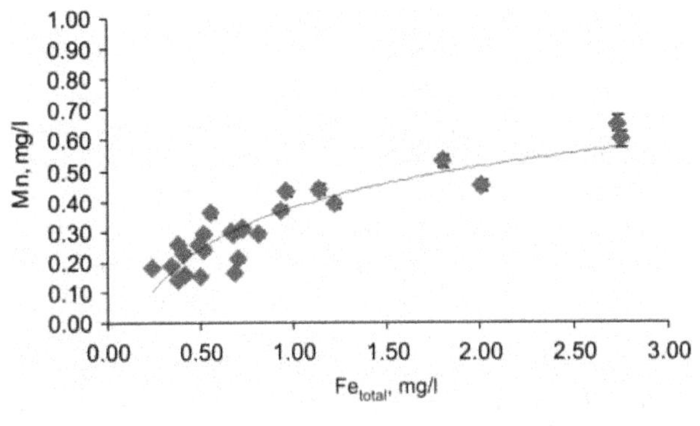

Fig. 6. Dependency between total iron and manganese concentrations in Vilnius network sampling points

Table 3. Dependency between nitrogen compounds and metals in Warsaw water network (Poland)

R^2	NH_4	NO_3	NO_2	Fe_{total}	Mn
NH_4	–	–	–	0.25	–
NO_3	–	–	–	0.52	–
NO_2	–	–	–	–	0.21
Fe_{total}	0.25	0.52	–	–	0.78
Mn	–	–	0.21	0.78	–

consumption by the denitrification reactions and Mn^{2+} is also potential electron donor for authotrophic or abiotic denitrification reactions (Rivett *et al.* 2008). The dissimilatory nitrate reduction to ammonium reaction occurs under much the same conditions as denitrification (Rivett *et al.* 2008).

Dependency between total iron and manganese concentrations was established as second order polynomial regression type ($R^2 = 0.84$):

$$C_{Mn} = -0.042 \cdot C^2_{Fe_{total}} + 0.304 \cdot C_{Fe_{total}} + 0.095. \quad (4)$$

As it was mentioned, no dependency between different compounds was defined in water network of Warsaw (Poland). Estimated dependencies are below significant level and unreliable (Table 3). The causality of obtained results is in inauspicious circumstances e.g. low concentrations of nitrogen compounds, iron and manganese.

The obtained results also showed that the potential nitrogen substances increasing induced by the distribution of groundwater could be reduced if: nitrogen compounds concentrations were extremely low (Warsaw water supply) at the outlet of the water treatment plant; water treatment implementation should reduce the levels of the nitrates and nitrites of the treated supplied water. These differences from Lithuanian case can be explained as follows. Drinking water in Warsaw is prepared from surface water sources. Water treatment technologies are strongly influenced by different chemical compounds that can completely change initial water composition (Langmark *et al.* 2005). Lithuanian ground water is used for drinking purposes and do not require complicated water treatment technologies. In this particular case the explored part of water supply system in Vilnius supplies ground water without treatment.

Different water quality changes were obtained using research results from groundwater and surface water sources of Vilnius and Warsaw respectively.

Conclusions

1. Results showed that groundwater (Vilnius water supply) was extremely susceptible to different nitrogen compounds concentration changes in the

pipelines, the mostly significant range of nitrate from 0.1 to 1.5 mg/L, in case ammonium concentrations from 0.09 to 0.25 mg/L. The occurrence of ammonium and nitrates in drinking water samples correlated positively ($R^2 = 0.76$ and $R^2 = 0.76$ respectively) with total iron concentration and negatively with the content of different nitrogen compounds (ammonium ions, nitrite, nitrate).

2. Significant increase data were obtained for manganese compounds in drinking water supplied to Vilnius, Lithuania. Nitrite concentration was slightly increasing during the research with positive correlation ($R^2 = 0.86$) on high manganese concentrations because of occurring micronutrient decomposition processes.

3. Comparing ammonium ions concentrations from two water sources (groundwater from Vilnius and surface water from Warsaw), it was noticed that ammonium ions concentration from Vilnius water supply was decreased to the maximum of 64%; the most significant maximum decrease of ammonium compounds from Warsaw water supply was nearly 91%.

4. Obtained results showed that different concentrations of nitrate up to ten times higher in Warsaw water supply comparing with Vilnius are found in the water supply of different origin (open water sources and groundwater respectively).

5. The obtained characteristics of iron and ammonium ions dependency can be useful for the estimation, project design, and the equipment start up procedure, renovation and maintenance of existing water supply systems.

Acknowledgements

On behalf of Lithuanian authors we would like to express our gratitude to the Laboratory of Ecotechnology – Water Center (SGGW Warsaw, Poland).

We are grateful to all JSC "Vilniaus vandenys" support in research activities and personally to the Director of Production Vytautas Kisielis who helped with practical consultancy.

We are thankful to Lithuanian Scientific Council for financial support during the research activities under the frame of students' scientific practice.

References

Agatemor, C. P.; Okolo, O. 2007. University of Benin water supply system: microbiological and physico-chemical assessment, *Environmentalist* 27: 227–239. http://dx.doi.org/10.1007/s10669-007-9000-4

Agatemor, C. P.; Okolo, O. 2008. Studies of corrosion tendency of drinking water in the distribution system at the University of Benin, *Environmentalist* 28: 379–384. http://dx.doi.org/10.1007/s10669-007-9152-2

Amosenkiene, A.; Valentukeviciene, M.; Mazeikiene, A.; Kanapickas, R. 2009. An assessment of factors having impact on water quality in water supply pipelines, *Science – Future of Lithuania, Environmental Protection Engineering* 1(4): 65–70.

Berry, D.; Xi, C.; Raskin, L. 2006. Microbial ecology of drinking water distribution systems, *Current Opinion in Biotechnology* 17: 297–302. http://dx.doi.org/10.1016/j.copbio.2006.05.007

Bonadonna, L.; Briancesco, R.; Libera, S. D.; Lacchetti, I.; Paradiso, R.; Semproni, M. 2009. Microbial characterization of water and biofilms in drinking water distribution systems at sport facilities, *Central European Journal of Public Health* 17: 99–102.

Cerrato, J. M.; Reyes, L. P.; Alvarado, C. N.; Dietrich, A. M. 2006. Effect of PVC and iron materials on Mn(II) deposition in drinking water distribution systems, *Water Research* 40: 2720–2726. http://dx.doi.org/10.1016/j.watres.2006.04.035

Gerke, T. L.; Maynard, J. B.; Schock, M. R.; Lytle, D. L. 2008. Physiochemical characterization of five iron tubercles from a single drinking water distribution system: possible new insights on their formation and growth, *Corrosion Science* 50: 2030–2039. http://dx.doi.org/10.1016/j.corsci.2008.05.005

Gladwin, M. T.; Crawford, J. H.; Patel, R. P. 2004. Biomedical implications for hemoglobin interaction with nitric oxide, *Free Radical Biology & Medicine* 36: 442–453.

Jaskelevicius, B.; Lynikiene, V. 2009. Investigation of influence of Lapes landfill leachate on ground and surface water pollution with heavy metals, *Journal of Environmental Engineering and Landscape Management* 17(3): 131–139. http://dx.doi.org/10.3846/1648-6897.2009.17.131-139

Langmark, J.; Storey, M. V.; Ashbolt, N. J.; Stenstro, T. A. 2005. Biofilms in an urban water distribution system: measurement of biofilm biomass, pathogens and pathogen persistence within the Greater Stockholm area, Sweden, *Water Science & Technology* 52: 181–189.

Lee, D.; Lee, J.; Kim, S. 2005. Diversity and dynamics of bacterial species in a biofilm at the end of the Seoul water distribution systems, *World Journal of Microbiology & Biotechnology* 21: 155–162. http://dx.doi.org/10.1007/s11274-004-2890-0

Lehtola, M. J.; Miettinen, I. T.; Lampola, T.; Hirvonen, A.; Vartiainen, T.; Martikainen, P. J. 2005. Pipeline materials modify the effectiveness of disinfectants in drinking water distribution systems, *Water Research* 39: 1962–1971. http://dx.doi.org/10.1016/j.watres.2005.03.009

Lipponen, M. T. T.; Suutari, M. H.; Martikainen, P. J. 2002. Occurrence of nitrifying bacteria and nitrification in Finnish drinking water distribution system, *Water Research* 36: 4319–4329. http://dx.doi.org/10.1016/S0043-1354(02)00169-0

Markku, J.; Lehtola, M. J.; Juhna, T.; Miettinen, I. T.; Vartiainen, T.; Martikainen, P. J. 2004. Formation of biofilms in drinking water distribution networks, a case study in two cities in Finland and Latvia, *Journal of Industrial Microbiology and Biotechnology* 31: 489–494. http://dx.doi.org/10.1007/s10295-004-0173-2

Mazeikiene, A.; Valentukeviciene, M.; Jankauskas, J. 2010. Laboratory study of ammonium ion removal by using zeolite (clinoptilolite) to treat drinking water, *Journal of Environmental Engineering and Landscape Management* 18(1): 54–61. http://dx.doi.org/10.3846/jeelm.2010.07

McNeill, L. S.; Edwards, M. 2002. The importance of temperature in assessing iron pipe corrosion in water distribution systems, *Environmental Monitoring and Assessment* 77: 229–242. http://dx.doi.org/10.1023/A:1016021815596

Nawrocki, J.; Raczyk-Stanislawiak, U.; Swietlik, J.; Olejnik, A.; Sroka, M. J. 2010. Corrosion in a distribution system: Steady water and its composition, *Water Research* 44: 1863–1872. http://dx.doi.org/10.1016/j.watres.2009.12.020

Neville, A. 2001. Effect of cement paste on drinking water, *Materials and Structures* 34: 367–372.

Ranieri, E.; Swietlik, J. 2010. DBPs control in European drinking water treatment plants using chlorine dioxide: two case studies, *Journal of Environmental Engineering and Landscape Management* 18(2): 85–91. http://dx.doi.org/10.3846/jeelm.2010.10

Rivett, M. O.; Buss, S. R.; Morgan, P.; Smith, J. W. N.; Bemment, C. D. 2008. Nitrate attenuation in groundwater: a review of biogeochemical controlling processes, *Water Research* 42: 4215–4232. http://dx.doi.org/10.1016/j.watres.2008.07.020

Sarin, P.; Snoeyink, V. L.; Lytle, D. A.; Kriven, W. M. 2004. Iron corrosion scales: model for scale growth, iron release, and colored water formation, *Journal of Environmental Engineering* 130(4): 364–373. http://dx.doi.org/10.1061/(ASCE)0733-9372(2004)130:4(364)

Srinivasan, S.; Harringto, G. W.; Xagoraraki, I.; Goel, R. 2008. Factors affecting bulk to total bacteria ratio in drinking water distribution systems, *Water Research* 42: 3393–3404. http://dx.doi.org/10.1016/j.watres.2008.04.025

Venkatesan, G.; Swaminathan, G. 2009. Review of chloride and sulphate attenuation in ground water nearby solid-waste landfill sites, *Journal of Environmental Engineering and Landscape Management* 17(1): 1–7. http://dx.doi.org/10.3846/1648-6897.2009.17.Ia-Ig

Marina VALENTUKEVIČIENĖ. Associated Professor of Vilnius Gediminas Technical University and European projects evaluator for EC. Research interests: eco-friendly water treatment technologies, sustainable use of water resources, environmental impact assessment, water reuse technologies, sustainable living environment.

Agnieszka KARCZMARCZYK. Adjunct at Faculty of Civil and Environmental Engineering and Dean's attorney for International Cooperation and Erasmus Exchange Programme at Warsaw University of Life Science (Poland). Research interests: biological wastewater treatment technologies, protection of water resources, environmental impact assessment, sustainable development.

Anželika JURKIENĖ. MSc of Hydrogeology and Engineering Geology, Laboratory assistant at Water Management Department of Vilnius Gediminas Technical University. Research interests: design of water network, water treatment technologies, ground water chemistry.

Auksė GRIGAITYTĖ. Doctoral student of Vilnius Gediminas Technical University and Manager at Water Supply Department of JSC "Vilniaus vandenys". Research interests: water supply, improvement of water quality, sustainable water use, sustainable pipelines materials.

23

INFLUENCE OF HYDROPOWER DAMS ON THE STATE OF MACROINVERTEBRATES ASSEMBLAGES IN THE VIRVYTE RIVER, LITHUANIA

Saulius Vaikasas[a], Kęstutis Palaima[a], Virginija Pliūraitė[b]

[a]*Water Research Institute, Aleksandras Stulginskis University, Parko g. 6, 58102 Vilainiai, Kėdainiai distr, Lithuania*

[b]*Nature Research Centre, Akademijos g. 2, 08412 Vilnius, Lithuania*

Abstract. Macroinvertebrate organization along the Virvytė River was examined to relate biological response in environmental changes. Benthic macroinvertebrates and a range of environmental variables were sampled from three study sections: control, HPP dams and below HPP dams. Total macroinvertebrate taxon number (NT = 28), EPT taxon number (NT = 13) and Shannon-Wiener diversity index (H′ = 3.85) were the highest in control sites of the river and the smallest in HPP dams (17, 5, 2.75, respectively). Taxon number and total abundance of macroinvertebrates declined significantly both in HPP dams and below them in comparison with the control sites. Total macroinvertebrate abundance in the control sites (698±41 ind.m^{-2}) was significantly higher than in HPP dams and below them (430± 21 ind.m^{-2} and 425±22 ind.m^{-2}, respectively). The obtained data show that the investigated sections differed in terms of composition and relative abundance of different taxa of macroinvertebrates. Compared to reference conditions (control sites), macroinvertebrate samples from the sites below the dams and in them had relatively more Chironomidae larvae, Oligochaeta and Mollusca and fewer of the more sensitive taxa Plecoptera, Ephemeroptera and Trichoptera. According to Danish Stream Fauna Index and Hilsenhoff biotic index, water quality in the Virvytė River control sites was higher in comparison with water quality in HPP dams and below-dam sites.

Keywords: Virvytė River, HPP dams, benthic macroinvertebrates, water pollution.

Reference to this paper should be made as follows: Vaikasas, S.; Palaima, K.; Pliūraitė, V. 2013. Influence of hydropower dams on the state of macroinvertebrates assemblages in the Virvyte River, Lithuania, *Journal of Environmental Engineering and Landscape Management* 21(4): 305–315. http://dx.doi.org/10.3846/16486897.2013.796956

Introduction

Modification of natural flow regimes alters the abundance and composition of biological conditions in aquatic environments (Wright *et al.* 2002; Winterwerp, Kesteren 2004; Ambers 2007; Soulsby *et al.* 2010). Dams affect ecosystems in a number of ways: by altering the natural flow fluctuations, modifying the physicochemical characteristics of riverine water and fragmenting the continuity of rivers (Grzybkowska, Dukowska 2002; Bahuguna *et al.* 2004; Lamsodis, Vaikasas 2005; Penczak, Kruk 2005; Pietraszewski *et al.* 2008; Zdankus *et al.* 2008; Vaikasas 2010).

Hydropower dams (HPP) block rivers, reducing connectivity and resulting in negative effects on stream biota upstream and downstream from the impoundment (Céréghino *et al.* 2002; Grubbs, Taylor 2004; Greathouse *et al.* 2006; WFD & Hydropower 2007; Hu *et al.* 2008). Dams also block movement of fishes and affect habitat and physicochemical conditions of streams by converting lotic habitats to lentic, changing stream flow, altering water quality and modifying channel morphology and bed structure by increasing siltation upstream and erosion downstream (Sear *et al.* 2004; Malik, Richardson 2009). These alterations cause changes in assemblage structure of fishes and macroinvertebrates via shifts in composition, abundance and diversity both upstream and downstream from the impoundment.

The responses of macroinvertebrates to regulated river systems are often complex and variable. Considerable evidence, however, exists that hydraulic conditions are the driving forces affecting distribution and abundance patterns of benthic invertebrates (Christopher *et al.* 2007).

Some authors have reported changes in taxonomic composition of macroinvertebrate communities due to alteration of fluvial habitats by effect of large dams (Lessard, Hayes 2003), while others have found a relatively minor impact of small dams on the river

Corresponding author: Saulius Vaikasas
E-mail: s.vaikasas@delfi.lt

biota in downstream reaches (Sharma *et al.* 2005; Ambers 2007). According to Principe (2010), differences in taxonomic composition between upstream and downstream reaches have not been detected; however, analysis of variance (ANOVA) showed some changes in richness and diversity, in the density of filtering collectors and shredders and in the abundance of some species. These results indicate that small dams have only a moderate impact compared to the documented impact of large dams.

Studies that focused on benthic macroinvertebrates usually found that sampling sites immediately downstream of dams were characterised by lower taxonomic diversity than unaffected sites upstream or sites further downstream where dam effects have attenuated (Céréghino *et al.* 2002; Camargo *et al.* 2005). Other studies have documented either higher diversity just below a dam than at sites further downstream or decreases only in certain taxonomic groups like Ephemeroptera, Plecoptera, Trichoptera (EPT), with little difference in total diversity between upstream and downstream sites (Lessard, Hayes 2003). Likewise, total abundance has been found either to decrease or to increase in dam tailwaters, the latter occurring when dominance of certain taxa like Chironomidae becomes pronounced (Munn, Brusven 1991).

Ambers (2007) pointed out that the impact of small dams declined to negligible ecological levels a few kilometres downstream. Large dams produce notable changes in macroinvertebrate communities by elimination of habitats, flow regulation and alteration of thermal regime (Lessard, Hayes 2003; Grubbs, Taylor 2004; Doyle *et al.* 2005; Marčiulionienė *et al.* 2011). As a result, sensitive species disappear and density of entropic species increases.

There is a large number of publications dealing with the impact of hydropower ponds on macroinvertebrate communities (Cortes *et al.* 2002; Maiolini *et al.* 2007; Katano *et al.* 2009; Tszydel *et al.* 2009; Principe 2010). Until now the impact of hydropower ponds on macroinvertebrate fauna has not been studied in Lithuania streams or rivers.

The aim of this study was to assess the effect of small HPP dams on stream habitats and on the structure of benthic macroinvertebrate community of the Virvytė River in Lithuania.

1. Materials and methods

The study was conducted in April 2010 in the Virvytė River. The length of the river is 99.7 km, draining a catchment of 1.134 km^2 with an average slope of 0.035% and an average discharge of 9.93 m^{3}s^{-1} (Gailiušis *et al.* 2001). Ten hydropower plants are built on the Virvytė River (Fig. 1).

The macroinvertebrates for our study were collected in 23 sampling sites differing in morphology along the Virvytė River (Table 1): 3 control sites (dam-unaffected), 10 sites in HPP dams and 10 sites below HPP dams. Gravel and pebble were dominating bottom substrate.

The samples of macroinvertebrates were collected by the kick-sampling method in three 0.1 m^2 areas at each study site (Kleemola, Söderman 1993). Additional samples of macroinvertebrates (2 samples per site) were taken from plants, stones or stumps to determine the Danish Stream Fauna Index (DSFI). A total of 115 samples were collected, sieved using a 500 µm mesh, transferred into plastic flasks and stored in a 4% formaldehyde solution. In the laboratory, all animals were separated, counted and identified to the species or genus level (except Oligochaeta) under a binocular dissecting microscope.

We calculated total number of macroinvertebrate taxa (NT), number EPT taxa, Shannon-Wiener diversity index (H′) as well as the total abundance (ind. m^{-2}) and relative abundance (%) of indicatory taxonomic groups (EPT, Chironomidae, Oligochaeta, Mollusca, Coleoptera and Isopoda).

DSFI (Arbačiauskas 2009) and Hilsenhoff biotic index (HBI) (Hilsenhoff 1988) were calculated to assess the ecological status of investigated river sites. HBI tolerance values were taken from Mandaville (2002).

The relationships between macroinvertebrate metrics and water quality values were determined using Sperman rank correlation. The General Linear Model ANOVA and Fisher Least Significant Difference (LSD) test were used to determine differences in macroinvertebrate metrics and morphological characteristics among groups of river sites. All species data were log (1 + x) transformed prior to analysis. Calculations were done with Statistica for Windows, version 6.0 (Sakalauskas 2003).

2. Results and discussion

The measured values of the main physiochemical variables in the investigated sites of the Virvytė River are presented in Table 2. Total nitrogen concentrations in water exceeded the maximum permissible concentration, approved by the Ministry of Environment of Lithuania (Lithuanian Standards 2010), in and below hydropower dams of Kapėnai, Skleipiai and Gudai. The values of N total in sediments of HPP dams were higher in comparison with the values below HPP dams and in control sites.

A total of 69 macroinvertebrate taxa belonging to 37 families were identified in the investigated river. The greatest taxa richness was recorded for Trichoptera (11 taxa: 10 species and 1 genus), Mollusca (11 species) and Ephemeroptera (11 species). Chironomids

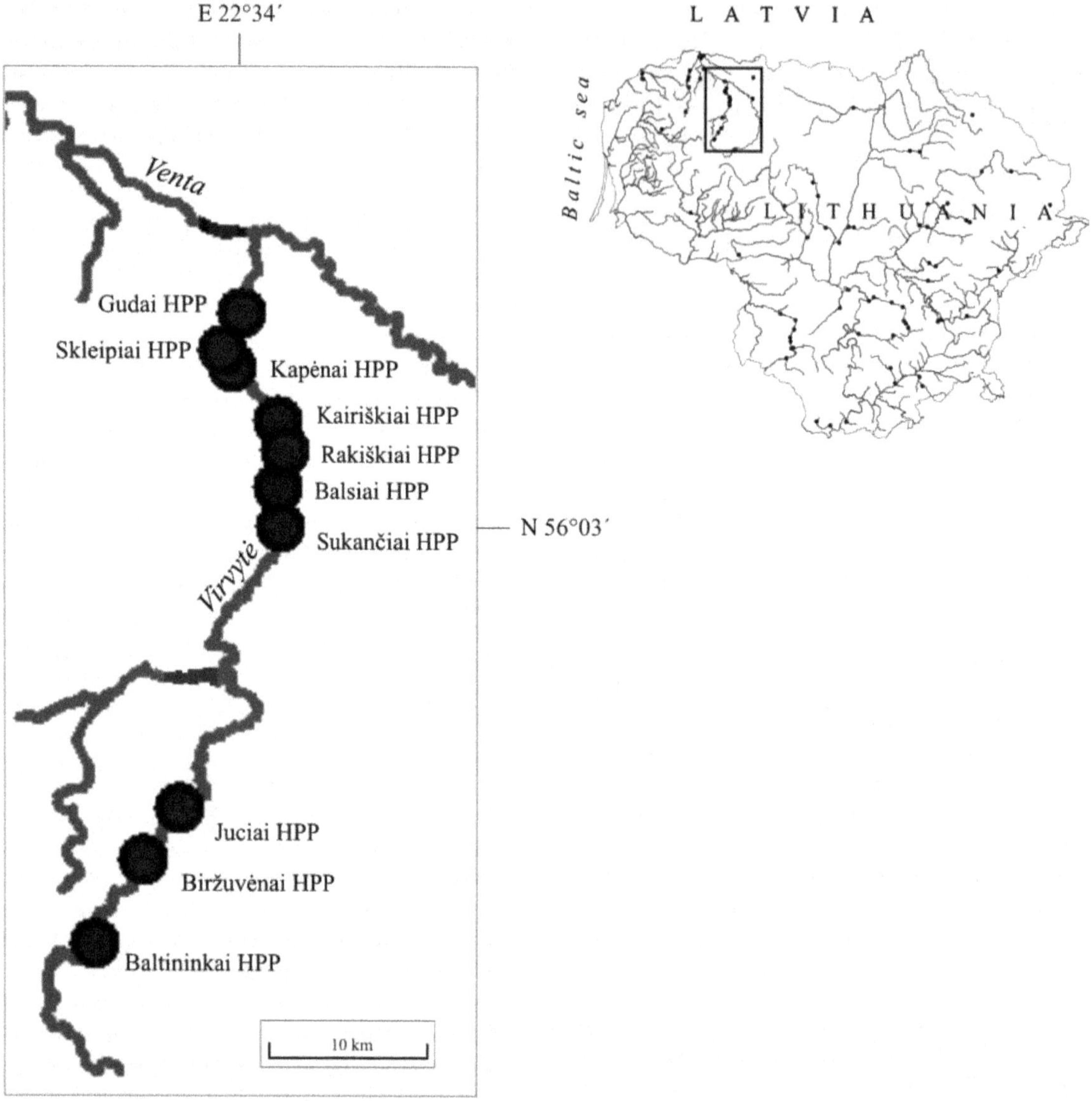

Fig. 1. Location of HPP in the river Virvytė

Cricotopus algarum and oligochaetes species (sp.) were detected in all investigated sites.

The total number of macroinvertebrate taxa (NT) and the number of EPT taxa were highly variable across the river, ranging within 12–30 and 2–14, respectively (Fig. 2). According to Munn and Brusven (1991), the major factors contributing to the simplified macroinvertebrate community are the reduced habitat diversity, fluctuating water levels, altered thermal regime, and possibly altered food supply.

The results obtained show that macroinvertebrate community compositions from the three sections – control sites, sites in HPP dams and sites below HPP dams – were different. Caddisflies *Notidobia ciliaris, Ithytrichia lamellaris, Hydroptila* sp., mayflies *Heptagenia sulphurea*, water beetles *Elmis* sp., molluscs *Ancylus fluviatilis* and dipterans *Dicranota bimaculata* species were absent from the sites in and below dams. The sites in and below the dams of the Virvytė River were dominated by Diptera and Oligochaeta in abundance. Ephemeroptera, Trichoptera and Plecoptera dominated in control sites of the Virvytė River.

The Shannon-Wiener diversity index (H′) ranged from 1.59 to 3.87 in the Virvytė (Fig. 2). The lowest H′ was detected in the Baltininkai HPP dam, where the amount of N total in sediments was the highest. The total number of macroinvertebrate taxa was the lowest (12) in this HPP dam, too.

The total abundance of benthic macroinvertebrates was found to range from 250 to 770 ind. m^{-2} in the Virvytė River (Fig. 3).

Relative abundances of different taxonomic groups of macroinvertebrates also ranged within a broad scale. Stoneflies were not found in HPP dams, and constituted only 0% to 8.7% below HPP dams and 15.1% to 20.5% in control sites (Fig. 4). According to Bahuguna *et al.* (2004), stoneflies were absent from the regulated stretch. Mayflies were the most abundant

Table 1. General morphological characteristics in studied sites of Virvytė River (Nemunas basin, Lithuania)

Site no			Longitude	Latitude	Distance from mouth, in km	Bottom substrate
1	Janapole	control	55°48′59′′	22°22′11′′	88.8	pebble
2	Baltininkų HPP	HPP dam	55°50′53′′	22°24′30′′	77.7	gravel
3		below HPP dam	55°50′56′′	22°24′40′′		pebble
4	Biržuvėnų HPP	HPP dam	55°53′35′′	22°27′21′′	72.0	gravel
5		below HPP dam	55°53′41′′	22°27′24′′		gravel
6	Jucių HPP	HPP dam	55°55′18′′	22°29′24′′	65.2	gravel
7		below HPP dam	55°55′21′′	22°29′24′′		gravel
8	Mažikai	control	55°56′43′′	22°30′18′′	58.1	pebble
9	Sukončių HPP	HPP dam	56°4′21′′	22°34′57′′	29.8	silt-sand
10		below HPP dam	56°4′24′′	22°35′1′′		gravel
11	Balsių HPP	HPP dam	56°5′38′′	22°34′47′′	27.7	gravel
12		below HPP dam	56°5′43′′	22°34′53′′		pebble
13	Rakiškių HPP	HPP dam	56°6′50′′	22°35′11′′	23.2	gravel
14		below HPP dam	56°6′54′′	22°35′17′′		gravel
15	Kairiškių HPP	HPP dam	56°7′48′′	22°34′39′′	20.2	gravel
16		below HPP dam	56°7′50′′	22°34′46′′		pebble
17	Kapėnų HPP	HPP dam	56°9′29′′	22°31′56′′	13.4	silt-sand
18		below HPP dam	56°9′32′′	22°31′53′′		pebble
19	Skleipių HPP	HPP dam	56°9′58′′	22°31′21′′	10.8	silt-sand
20		below HPP dam	56°10′3′′	22°31′29′′		pebble
21	Gudų HPP	HPP dam	56°11′12′′	22°32′23′′	6.6	gravel
22		below HPP dam	56°11′15′′	22°32′30′′		pebble
23	Gyvuoliai	control	56°12′40′′	22°33′23′′	2.8	pebble

Table 2. The values of chemical variables (*N* – nitrogen, *P* – phosphorus) and suspended (sediments SS) in studied sites of Virvytė River

Site no.	*N* total in water, mg l^{-1}	*P* total in water, mg l^{-1}	SS, mg l^{-1}	*N* total in sediments, mg kg^{-1}	*P* total in sediments, mg kg^{-1}
1	1.34	0.040	5.2	350	290
2	1.54	0.052	3.6	5023	650
3	1.33	0.041	6.2	716	342
4	1.34	0.039	2.2	1051	224
5	1.55	0.039	7.4	390	189
6	1.59	0.040	4.8	2745	638
7	1.29	0.053	3.0	257	147
8	1.66	0.045	13.4	228	239
9	2.10	0.050	9.0	2774	425
10	2.36	0.047	10.0	650	254
11	2.36	0.042	22.4	744	247
12	2.10	0.040	10.0	375	187
13	2.10	0.046	8.4	2190	606
14	2.50	0.044	12.6	1577	376
15	2.48	0.049	40.8	1022	247
16	2.36	0.052	9.6	738	333
17	**2.72**	0.053	14.2	2628	1063
18	**2.60**	0.044	8.0	467	217
19	**2.77**	0.059	57.8	2920	472
20	**2.71**	0.059	20.0	686	244
21	**2.59**	0.051	14.1	423	301
22	**2.65**	0.052	9.0	248	254
23	1.51	0.042	7.8	291	250

Note: In bold – exceeded means of maximum permissible concentrations approved by Ministry of Environment of Lithuania (Lithuanian Standards 2010).

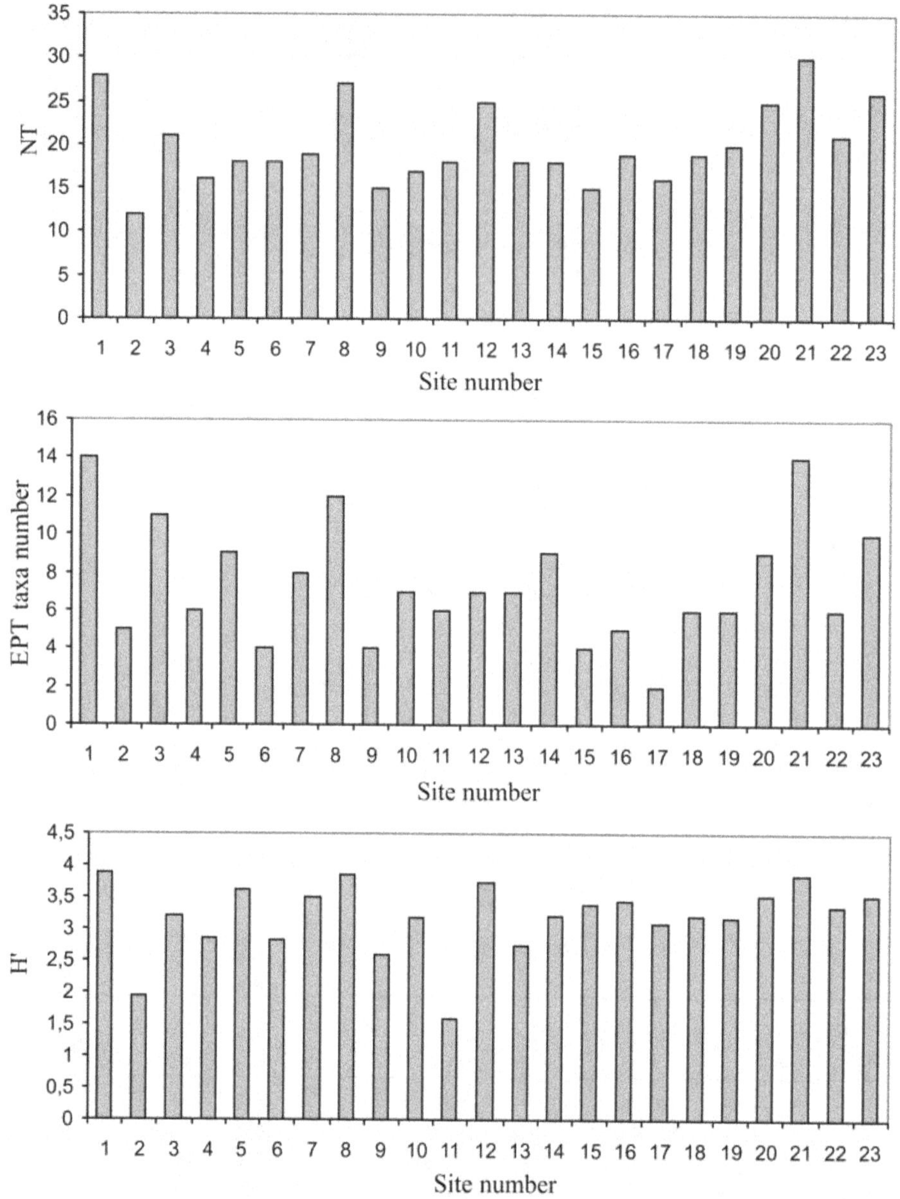

Fig. 2. Distribution of NT, EPT taxa number, H´ in studied sites of Virvytė River

group of macroinvertebrates in control sites (sites 1, 8, 23) and below Rakiškės (site 14) and Kairiškės HPP dams (site 16). We found that mayflies *Baetis rhodani*, *Siphlonurus alternatus* and *Caenis macrura* were more abundant in and below HPP dams. The mayfly *Heptagenia sulphurea* was absent. Other studies revealed increase of Baetidae and Caenidae and decrease of Heptageniidae due to flow reduction (Mantel *et al.* 2010).

Chironomids were the most abundant in HPP dams in our study, while caddisflies were found in smaller numbers (Fig. 4). The relative abundance of mayflies in and below HPP dams was lower, too.

We found that NT was negatively correlated with N total in sediments (Table 3). The total abundance of individuals and relative abundance of caddisflies, EPT and water beetles larvae was negatively correlated with N total in water, and molluscs showed an opposite trend in relation to N total in water (Table 3).

The relative abundance of caddisflies was negatively correlated with *P* total in water (Table 3). Relative abundance of EPT, water beetles larvae, and DSFI were negatively correlated with *P* total in sediments, whereas relative abundance of oligochaetes and molluscs showed an opposite trend in relation to *P* total in sediments (Table 3). Relative abundance of molluscs was positively correlated with N total in water and N total in sediments (Table 3). No reliable correlation was found between *P* total in water and macroinvertebrate metrics (Vaikasas, Lamsodis 2011).

After rearrangement of sites to groups according to differentiating environmental descriptors (control, HPP dam, below HPP dam), statistically significant differences were found in NT, EPT taxa number, H´, total

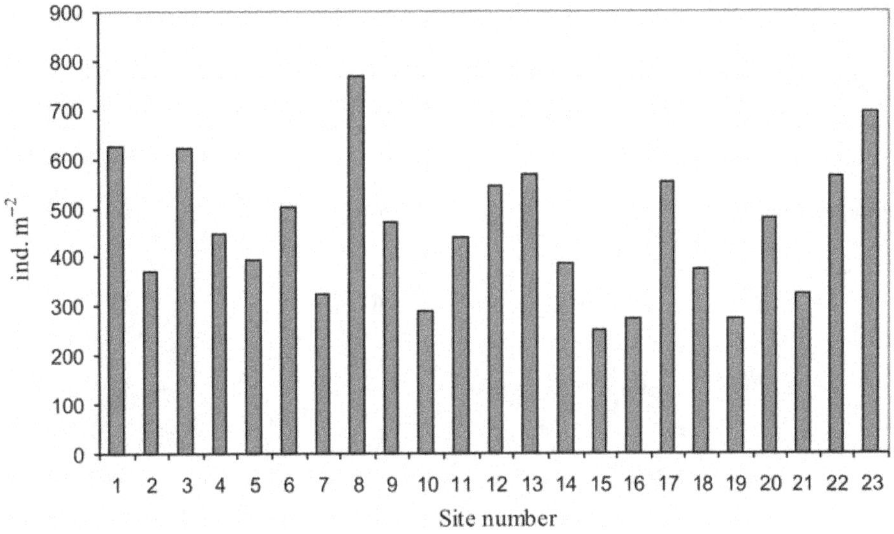

Fig. 3. Total macroinvertebrate abundance in studied sites of Virvytė River

abundance of macroinvertebrates (ind.m^{-2}) and relative (%) abundance of Plecoptera, Trichoptera EPT, Chironomidae and Oligochaeta in the Virvytė River (Figs 5, 6; Table 4). According to Greathouse *et al.* (2006), significant differences were found between dammed and undammed sites for many macroinvertebrate taxa.

The values of NT (ANOVA $F = 6848.83$, $p = 0.0004$), EPT taxa number (ANOVA $F = 979.27$, $p = 0.002$), H′ (ANOVA $F = 2566.4$, $p = 0.002$), total abundance of macroinvertebrates (ANOVA: $F = 25214.17$, $p = 0.00001$), percentage of Trichoptera and percentage of EPT were significantly higher in the control sites than in the sites in and below HPP dams in the investigated river (Figs 5, 6; Table 4) (Fisher's LSD test). The relative abundance of Plecoptera was significantly higher in the control sites (17.2) than in the sites below HPP dams (2.5), and the relative abundance of Ephemeroptera were significantly higher in the control sites (48.7) than in HPP dams (24.4) in the investigated River (Fisher's LSD test) (Table 4). Ecosystem health has an inverse relationship with human impacts resulting in the presence of few tolerant species in very large numbers (Brown 2005; Novotny *et al.* 2005). According to Nichols *et al.* (2006), macroinvertebrate samples from sites below dams had relatively more Chironomidae larvae and Oligochaeta and fewer of the more sensitive taxa, Plecoptera, Ephemeroptera and Trichoptera.

The Shannon-Wiener diversity index (H′) was significantly lower (ANOVA $F = 2566.4$, $p = 0.002$) in

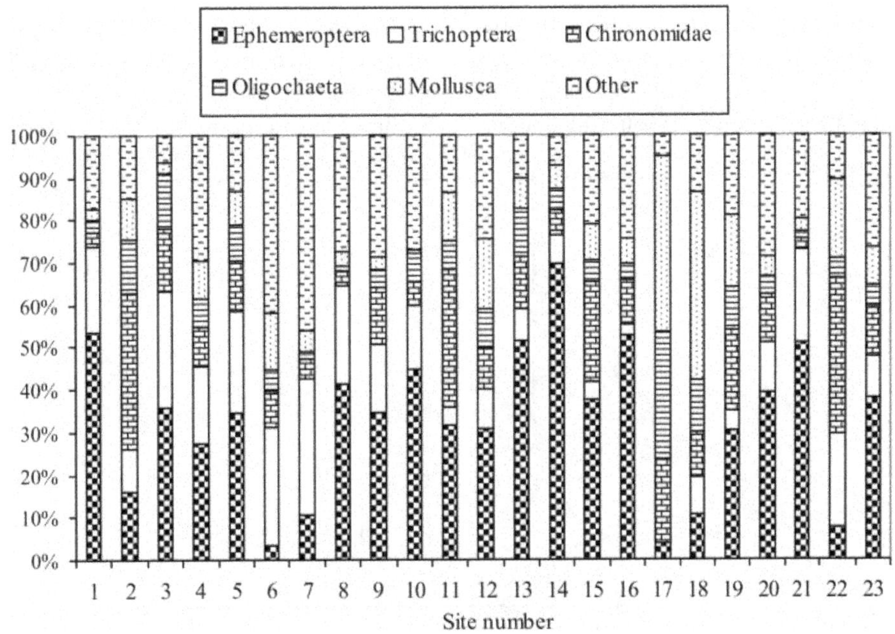

Fig. 4. Relative abundances of the main benthic macroinvertebrate groups in studied sites of Virvytė River

Table 3. Spearman's rank correlation coefficient between macroinvertebrate metrics and chemical parameters in the Virvytė River

	N total, mg l^{-1} in water	*P* total, mg l^{-1} in water	*N* total in sediments	*P* total in sediments
NT	ns	ns	−0.46**	ns
EPT	ns	ns	ns	ns
Total abundance of individuals (ind m^{-2})	−0.26*	ns	ns	ns
Percentage of individuals (%)				
Plecoptera	ns	ns	−0.33*	ns
Trichoptera	−0.74**	−0.30*	ns	ns
EPT	−0.34*	ns	ns	−0.25*
Coleoptera	−0.18*	ns	−0.36*	−0.39*
Chironomidae	ns	Ns	ns	Ns
Oligochaeta	ns	ns	0.45*	0.63**
Mollusca	0.40 *	ns	ns	0.35
Biotic index				
DSFI	ns	ns	ns	−0.55*
HBI	ns	ns	ns	ns

* $p<0.05$, ** $p<0.01$.

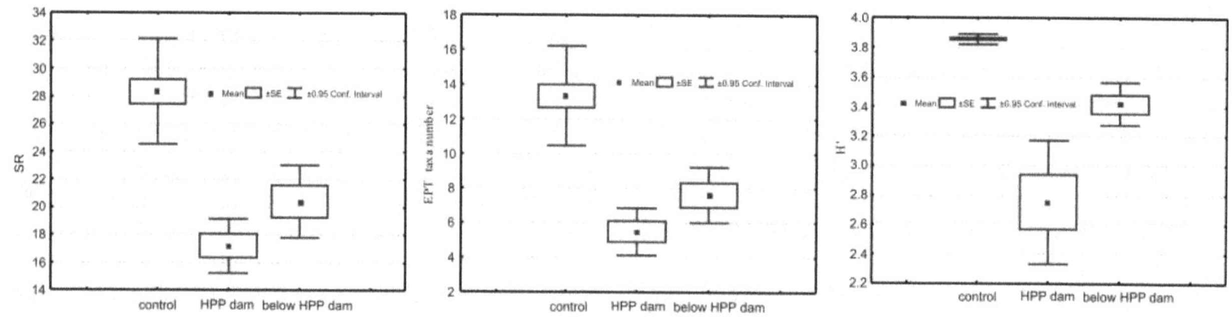

Fig. 5. The number of total macroinvertebrate taxa (NT), sensitive to pollution EPT taxa and the means of Shannon-Wiener diversity index (H′) in studied sites of Virvytė River

HPP dams in comparison with control sites and sites below HPP dams (Fig. 5).

Dam building has significant impacts on the composition of macroinvertebrates just above the dam site, probably as a result of deposition of inorganic material within a small reservoir and changes in water velocity (Doyle *et al.* 2005; Sharma *et al.* 2005). We found that both of the disturbed sites (in HPP dams and below dams) had lower taxa number of macroinvertebrates (both total and EPT taxa) than the control sites (Fig. 5, Table 4) (Fisher's LSD test). These findings are supported by the study of Jesus *et al.*

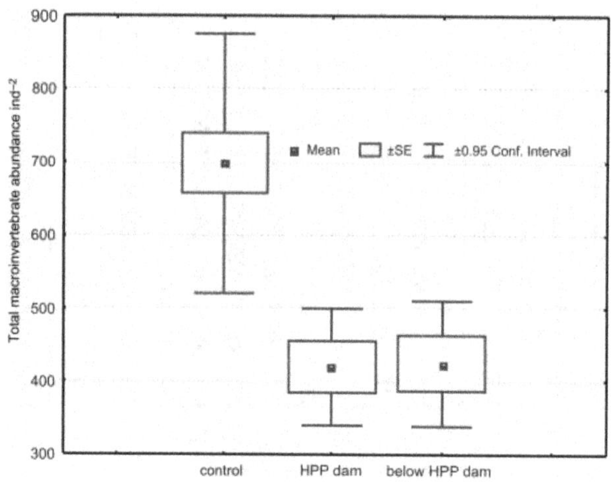

Fig. 6. The means of total macroinvertebrate abundance in studied sites of Virvytė River

Table 4. Mean±SE characteristics of macroinvertebrate metrics at control, HPP dam, below HPP dam reaches of Virvytė River

Metric	Control	HPP dam	Below HPP dam
Total macroinvertebrate taxa number	28±1*	17±1	20±2
EPT taxa	13±1*	5±1	8±1
Percentage of individuals (%)			
Ephemeroptera	48.7±3.8*	24.4±5.1*	36.7±3.6
Trichoptera	21.8±0.9*	11.4±2.9*	14.7±3.1
Plecoptera	17.2±1.6*	0	2.5±0.9*
EPT	87.7±1.4*	35.8±4.8*	53.9±4.6*
Coleoptera	4.7±1.1	4.3±1.4	10.0±2.6
Chironomidae	3.0±0.3*	21.2±3.4*	9.8±1.0*
Oligochaeta	2.1±0.4*	9.7±2.4	6.9±1.1
Mollusca	2.9±0.1*	13.8±3.4*	10.2±4.0
Isopoda	0	6.9±2.5	4.8±2.5

*Statistically significant difference from all other study reaches Fisher's LSD test, $p<0.05$.

(2004), who described the use of EPT taxa richness (in %) as the most descriptive tool in analyzing biological data to ensure the most effective investigation of water quality.

In a study by Katano *et al.* (2009), the change in assemblage structure did not involve a substantial loss of taxa or disappearance of EPT taxa; instead, a shift occurred in the dominant taxa. An increase in macroinvertebrate density below dams compared to above-dam sites is commonly accompanied by a significant reduction in taxonomic diversity, disappearance of EPT taxa or dominance of dipterans (Munn, Brusven 1991). The values of relative abundance of Chironomidae, Oligochaeta and molluscs were significantly lower in the control sites in comparison with the sites in and below the dams (Table 4) (Fisher's LSD test). The proportion of tolerant chironomids and non-insect individuals (oligochaetes) increased at degraded sites and were dominating in organically polluted sites (Masese *et al.* 2009).

Our results suggest that hydropower plants have a negative impact on water quality. DSFI value was the highest in control sites and reliably differed from the value in HPP dams (ANOVA $F = 11336.6, p = 0.0001$). Water quality was classified as very good in control sites, good in sites below HPP dams and moderate in HPP dams (Fig. 7). The HBI showed a significant variation between control river segments and segments in and below HPP dams (ANOVA $F = 4113.28, p = 0.0001$). According to HBI, water quality was excellent in the Virvytė River control sites, fair below HPP dams and fairly poor in HPP ponds. The results of investigations by Thomson *et al.* (2005) also showed that sites downstream of the dam had poorer HBI scores than upstream (control) sites. Cortes *et al.* (2002) found that in the regulated segment of the river, poor water quality and lack of litter input affected mainly the shredders group.

In conclusion, construction of hydropower plants induces cardinal changes in species composition of macrozoobenthos and water quality of the river. Not only most reophylic species of macrozoobenthos become extinct, but also distribution and abundance of the surviving spices changes, which affects distribution and life quality of other vertebrate and invertebrate animals.

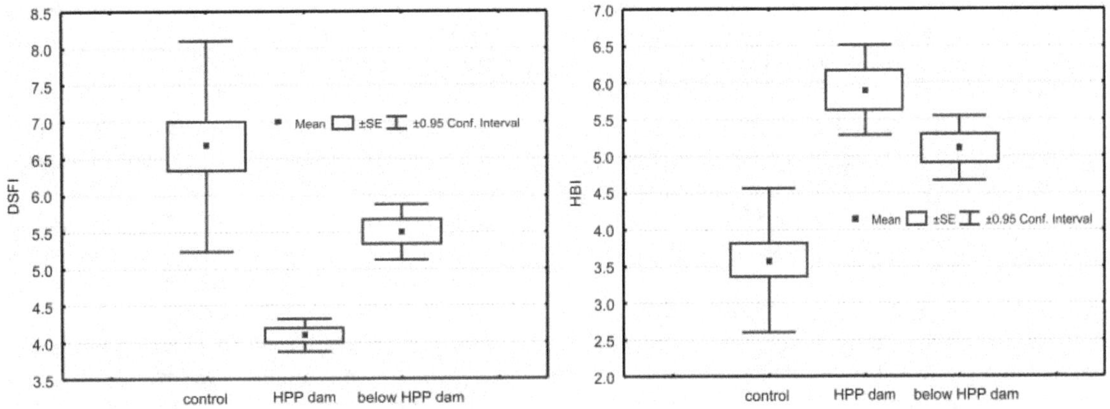

Fig. 7. The means of biotic indices score in studied sites of Virvytė River

Conclusions

1. Dam building has significant impacts on macroinvertebrate composition in the Virvytė River. Both of the disturbed sites (in and below HPP dams) have significant lower taxa number of macroinvertebrates (both total and EPT taxa) compared to the control sites. The total abundance of macroinvertebrates was significantly higher in control sites of river than in the sites in and below HPP dams.

2. The macroinvertebrate assemblages in control sites dominated pollution sensitive EPT (87.7% of total abundance) and in HPP dams pollution tolerant chironomids (21.2% of total abundance).

3. According to DSFI, water quality was classified as very good in the control sites, good below HPP dams and moderate in HPP dams. Water quality according to HBI was excellent in the Virvytė River control sites, fair below HPP dams and fairly poor in HPP dams.

Acknowledgments

The authors would like to thank the reviewers for their helpful comments to improve the article.

References

Ambers, R. K. R. 2007. Effects of a small, century-old dam on a second order stream in Virginia Piedmont, *Southeastern Geographer* 47: 181–201. http://dx.doi.org/10.1353/sgo.2007.0018

Arbačiauskas, K. 2009. Bentoso bestuburiai, in *Gyvunijos monitorinio metodai* [Fauna monitoring methods]. Vilnius: VU Ekologijos Institutas, 22–45.

Bahuguna, B. K.; Nautiyal, R.; Nautiyal, P.; Singh, H. R. 2004. Stream regulation: variations in the density, composition and diversity of benthic macroinvertebrates occurring in the up and downstream sections of the impounded zone of the river Ganga in the foothills, *Tropical Ecology* 45(2): 251–261.

Brown, L. R. 2005. Aquatic assemblages of the highly urbanized Santa Ana river basin, California, *American Fisheries Society Symposium* 47: 263–287.

Camargo, J. A.; Alonso, A; de la Puente, M. 2005. Eutrophication downstream from small reservoirs in mountain rivers of central Spain, *Water Research* 39: 3376–3384. http://dx.doi.org/10.1016/j.watres.2005.05.048

Céréghino, R.; Cugny, P.; Lavandier, P. 2002. Influence of intermittent hydropeaking on the longitudinal zonation patterns of benthic invertebrates in a mountain stream, *International Review of Hydrobiology* 87: 47–60. http://dx.doi.org/10.1002/1522-2632(200201)87:<47::AID-IROH47>3.0.CO;2-9

Christopher, A. T.; Granata, T. C.; Murphy, R. P.; Livchak, C. J. 2007. Using a coupled eco-hydrodynamic model to predict habitat for target species following dam removal, *Ecological Engineering* 30(3): 215–230. http://dx.doi.org/10.1016/j.ecoleng.2006.11.006

Cortes, R. M. V.; Ferreira, M. T.; Oliveira, S. V.; Oliveira, D. 2002. Macroinvertebrate community structure in a regulated river segment with different flow conditions, *River Research and Applications* 18(4): 367–382. http://dx.doi.org/10.1002/rra.679

Doyle, M. W.; Stanley, E. H.; Strayer, D. I.; Jacobson, R. B.; Schmidt, J. C. 2005. Effective discharge analysis of ecological processes in streams, *Water Resources Research* 41: 16. http://dx.doi.org/10.1029/2005WR004222

Gailiušis, B.; Jablonskis, J.; Kovalenkovienė, M. 2001. *Lietuvos upės. Hidrografija ir nuotėkis* [Lithuanian rivers hydrograph and runoff]. Kaunas: Lietuvos energetikos institutas. 792 p.

Greathouse, E. A.; Pringle, C. M.; McDowell, W. H.; Holmquist, J. G. 2006. Indirect upstream effects of dams: consequences of migratory consumer extirpation in Puerto Rico, *Ecological Applications* 16: 339–352. http://dx.doi.org/10.1890/05-0243

Grubbs, S. A.; Taylor, J. M. 2004. The influence of flow impoundment and river regulation on the distribution of riverine macroinvertebrates at Mammath cave National Park, Kentucky, U.S.A, *Hydrobiologia* 520: 19–28. http//dx.doi.org/10.1023/B:HydR0000027722.23374.dc.

Grzybkowska, M.; Dukowska, M. 2002. Communities of Chironomidae (Diptera) above and below a reservoir on a lowland river: long-term study, *Annales Zoologici* 52: 235–247. http://dx.doi.org/10.3161/000345413X666066

Hilsenhoff, W. L. 1988. Rapid field assessment of organic pollution with a family-level biotic index, *Journal of North American Benthological Society* 7: 65–68. http://dx.doi.org/10.2307/1467832

Hu, W.; Wang, G.; Deng, W.; Li, S. 2008. The influence of dams on ecohydrological conditions in the Huaihe River basin, China, *Ecological Engineering* 33(3–4): 233–241. http://dx.doi.org/10.1016/j.ecoleng.2008.04.003

Jesus, T.; Formingo, N.; Santos, P.; Tavares, G. R. 2004. Impact evaluation of Vila Vicosa small hydroelectric power plant (Portugal) on the water quality and the dynamics of the benthic macroinvertebrate communities of the Ardena river, *Limnetica* 23(3–4): 241–256.

Katano, I.; Negishi, J.; Minagawa, T.; Doi, H.; Kawaguchi, Y.; Kayaba, Y. 2009. Longitudinal macroinvertebrate organization over contrasting discontinuities: effects of a dam and tributary, *Journal of North American Benthological Society* 28(2): 331–351. http://dx.doi.org/10.1899/08-010.1

Kleemola, S.; Söderman, G. 1993. *Manual for integrated monitoring. programme phase 1993–1996. Hydrobiology of streams.* Helsinki: National board of water and the environment, 67–68.

Lamsodis, R.; Vaikasas, S. 2005. The potential to retain nitrogen in beaver ponds and delta floodplains of the River Nemunas, *Large Rivers* 15(1–4) *Archiv fur Hydrobiologie Suppl.* 155/1: 225–241.

Lessard, J. L.; Hayes, D. B. 2003. Effects of elevated water temperature on fish and macroinvertebrate communities below small dam, *River Research and Applications* 19: 721–732. http://dx.doi.org/10.1002/rra.713

Lithuanian Standards. 2010. Paviršinių vandens telkinių būklės nustatymo metodika [Methods for Assessment of the Status of Surface Water Bodies], *Valstybės žinios* [The Official Gazette], 29–1363.

Maiolini, B.; Silveri, L.; Lencioni, V. 2007. Hydroelectric power generation and disruption of the natural stream flow: effects on the zoobenthic community, *Studi Trentini di Scienze Naturali Acta Biologica* 83: 21–26.

Malik, A. U.; Richardson, J. S. 2009. Riparian vegetation change in upstream reaches of three temperate rivers dammed for hydroelectric generation in British Columbia, Canada, *Ecological Engineering* 35(5): 810–819. http://dx.doi.org/10.1016/j.ecoleng.2008.12.005

Mandaville, S. M. 2002. *Benthic macroinvertebrates in freshwaters. Taxa tolerance values, metrics, and protocols.* Halifax: Soil and water cons. soc. of metro Halifax.

Mantel, S. K.; Muller, W. J.; Hughes, D. A. 2010. Ecological impact of small dams on South African rivers: part 2: biotic response – abundance and composition of macroinvertebrate communities, *Water SA* 36(3): 361–370.

Marčiulionienė, D.; Montvydienė, D.; Kazlauskienė, N.; Kesminas, V. 2011. Changes in macrophytes and fish communities in the cooler of Ignalina nuclear power plant (1988–2008), *Journal of Environmental Engineering and Landscape Management* 19(1): 21–33. http://dx.doi.org/10.3846/16486897.2011.557273

Masese, F. O.; Raburu, P. O.; Muchiri, M. 2009. A preliminary benthic macroinvertebrate index of biotic integrity (B-IBI) for monitoring the Moiben River, Lake Victoria Basin, Kenya, *African Journal of Aquatic Science* 34(1): 1–14. http://dx.doi.org/10.2989/AJAS.2009.34.1.1.726

Munn, M. D.; Brusven, M. A. 1991. Benthic macroinvertebrate communities in nonregulated and regulated waters of the Clearwater river, Idaho, U.S.A, *Regulated Rivers: Research and Management* 6(1): 1–11. http://dx.doi.org/10.1002/rrr.3450060102

Nichols, S.; Norris, R.; Maher, W.; Thoms, M. 2006. Ecological effects of serial impoundment on the Cotter River, Australia, *Hydrobiologia* 572: 255–273. http://dx.doi.org/10.1007/s10750-005-0995-6

Novotny, V.; Batošová, A.; O'Reily, N.; Ehlinger, T. 2005. Unlocking the relationship of biotic integrity of impaired waters to anthropogenic stresses, *Water Research* 39: 184–198. http://dx.doi.org/10.1016/j.waters.2004.09.002

Penczak, T.; Kruk, A. 2005. Patternizing of impoundment impact (1985–2002) on fish assemblages in a lowland river using the Kohonen algorithm, *Journal of Applied Ichtyology* 21(3): 169–177. http://dx.doi.org/10.1111/j.1439-0426.2005.00649.x

Pietraszewski, D.; Marszał, L.; Zięba, G.; Przybylski, M.; Zieliński, P. 2008. Fish fauna of the Sanna River system, *Scientific Annual of the Polish Angling Association* 21: 129–146.

Principe, R. 2010. Ecological effects of small dams on benthic macroinvertebrate communities of mountain streams (Córdoba, Argentina), *Annales de Limnoligie – International Journal of Limnology* 46: 77–91. http://dx.doi.org/10.1051/limn/2010010

Sakalauskas, V. 2003. *Duomenų analizė su STATISTIKA* [Data analysis with STATISTIKA]. Vilnius: Margi raštai. 235 p.

Sear, D. A.; Carling, P. A.; Greig, S. M. 2004. Fine sediment accumulation in spawning gravels and the effects on interstitial flow, in *Proceedings of the fifth Symposium on Ecohydraulics. Aquatic Habitats: Analysis & Restoration*, 12–17 September, 2004, Madrid, Spain, 808–812.

Sharma, Ch. M.; Sharma, S.; Borgstrom, R.; Bryceson, I. 2005. Impact of a small dam on macroinvertebrates: a case study in the Tinau River, Neoal, *Aquatic Ecosystem Health and Management* 8: 267–275. http://dx.doi.org/10.1080/14634980500218332

Soulsby, R. L.; Manning, A. J.; Whitehouse, R. J. S.; Spearman, J. R. 2010. *Development of a generic physically-based formula for the settling flux of natural estuarine cohesive sediment*: Final Report-summary. HR Wallingford company research project DDY0409. 1 p.

Thomson, J. P.; Hart, D. D.; Charles, D. F.; Nightengale, T. L.; Winter, D. M. 2005. Effects of removal of a small dam on downstream macroinvertebrate and algal assemblages in a Pennsylvania stream, *Journal of North American Benthological Society* 24(1): 192–207. http://dx.doi.org/10.1899/0887-3593(2005)024<0192:EOROAS>2.0.CO;2

Tszydel, M.; Grzybkowska, M.; Kruk, A. 2009. Influence of dam removal on trichoptera assemblages in the lowland Drzewiczka River, Poland, *Hydrobiologia* 630(1): 75–89. http://dx.doi.org/10.1007/s10750-009-9781-1

Vaikasas, S. 2010. Mathematical modelling of sediment dynamics and their deposition in Lithuanian rivers and their deltas, *Journal of Environmental Engineering and Landscape Management* 18(3): 207–216. http://dx.doi.org/10.3846/jeelm.2010.24

Vaikasas, S.; Lamsodis, R. 2011. Sediment and nutrient retention in reservoirs of small Hydropower Plants, in *Proceedings of the 8th International Conference "Environmental Engineering"*, 22–23 May, 2011, Vilnius Gediminas Technical University, Vilnius, 684–691.

Winterwerp, J. C.; Kesteren, W. G. M. 2004. Introduction to the physic of cohesive sediment in the marine environment, in van Loon, T. (Ed.). *Developments in sedimentology 56*. Amsterdam: Elsevier. 466 p.

Wright, J. F.; Gunn, R. J. M.; Winder, J. M.; Wiggers, R.; Vowles, K.; Clarke, R. T.; Harris, I. 2002. A comparison of the macrophyte cover and macroinvertebrate fauna at three sites on the River Kennet in the mid-1970s and late 1990s, *Science of the Total Environment* 282: 121–142. http://dx.doi.org/10.1016/S0048-9697(01)00948-2

WFD & Hydropower. 2007. Water framework directive & hydropower, in *Proceedings of the Common Implementation Strategy Workshop*, 4–5 June, 2007, Berlin, 1–4.

Zdankus, N.; Vaikasas, S.; Sabas, G. 2008. Impact of a hydropower plant on the downstream reach of a river, *Journal of Environmental Engineering and Landscape Management* 16(3): 128–134. http://dx.doi.org/10.3846/1648-6897.2008.16.128-134

Saulius VAIKASAS. Prof., Dr Habil at Department of Hydraulics, Vilnius Gediminas Technical University (VGTU), 2007. Principle researcher of Water Research Institute of ASU, 2003. Doctor Habil of Science at Lithuanian University of Agriculture and Kaunas Technological University (Environmental Engineering and Landscape Management), 2001. Doctor of Science at WMI LUA, 1993. Publications: author of 1 monograph, 2 educational books, over 250 research papers. Research interests: hydraulic modelling, floods, sedimentation and riverbed processes, ecohydraulics.

Kęstutis PALAIMA. PhD student at the Water Research Institute of ASU, Technology Science of Environmental Engineering and Landscape Management direction.

Virginija PLIŪRAITĖ. Dr, Sector of Freshwater Ecology, Laboratory of Ecology and Physiology of Hydrobionts, Department of Aquatic Ecosystems, Nature Research Centre. Research interests: ecology of aquatic invertebrates, investigation of seasonal dynamics of macrozoobenthos communities and changes in rivers affected by anthropogenic activity.

24

COMPARING ENVIRONMENTAL IMPACTS OF NATURAL INERT AND RECYCLED CONSTRUCTION AND DEMOLITION WASTE PROCESSING USING LCA

Isabela Maria Simion[a], Maria Emiliana Fortuna[a, c], Alessandra Bonoli[b], Maria Gavrilescu[a, d]

[a]*Department of Environmental Engineering and Management, Faculty of Chemical Engineering and Environmental Protection, "Gheorghe Asachi" Technical University of Iasi, 73 Prof Dr Docent Dimitrie Mangeron Street, 700050 Iasi, Romania*

[b]*DICAM, Alma Mater Studiorum University of Studies of Bologna, 2 Risorgimento Street, 40136 Bologna, Italy*

[c]*Petru Poni Institute of Macromoecularl Chemistry, 41A Grigore Ghica Voda Alley, 700487 Iasi, Romania*

[d]*Academy of Romanian Scientists, 54 Splaiul Independentei, 050094 Bucharest, Romania*

Abstract. Construction and demolition wastes (C&DW) are usually recognized as not dangerous, but their accumulation can generate serious environmental problems. In spite of C&DW high potential to be reused/recycled, the practical procedures need to be assessed in terms of environmental consequences. The objective of this study is to quantify the environmental impacts of C&DW recycling/reuse, specifically in the production of aggregate 0/30 mm, comparative to those generated during the natural inert processing, in terms of global impacts addressing the whole process and for each technological phase. The analysis was carried out using Life Cycle Assessment methodology, assisted by SimaPro software, and based on primary data collected directly from the Italian Emilia Romagna region. Three methods were used for impact quantification: *Eco-Indicator 99*, *EDIP/UMIP* and *Cumulative Energy Demand*. The analysis revealed that the environmental impacts generated by C&DW recycling/reuse accounting for about 40% of the impacts induced by natural inert processing.

Keywords: C&DW, energy, environmental impact, Life Cycle Assessment, natural inert, recycling.

Reference to this paper should be made as follows: Simion, I. M.; Fortuna, M. E.; Bonoli, A.; Gavrilescu, M. 2013. Comparing environmental impacts of natural inert and recycled construction and demolition waste processing using LCA, *Journal of Environmental Engineering and Landscape Management* 21(4): 273–287. http://dx.doi:10.3846/16486897.2013.852558

Introduction

Waste generation is the result of inefficient and unsustainable uses of natural resources and energy in production processes, which leads to economic and environmental problems, such as large additional expenses for the collection, processing and waste landfilling and elimination (Dall'Ara *et al.* 2012; Schiopu *et al.* 2012; Simion *et al.* 2012; Taboada-González *et al.* 2012; Yeheyis *et al.* 2012). The enormous production of waste generated during various anthropogenic activities induces the need to extend the life cycle of the products as well as the producer responsibilities, which is a subject of great interest for scientific research, as well as for various actors involved in production processes and waste treatment and management (del Rio Merino *et al.* 2010; Vlase *et al.* 2012; Wiesmeth, Hackl 2011). One of the key issues of a continuous progress is ensuring the sustainable management of prevention, control and remediation processes relative to the environmental components, associated in particular to waste minimization and valorization (Gavrilescu 2004; Simion *et al.* 2013).

In this context, a sustainable approach should address various procedures for reducing waste and/or for the optimization of waste recovery, recycle and reuse as well as its processing in various ways (Agamuthu 2008; del Rio

Corresponding author: Maria Gavrilescu
E-mail: mgav@ch.tuiasi.ro; mgav_ro@yahoo.com

Merino *et al.* 2010; Fortuna *et al.* 2012; Ghinea, Gavrilescu 2010; Petraru, Gavrilescu 2010; Rigamonti *et al.* 2009; Schiopu, Gavrilescu 2010).

Construction and demolition wastes (C&DW) include diverse categories of materials that, when recovered, can represent high-value materials and resources for new construction (EPA 2002; Simion *et al.* 2013). However, C&DW is a significant part of the total solid waste streams, since building materials account for almost half of all materials used and about half the solid waste generated worldwide (Cochran *et al.* 2007; Dosal *et al.* 2012; Edge Environment 2011; Kourmpanis *et al.* 2008; Vasquez 2013). An Australian report shows that C&DW generation in Australia during 2004–2005 "was 15.1 million tonnes, of which 7.6 million tonnes was recycled materials (timber, steel, concrete, rubble and soil) and 7.5 million tonnes was residual waste to landfill. In 2006–2007, 43,777,000 tonnes of waste was generated, 38 per cent of which was from C&DW" (Edge Environment 2011). C&DW represents 10–30% of total waste generation in Europe, which means about 180 million tonnes per year, approximated at 480 kg per person. While only 28% is recycled, the rest of 72% is disposed (Bressi, Puia 2000; EPA 2002; Pittalis 2009; Poon 2007). Table 1 shows the amounts of C&DW generated yearly in European countries in 2006–2010 (Eurostat 2013). On the other hand, large quantities of natural primary aggregates and building materials are extracted each year due to increasing demands. At a global level, 60% of the raw materials extracted from the lithosphere are used for civil works and building construction. In Europe, the extraction of minerals for building is about 4.8 tonnes per inhabitant per year, which is 64 times the average footprint per person (Bribián *et al.* 2011; Simion *et al.* 2013).

Unfortunately, materials resulting from construction and demolition are often considered as **waste** for disposal, rather than **resources** for processing and reuse. They have negative impacts on environment, economy, public health and social life (Ionescu *et al.* 2011; Robu *et al.* 2007; SARMa 2011; Yeheyis *et al.* 2012). Governmental regulation and legislation worldwide can be considered as positive reactions to the need of waste minimization, since they often generated the market framework for building materials and products resulting from the construction and

Table 1. C&DW generation in the European countries during 2004–2010 (Eurostat 2013)

Country \ Time	2004	2006	2008	2010
France	3.461	3.680	3.942	4.011
Austria	3.418	3.788	3.765	1.074
Netherlands	3.047	3.464	3.617	4.698
Ireland	2.773	3.896	0	360
Germany (until 1990 former territory of the FRG)	2.322	2.386	2.402	2.336
United Kingdom	1.658	1.808	1.645	1.695
European Union (27 countries)	1.576	1.708	1.721	1.713
Sweden	1.142	985	359	1.000
Spain	1.085	1.073	986	824
Belgium	1.059	1.241	1.1442	1.667
Italy	845	888	1.165	981
Czech Republic	796	816	1.022	889
Denmark	791	1.067	1.033	572
Cyprus	660	386	544	1.288
Slovenia	455	496	681	737
Bulgaria	385	133	240	10
Estonia	362	534	820	326
Greece	300	613	608	184
Slovakia	261	170	241	329
Portugal	250	341	761	1.041
Norway	240	269	314	316
Hungary	172	302	323	307
Croatia	145	–	29	2
Lithuania	104	103	123	109
Iceland	63	–	–	–
Poland	52	371	182	545
Latvia	4	8	5	10
Romania	4	2	15	11
European Union (15 countries)	–	–	–	–
Turkey	–	–	0	0

demolition (C&DW) waste streams (Edge Environment 2011).

However, it is very difficult to know correct data and establish statistics with any degree of confidence because this form of waste is deposited both legally and illegally. Also C&DW is often not recorded as a separate waste stream, or is recorded incorrectly. Besides, there seems to be confusion regarding the status of clean fill materials that can be accepted for landfills establishment according to European regulations (Bonoli, Garfi 2008; Dosal *et al.* 2012).

Even though this material is derived from construction or building projects, it is seldom recorded as C&DW (EPA 2002; Simion *et al.* 2013).

The demolition of walls or renovation in course of restructuring activities are perhaps the most immediately reminded of to the memory of the ordinary citizen, in connection with the mounds of rubble often improperly abandoned on roads and bank river (Balázs *et al.* 2010; D'Andrea 1999).

1. Structure of C&DW and recycling/reuse framework

Most of the materials obtained during dismantling can be recycled and, if they receive an appropriate treatment process, can become recycled raw materials that find their target market materials for building and for construction in civil engineering works. These products can be used for many applications often quite comparable to performance those of natural aggregates (Bonoli *et al.* 2006; Taranu *et al.* 2012). In order to increase the recycling amounts of C&DW it is necessary to know the specific composition of this waste (Dosal *et al.* 2012; Simion *et al.* 2013). The typical components of C&DW include concrete, asphalt, wood, metal, drywall, and smaller amounts of packaging materials, such as paper and plastic (Coronado *et al.* 2011). However, C&DW has a very inconsistent composition according to local building techniques, climate, economic activities, technological development in the area, available raw materials (Table 2). Their characteristics depend in a high degree on the region and involve large numbers of stakeholders (Coelho, de Brito 2013; Taranu *et al.* 2012).

The enormous quantities of C&DW produced by many anthropogenic activities generate the need of its recycling: this is to reduce withdrawals from quarries of natural materials and at the same time to reduce the quantities of waste disposed in landfills. That's in accordance with European directives and national laws envisage the achievement of important objectives of recycling and recovery. Directive 2008/98/EC on waste incorporates the need to quantify the waste stream and to improve the C&DW recovery efficiency in the European Union (EC Directive 2008). Moreover, the European Commission considers C&DW as a priority in waste stream management due to large quantities produced yearly and, above all, following its high potential for reuse and recycling embedded in the waste structure (Coronado *et al.* 2011; Solís-Guzmán *et al.* 2009).

Table 2. Compositions of C&DW in Europe (EPA, 2002)

Material	Percentage of total fill (%)
Concrete	20–50
Bricks	5–20
Timber	5–20
Steel	5–15
Soil	15–70
Green Waste	5–20
Plastic	5

Certainly, recycling and waste reprocessing are highly affected by this situation and their own impacts, mostly due to transporting and reprocessing the materials recovered could be increased by any inconsistence in waste composition. All these can weaken the benefits from recycling, so as it is necessary to examine the lifecycle as a whole and quantify the environmental impacts, based on an adequate knowledge of the inputs to material recovery facilities, even more the Landfill Directive asks the recycling target of C&DW by 2020 to be 70% (Bribian *et al.* 2011; Coronado *et al.* 2011). The optimization of recycling operations in dedicated facilities is necessary in order to achieve this target, so as it is necessary to group the waste into homogeneous fractions by (Bonoli *et al.* 2006):

- selection of raw materials and their appropriate storage containers;
- separation of waste to facilitate the treatment.

The recycling of building materials has its origin in the time of the total or partial demolition of a manufactured housing (Mymrin, Correa 2007; Rigamonti *et al.* 1996). Demolition technique adopted is closely related with waste valorization: in fact, it has a direct influence on the quality of waste produced and of recycled materials. Those recycled materials deriving from treated homogeneous waste possess a higher quality than those coming from a heterogeneous mix (Pitalis 2009; Taranu *et al.* 2012).

These actions are feasible if the building processes use less natural resources and energy, and do not generate more waste than the production processes of similar products using virgin resources. Furthermore, at present the cost implications of preparing recycled or reused materials are often unclear (Gorgolewski *et al.* 2006; Bressi 1999).

In particular, the use of C&DW after adequate treatment for civil engineering purposes may bring economic and environmental advantages due to the lesser need for their transportation to landfill, reduced spaces designed for authorized dumps, and to the considerable savings in traditional quarry materials (Cupo-Pagano *et al.* 1994). The Life Cycle Assessment (LCA) tool makes possible a specific examination of particular construction waste streams, taking into account impacts occurring over the entire life cycle. There are numerous studies published in which the LCA is applied to evaluate the impact of different construction materials and solutions (Bribián *et al.* 2011; Khasreen *et al.* 2009; Koroneos, Dompros 2007).

The aim of this paper is to analyze the processes and materials involved in the Italian C&DW management system and the environmental benefits of C&DW recycling/reuse. Based on the above consideration, in this study we developed a life cycle assessment approach to compare the recycling/reuse scenario of C&DW management with the conventional scenario, which processes natural inert, against a range of environmental criteria. Two scenarios were considered:

- *scenario 1*: production of *aggregate* 0/30 mm from natural inert quarries (Fig. 1);
- *scenario 2*: production of *aggregate* 0/30 mm from C&D inert waste (Fig. 2).

The two scenarios specific for the Italian Emilia Romagna region are discussed and evaluated based on their environmental impact, exploiting the facilities offered by simulation software SimaPro 7.3.3, offered by the DICAM University of Bologna.

2. Materials and method

2.1. LCA methodology

Customarily related to the analysis of products and materials, the use of LCA has extended in recent years to cover larger processes and services such as waste management (Curan 1996; Powell *et al.* 1996). In this study the methodological standards ISO 14040:2006 and ISO 14044:2006 were used. According to these standards, Life Cycle Assessment (LCA) is an internationally standardized methodology for environmental assessment, applied to evaluate the environmental impact of a product or system (ISO 2006a, b; Comandaru *et al.* 2012; Georgakellos 2011; Ghinea *et al.* 2012). LCA consists of four steps: goal and scope definition, inventory analysis, impact assessment and interpretation (Curan 1996; Ghinea, Gavrilescu 2010b; Ghinea *et al.* 2012; Iosip *et al.* 2010; ISO 2006a, b). This methodology can be used for modeling and simulation of waste management scenarios, supported by SimaPro software, while all of the data needed for the life cycle inventory are collected from the literature, the database of the software and the municipal waste services (Kazemi *et al.* 2012). With SimaPro software tool, life cycle balances were elaborated and analyzed in specific ways.

2.2. Goal and scope definition

The aim of the present analysis was to provide multiple criteria for decision-making, according to the values of three different methods to assess impact categories for the two scenarios: **natural inert processing** (*scenario 1*) and **C&DW recycling/reuse** (*scenario 2*). The three methods applied for impact assessment are: *Eco-Indicator 99, EDIP/UMIP* and *Cumulative Energy Demand.*

The focal point is the inert waste, and the data are compared to those for natural inert materials, related to 1 tonne of processed material as the functional unit. The boundaries of the systems are drawn so as to comprise all significant impacts.

2.3. Production and reuse of C&DW – system boundaries

In the first step of LCA the boundaries of systems are defined to identify inputs and outputs, so as to consider all processes, the input data on energy flows and material flows and output data related to specific issues. The necessary treatment for C&DW recycling/reuse as raw materials in new processes consists in: preliminary separation, primary crushing, direct classification or screening, secondary crushing and new separation. An efficient plant is equipped with a technological line, which allows the separation the incoming material into three streams: a stone material, a light fraction (paper, plastic, wood), a metal fraction (Bonoli *et al.* 2006).

2.4. Inventory analyses phase

The inventory phase is conducted by collecting all information, using specific methods that were analyzed comparatively with studies from literature and software libraries, for all the sectors involving materials, energy and fuels. Natural inert and inert wastes are analyzed in all their phase of processing, normalized to the functional unit (1 tonne) of material entering in the systems. The database BUWAL 250 was used for all stages of assessment, excepting the distribution of waste and iron extraction, for which the Ecoinvent database has been used.

After data inventory collection and data normalization to the functional unit, the environmental impact was evaluated.

2.5. Life cycle impact assessment

Life cycle impact assessment is the phase in the LCA aiming at understanding and evaluating the magnitude and significance of the potential environmental impacts of a product system. The life cycle of a product ranges from resource extraction via material processing, manufacturing, and product use or service delivery, to recycling, and to the disposal of any remaining waste.

For the production system *aggregate* 0/30 mm from quarry material, the LCA is conducted based on the *cradle-to-gate* approach. In fact, the study begins with the procurement of raw materials and ends with placing the product on the market. For the production *aggregate* 0/30 mm system by processing C&DW, the LCA is conducted with *gate-to-gate* approach, which means that the study examines only the treatment of the waste in the recycling plant. The inputs and outputs taken into consideration are reported, both in the case of production of the natural inert or the recycling of inert waste. The impact categories to analyze in this study were selected considering the current energy and environmental problem in the European area (Bribián *et al.* 2011).

3. Results and discussions

3.1. Impact assessment for processing scenarios

Impacts evaluation for the two scenarios (**natural inert processing** – *scenario 1,* and **C&DW recycling/reuse** – *scenario 2*), achieved with SimaPro software according to Life Cycle Assessment methodology and based on the three methods (*Eco-Indicator 99, EDIP/UMIP* and *Cumulative Energy Demand*) led to interesting results as decision making support, presented below.

3.1.1. Eco-Indicator 99 method

The *Eco-Indicator 99* method is used for the characterization of the impact on *human health, ecosystem quality and resources*. These three categories of impact are denoted as "macro categories", but these impact categories include

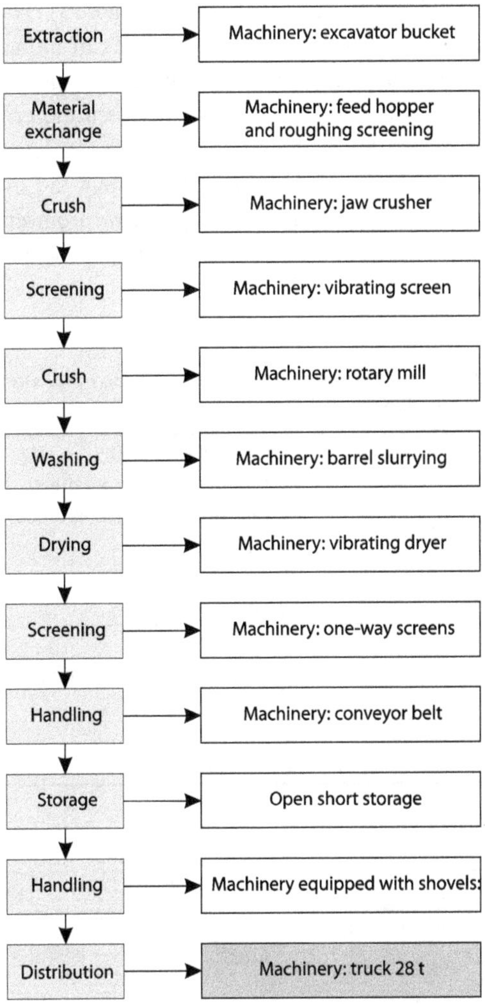

Fig. 1. Scheme of the production process of *aggregate* 0/30 mm from natural inert material

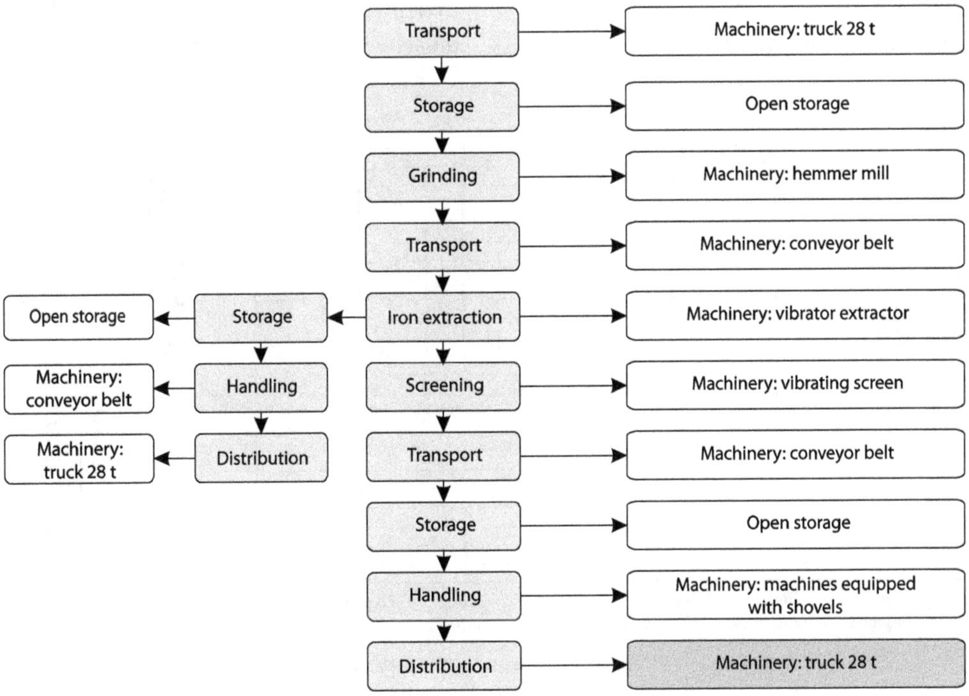

Fig. 2. Scheme of the production process of *aggregate* 0/30 mm from inert waste

some individual categories of impacts (Table 3). The characterization of the three main categories of impact and of the individual categories is illustrated in Figs 3 and 4, which allow making comparisons between the environmental effects due to the production of *aggregate* in the two analyzed scenarios. It can be seen that the impacts generated by the production of *aggregate* from crushed natural inert are important in terms of human health, ecosystem quality and resources.

Table 3. Impact categories determined with *Eco-Indicator 99* method (Goedkoop *et al.* 2010)

Human health	Carcinogens Respiratory effects (organics) Respiratory effects (inorganics) Climate changes Radiation Ozone layer
Ecosystem quality	Ecotoxicity Acidification/ Eutrophication Land use
Resources	Minerals Fossil fuels

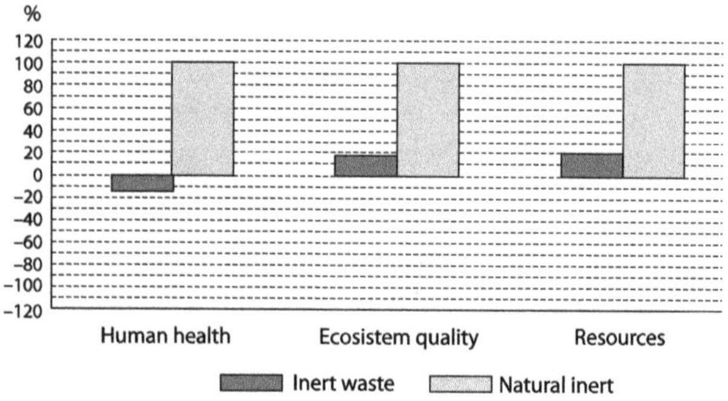

Fig. 3. Characterization of impact with *Eco-Indicator* method

On the other hand, C&DW recycling generates positive impacts on human health, and negative impacts on ecosystem quality and resources but accounting for about 20% of the impacts generated by natural inert processing.

If the analysis is detailed by taking into account the individual categories of impacts, it is evident that the two scenarios induce negative impacts for all these subcategories excepting *Carcinogens* and *Minerals*. In these two cases, C&DW recycling generate positive impacts.

However, unlike the situation presented in Fig. 3, C&DW recycling induce high negative impacts in terms of ionizing *Radiation* and *Land Use* (since the temporary storage of waste involves occupation of large areas), while for the other impact categories, the effects of C&DW recycling account for maximum 30% of the impact of *scenario 1*.

The recycling of C&DW allows, therefore, obtaining the same product as from the natural inert quarry, but with much lower impact on humans and the environment, favorizing the conservation of mineral resources, since their extraction generates extensively environmental damages.

3.1.2. EDIP/UMIP method

This method allows the evaluation of various indicators, the most important of which is the Global Warming Potential (GWP), due to greenhouse gas emissions. Other impact categories able to be assessed by the *EDIP/UMIP* are as follows: ozone depletion; acidification; eutrophication; photochemical smog; ecotoxicity (water, chronic); ecotoxicity (soil chronic); human toxicity (air); human toxicity (water); human toxicity (soil); bulk waste; radioactive waste; slags/ashes; resources; hazardous waste.

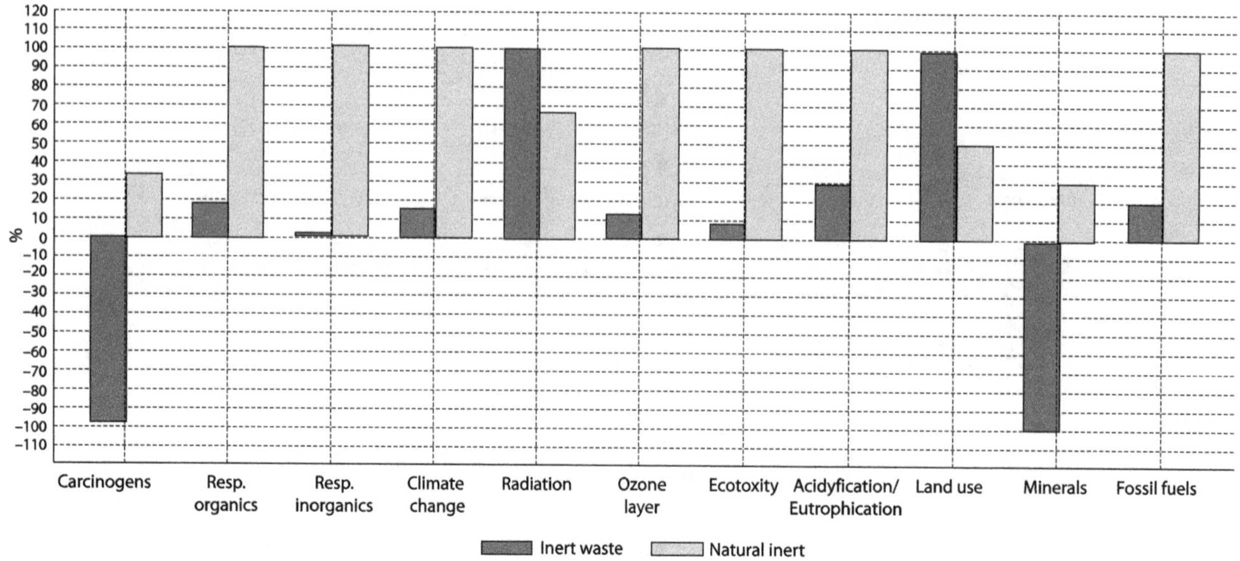

Fig. 4. Characterization of impact categories with *Eco-Indicator* method

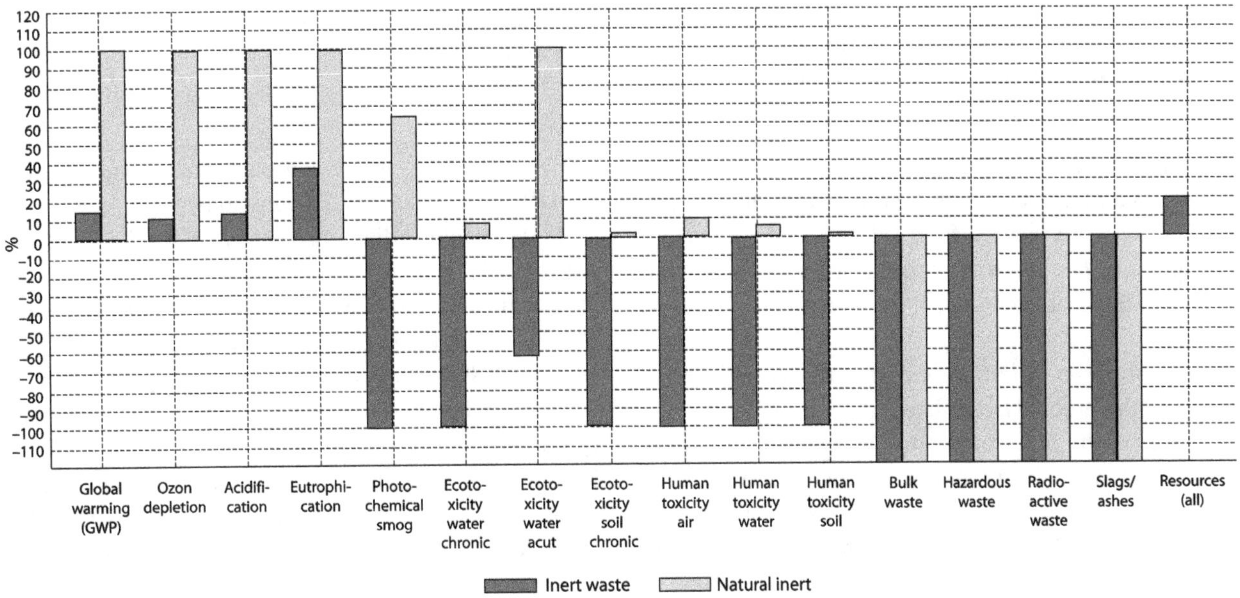

Fig. 5. Characterization of impacts with *EDIP/UMIP* method

Figure 5 demonstrates that C&DW recycling and reuse (*scenario 2*) generates positive impacts in terms of Photochemical smog, Ecotoxicity of water (chronic, acute), Ecotoxicity of soil, Human toxicity (air, water, soil), bulk waste, hazardous *waste, Radioactive waste, Slag/ashes,* while the negative impacts account for maximum 40% of those of *scenario 1*.

3.1.3. Cumulative Energy Demand method

This method considers the energy consumption, referring to the different types of energy utilized in all phases of life cycle, as extraction, processing, recycling, etc. measured in MJ (Fig. 6). Overall, the energetic consumption for *scenario 2* is 17% of that for *scenario 1*. Energy consumption for the production of *aggregate* from C&DW recycling or extracting natural inert material is presented in Figure 6.

It's important to observe that the energy necessary for the production of *aggregate* from natural inert is higher (1664.11 MJ) than that for inert waste (246.41 MJ). Most of the energy comes from burning fossil fuels, which have a high impact on health and the environment.

The categories of energy taken into account in this analysis are obtained from: fossil fuels, nuclear, biomass, renewable (wind, solar, geothermal, etc.), hydroelectric sources (Table 4). Data from Table 4 shows that there are significant differences between the energy consumption for C&DW recycling/reuse (*scenario 2*) and natural inert processing, as follows:

- the potential consumption of energy from fossil fuel for *scenario 2* is 16% from that associated to *scenario 1*;
- the potential consumption of nuclear energy associated to *scenario 2* is 11% from that for *scenario 1*;
- the energy from biomass is coupled with energy production in *scenario 2*, unlike the situation for *scenario 1*, which consumes energy equivalent to that from biomass;
- the potential consumption of hydroelectric energy for *scenario 2* is 9% from that for *scenario 1*;
- the potential consumption of wind, solar, geothermal energy is similar in the two scenarios.

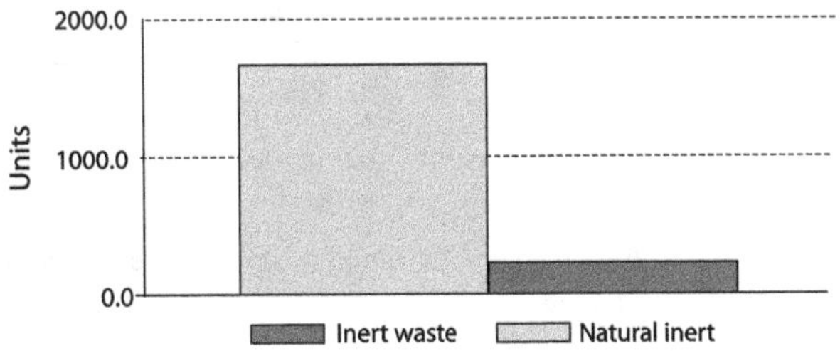

Fig. 6. Energy consumption with *Cumulative Energy Demand* method

Table 4. Energy consumption for different type of produced energy

	Natural inert (MJ)	Inert waste (MJ)
Fossil fuels	1300	210
Nuclear	185	20.6
Biomas	0.06	–0.09
Wind, solar, geothermal	0.05	0.04
Hydroelectric	178	16.4

3.2. Assessment of impacts associated to various phases for the production of aggregate 0/30 mm by natural inert processing (sce-nario 1)

3.2.1. Eco-Indicator 99 method

The analysis shows that the most impact is generated due to the usage of aggregates for extraction, such as excavator buckets and this is because they have high energy consumption. The resulting impacts are high on human health, quality ecosystems and the consumption of resource (Fig. 7).

Handling and distribution of materials, as other phases of the process have also high environmental impact, even if the order of magnitude is lower than that for extraction, both due to the use of machines such as power shovels and trucks that have substantial fuel consumption and consequently high impact (Fig. 7). However, the primary and secondary crushing have smaller impacts. The same is true in the assessment of impacts for the individual categories expressed in Fig. 8. In this case it is necessary to emphasize that, as regards the categories relating to radiation, the use of the territory and the consumption of minerals play fundamental roles in the distribution phase of the material, while the other does not contribute to the formation of this kind of impacts.

3.2.2. EDIP/UMIP method

This method takes in account many indicators, including the GWP, as a result of CO_2 emissions.

The Sankey diagram presented in Fig. 9 shows that high GWP due to the production of CO_2 is associated with the extraction of aggregate from the quarry (84,800 g), followed by handling through mechanical shovels (6,890 g) and distribution (4,330 g) respectively. The contribution of the stages of primary and secondary crushing is reduced, respectively 2,030 g of CO2 and 3,200 g of CO2, while the steps of screening and washing do not involve the production of carbon dioxide (Fig. 10).

Fig. 10 shows that the extraction of aggregate from the quarry and the distribution of the *aggregate* output from the processing plant induce higher environmental impacts. The first one contributes mostly to the Global Warming Potential, Ozone Depletion, Acidification, Eutrophication, *Photochemical smog*, *Human toxicity* (air) and consumption of *Resources*, while the latter is mainly responsible for *Human toxicity* (water, soil).

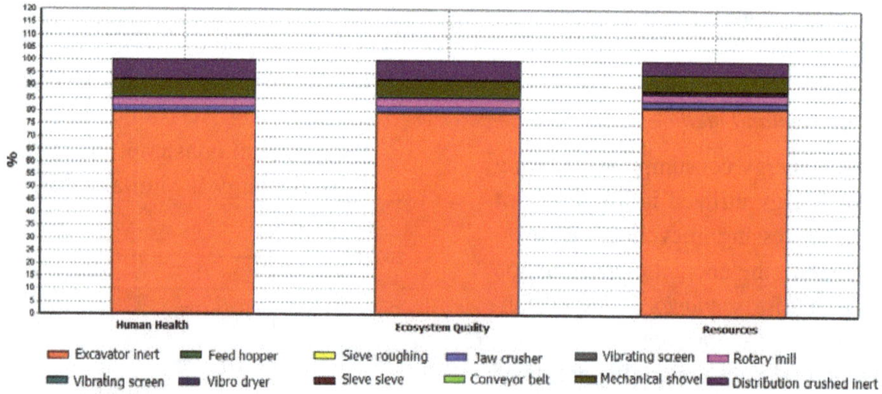

Fig. 7. Characterization of impact category for each phase of *scenario 1* with *Eco-Indicator 99* method

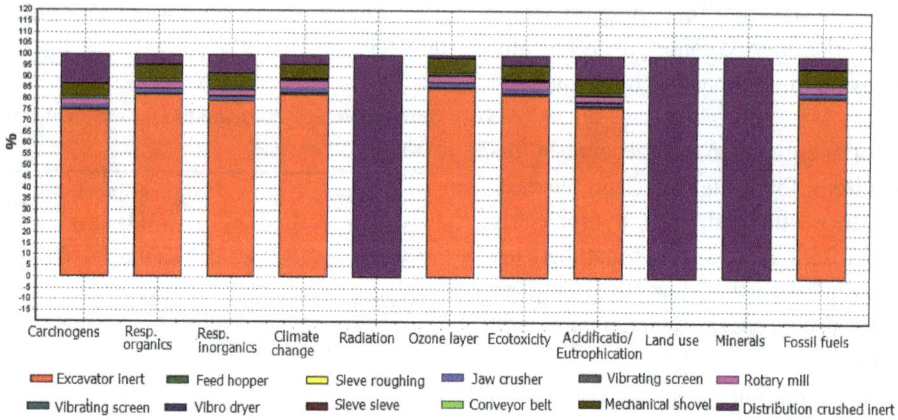

Fig. 8. Characterization of individual impacts for each phase of *scenario 1* with *Eco-Indicator 99* method

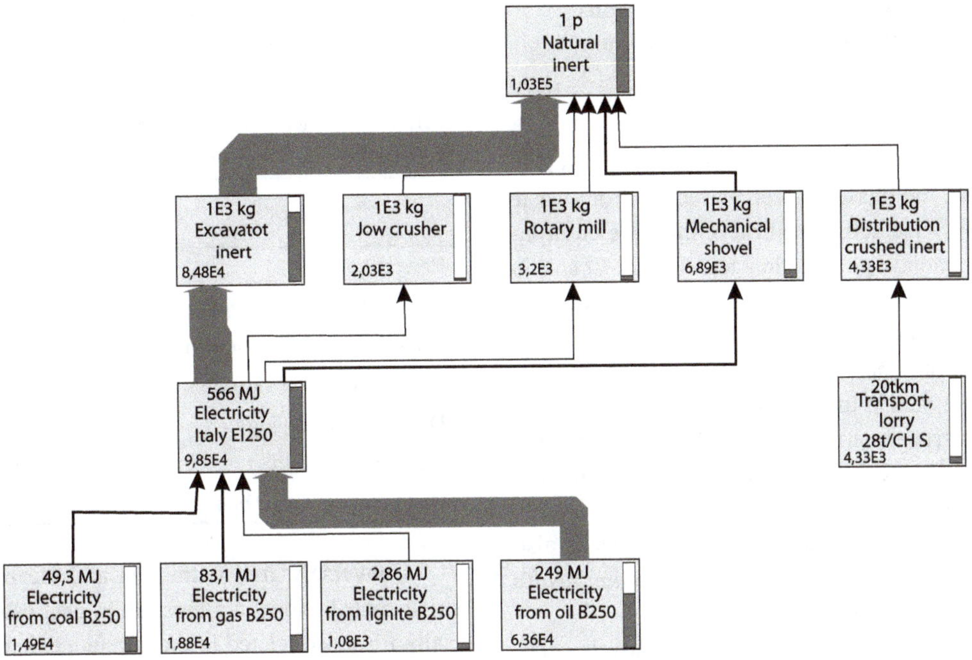

Fig. 9. Mass balance flow sheet (Sankey diagram) for CO_2 produced during natural inert processing, associated to GWP impact category

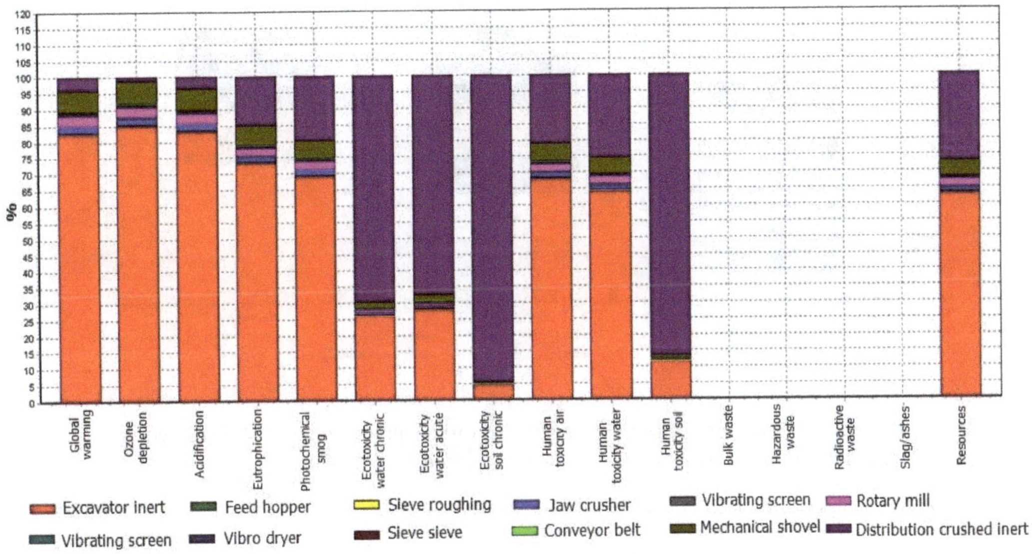

Fig. 10. Characterization of individual impacts for each phase of *scenario 1* with *EDIP/UMIP* method

3.2.3. *Cumulative Energy Demand method*

From Fig. 11 it can be seen that higher energy consumption (1380 MJ equivalent) is attributable to the extraction stage (the excavation process), followed by material handling by mechanical shovels (111.6 MJ equivalents), transport and distribution by truck (70.95 MJ equivalents), the secondary crusher (51.77 MJ equivalent) and the jaw crusher (33.02 MJ equivalent). In Fig. 11 the steps with lower energy consumption equivalent to 10 MJ were ignored because they contribute to the expenditure of energy to a less significantly degree.

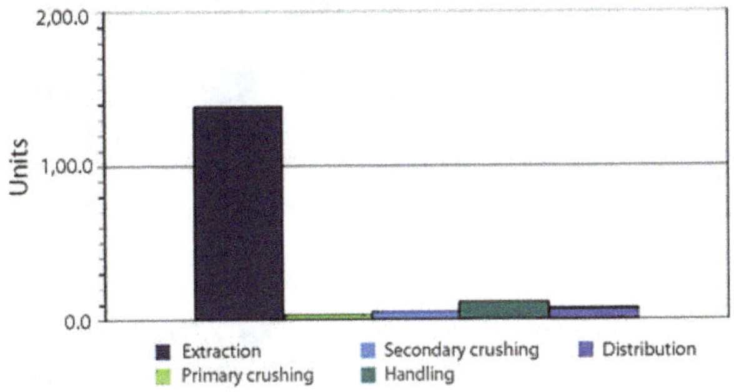

Fig. 11. Energy of individual phases for inert crushed processing

3.3. Assessment of impacts associated to various phases for the production of aggregate 0/30 mm by C&DW recycling/reuse (scenario 2)

3.3.1. Eco-Indicator 99 method

In the production of *aggregate* 0/30 mm by C&DW recycling/reuse, the phases of iron extraction from waste and, respectively the subsequent recycling have environmental and economic positive impacts. In fact, *scenario 2* allows the recovery and sale of various materials that would otherwise be disposed of in landfills (Fig. 12). Table 5 shows the energy consumption related to other phases that take part in the preparation process of *aggregate* 0/30 mm from quarry material. On the other hand, transportation of materials from the scrap yard generates the highest negative impact in terms of *Human health*, followed by materials distribution phase, which is also characterized by high impacts in terms of *Human health and Ecosystem quality* (Fig. 13).

Figure 13 also shows that mechanical shovels generate an impact almost equal to that of transport and distribution, if not superior, in the impact categories relating to the destruction of ozone layer, ecotoxicity and the consumption of fossil fuels.

Table 5. Machinery used in the manufacture of grinding with the consumption less than 10 MJ equivalents

Phases	Energy consumption (MJ equivalent)
Power (hopper)	0.87
Sieving (sieve roughing)	4.39
Screening (vibrating screen)	2.44
Washing (beating slurrying)	8.04
Drying (vibro drying)	1.82
Sieving (sieve sieve)	5.76
Handling (conveyor belt)	1.07

Conversely, iron extraction and recycling have positive contributions in almost all impact categories, especially for those related to the carcinogenic effects on respiration due to inorganic substances, ecotoxicity and the consumption of minerals.

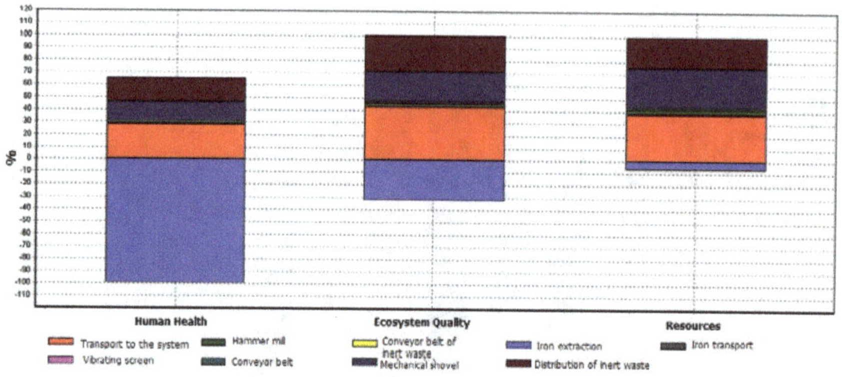

Fig. 12. Characterization of impacts generated by various processing phases involved in C&DW recycling/reuse, based on *Eco-Indicator* method

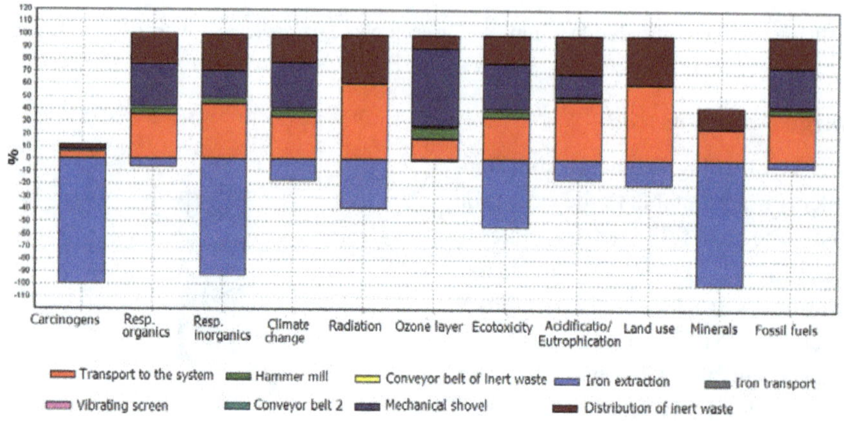

Fig. 13. Characterization of impacts generated in various phases of C&DW recycling/reuse process assessed with the *Eco-Indicator* method

3.3.2. EDIP/UMIP method

The flow diagram of CO_2 balance (Sankey diagram, Fig. 14), closely related to the Global Warming Potential (GWP) highlights that the recycling of iron allows to avoid CO_2 emissions, equivalent to 3430 g. The material handling with mechanical shovels results in emissions of CO_2 equivalent to 6890 g, while the transport system and the distribution of the *aggregate* respectively are associated with CO_2 emissions equivalent to 6,500 g and 4,330 g, respectively (Fig. 14).

Similar arguments apply to sulfur dioxide (SO_2) and nitrogen peroxide (N_2O_4), respectively, responsible for acidification and eutrophication potentials (Fig. 15). SO_2 is generated in particular during handling with mechanical shovels (60.1 g), transport to the waste plant (48 g), distribution of aggregate (32 g) and grinding (8.6 g). The equivalent quantities of N_2O_4 generated during recycling process are of 80.5 g in the transportation of waste, 53.7 g in the distribution of aggregate, 21.1 g in handling with shovels and 3.02 g in grinding. The quantities of these

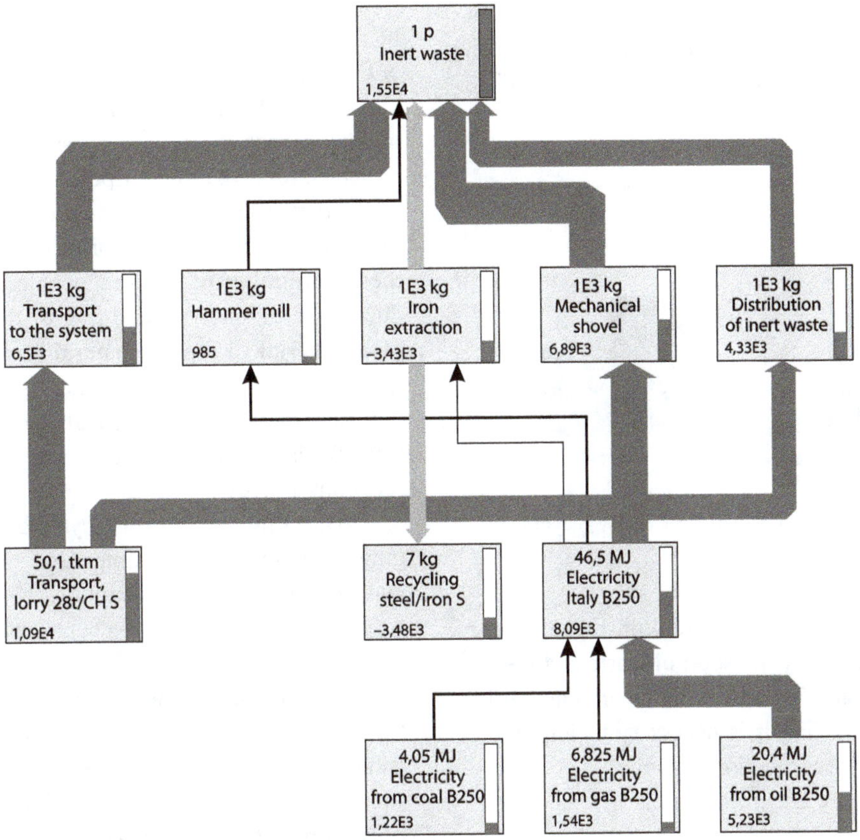

Fig. 14. Mass balance flow sheet (Sankey diagram) for CO_2 produced during C&DW recycling/reuse, associated to GWP impact category

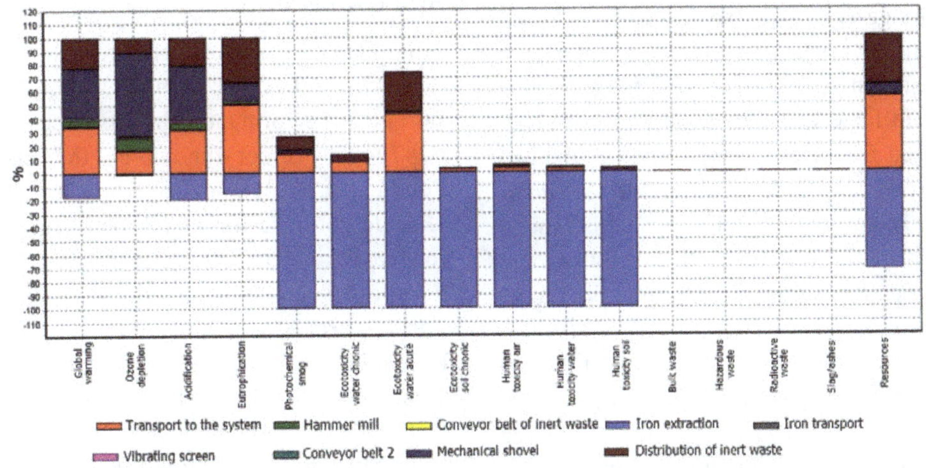

Fig. 15. Characterization of impacts generated in various phases of C&DW recycling/reuse process assessed with the *EDIP/UMIP* method

gases emitted in the other phases of C&DW recycling are relative small and therefore can be neglected.

3.3.3. Cumulative Energy Demand method

According to this method, an equivalent of 246.41 MJ is necessary to recycle **one tonne** of C&DW. The extraction of iron saves 60.78 MJ equivalents, while the transport to the recycling plant involves a consumption of 0.38 MJ equivalents, so that we have a positive balance of mining. In Fig. 16 the technological phases with higher energy consumption were compared, neglecting phases that consume less (screening and handling through the web site immediately after the screening, which needs 0.67 MJ equivalents each).

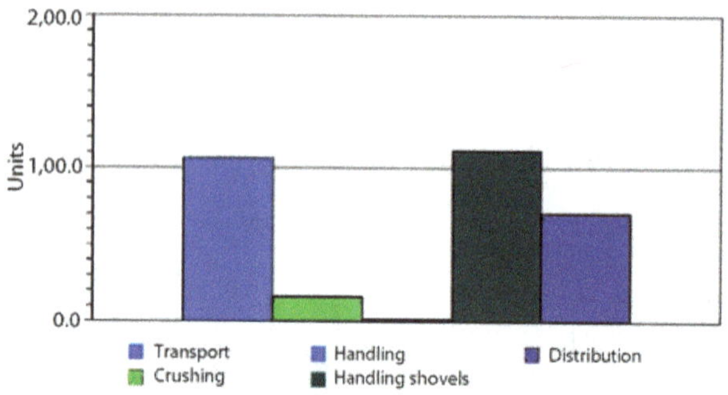

Fig. 16. Energy of individual phases for C&DW recycling/reuse

Handling shovels is the phase with higher energy consumption, followed by transport of waste and the distribution of aggregate. To reduce the environmental impact of C&DW recycling it is necessary to improve or remove the three phases mentioned above.

Conclusions

In this study, we applied LCA to quantify the environmental impacts in C&DW recycling/reuse, specifically in the production of *aggregate* 0/30 mm. A comparison was performed with the environmental performance of natural inert processing, in terms of global impacts addressing the whole process and for each technological phase. Three methods were used for impact quantification of the two scenarios: *Eco-Indicator 99*, *EDIP/UMIP* and *Cumulative Energy Demand*, according to Life Cycle Assessment methodology, supported by SimaPro 7.3.3 software, and based on primary data collected directly from the Italian Emilia Romagna region.

The comparison of the two scenarios – the production of *aggregate* from crushed inert (*scenario 1*) and C&D waste inert (*scenario 2*), with the mentioned methods showed that the recycling process generates a significantly lower environmental impact compared to that resulted from the natural inert processing, as regards the emission of pollutants, the consequences on human and ecosystems health, the amount of energy consumed. It should be noted that, in the process of recycling, recovery of iron has a positive impact on reducing the consumption of raw materials and energy consumed. Substantial differences appear between the two processes in terms of global warming potential, since the production of 1 tonne of *aggregate* from crushed inert generates 103,000 g of CO_2 equivalent, compared with 15,500 g of CO_2 generated from the recycling of C&DW (almost 7 times lower). A similar argument applies to the acidification and eutrophication, assessed in terms of SO_2 and N_2O_4 emissions.

The technological phase with the higher environmental impact and energy consumption is extraction of minerals, followed by transport and handling.

Apart from the above mentioned aspects, recycling of C&DW has several advantages, including: reducing the exploitation of non-renewable raw materials; reduction in the number and size of the landfill; creation of a material to replace natural raw materials such as gravel and sand, with equivalent performance. However, the selective collection of various components of C&DW should be helpful to ensure the efficiency of the recycling process.

Assessing the costs of recycling technologies, emerging affordability to promote these solutions rather than conventional disposal in landfills, the eco-tax introduced for the disposal of waste has contributed to this condition benefit of recycling.

Considering the mentioned positive effects associated to the C&DW recycling/reuse, from both environmental and economical perspectives, it would be necessary to strengthen the market for recycled aggregates, especially in the perspective of sustainable development in the construction sector.

Acknowledgement

This paper was carried out with the support of: POSDRU CUANTUMDOC "Doctoral Studies for European Performances in Research and Innovation" ID_79407 project funded by the European Social Found and Romanian Government; PERFORM-ERA "Postdoctoral Performance for Integration in the European Research Area" (ID-57649), financed by the European Social Fund and the Romanian Government. The research was also supported by the grant of the Romanian National Authority for Scientific Research, CNCS – UEFISCDI, project number PN-II-ID-PCE-2011-3-0559, contract 265/2011. Part of the research was supported by the European Union's Seventh Framework Programme (FP7/2007-2013) under grant agreement n°264115 – *STREAM.*

References

Agamuthu, P. 2008. Challenges in sustainable management of construction and demolition waste, *Waste Management & Research* 26: 491–492. http://dx.doi.org/10.1177/0734242X08100096

Balázs, S.; Antonini, E.; Tarantinic, M. 2012. *Application of Life Cycle Assessment (LCA) methodology for valorization of building demolition materials and products* [online], [cited 10 September 2012]. Available from Internet: http://www1.coe.neu.edu/~smgupta/4193-56-SPIE.PDF

Bonoli, A.; Garfi, M. 2008. The recycling of construction and demolition waste: the quality standards for the reuse in construction field, in *REWAS 2008, Global Simposium on Recycling, Waste Treatment and Clean Technology proceedings, WARRENDALE PA, TMS*, 1163–1169.

Bonoli, A.; Fabbri, S.; Goldoni, S. 2006. Demolition techniques and enhancement of waste from C&D, in *Recycling* 115–122 (in Italian).

Bressi, G. 1999. Recycling technologies of waste from C&D, in *The Recycling in Construction Aggregates*. Palermo, 4 June, 1999 (in Italian).

Bressi, G.; Puia, P. 2000. Aggregates from recycling: logistics and environmental issues in the economic evaluation of their use, in *Magenta*. Milano, 30 November, 2000 (in Italian).

Bribián, I. Z.; Capilla, A. V.; Uson, A. A. 2011. Life cycle assessment of building materials: comparative analysis of energy and environmental impacts and evaluation of the eco-efficiency improvement potential, *Building and Environment* 46: 1133–1140. http://dx.doi.org/10.1016/j.buildenv.2010.12.002

Coelho, A.; de Brito, J. 2013. Environmental analysis of a construction and demolition waste recycling plant in Portugal – Part I: energy consumption and CO_2 emissions, *Waste Management* 33(5): 1258–1267. http://www.dx.doi.org/10.1016/j.wasman.2013.01.025

Cochran, K.; Townsend, T.; Reinhart, D.; Heck, H. 2007. Estimation of regional building-related C&D debris generation and composition: case study for Florida, *Waste Management* 27: 921–931.

Comandaru, I. M.; Barjoveanu, G.; Peiu, N.; Ene, S. A.; Teodosiu, C. 2012. Life cycle assessment of wine: focus on water use impact assessment, *Environmental Engineering and Management Journal* 11: 533–543.

Coronado, M.; Dosal, E.; Coz, A.; Viguri, J. R.; Andres, A. 2011. Estimation of construction and demolition waste (C&DW) generation and multicriteria analysis of C&DW management alternatives: a case study in Spain, *Waste and Biomass Valorization* 2: 209–225. http://dx.doi.org/10.1007/s12649-011-9064-8

Cupo-Pagano, M.; D'Andrea, A.; Giavarini, C.; Marro, C. 1994. Use of building demolition waste for asphalt mixes: first results, in *Energy, Environment and Technological Innovation – Proceedings of III International Congress*, 5–11 November, 1994, Caracas – Venezuela, 203–208.

Curan, M. A. 1996. *Environmental Life-Cycle Assessment*. McGraw-Hill Professional Publishing.

D'Andrea, A. 1999. Aggregates recycling: characteristics, fields use, recycling of inert waste in construction, in *Proceedings of the National Conference*, Palermo, 81–103 (in Italian).

Dall'Ara, A.; Bonoli, A.; Serranti, S. 2012. An innovative procedure to characterize properties from tailored composts, *Environmental Engineering and Management Journal* 11: 1825–1832.

Dosal, E.; Coronado, M.; Munoz, I.; Viguri, J. R.; Andres, A. 2012. Application of multi-criteria decision-making tool to locate construction and demolition waste (C&DW) recycling facilities in a northern Spanish region, *Environmental Engineering and Management Journal* 11: 545–556.

EC Directive. 2008. Directive 2008/98/EC of the European Parliament and of the Council of 19 November 2008 on Waste and Repealing Certain Directives, *Official Journal of the European Union* L 312: 3–30.

Edge Environment. 2011. *Construction and demolition waste guide – recycling and re-use across the supply chain, Department of Sustainability, Environment, Water, Population and Communities* [online]. Australian Government, Canberra, Australia. Available from Internet: http://www.environment.gov.au/wastepolicy/publications/pubs/case-studies.pdf

EPA, 2002. *Construction and Demolition Waste, Waste Management and Resource Use Opportunities, Queensland.* Environmental Protection Agency, Brisbane, Australia.

Eurostat. 2013. *Statistics Database 2013, Waste Statistics* [online]. Available from Internet: http://epp.eurostat.ec.europa.eu/portal/page/portal/waste/waste_generation_management/generation/construction

Fortuna, M. E.; Simion, I. M.; Ghinea, C.; Petraru, M.; Cozma, P.; Apostol, L. C.; Hlihor, R. M.; Fertu, D. T.; Gavrilescu, M. 2012. Analysis and management of specific processes from environmental engineering and protection based on sustainability indicators, *Environmental Engineering and Management Journal* 11: 333–350.

Gavrilescu, M. 2004. Cleaner production as a tool for sustainable development, *Environmental Engineering and Management Journal* 3: 45–70.

Georgakellos, D. A. 2011. Determining the level of subsidies for environmental friendly technology vehicles, *Environmental Engineering and Management Journal* 10: 1495–1502.

Ghinea, C.; Gavrilescu, M. 2010. Decision support models for solid waste management – an overview, *Environmental Engineering and Management Journal* 9: 868–880.

Ghinea, C.; Petraru, M.; Bressers, H. Th. A.; Gavrilescu, M. 2012. Environmental evaluation of waste management scenarios – significance of the boundaries, *Journal of Environmental Engineering and Landscape Management* 20(1): 76–85. http://dx.doi.org/10.3846/16486897.2011.644665

Goedkoop, M.; Oele, M.; De Schryver, A.; Vieire, M.; Hegger, S. 2010. *SimaPro Database Manual. Methods library.* The Netherlands: PRé Consultants.

Gorgolewski, M.; Straka, V.; Edmonds, J.; Sergio, C. 2006. *Facilitating greater reuse and recycling of structural steel in the construction and demolition process*: Final report. Canadian Institute of Steel Construction, 3–63.

Ionescu, S.; Deac, V.; Corpodean, C. 2011. Risk of exposure to asbestos in case of the mechanical degradation of some asbocement composites, *Environmental Engineering and Management Journal* 9: 1773–1777.

Iosip, A.; Hortal, M.; Dobón, A.; Bobu, E. 2010. Comparative environmental impact assessment of corrugated board production, *Environmental Engineering and Management Journal* 9: 1281–1287.

ISO 14040: 2006a. *Environmental management – Life cycle assessment – Principles and framework.* European Standard

EN ISO 14040. The International Organization for Standardization, Geneva, Switzerland.

ISO 14044: 2006b. *Environmental management – Life cycle assessment – Requirements and Guidelines.* European Standard EN ISO 14040. The International Organization for Standardization, Geneva, Switzerland.

Kazemi, S.; Peiu, N.; Macoveanu, M. 2012. Waste management: A Life Cycle perspective, *Environmental Engineering and Management Journal* 11: 1639–1648.

Khasreen, M.; Banfill, P.; Menzies, G. 2009. Life-cycle assessment and the environmental impact of buildings: a review, *Sustainability* 1: 674–701. http://dx.doi.org/10.3390/su1030674

Koroneos, C.; Dompros, A. 2007. Environmental assessment of brick production in Greece, *Building and Environment* 42: 2114–2123. http://dx.doi.org/10.1016/j.buildenv.2006.03.006

Kourmpanis, B.; Papadopoulos, A.; Moustakas, K.; Stylianou, M.; Haralambous, K. J.; Loizidou, M. 2008. Preliminary study for the management of construction and demolition waste, *Waste Management & Research* 26: 267–275. http://dx.doi.org/10.1177/0734242X07083344

Mymrin, V.; Correa, S. M. 2007. New construction material from concrete production and demolition wastes and lime production waste, *Construction and Building Materials* 21: 578–582. http://dx.doi.org/10.1016/j.conbuildmat.2005.09.007

Petraru, M.; Gavrilescu, M. 2010. Pollution prevention: a key to economic and environmental sustainability, *Environmental Engineering and Management Journal* 9: 597–614.

Pittalis, C. 2009. *Inert waste from C & D: technologie recycling and applicazionie of LCA methodology for the production of stabilized inert 0/30 mm*: PhD Thesis. Coordinator Bonoli, A. Faculty of Inginery, Alma Mater Studiorum Universita di Bologna.

Poon, C. S. 2007. Management of construction and demolition waste, *Waste Management* 27: 159–160. http://dx.doi.org/10.1016/j.wasman.2007.08.013

Powell, J. C.; Craighill, A. L.; Parfitt, J. P.; Turner, R. K. 1996. A lifecycle assessment and economic valuation of recycling, *Journal of Environmental Planning and Management* 39: 97–112. http://dx.doi.org/10.1080/09640569612697

Rigamonti, L.; Grosso, M.; Giugliano, M. 2009. Life cycle assessment for optimizing the level of separated collection in integrated MSW management system, *Waste Management* 29: 934–944.

del Rio Merino, M.; Izquierdo Gracia, P.; Salto Weis Azevedo, I. 2010. Sustainable construction: construction and demolition waste reconsidered, *Waste Management and Research* 28: 118–129. http://dx.doi.org/10.1177/0734242X09103841

Robu, B. M.; Caliman, F. A.; Betianu, C.; Gavrilescu, M. 2007. Methods and procedures for environmental risk assessment, *Environmental Engineering and Management Journal* 6: 573–592.

SARMa, 2011. *Life Cycle Assessment (LCA) guidelines to be used in the SARMa project. Definition of a common methodology to boost use of LCA tools in sustainable production and recycling of aggregates.* 15 February, 2011. 60 p.

Schiopu, A.-M.; Gavrilescu, M. 2010. Municipal solid waste landfilling and treatment of resulting liquid effluents, *Environmental Engineering and Management Journal* 9: 993–1019.

Schiopu, A. M.; Piuleac, G. C.; Cojocaru, C.; Apostol, I.; Mamaliga, I.; Gavrilescu, M. 2012. Reducing environmental risk of landfills: leachate treatment by reverse osmosis, *Environmental Engineering and Management Journal* 11: 2319–2331.

Simion, I. M.; Fortuna, M. E.; Gavrilescu, M. 2012. *Use of sustainable indicators in solid waste management – a review*, Bulletin of the Polytechnic Institute of Iasi, Tome LVIII (LXII), Fasc. 1, 57–71.

Simion, I. M.; Ghinea, C.; Maxineasa, S. G.; Taranu, N.; Bonoli, A.; Gavrilescu, M. 2013. Ecological footprint applied in the assessment of construction and demolition waste integrated management, *Environmental Engineering and Management Journal* 12: 779–788.

Solís-Guzmán, J.; Marrero, M.; Montes-Delgado, M. V.; Ramírez-de-Arellano, A. 2009. A Spanish model for quantification and management of construction waste, *Waste Management* 29: 2542–2548.

Taboada-González, P.; Aguilar-Virgen, Q.; Ojeda-Benítez, S.; Armijo, C. 2012. Waste characterization and waste management perception in rural communities in Mexico: a case study, *Environmental Engineering and Management Journal* 11: 1751–1759.

Taranu, N.; Oprisan, G.; Entuc, I.; Budescu, M.; Munteanu, V.; Taranu, G. 2012. Composite and hybrid solutions for sustainable development in civil engineering, *Environmental Engineering and Management Journal* 11: 783–793.

Vasquez, E. 2013. Overview regarding construction and demolition waste in several countries, in Vasquez, E. (Ed.). *Progress of Recycling in the Built Environment: Rilem State-of-the-Art Reports* 8: 37–174. London: Springer.

Vlase, R.; Iftimi, V.; Gavrilescu, D. 2012. Sustainable use of recovered paper – Romanian paper industry, *Environmental Engineering and Management Journal* 11: 2077–2082.

Wiesmeth, H.; Hackl, D. 2011. How to successfully implement extended producer responsibility: considerations from an economic point of view, *Waste Management and Research* 29: 891–901. http://dx.doi.org/10.1177/0734242X11413333

Yeheyis, M.; Hewage, K.; Shahria Alam, M.; Eskicioglu, C.; Sadiq, R. 2012. An overview of construction and demolition waste management in Canada: a lifecycle analysis approach to sustainability, *Clean Technical Environmental Policy* 15(1): 81–91. http://dx.doi.org/10.1007/s10098-012-0481-6

Isabela Maria SIMION. PhD student since October 2010 and from 2011 assistant researcher in a IDEI Project, at Faculty of Chemical Engineering and Environmental Protection, "Gheorghe Asachi" Technical University of Iasi. Holds a Master's degree in Environmental Engineering and Management (2010) and a Bachelor's degree in Environmental Engineering (2009), both from "Gheorghe Asachi" Technical University of Iasi. Her research focuses on solid waste management, life cycle assessment, sustainable development, sustainable indicators, ecological footprint.

Maria Emiliana FORTUNA. PhD, Researcher at Institute of Macromolecular Chemistry "Petru Poni" Iasi. PhD degree (Department of natural and synthetic polymers – 2008), Master's degree in Environmental Engineering and Management (2004) at "Gheorghe Asachi" Technical University of Iasi and a Bachelor's degree Chemical Engineering (2003) at "Al. I. Cuza" University of Iasi. Her research focuses on impact of paper recycling, electrochemical methods for industrial water treatment, sustainable development, ecological footprint.

Alessandra BONOLI. Associate Professor at the Department of Civil Engineering, Chemistry, Environmental and Materials, Alma Mater Studiorum University of Bologna, Italy. Teaching and research interests: mineral processing, solid materials and waste recycling, water supply, water resource sustainable utilization, waste water treatment and recycling, environmental management and sustainability, life cycle assessment, appropriate environmental technologies for developing countries.

Maria GAVRILESCU. Professor at the Department of Environmental Engineering and Management of the "Gheorghe Asachi" Technical University of Iasi, Romania, where she teaches courses in Chemical and Biological Process Engineering, Process Synthesis and Analysis, Integrated Pollution Prevention and Control, Sustainable Industrial Production and Environmental Risk Assessment and Management. She has published numerous books and articles in the fields of chemical and environmental engineering, pharmaceutical and environmental biotechnology, pollution prevention and cleaner production, risk assessment and management, which include over 400 titles.

PERMISSIONS

 Every chapter published in this book has been scrutinized by our experts. Their significance has been extensively debated. The topics covered herein carry significant findings which will fuel the growth of the discipline. They may even be implemented as practical applications or may be referred to as a beginning point for another development.

The contributors of this book come from diverse backgrounds, making this book a truly international effort. This book will bring forth new frontiers with its revolutionizing research information and detailed analysis of the nascent developments around the world.

We would like to thank all the contributing authors for lending their expertise to make the book truly unique. They have played a crucial role in the development of this book. Without their invaluable contributions this book wouldn't have been possible. They have made vital efforts to compile up to date information on the varied aspects of this subject to make this book a valuable addition to the collection of many professionals and students.

This book was conceptualized with the vision of imparting up-to-date information and advanced data in this field. To ensure the same, a matchless editorial board was set up. Every individual on the board went through rigorous rounds of assessment to prove their worth. After which they invested a large part of their time researching and compiling the most relevant data for our readers.

The editorial board has been involved in producing this book since its inception. They have spent rigorous hours researching and exploring the diverse topics which have resulted in the successful publishing of this book. They have passed on their knowledge of decades through this book. To expedite this challenging task, the publisher supported the team at every step. A small team of assistant editors was also appointed to further simplify the editing procedure and attain best results for the readers.

Apart from the editorial board, the designing team has also invested a significant amount of their time in understanding the subject and creating the most relevant covers. They scrutinized every image to scout for the most suitable representation of the subject and create an appropriate cover for the book.

The publishing team has been an ardent support to the editorial, designing and production team. Their endless efforts to recruit the best for this project, has resulted in the accomplishment of this book. They are a veteran in the field of academics and their pool of knowledge is as vast as their experience in printing. Their expertise and guidance has proved useful at every step. Their uncompromising quality standards have made this book an exceptional effort. Their encouragement from time to time has been an inspiration for everyone.

The publisher and the editorial board hope that this book will prove to be a valuable piece of knowledge for researchers, students, practitioners and scholars across the globe.

LIST OF CONTRIBUTORS

Mattia Cai and Davide Pettenella
Department of Land, Environment, Agriculture and Forestry, University of Padova, Agripolis, V.le dell'Università, 16, 35020 Legnaro (PD), Italy

Antanas Sigitas Šileika
Water Research Institute, Aleksandras Stulginskis University, Parko g. 6, Vilainiai, LT-58102 K'dainiai, Lithuania

Mats Wallin
Department of Aquatic Science and Assessment, Swedish University of Agricultural Sciences, 7050, 750 07 Uppsala, Sweden

Kazimieras Gaigalis
Alekandras Stulginskis University, Water Research Institute, Mituvos g. 9, LT-50134 Kaunas, Lithuania

Dainius Jasaitis and Aloyzas Girgždys
Department of Physics, Vilnius Gediminas Technical University, Saulėtekio al. 11, LT-10223 Vilnius, Lithuania

Sławomir Żak and Terese Rauckyte-Żak
Faculty of Chemical Technology and Engineering, University of Technology and Life Sciences, 3 Seminaryjna Street, 85-326 Bydgoszcz, Poland

Alfredas Laurinavičius
Faculty of Environmental Engineering, Vilnius Gediminas Technical University, Sauletekio al. 11, LT-10223 Vilnius, Lithuania

Gediminas Marčiukaitis and Gintaras Jurkėnas
Department of Reinforced Concrete and Masonry Structures, Vilnius Gediminas Technical University, Sauletekio al. 11, 10223 Vilnius, Lithuania

Himanshu Patel and R. T. Vashi
Department of Chemistry, Navyug Science College, Rander Road, Surat 395009, Gujarat, India

Alfonsas Rimkus and Saulius Vaikasas
Water Research Institute, Aleksandras Stulginskis University, Parko g. 6, LT-58102 Vilainiai, Ke˙dainiai, Lithuania

Boriss Gjunsburgs
Department of Water Engineering and Technology, Riga Technical University, Azenes str. 16/20, LV-1048 Riga, Latvia

Yuan Li, Yanqun Zu and Guoqing Gao
College of Resources and Environment, Yunnan Agricultural University, Kunming, China

Bozhi Wu and Fenggen Guo
College of Agronomy and Biotechnology, Yunnan Agricultural University, Kunming, China

Junqin Wu, Yuechun Zhao, Lu Liu, Biao Fan and Minghua Li
Department of Applied Chemistry, South China Agricultural University, Guangzhou 510642, China

Vitas Marozas
Department of Ecology, Aleksandras Stulginskis University, Studentu g. 11, Akademija, LT-53361 Kaunas distr., Lithuania

Kestutis Armolaitis and Juratė Aleinikovienė
Department of Ecology Institute of Forestry, Lithuanian Research Centre for Agriculture and Forestry (LRCAF), Liepu g. 1, Girionys, LT-53101 Kaunas distr., Lithuania

Juratė Aleinikovienė
Department of Soil Science and Plant Nutrition, Aleksandras Stulginskis University, Studentu g. 11, Akademija, LT-53361 Kaunas distr., Lithuania

Mindaugas Gudas
Environment Protection Agency, Juozapaviciaus g. 9, LT-09311 Vilnius, Lithuania

Arvydas Povilaitis
Aleksandras Stulginskis University, Studentu g. 11, LT-53361 Kaunas, Lithuania

Ernestas Zaleckas and Eglė Sendžikienė
Institute of Environment, Aleksandras Stulginskis University, Studentų g. 11, LT-53361, Akademija, Kauno r., Lithuania

Valdas Paulauskas
Department of Chemistry, Aleksandras Stulginskis University, Studentų g. 11, LT-53361, Akademija, Kauno r., Lithuania

Huasai Simujide, Chen Aorigele, Bai Manda, Mei-Yun Wu, Yu Li and Tong-Ri-Ge Bai
College of Animal Science, Inner Mongolia Agricultural University, Hohhot 010018, PR China

Chun-Jie Wang
College of Veterinary Medicine, Inner Mongolia Agricultural University, Hohhot 010018, PR China

Ma Lina
College of Life Sciences, Inner Mongolia Agricultural University, Hohhot 010018, PR China

Vita Kiriliauskaitė and Vida Bendikienė
Department of Biochemistry and Biophysics, Faculty of Natural Sciences, Vilnius University, M. K. Ciurlionio g. 21, LT-03101 Vilnius, Lithuania

Benediktas Juodka
Vilnius University, Vilnius, Lithuania

Alfredas Rimkus, Saugirdas Pukalskas, Jonas Matijošius and Edgar Sokolovskij
Department of Automobile Transport, Vilnius Gediminas Technical University, J. Basanaviciaus g. 28, LT-03224 Vilnius, Lithuania

Susana López-Aparicio
NILU - Norwegian Institute for Air Research, Instituttveien 18, 2027-Kjeller, Norway

Rima Grašienė
Lithuanian Theatre, Music and Cinema Museum, Vilniaus g. 41, 01119 Vilnius, Lithuania

Albinas Lugauskas, Igoris Prosyčevas, Algirdas Narkevičius, Aušra Selskienė, Dalia Bučinskienė, Alma Ručinskienė and Elena Binkauskienė
State Research Institute, Center for Physical Sciences and Technology, A. Goštauto g. 9, LT-01108 Vilnius, Lithuania

Igoris Prosyčevas
Institute of Materials Science, Kaunas University of Technology, Savanorių pr. 271, LT-50131 Kaunas, Lithuania

Anil Kumar
National Institute of Immunology, Aruna Asaf Ali Marg, a10067 New Delhi, India

Purnima Dhall and Rita Kumar
Institute of Genomics and Integrative Biology, Mall Road, 110007 Delhi, India

Sanak Ray, Ayusman Mohanty, Torati Sri Ramulu and Gautam Roy Chaudhury
Department of Environment and Sustainability, Institute of Minerals and Materials Technology (CSIR), Bhubaneswar 751013 Odisha, India

Elena Berte and Bruno Zanon
Department of Civil and Environmental Engineering, University of Trento, Via Mesiano 77, 38123 Trento, Italy

Thomas Panagopoulos
Research Centre for Spatial and Organizational Dynamics (CIEO), University of Algarve, 8000 Faro, Portugal

Janina Šyvokienė and Milda Zita Vosylienė
Nature Research Centre, Akademijos g. 2, LT-08412 Vilnius, Lithuania

Marina Valentukevičienė, Anželika Jurkienė and Auksė Grigaitytė
Department of Water Management, Vilnius Gediminas Technical University, Saulėtekio al. 11, LT-10223 Vilnius, Lithuania

Agnieszka Karczmarczyk
Faculty of Civil and Environmental Engineering, Warsaw University of Life Science, ul. Nowoursynowska 159, 02-776 Warsaw, Poland

Saulius Vaikasas and Kęstutis Palaima
Water Research Institute, Aleksandras Stulginskis University, Parko g. 6, 58102 Vilainiai, Kėdainiai distr, Lithuania

Virginija Pliūraitė
Nature Research Centre, Akademijos g. 2, 08412 Vilnius, Lithuania

Isabela Maria Simion
Department of Environmental Engineering and Management, Faculty of Chemical Engineering and Environmental Protection, "Gheorghe Asachi" Technical University of Iasi, 73 Prof Dr Docent Dimitrie Mangeron Street, 700050 Iasi, Romania

Maria Emiliana Fortuna
Department of Environmental Engineering and Management, Faculty of Chemical Engineering and Environmental Protection, "Gheorghe Asachi" Technical University of Iasi, 73 Prof Dr Docent Dimitrie Mangeron Street, 700050 Iasi, Romania
Petru Poni Institute of Macromoecularl Chemistry, 41A Grigore Ghica Voda Alley, 700487 Iasi, Romania

Maria Gavrilescu
Department of Environmental Engineering and Management, Faculty of Chemical Engineering and Environmental Protection, "Gheorghe Asachi" Technical University of Iasi, 73 Prof Dr Docent Dimitrie Mangeron Street, 700050 Iasi, Romania
Academy of Romanian Scientists, 54 Splaiul Independentei, 050094 Bucharest, Romania

Alessandra Bonoli
DICAM, Alma Mater Studiorum University of Studies of Bologna, 2 Risorgimento Street, 40136 Bologna, Italy

Index

www.ingramcontent.com/pod-product-compliance
Lightning Source LLC
LaVergne TN
LVHW060520210926
848485LV00005B/1115